DYNAMICS OF STELLAR SYSTEMS

1 – D. Lynden-Bell, 2 – A. J. Kalnajs, 3 – G. Janin, 4 – S. J. Aarseth, 5 – C. Wilson, 6 – I. R. King, 7 – A. Hayli, 8 – W. C. Saslaw, 9 – J. Colin, 10 – H. Capaccioli, 11 – G. D. Illingworth, 12 – F. Bertola, 13 – K. C. Freeman, 14 – T. S. van Albada, 15 – J. E. Baldwin, 16 – C. Turon, 17 – C. Hunter, 18 – M. Lecar, 19 – L. Cohen, 20 – R. Wielen, 21 – P. Bouvier, 22 – S. M. Alladin, 23 – M. Mayor, 24 – R. André-Jeannin, 25 – P. O. Vandervoort, 26 – J.-P. Parisot, 27 – J. Peralta, 28 – E. Oblak, 29 – F. Puel, 30 – F. Hohl, 31 – P. Brosche, 32 – M. Schmidt, 33 – E. D. Fackerell, 34 – R. B. Larson, 35 – R. H. Miller, 36 – J. M. Bardeen, 37 – M. Roberts, 38 – C. Froeschlé, 39 – P. Biermann, 40 – K. A. Innanen, 41 – D. C. Heggie, 42 – M. J. Valtonen, 43 – B. Barbanis, 44 – G. Severne, 45 – M. Feix, 46 – F. House, 47 – M. Hénon, 48 – J. Delhaye, 49 – J. Ipser, 50 – J. P. Lafon, 51 – L. Martinet, 52 – A. Brahic.

INTERNATIONAL ASTRONOMICAL UNION
UNION ASTRONOMIQUE INTERNATIONALE

SYMPOSIUM No. 69
HELD IN BESANÇON, FRANCE, SEPTEMBER 9–13, 1974

DYNAMICS
OF STELLAR SYSTEMS

EDITED BY

AVRAM HAYLI

Observatoire de Besançon, France

D. REIDEL PUBLISHING COMPANY

DORDRECHT-HOLLAND / BOSTON-U.S.A.

1975

Library of Congress Cataloging in Publication Data

Main entry under title:

Dynamics of stellar systems.

(Symposium – International Astronomical Union ; no. 69)
"Held in Besançon, France, September 9–13, 1974."
Includes bibliographies and indexes.
1. Stars—Clusters—Congresses. 2. Galaxies—Congresses.
I. Hayli, Avram. II. Series: International Astronomical
Union. Symposium ; no. 69.
QB853.D9   523.1′12   75–12976
ISBN-13:978-90-277-0590-7   e-ISBN-13:978-94-010-1818-0
DOI: 10.1007/978-94-010-1818-0

Published on behalf of
the International Astronomical Union
by
D. Reidel Publishing Company, P.O. Box 17, Dordrecht, Holland

Sold and distributed in the U.S.A., Canada, and Mexico
by D. Reidel Publishing Company, Inc.
306 Dartmouth Street, Boston,
Mass. 02116, U.S.A.

TABLE OF CONTENTS

PREFACE

The idea of holding this Symposium has its origin in a conversation with G. Contopoulos during the winter of 1973. It was then clear that the progress realized in Stellar Dynamics since the Thessaloniki symposium had given a new shape to the field. Other meetings such as the C.N.R.S. colloquium held in Paris in 1967 or the I.A.U. colloquium held in Cambridge, England, in 1970 had in the meantime given opportunities to review the advances achieved in several branches of this field. We thought that time had come to organize a general confrontation of the new results obtained in the approach of the gravitational N-body problem by different methods, in the investigation of spherical and flattened systems, in the comparisons with observations and in more theoretical speculations on orbits, integrals of motion, dense nuclei or relativistic stellar dynamics.

Things were made easy by the support of Commissions 33 and 37 which welcomed the proposition and by I. R. King who accepted to act as Chairman of the Scientific Organizing Committee. The final decision to meet in Besançon was taken during the XVth General Assembly of the I.A.U. in Sydney in August 1973.

The members of the Organizing Committee were: I. R. King (Chairman), G. Contopoulos, A. Hayli, M. Hénon, G. Hori, D. Lynden-Bell, L. Perek, L. Spitzer, R. Wielen, Ya. Zel'dovich while the members of the Local Committee were: A. Hayli (Chairman), J. Colin, M. Crézé, E. Oblak, F. Puel, and Miss H. Taillardat (Secretary).

The symposium was sponsored by the following authorities: the International Astronomical Union, the Délégation Générale à la Recherche Scientifique et Technique, the Ministère de l'Education Nationale, Besançon Observatory, the University of Besançon, the Conseil Général du Doubs and the City of Besançon.

The sessions were held in the conference room of Besançon Observatory. There were 11 invited papers and 29 contributed papers.

The paper by G. Pels, J. H. Oort and Mrs H. A. Pels-Kluyver was read by I. R. King and was not followed by a discussion. The second paper by R. H. Miller was presented during the Symposium as a commentary to D. C. Heggie's paper; it has been published separately. I. R. King asked D. Lynden-Bell to present the General Conclusion: it was followed by a Final Discussion.

Discussions are edited almost in their entirety. They were essentially reproduced from the slips completed by participants but the tape recordings of the Proceedings were also of some help.

For the material organization of the Symposium I have been considerably helped by my colleagues of Besançon Observatory. I would like to mention especially Miss Hélène Taillardat who helped me also in the preparation of the manuscripts, Dr M. Crézé, MM. J. Colin, E. Davoust, E. Oblak, J.-P. Parisot, F. Puel, Mrs D. Chabod and M. B. Mougin.

A. HAYLI

LIST OF PARTICIPANTS

S. J. Aarseth, Institute of Astronomy, Cambridge, U.K.

S. M. Alladin, Osmania University, Hyderabad, India

R. André-Jeannin, Université de Paris VI, Paris, France

E. Athanassoula-Georgala, N.R.C. 'Democritos', Athens, Greece

J. E. Baldwin, Cavendish Laboratory, Cambridge, U.K.

B. Barbanis, University of Patras, Patras, Greece

J. M. Bardeen, Yale University, New Haven, Conn., U.S.A.

F. Bertola, Osservatorio Astronomico, Padova, Italy

P. Biermann, Universität-Sternwarte, Göttingen, F.R.G.

P. Bouvier, Observatoire de Genève, Genève, Switzerland

A. Brahic, Observatoire de Paris, Meudon, France

P. Brosche, Astronomisches Institut, Bonn, F.R.G.

P. Brunet-Crosa, Universitaria de Barcelona, Barcelona, Spain

M. Capaccioli, Osservatorio Astronomico, Padova, Italy

S. Clairemidi, Université de Besançon, Besançon, France

L. Cohen, Hunter College, New-York, N.Y., U.S.A.

J. Colin, Observatoire de Besançon, Besançon, France

S. Considère, Observatoire de Besançon, Besançon, France

G. Contopoulos, Institute of Astronomy, Thessaloniki, Greece

M. Crézé, Observatoire de Besançon, Besançon, France

E. Davoust, Observatoire de Besançon, Besançon, France

E. D. Fackerell, University of Sydney, Sydney, Australia

M. R. Feix, GRI/CNRS, Orléans, France

K. C. Freeman, Mount Stromlo Observatory, Canberra, Australia

C. Froeschlé, Observatoire de Nice, Nice, France

L. Galgani, Instituto de Fisica, Milano, Italy

J. R. Gott III, California Institute of Technology, Pasadena, Calif., U.S.A.

A. Hayli, Observatoire de Besançon, Besançon, France

D. C. Heggie, Institute of Astronomy, Cambridge, U.K.

M. Hénon, Observatoire de Nice, Nice, France

F. Hohl, NASA, Newport News, Va., U.S.A.

F. House, Astronomisches Institut, Bochum, F.R.G.

C. Hunter, Florida State University, Tallahassee, Fla., U.S.A.

G. Illingworth, Kitt Peak Observatory, Tucson, Ariz., U.S.A.

K. A. Innanen, York University, Downsview, Canada

J. R. Ipser, LASR, Chicago, Ill., U.S.A.

G. Janin, ESOC, Darmstadt, F.R.G.

A. J. Kalnajs, Mount Stromlo Observatory, Canberra, Australia

I. R. King, California University, Berkeley, Calif., U.S.A.

J.-P. Lafon, Observatoire de Paris, Meudon, France

R. B. Larson, Yale University Observatory, New Haven, Conn., U.S.A.

M. Lecar, Center for Astrophysics, Cambridge, Mass., U.S.A.

D. Lynden-Bell, Institute of Astronomy, Cambridge, U.K.

L. Martinet, Observatoire de Genève, Genève, Switzerland

M. Mayor, Observatoire de Genève, Genève, Switzerland

R. H. Miller, University of Chicago, Chicago, Ill., U.S.A.

E. Oblak, Observatoire de Besançon, Besançon, France

J.-P. Parisot, Observatoire de Besançon, Besançon, France

J. Peralta, Universidad del Zulia, Maracaibo, Venezuela

P. Pişmiş, Instituto de Astronomia, Mexico, Mexico

F. Puel, Observatoire de Besançon, Besançon, France

M. S. Roberts, National Radio Astronomy Observatory, Charlottesville, Va., U.S.A.

W. C. Saslaw, Jesus College, Cambridge, U.K.

J.-P. Scheidecker, Observatoire de Nice, Nice, France

M. Schmidt, California Institute of Technology, Pasadena, Calif., U.S.A.

A. Scotti, Istituto de Fisica, Parma, Italy

G. Severne, Universiteit Brussel, Brussel, Belgium

C. Simo, Universitaria de Barcelona, Barcelona, Spain

L. Spitzer, Princeton University Observatory, Princeton, N.J., U.S.A.

C. Turon, Observatoire de Paris, Paris, France

S. Vaghi, Observatoire de Paris, Meudon, France

M. J. Valtonen, Institute of Astronomy, Cambridge, U.K.

T. S. Van Albada, Kapteyn Laboratorium, Groningen, The Netherlands

P. O. Vandervoort, University of Chicago, Chicago, Ill., U.S.A.

R. Wielen, Hamburger Sternwarte, Hamburg, F.R.G.

C. Wilson, Hale Observatories, Pasadena, Calif., U.S.A.

PART I

SPHERICAL SYSTEMS

DYNAMICAL THEORY OF
SPHERICAL STELLAR SYSTEMS WITH LARGE N

L. SPITZER, JR.

Princeton University Observatory, Princeton, N.J., U.S.A.

1. The General Problem

The dynamical evolution of a spherical star system is attractive to the theorist for two reasons. In the first place the physical problem, with some idealization, appears to have an appealing, if somewhat deceptive, simplicity; a large number of point masses, attracting each other with inverse-square gravitational forces, are assumed to move in a quasi-steady system with spherical symmetry. In the second place, the observed globular clusters appear to conform rather closely to this ideal of a spherical star system; to analyze the changes of these clusters with time, resulting from the dynamical interaction of stars with each other, is surely an important aspect of the theory of galactic evolution.

Some aspects of this evolution may be deduced from general statistical principles. Clearly the total energy of the system, including the kinetic energy, T, and the potential energy, W, must be constant. From the second law of thermodynamics we know that the direction of any dynamical evolution must be such as to increase the total entropy. In terms of statistical mechanics when energy is exchanged back and forth between particles in a steady state, with no energy input from the outside, the system will tend to evolve towards the most probable state, where the distribution of particles, or stars, among different energy states is in accordance with the probability P_i, computed from the equation

$$P_i = Cg_i \exp(-\lambda E_i),\tag{1}$$

where E_i is the energy of a star in some particular state and g_i is the statistical weight of that state, proportional to the amount of phase space available; C is a normalization constant, while the parameter λ is inversely proportional to the mean energy of each particle, averaged over all particles.

The application of Equation (1) to a system of gravitationally interacting mass points runs into immediate difficulty because P_i increases steadily as E_i becomes more strongly negative, and there is no lower limit on E_i. Thus tightly bound gravitational subsystems may be expected to form and become even more strongly bound as a result of energy exchanges resulting from mutual encounters. Another difficulty is posed at the other end of the energy scale by stars for which E_i is zero or slightly positive; the total phase space available for stars in such states may be regarded as infinite. Thus it is clear that a self-gravitating cluster has difficulty reaching a permanent equilibrium state. One final state appears to be possible, – two tightly bound

Hayli (ed.), Dynamics of Stellar Systems, 3–26. All Rights Reserved.

stars revolving about each other in a permanent Keplerian ellipse, with all other stars escaping to infinity. This particular state is unchanging not because it has reached the maximum possible entropy, but rather because dynamical evolution has ceased, if we ignore gravitational radiation. In any case it is clear that in the evolution of a cluster a part of the system will expand, with perhaps some stars escaping, while another part will contract. If the simplicity of an isolated binary system is not reached, the contraction will continue until the initial assumptions are no longer valid. The finite stellar radius may be important if physical collisions occur, while general relativistic effects can be important if the escape velocity from the system becomes comparable with the velocity of light, as, for example, if material collapses toward a black hole.

We now list four different pathways by which a star system may evolve in conformity with Equation (1).

(1) Contraction of cluster as a whole, with escape or 'evaporation' of some stars.

(2) Contraction of inner core of cluster, with expansion at intermediate radii.

(3) Contraction of subsystem of heavier stars, with expansion for the lighter stars.

(4) Formation and contraction of binary systems.

A chief purpose of the dynamical theory summarized in the ensuing sections is to analyze the rate at which each of these processes occurs. In this introductory section we explain briefly the physical nature of each such process.

Process (1), first proposed by Ambartsumian (1938) and Spitzer (1940), is in many ways the simplest of the four, since it can be discussed quantitatively without any real evaluation of the internal structure of the cluster. Thus a moderately realistic estimate of the escape rate can be obtained with the entirely unrealistic assumption that the cluster is homogeneous and uniform. From the familiar virial theorem it is readily shown that the mean square value of the escape velocity, v_∞, is given for an isolated system in general by

$$\langle v_\infty^2 \rangle = 4 \langle v^2 \rangle. \tag{2}$$

Hence if a Maxwellian velocity distribution is established in the relaxation time, t_r, the fraction ξ_e of stars that will gain enough energy to escape during this time is the fraction of the particles in a Maxwellian distribution, $P_M(v)$, which have velocities twice the rms value, v_m, giving

$$\xi_e = \int_{2v_m}^{\infty} P_M(v) \, dv = 7.4 \times 10^{-3}. \tag{3}$$

As a result of this evaporation of stars, both the mass and energy of the cluster must decrease, with the cluster becoming progressively more tightly bound. For a homogeneous uniform cluster the evolutionary history can be traced (King, 1958b) until the system collapses entirely after a life of about 40 initial relaxation times. More elaborate dynamical investigations, using the Fokker-Planck equation for exact analytical computations of dynamical relaxation gave only slightly different results for homogeneous clusters (Spitzer and Harm, 1958; King, 1965, 1966); for actual clusters the variations of density and relaxation time with position must also be considered (King, 1958a).

Process (2) can take at least two forms, depending on whether the energy released from the contraction of the core appears in the outer part of the cluster, or 'halo', or somewhere in the intermediate region. If the former, then an increasing number of stars will populate the halo with low binding energies and large apocentric distances. Such stars have relatively long periods and return only infrequently to the denser regions of the cluster. As we shall see in Section 3, the accumulation of stars in the outer halo is closely related to the escape of stars associated with process (1), and these two effects may conveniently be discussed together.

If the energy released by the core appears in the intermediate region of the cluster, process (2) takes a somewhat different form. To discuss this mechanism quantitatively requires rather detailed knowledge of the internal structure of the cluster. Analyses to date have been confined to rather idealized systems. In particular, Antonov (1962) and Lynden-Bell and Wood (1968) have considered this process in isothermal spheres confined within hypothetical rigid boundaries. The analysis assumed that the mean free path was short and thus, unlike the stellar case, the velocity distribution remained everywhere isotropic during the evolution. Their work demonstrated that such spheres would be thermally unstable if the central density exceeded the density at the boundary by a factor greater than 708. In this situation, the system has the remarkable property that the entropy is increased by a development of a thermal gradient, with heat flowing from the contracting core to the expanding outer region. Since the kinetic energies of particles in a self-gravitating system increase as the system gives up energy and contracts, the contracting core becomes steadily hotter as it loses energy, while the outer region cools.

In actual clusters the physical situation is somewhat different from that assumed in these analyses – see Section 3 – but one might expect that perhaps a similar collapse of the central core – called the 'gravothermal catastrophe' by Lynden-Bell and Wood – may occur if the density change within the isothermal region of the cluster amounts to more than some three orders of magnitude. There have been no computations of the rate to be expected for this type of collapse, but one might expect an accelerated collapse, since the relaxation time gets markedly shorter as the central core becomes denser.

Process (3), which Spitzer (1969) pointed out could lead to contraction without limit, results in a straightforward way from the tendency towards equipartition of energy between stars of different masses. During one relaxation time, t_{rh}, a heavier star should lose an appreciable fraction of its kinetic energy to the lighter stars and fall towards the center of the system, creating appreciable mass stratification. If there are only a few such heavy stars, they will reach equilibrium near the cluster center with lower velocities than the lighter stars and about the same kinetic energies. If the amount of mass in this concentration of heavy stars is appreciable, however, the self-attraction of these stars will require increased random velocities in equilibrium, equipartition with the lighter stars becomes impossible, and the subgroup of heavy stars will continue to lose energy to the lighter stars, continually contracting with increasing velocities and increasing departure from equipartition. The initial time scale for this type of collapse should be of the order of the relaxation time in the inner regions of the cluster, but in contrast to the evaporation and gravothermal catastrophes there is no reason to expect any acceleration of the collapse. Indeed, if kinetic energy is lost by the heavy stars at a constant rate, which would be expected if the density and rms velocity of the light stars remained constant, the mean value of M/r for these stars might be expected to increase at a constant rate, giving dr_1/dt varying as $-r_1^2$, if r_1 is the mean radius for this subgroup.

Process (4) was pointed out by Aarseth (1968) and van Albada (1968) as a major influence on the evolution of systems with some 25 to 250 stars. Numerical integration of the exact equations of motion for such systems, following von Hoerner (1960, 1963), showed that binary stars were formed and disrupted, but that one or more tightly bound binaries generally accounted for most of the binding energy of the cluster at the end of the evolutionary period followed. General arguments – Spitzer and Hart (1971a), referred to as Paper I – show that the three-body encounters which form tightly bound binary systems become less and less important as the number of stars, N, increases. A detailed analysis by Heggie (1974) verifies this result and shows that in a cluster of 10^5 stars, for example, the formation of binaries should be quite unimportant for dynamical evolution. We shall therefore neglect process (4) throughout most of the discussion below.

As emphasized by Heggie (1974), there are two situations in which absorption of energy by binary systems can be important even in clusters with large N. Firstly, if the core of the cluster collapses, the formation of binaries may become important when a relatively small number of stars is involved in a high-density core; we return to this subject again in Section 6. Secondly, if a sufficient fraction of the stars are binary from the beginning, with a separation of a few astronomical units between components, the dynamical evolution may be profoundly affected, with the cluster as a whole perhaps even expanding rather than contracting. It has often been assumed that binaries are relatively infrequent in systems of population type II, and indeed very few binaries have been detected in globular clusters. However in van de Kamp's (1971) list of stars within 5.2 pc, the fraction which are in fact multiple systems is about 40% independent of space velocity. Thus of the 13 stars with a heliocentric velocity known to exceed 70 km s^{-1}, 3 are double and 2 are triple systems. Of the remaining 32 stars, with lower (or uncertain) velocities, 13 are double stars, with 1 triple. The possible presence of such multiple systems in globular clusters has so far been generally ignored and must certainly be considered in future work.

2. Dynamical Methods for Large-N Systems

Since it is not feasible to integrate numerically the exact equations of motion for a system with a very large number of stars, an approximate solution must be sought. As the basis for such an approximation in spherically symmetric systems we may express the gravitational acceleration $\mathbf{g}_i$, for any star i, as the sum of two terms;

$$\mathbf{g}_i = \frac{d\mathbf{v}_i}{dt} = -\mathbf{r}_1 \frac{\partial \varphi(r_i, t)}{\partial r} + \mathbf{F}_i(t), \tag{4}$$

where $\mathbf{r}_1$ is a unit vector in the r direction. The radial potential function, $\varphi(r, t)$, may be computed from the smoothed distribution of stars; if the system is in a quasi-steady state, $\varphi(r, t)$ will change only slowly with time. The function $\mathbf{F}_i(t)$ represents for each star the difference between the actual acceleration and the smoothed spherically symmetric value, and results from the granularity of the actual distribution of mass points. The basic approximation made is that the effects produced by this granularity may be computed from an analysis of successive two-body encounters. This approximation is believed to be relatively accurate, although for some effects, such as the formation of binaries, one must consider encounters of three bodies and other higher-order correlations.

The analytical theory of two-body encounters has been thoroughly explored by Chandrasekhar (1942), and by Rosenbluth et al. (1957). However, to apply this theory to the complex configuration of a stellar cluster seems discouragingly complicated, – see Hénon (1973). Instead, Monte Carlo techniques have been used to follow numerically the orbits of some thousand 'test stars', taken to be a representative sample of a larger group. In these techniques the orbit of each star in the potential field, $\varphi(r, t)$, is taken into account, and velocity perturbations are applied with suitable random sampling techniques so that the net effect of two-body encounters is correctly given. Two different methods have been published, the first by Hénon (1967, 1971, 1973), the second by a group at Princeton (see Spitzer and associates, Papers I–VII). These methods differ both in the consideration of stellar orbits in the potential field $\varphi(r, t)$ and in the choice of velocity perturbations resulting from $\mathbf{E}_i(t)$.

In the work by Hénon the unperturbed stellar orbits are not actually computed in detail. For his Monte Carlo computations the only information required from the orbits is the resultant probability distribution of r_i, the star's distance from the center, and this can be determined from the star's energy, E, and angular momentum J (both taken per unit mass) if the potential field $\varphi(r, t)$ is known. For each of the test stars a particular position in its orbit is determined, in accordance with this probability distribution. Adjacent stars are then assumed to have a mutual encounter, changing the velocity of each. All the geometrical angles in this encounter are chosen at random, but the collision parameter, p, (the distance of closest approach in the absence of mutual forces) is computed to give the correct total mean square deflection, considering the local density, the time step used and the actual relative velocity of the two stars. After each such collision, new random positions are chosen for each of the two stars, taking into account the new E, J of each and the potential $\varphi(r, t)$ is recomputed; spherical symmetry is assured by the assumption that each star considered is one of a large number of stars all with identical radial and transverse velocities and all arranged in a spherical shell (called a 'superstar' by Hénon).

In the computations at Princeton, (see Spitzer and associates, esp. Papers I, IV and VII) Equation (4) is integrated numerically with time to give $r_i(t)$ for each of the 1000 test stars; again, for the computation of $\varphi(r, t)$ each test star is assumed to represent a large number of similar stars arranged in a uniform spherical shell. The velocity perturbations are computed in groups rather than from individual two-body collisions as in Hénon's method. The cluster is divided into 25 spherical shells or regions, each one containing 40 test stars; the mean density ϱ and rms velocity v_m are determined in each region, with separate values for stars of each mass group. For each star of velocity $\mathbf{v}$, the velocity perturbations are assigned according to a Gaussian distribution formula which gives the correct first and second moments of $\Delta\mathbf{v}$ for a star of that mass and velocity interacting separately with the stars of each mass group; each such group is assumed to have a Maxwellian distribution of velocities, with the rms velocities and densities computed for that group of field stars in the region considered.

In the Princeton computations the ratio of the relaxation time, t_{rh}, to the mean crossing time, r_h/v_m, is taken to be in the range from 10 to 50, corresponding to a cluster with N between 500 and 4000. For clusters with so small a value of N the use of a smoothed radial potential becomes questionable, and the Monte Carlo computations are not completely valid. For clusters with N equal to 10^5, the actual ratio of t_{rh} to r_h/v_m (which varies as $N/\log(0.4\ N)$ – see Paper I) is about 800, and the computing time involved in following stars for so many orbits would be prohibitively large. Fortunately, most aspects of cluster evolution, as measured in units of t_{rh}, are independent of the crossing time, a fact which forms the basis of Hénon's method, and hence the Princeton computations can be taken as representing the dynamical history of clusters with large N. In considering the escape of stars from the cluster, an effect whose detailed mechanism depends on the ratio of relaxation to crossing time and hence on N, a separate theoretical study indicates the modifications needed in the computed models (see Section 3.3.).

Comparing these two methods, the one developed by Hénon has the advantage that it requires somewhat less computing time. In addition, the computation of velocity perturbations takes the exact distribution of stellar velocities into account; this latter advantage is not too significant, since in the denser regions where encounters are significant the actual distribution is generally rather closely Maxwellian. The Princeton method has some advantage in that individual orbits are followed. Initial dynamical collapse can be considered and, as pointed out above, a basis is given for analyzing the effect of distant encounters in producing escape of stars, which cannot be considered directly with Hénon's method.

3. Results for Isolated Systems with Stars of Same Mass

3.1. Reference relaxation time

In this section we shall give some results for the simplest possible system, obtained with Monte Carlo calculations, mostly by the Princeton group. It is often convenient to express time in units of a reference relaxation time, defined as the relaxation time at the mean density for the inner half of the cluster mass, M, with stars assumed to have the rms velocity of the cluster as a whole. With some simplifications and approximations, – discussed in Paper I – we obtain

$$t_{rh} = \frac{0.060 \, M^{1/2} r_h^{3/2}}{m G^{1/2} \log(0.4 \, N)},$$ (5)

where r_h is the radius containing half the mass, N is the total number of stars, and m is the mean mass, equal to M/N. The median radius, r_h, changes relatively little during the evolution of most of the models; in any case it is simplest to use the initial value of t_{rh} as a reference parameter throughout the life of the cluster. Because of the tremendous central concentration of most clusters, the local relaxation time can be several orders of magnitude shorter than t_{rh} at the center and enormously greater in the far halo.

3.2. DEVELOPMENT OF ISOTHERMAL SPHERE AND SURROUNDING HALO

As noted above, one would expect collisions to establish a velocity distribution close to the Maxwellian in the central regions of the cluster. The Monte Carlo models may be used to show the actual value of $f(E)$, the density of test stars in phase space, as a function of E, the star's energy per unit mass. Figure 1, taken from Paper IV, shows the values obtained in the inner three regions (containing the inner 12% of the mass) in one of the model one-component systems. The energy scale is defined so that a star at rest at infinity has zero energy. In the outer regions, the phase density should be a function of J as well as of E, but in the central regions the velocity distribution is isotropic and f is a function of E only.

The points mostly lie close to the solid line, representing an exponential decrease of $f(E)$ with increasing E. However, this distribution would give a finite value for $f(0)$, giving much too large an escape rate. For E only slightly negative, the points lie reasonably close to the dashed line, representing the 'lowered Maxwellian'

$$f(E) = C \exp[(-3E/v_m^2) - 1],$$ (6)

proposed by Michie (1963) and King (1965); C is a constant of proportionality. Equation (6) is based on the assumption that the energy required for escape is zero. This lowered Maxwellian distribution seems a reasonably good approximation for the cluster models. At values of E between the last plotted point and zero, the phase space density drops to a value much below that given in Equation (6). The statistical weight of these states of slight negative energy is relatively very large, and the gradual increase of $f(E, 0)$ in these states as time goes on is an important part of the cluster evolution, which leads, as we shall see below, both to the development of the halo and to escape of some stars.

As a result of this tendency towards an isotropic Maxwellian distribution, the inner half of the cluster, within the radius r_h, approximates the inner region of an isothermal sphere. There are slight deviations from strictly isothermal conditions because of the drop in the velocity distribution function at energies near zero, discussed above.

However, in the central regions this drop is way out in the wings of the velocity distribution function and produces little direct effect locally, except for a slight decrease of the rms velocity with increasing radius.

In the halo a quite different situation prevails. Since the local dynamical relaxation time is very long, the velocity distribution is anisotropic. The phase space density is

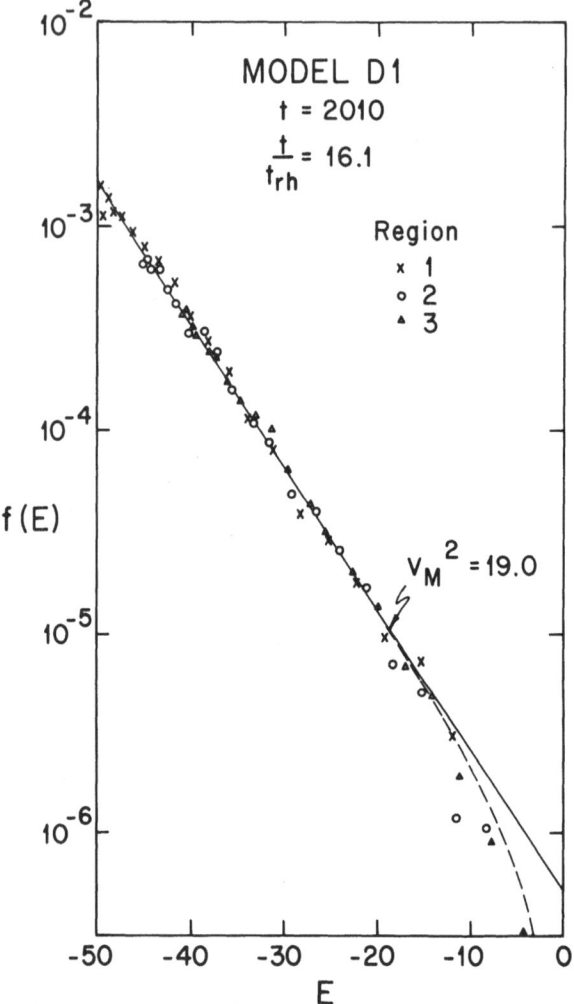

Fig. 1. Density in phase space, $f(E)$, as a function of stellar energy, E (potential plus kinetic), in the inner regions of Model D1 fairly late in its development. The straight line represents the usual Maxwellian distribution, while the dashed line represents the lowered Maxwellian distribution (Equation (6) in text), which decreases linearly to 0 as the energy approaches 0, the value for a star motionless at infinity.

constant along a dynamical trajectory when collisions are infrequent, and hence along orbits which intersect the core, $f(E, J)$ has the same values as in the core. For J greater than a certain critical value, the orbits are restricted to regions of low density, outside the isothermal core, and if these orbits are not populated originally, they will remain unpopulated. Thus the halo contains stars on predominantly radial orbits, with transverse velocities varying as $1/r$ since the angular momenta remain constant with increasing distance.

To obtain the density in the halo we integrate $f(E, J)$ over $2\pi v_t \, dv_t \, dv_r$, where v_t and v_r are the tangential and radial velocities, respectively. As a reasonable first approximation we may set $f(E, J)$ equal to $f(E, 0)$ for J less than some upper limit, J_1, and zero outside. Thus $f(E, J)$ differs from zero only within a range* Δv_t from 0 to J_1/r, and we have

$$\varrho(r) = \pi(\Delta v_t)^2 (\Delta v_r) \, \bar{f}, \qquad (7)$$

where $\bar{f}$ is a mean value of $f(E, 0)$ over v_r, taken over the range* Δv_r from 0 to the escape velocity $(2GM/r)^{1/2}$. According to Equation (6), expanding the exponential for

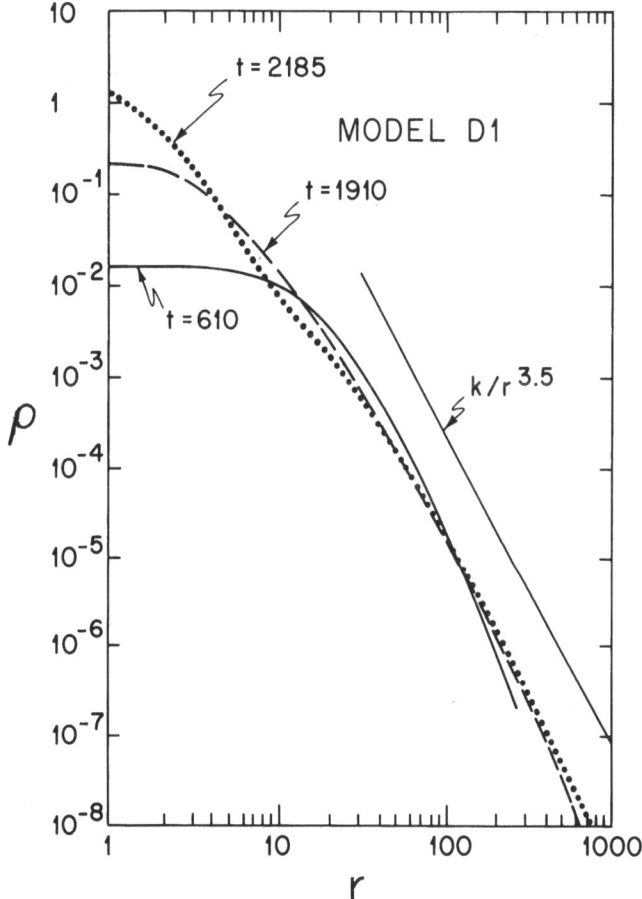

Fig. 2. Density distribution, $\varrho(r)$, as a function of distance r from the center at different times in the development of Model D1. The straight line at the right represents the theoretical density distribution in a well developed halo. This model starts with an initial collapse, which generates a substantial halo at the beginning. The dimensionless units for ϱ and r are explained in Paper II.

---

* The velocity ranges Δv_t and Δv_r in Equation (7) should not be confused with the velocity perturbations which are denoted by these same symbols in Paper IV.

small E, f decreases from $-3CE/v_m^2$ to zero as v_r increases over this range; since $-E$ equals GM/r when v_r vanishes, $\bar{f}$ is proportional to $1/r$. Since, as we have seen, Δv_t and Δv_r vary as $1/r$ and $1/r^{1/2}$, respectively, Equation (7) yields

$$\varrho(r) = k/r^{3.5}, \tag{8}$$

where k is a constant of proportionality.

The density distributions in two different models at successive times are plotted in Figures 2 and 3, which show the isothermal central region and the outlying halo. The

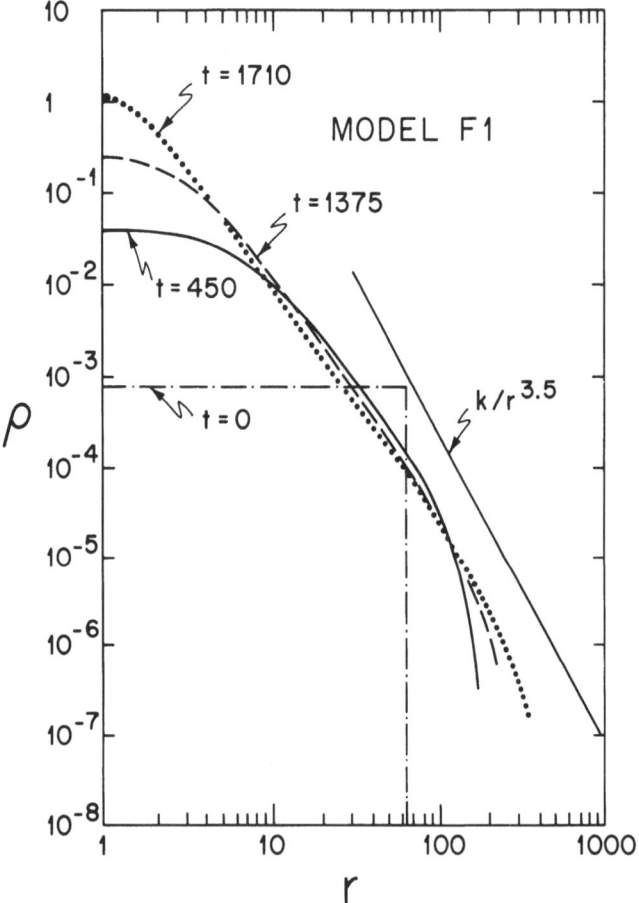

Fig. 3.   Density distribution $\varrho(r)$ at different times in the development of Model F1. This model starts as a sphere of constant density in equilibrium, with all stars in circular orbits. There is no appreciable halo at the beginning, and the halo develops much later than in Model D1, shown in Figure 1.

upper right solid lines represent Equation (8), which seems to provide a good fit for the halo. The progressive contraction of the isothermal core, which is discussed in more detail below, is evident in these figures. The outward growth of the halo is also evident. For stars in the halo the mean value of J, the angular momentum of each star, does not change appreciably with time, nor does the value of the constant k in Equation (8).

Halo growth consists of extending the range of r over which Equation (8) is valid; at greater r the density drops off very sharply. In terms of $f(E, 0)$ the development consists of increasing the phase-space density at energies only slightly negative up to the value given in Equation (6). Halo growth is particularly evident for Model F, for which the initial equilibrium configuration is a sphere of constant density, with no halo whatever, and therefore the entire halo must develop as a result of two-body encounters. Model D starts off with an initial collapse, which produces a substantial halo to begin with, and the outward growth of the halo is less noticeable.

One way of presenting the Monte Carlo data on evolution of a cluster is to plot the radii containing different fractions of the total mass as functions of the time. Such a plot for Model F is given in Figure 4, where the solid line gives the results obtained by Hénon, while the individual points show Princeton results from Paper IV. The close agreement is very gratifying. The outward extension of the halo and the contraction of the core are both evident.

3.3. ESCAPE OF STARS

Stars can escape from a cluster by a close encounter with another star at any position within the cluster. Such encounters produce velocity changes comparable with the relative velocity of the two stars, and in view of Equation (2), velocities exceeding v_∞ can result. The resultant rate of escape for the cluster as a whole has been computed by Hénon (1969).

In clusters with large N the cumulative effect of many distant encounters, producing small deflections, is more important in leading to escape of stars. In any one orbit the change of energy is small, leading in the first instance to outward diffusion of the stars and growth of the halo rather than to escape. We analyze the conditions under which halo stars can actually escape. We may define ε_2 as the root mean square change of energy of a halo star during one orbit from apocenter, where $r = r_a$, through the isothermal core and back again to apocenter. Evidently ε_2 will increase as J decreases, increasing the density experienced by the star at pericenter. For each J, ε_2 will not depend significantly on E for halo stars, when E is only slightly less than zero. There will also be a mean change of energy per orbit, denoted by ε_1, but this is smaller and we may neglect it. Evidently when the energy of the halo star, closely equal to $-GM/r_a$, equals $-\varepsilon_2$, escape is somewhat likely. We define this value of r_a as the critical distance, r_f. As r_a increases above r_f, the probability of escape per orbit approaches 0.5.

Let us now consider what happens when N, the total number of stars, increases. We keep the total mass M and the cluster dimensions constant, varying only the mean mass per star: thus the stellar energy per unit mass and the potential function $\varphi(r)$ remain unaffected. In this case it may be shown – see Paper II – that ε_2 decreases as $1/N^{1/2}$, while ε_1 varies as $1/N$. Hence the critical radius r_f increases as $N^{1/2}$, and the halo development must proceed to greater radial distances before escape becomes possible. This development to large r is limited by the long periods of the extended halo orbits. The period of a star with apocenter distance r_a is proportional to $r_a^{-3/2}$, and thus P_f, the period at the critical distance, increases as $N^{3/4}$. Since the time required for evolution by encounters varies as N (the small variation of $\log N$ is ignored here), diffusion of halo stars out to the critical distance occurs more readily as N is increased, even though the critical distance increases with N. It may be shown that after the halo has grown out to the critical radius, the rate of escape per unit t_{rh}

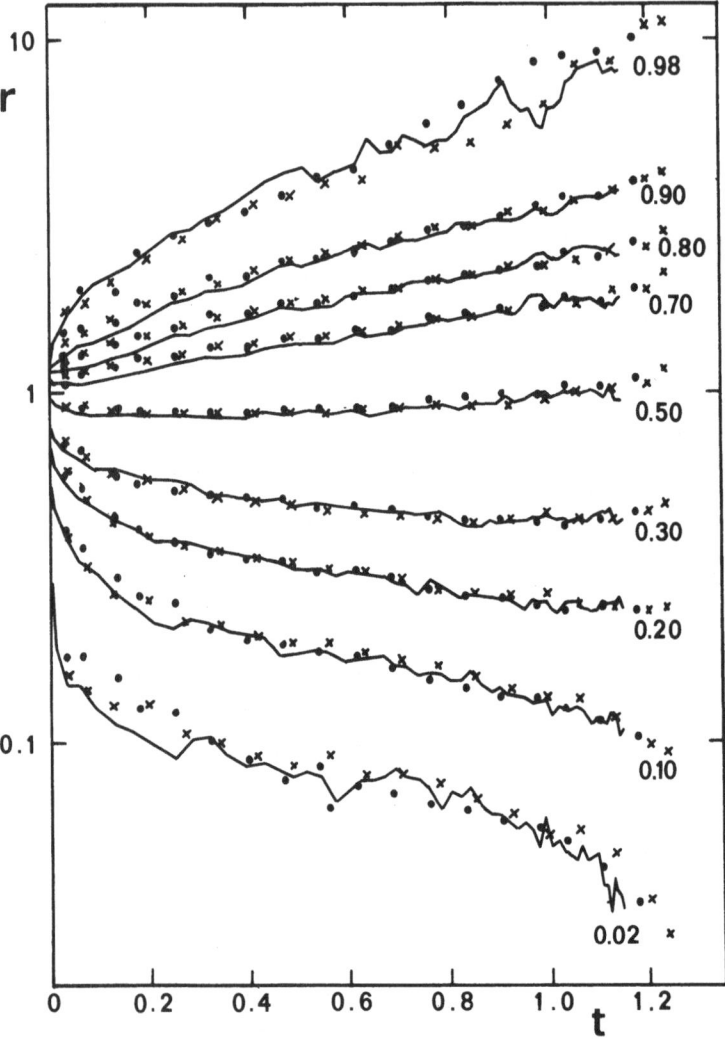

Fig. 4. Values of the radius containing various fractions of the total mass of Model F, plotted against the dimensionless time used by Hénon. The solid line represents the computations by Hénon (1973) for this model, while the dots and crosses represent values obtained by Spitzer and Thuan (1972) for models F1 and F2, respectively; these two models differ only in the ratio of the relaxation time, t_{rh}, to the dynamical crossing time, r_h/v_m, which equals 10 and 20, respectively, for F1 and F2.

in independent of N, although the mean energy of the escaping stars, which equals $0.58\,\varepsilon_2$ – see Paper III – will decrease as $1/N^{1/2}$.

The Monte Carlo calculations at Princeton give quantitative results in accord with these theoretical expectations. Thus in Model F, where there is no halo to start with and the halo does not become fully developed until near the end of the computations, no stars escape during the first 60% of the period covered, and during the last 10% of the time the escape rate reaches about 0.3% per time interval t_{rh}. In Model D, where the halo develops much earlier, the escape rate reaches this same value about 55% of the way through the evolutionary time followed. The value of ε_2 in Model D1 is about 2 for $J = 100$, a representative mean value for the halo, giving a highly eccentric orbit. Thus r_f is about 500 for this model, with about the same value for Model F1 also. The corresponding period, P_f, is 800 time units (for $G = 1$ and $M = 10^3$, in the dimensionless units used). The halo in Model D1 becomes well developed out to this radius well before the end of the computations, while in Model F1, this stage is reached only very near the end. Separate computations in which N was varied, keeping all other quantities the same, confirm that the mean energy of escaping stars varies about as $N^{-1/2}$, and is roughly equal to $\frac{1}{2}\varepsilon_2$.

The measured rate of escape, ξ_e, of about 3×10^{-3} per interval of t_{rh} is less than computed from Equation (3) by a factor 0.4. If somewhat more sophisticated methods are used for predicting ξ_e, the reduction factor is slightly smaller, about 0.3. This discrepancy is largely accounted for by the rate at which stars accumulate in the outermost halo; in Model D1 this rate is about twice the measured ξ_e. These stars return to the isothermal core so infrequently that they are very similar to escaping stars and should really be grouped with them for some purposes. For systems with N as large as 10^5, the halo would be more fully developed, the ratio of stars accumulating in the far halo to those escaping, per unit time interval, would fall below unity late in the evolutionary development, and ξ_e would be closer to the predicted value of about 0.01.

In terms of these results it is clear why Hénon's method does not give the rate of escape. The dynamical time in which a star moves around its orbit from apocenter to apocenter does not appear in his method, which corresponds to the limiting case of infinite N. Thus the rms change of energy in a single orbit is not considered, and there is no way of determining the value of r_f beyond which stars will begin to escape. His technique gives correctly the total rate at which stars diffuse up towards zero energy, but cannot distinguish evaporation at slight positive energy from halo buildup at slight negative energy. The Fokker-Planck diffusion equation, based on the limit of infinitesimal velocity changes, gives no escape of stars from an isolated cluster. As shown in detail in Paper III, consideration of finite velocity changes in the neighborhood of zero E is required to give correctly the escape rate which is, in fact, present.

One may ask what the present theory should predict for the process of escape from systems with smaller N. If $N \approx 250$, ε_2 is about 60% of the mean kinetic energy per unit mass of the cluster stars. If the present theory is still applicable for systems of such low N, which it may conceivably be at times before the presence of binaries dominates the exchange of kinetic energy between stars, the mean energy of the escaping stars (their kinetic energy at infinity) should be about a third of the mean kinetic energy of the cluster stars, in rough agreement with the results obtained by Aarseth (1975) for one-component systems with $N = 250$. Detailed numerical agreement is not to be expected, since the standard theory for large N ignores terms whose relative order is $1/\ln N$. However it is evident that for small N the exchange of energy in one orbit is relatively large, in contrast with the conventional velocity diffusion picture which must be applicable for large N.

3.4. Contraction of Isothermal Core

In all the Monte Carlo models the core contracts steadily throughout the evolution. At the beginning, the driving force behind this contraction for the one-component models is apparently the diffusion of stars toward zero energy in response to the drop of $f(E)$ below the Maxwellian value for slightly negative energies (shown in Figure 1). A detailed energy accounting in Paper IV shows that during most of the cluster's evolution the energy gained by stars which accumulate in the halo at slight negative energies or which escape at slight positive energies is just about balanced by the energy given up by the contraction of the inner core. Furthermore, the rate of contraction is about that predicted from the simple evaporation theories. As we have seen, these neglect halo formation, but they give correctly the rate at which stars accumulate at energies near zero; whether a star ends up orbiting in the halo or escaping from the system does not affect much the energy it has required to raise it above the average stars.

Towards the end of the evolutionary calculation the contraction seems to change its character. The rate of contraction, as measured at the radius containing 2% of the mass, actually accelerates, although the rate at which stars escape and accumulate in the halo does not seem to change very much. In Figure 5 the radii containing 2% and 10% of the mass of Model F are plotted against time. Evidently the squares of these radii, proportional to the spherical areas at these radii, decrease linearly with time, and seem destined to reach zero in a finite time. The time interval between the start of the cluster's evolutionary history and this final collapse of the central region is denoted by t_{coll}, and is from 12 to 19 times t_{rh} in different isolated clusters. Somewhat similar results were obtained by Larson (1970) with an approximate analytical theory.

A detailed energy accounting for Model D (see Paper IV) confirms that the rate of energy release from the core very late in the evolution is greater than can be accounted for by diffusion of particles up to small positive or negative energies. Instead the energy released near the center goes into readjustment of intermediate regions of the cluster. In view of the relatively small numbers involved, this result is not very secure statistically. Just what is happening within the cluster is not clear in detail, but one is tempted to identify this final collapse with the gravothermal instability described as process (2) in Section 1. On this basis one would expect the range of density variation within the isothermal sphere to surpass some critical value before the final collapse. Since there is a gradual transition between the isothermal central regions and the halo, this expectation cannot be tested precisely. However, in the later stages of development the median radius, r_h, provides a rough approximation for the radius of the isothermal zone. The ratio of the central density, $\varrho(0)$, to $\varrho(r_h)$, the density for r equal r_h, attains the value 10^3 relatively late in the collapse of the one-component models analyzed, and a gravothermal catastrophe under these conditions would seem a logical inference.

On the other hand, it must be admitted that the physical situation in a cluster of large N is quite different from that envisaged in the theoretical analysis of bounded isothermal spheres of gas. In addition to the lack of a rigid boundary, the local dynami-

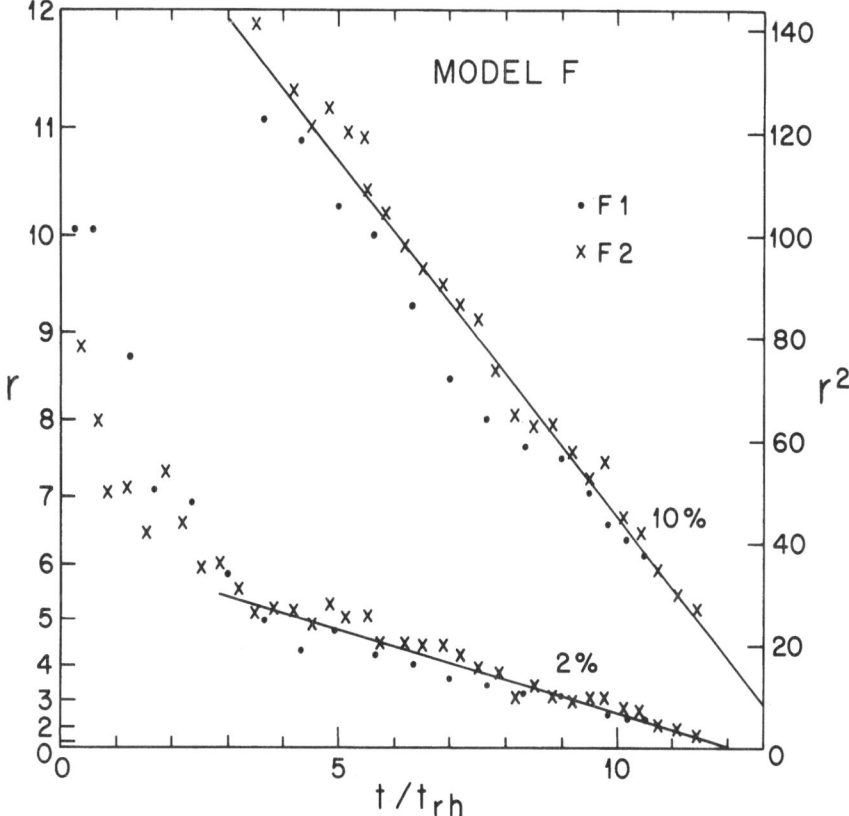

Fig. 5. Values of the radius containing 2% and 10% of the total mass of Model F, replotted from Figure 4. As before, dots and crosses represent the results for Models F1 and F2, respectively. The straight lines represent empirical fits to the data.

cal relaxation time is much longer than the dynamical crossing time, or the comparable orbital period. Thus the phase space density will be constant along each dynamical trajectory. Any departure from thermal equilibrium in the intermediate region of the cluster will therefore appear not in a perturbation of a local temperature, with radial temperature gradients appearing, but rather in a difference of $f(E, J)$ between orbits of the same E but different J; i.e., in a developing anisotropy of the velocity distribution. More generally, Miller (1973) has argued that the thermodynamic considerations which lead to the gravothermal instability are not directly applicable to stellar systems. More analysis is clearly required, and the present tentative identification of the gravothermal instability is suggestive rather than conclusive.

If this identification were correct, one would naturally question what role the gravothermal instability should play in systems of relatively small N, which have been followed by direct integration of the exact equations. There is some question whether this effect should appear when N is between 100 and 250. In the Monte Carlo models, this instability appears when the density within the inner 2% of the cluster mass is some 1000 times as great as the lowest density in the isothermal region. Such a mean density becomes difficult to define when the number of stars involved is as low as 2, and even with as many as 5 stars, there is some question as to how good the correspondence should be between an actual system and a model based on an essentially continuous distribution of matter.

Quite apart from the detailed mechanisms involved, one may ask why the area of the spheres containing a small fixed percentage of the total mass should decrease linearly with time during the final collapse, as indicated in Figure 5. It was shown in Paper IV that this result follows approximately if one assumes that the contracting core loses a certain fraction, η, of its kinetic energy during each relaxation time at the center. The energy lost is presumably conducted away to more distant regions of the cluster; the core contracts, of course, as a result of the energy drain, and the gravitational energy so released heats the core to higher rms velocities than before. Numerical values in Paper IV for the collapse rates indicate that on the average

$$\langle \eta \rangle = 2.2 \times 10^{-3}, \tag{9}$$

with an average deviation of about 10% for the three models of isolated systems.

4. Results for Systems with a Distribution of Stellar Mass

4.1. Development of Mass Stratification

If two groups of stars are present, each with a Maxwellian velocity distribution with rms velocities v_{m1} and v_{m2}, and with masses m_1 and m_2, then encounters between the two groups of stars in a region where the particle densities are n_1 and n_2 will change the mean energy, E_{m1} of group 1 stars at a rate (Spitzer, 1969)

$$\frac{dE_{m1}}{dt} = (\tfrac{1}{2}m_2 v_{m2}^2 - \tfrac{1}{2}m_1 v_{m1}^2)/t_{eq}(1, 2), \tag{10}$$

where

$$t_{eq}(1, 2) = \frac{(v_{m1}^2 + v_{m2}^2)^{3/2}}{8(6\pi)^{1/2} n_2 m_1 m_2 G^2 \ln(0.4N)}. \tag{11}$$

Conservation of energy requires that

$$\frac{t_{eq}(1, 2)}{n_1} = \frac{t_{eq}(2, 1)}{n_2}. \tag{12}$$

The ratio of the 'equipartition time' t_{eq} to the local relaxation time $t_r(2, 2)$, for encounters among stars of group 2, is given by

$$\frac{t_{eq}(1, 2)}{t_r(2, 2)} = 0.44 \frac{m_2}{m_1}\left(1 + \frac{v_{m1}^2}{v_{m2}^2}\right)^{3/2}.$$

(13)

If the lighter stars, of mass m_2, predominate, the relaxation time t_r for the entire system is close to $t_r(2, 2)$, and the equipartition time for the heavies will be somewhat shorter than t_r.

One would expect that in a system with an initial distribution of masses the tendency towards equipartition would produce a marked stratification within a time comparable to t_{rh}. Early Monte Carlo computations reported in Paper II demonstrated this effect. Several rather serious approximations made in this work have now been improved, – see Paper VII. Figure 6 shows the mass fractionation in a cluster

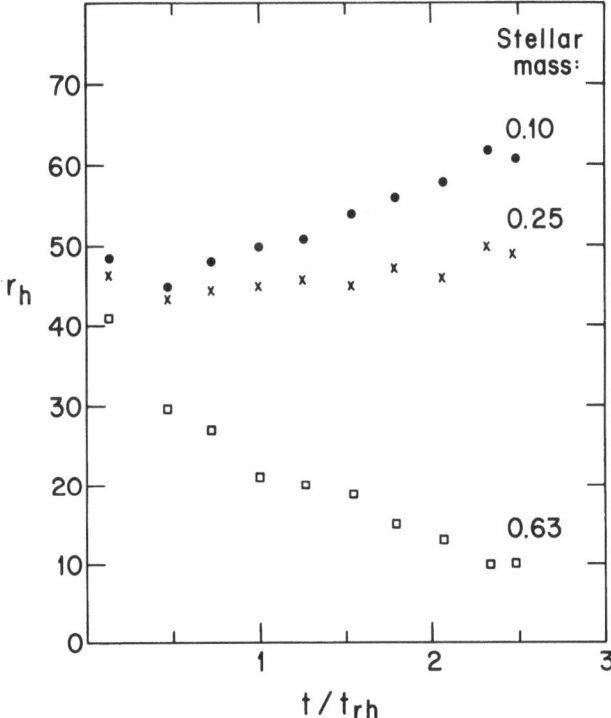

Fig. 6.   Radius containing half the mass for each of three mass groups in Model La, plotted as a function of time in units of t_{rh}. The stellar masses given are ten times the values used in the computer program and, in solar masses, give representative values for old systems of population type II.

with three components, (Model La) with the mass distribution spectrum slightly steeper than in the familiar Salpeter function; the stellar masses are in the ratio 1:0.4:0.16, with the total mass in each group in the ratio 1:2:3. This figure shows clearly how the heavier stars settle towards the center in a time of order t_{rh}. The rate at which stars accumulate in the halo or escape from the system per time interval equal to t_{rh}, is much the same as in the one-component case, except that this rate is about twice as high for the lighter stars as for the heavy ones. The stars of intermediate mass escape at about the same rate (within about 20%) as the lightest stars, in general agreement with the results obtained for small-N systems by Wielen (1968) and by Aarseth and Woolf (1972).

4.2. CONTRACTION OF ISOTHERMAL CORE

The initial contraction of the isothermal core for the three-component models is substantially faster than in the one-component system, as a result of the mass stratification. In all these models the contraction accelerates at the end, with the area of the spherical shells containing a fixed percentage of the total mass decreasing linearly with t as before. The time, t_{coll}, from beginning to final collapse is shorter than before, with t_{coll}/t_{rh} equal to about 3, as compared with values between 12 and 19 for the one-component models.

For one-component models the mechanism producing contraction changed during the evolution of the system, with evaporation and halo buildup important initially and the gravothermal instability apparently dominant at the end. A similar transition apparently occurs for the three-component models. The rate of collapse at the end is about an order of magnitude too great to attribute to the energy transferred away from the heavies through equipartition, at a rate given by Equation (10). This result follows from the fact that the total fraction of the mass present in the form of light and intermediate stars together is very small in the central regions, – less than about 10% within the sphere containing 4% of the system's mass. However, the rate of collapse found in these models is consistent with that observed in the one-component models; the values of $1000\,\eta$ for the three-component models range from 1.9 to 5.1, averaging 3.0, in rough accord with Equation (9). Furthermore, the density range within the inner half of the mass of the system again reaches some three orders of magnitude late in the collapse phase.

Important evidence on this question of collapse mechanism is obtained by a computer run in which encounters of heavy stars with each other were suspended arbitrarily at a later stage in the collapse. One would expect that this modification in the program would have no great effect on the tendency towards energy loss to the lighter stars, but should slow down the outwards energy flux important in the gravothermal instability. When this modification was made in one of the models, at a relatively advanced stage of evolution, the collapse of the very inner region (containing 2% of the mass) slowed down to about half its previous rate. Further out, the radius containing 10% of the mass stopped contracting entirely. While these results have not been explained in detail, they indicate that the transfer of energy from the heavy stars to the light stars is not the predominant mechanism responsible for the final collapse. In all probability, whatever detailed process accounts for the final collapse of one-component systems also plays an important part in the similar collapse of three-component clusters.

5. Results for Systems in Perturbing External Fields

While the computations for isolated systems are of great physical interest, the attempt to explain actual clusters in our own Galaxy must take into account the external gravitational field to which these clusters are subject. The smoothed gravitational potential of the Galaxy itself is the largest such field, and the primary one which we shall consider. This field produces two effects. First, the cluster will undergo a quasi-steady tidal distortion, with stars escaping entirely whenever they enter certain 'escape regions'. Second, the cluster will gain internal energy as a result of changes in the local galactic field resulting from cluster motion. Neither of these effects is consistent with a spherically symmetric model of a cluster, but the resultant modifications of evolutionary development can be computed approximately with such symmetric models.

If we consider first the quasi-steady effects of the field, the tidal distortion can obviously not be taken into account when spherical symmetry is assumed. However, the effect of the tidal field on the rate of escape can be computed if we assume that any star reaching the distance r_e from the cluster center will escape; the escape radius, r_e, can then be set equal to the minimum distance from the cluster center to the actual escape regions. In fact, stars in some directions can be stably bound to the cluster for substantial times at distances much exceeding r_e, but should escape in time when encounters deflect them into the true escape regions.

The most serious approximation associated with the assumption of spherical symmetry is that the angular momentum of each star remains constant in the absence of gravitational encounters between stars. The tidal field would be expected to make the velocities of the halo star more nearly isotropic. Hence a star diffusing towards zero energy, as a result of encounters during periodic passages through the isothermal core, might penetrate progressively less deeply into the core in its successive orbits; as pointed out by Aarseth and Woolf (1972), this effect would be expected to decrease the escape rate somewhat, offsetting at least partially the effect of reduced r_e. On the other hand, numerical orbit computations by Prata (1971) showed that an external tidal field did not in fact produce an isotropic distribution of velocities for r less than about $\frac{1}{2}r_e$; possibly this result is produced by some third integral which constrains the orbit. The only firm conclusion indicated at present is that one must question the validity of results obtained with spherically symmetric models of tidally perturbed clusters.

We consider next effects produced by changes in the external field experienced by the cluster. As pointed out by Ostriker et al. (1972) rapid motion of a globular cluster through the galactic disc produces a transient compressive force perpendicular to the galactic plane, which can effectively heat a cluster; this effect is called a 'compressive shock'. This increase of internal energy can be computed for a spherically symmetric model, neglecting the transitory nonspherical distortions. The problem of the constancy of angular momentum in spherically symmetric models is unimportant in this case, since shock heating destroys this constancy in any case, and the computed models should be reasonably realistic.

Monte Carlo models of systems with tidal cut-offs and with shock heating have been computed both for one-component (Spitzer and Chevalier, 1972 – Paper V) and three-component (Spitzer and Shull, 1975 – Paper VII) systems. Because of the many parameters involved, a full analysis of these results would be outside the scope of the present paper, and only a brief general review will be given.

As would be expected, the assumption of a finite escape radius, r_e, increases the escape rate, and also accelerates the initial contraction of one-component models, decreasing the time until the core collapses. If shock heating is assumed, the rate of escape is increased even further. This effect becomes particularly marked when a distribution of masses is present, since mass stratification forces the lighter stars out to larger radii, where shock heating is most effective. Thus imposing an escape radius as small as about $3r_h$ increases the escape rate by about an order of magnitude above that for an isolated system. If moderate shock heating is assumed for a system with three mass components, another order of magnitude increase in ξ_e is produced towards the end of the evolutionary development, leading to escape of about three fourths of the lightest stars during the time required to produce collapse of the core.

Since the effectiveness of shock heating for a particular star varies as the square of the distance from the cluster center, reasonable amounts of shock heating in existing globular clusters generally have no direct effect on collapse of the inner isothermal core once this collapse has started. However, shock heating can sometimes hasten the rate of collapse by affecting the initial rate of contraction. Thus a small amount of shock heating, by getting rid of stars in the outer part of a one-component cluster, can accelerate the rate of diffusion of stars from the core up towards small total energy, increasing the initial rate of contraction and decreasing t_{coll}. In a multi-component system, on the other hand, when shock heating increases the rate of escape of the lighter stars, the number of these stars in the central core is diminished, and the rate at which the heavier stars lose energy through equipartition is somewhat reduced; this effect tends to offset the other, and shock heating has no marked effect on t_{coll} in multi-component systems. With strong shock heating, such as may have existed at an earlier stage when clusters were more extended than they are now, complete dissolution of the system can be produced. Evidently with so many physical processes occurring a wide range of evolutionary histories becomes possible.

6. Evolution of Globular Clusters

To compare the theoretical models with observed clusters we must know the values of t_{rh}, the reference relaxation time (see Equation (5)). These values depend on the cluster mass, which has been determined from velocity dispersions measured in four globular clusters (Illingworth and Freeman, 1974, and references cited therein). If we adopt the value of $1.0\ M_\odot/L_\odot$ for M/L and use the observational data on clusters obtained by Peterson and King (1975) together with the analysis of these data by Ostriker et al. (1972), values of t_{rh} can be computed for 32 clusters. The distribution of these values is shown in Figure 7. While these relaxation times cover a large spread, only one value exceeds 10^{10} yr. It seems likely that most clusters have existed for times substantially exceeding t_{rh}.

Two consequences seem to follow directly. Firstly, pronounced stratification of mass must have developed during the life of most of the clusters, provided that a wide initial distribution of masses was present initially. Secondly, if we continue to

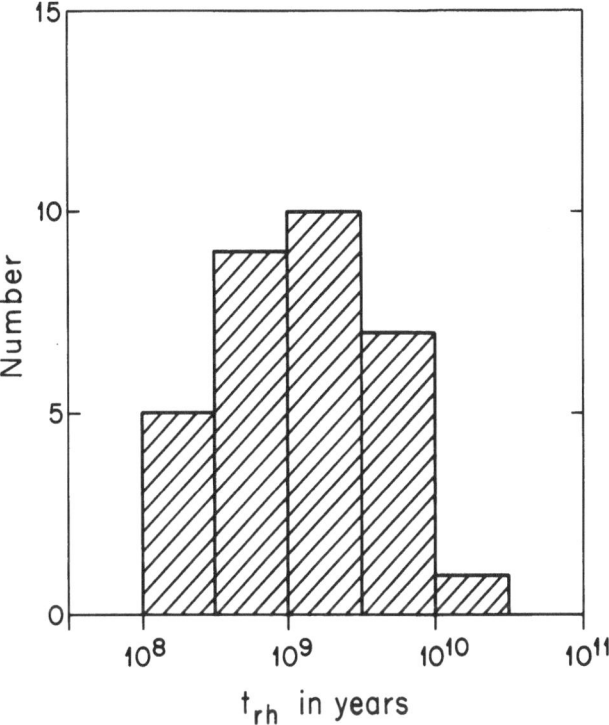

Fig. 7. Histogram showing the distribution of values of t_{rh} in 32 clusters. The quantity t_{rh} is the relaxation time for a star whose velocity equals the rms value for the cluster, and which is moving through a region of particle density equal to the mean value for the inner half of the cluster's mass.

ignore possible effects produced by an initial distribution of binaries, we must conclude that the cores of most clusters have gone through the collapse phase resulting from the gravothermal instability. We discuss each of these processes in turn.

Any stars of relatively low mass in a cluster should be relatively concentrated in the outer regions of each cluster. The prominent stars, all those above the red-giant turn-off point, all have presumably about the same mass, except perhaps for a short period before their death when mass loss is significant. The mass of these stars in the upper part of the HR diagram is estimated by Ostriker *et al.* (1972) as 0.8 $M_\odot$. Stars less massive than about 0.3 $M_\odot$ will be fainter than the 26th magnitude even in the closest clusters $(m - M \geqslant 14)$, and hence too faint to detect. Whether these faint stars have escaped entirely, as suggested by Ostriker *et al.* (1972), or whether they are present in an extended halo around each cluster cannot be determined from the present evidence. Nearly complete escape of the lighter stars might have been possible at an earlier epoch, when the cluster was considerably more diffuse and the rate of shock heating was relatively more rapid. On the other hand, in the three-component Monte Carlo computations, the distribution of the heavier stars seems not unlike the observed clusters. If the presence of the lighter stars in these models is ignored and the virial theorem is applied to the heavier stars only, the derived mass exceeds only slightly the true total mass in the heavy stars; i.e., the self-attraction of the heavy stars is the dominant influence in accounting for their velocity dispersion. If a relatively massive halo of light stars is actually present in a globular cluster, the variation of velocity dispersion of the heavy stars with distance from the center should clearly be influenced, but observational evidence on so refined a point is entirely lacking. In any case it is clear that mass stratification, plus enhanced escape of the lighter stars, would seem to offer a reasonable interpretation for the observed clusters and their low apparent M/L ratio.

Since collapse of an inner core appears to be an entirely possible event in the life of a cluster, one may inquire what the consequences of such a collapse might be. There seem to be two general possibilities for the final development of a collapsing core. On the one hand, the collapse may be terminated by the presence of binaries, which sink to lower and lower energies, ejecting other stars from the core if not from the cluster. As suggested by Heggie (1974) such binaries may form naturally as fewer and fewer stars become involved in the collapsing region. In this connection it is conceivable that a few tightly bound binaries present in the cluster may sink to the center of the system because of the higher combined mass of each binary; such binaries would then interact increasingly with the single stars composing the bulk of the contracting core.

On the other hand, if binaries are not present it would appear to be impossible to stop the collapse until the finite sizes of the stars become important. Physical collisions between stars will then occur. Since the stellar velocities at this time will be small compared to escape velocities from a stellar surface, coalescence between two colliding stars (Colgate, 1967; Sanders, 1970) will be more likely than disruption. Thus a few massive stars will form, and may either explode as supernovae or form binary systems which will tend to give up energy to single stars. If only a few supernovae were produced in each core collapse, the chance of observing such an event either in our own Galaxy or in neighboring systems would be very small, since supernovae formed in this way would constitute an extremely small addition to the overall supernova rate per galaxy.

In either case some of the mass of the core is likely to be ejected either as gas or as stars, and it is possible that the entire cluster may expand somewhat. It is difficult to see, however, how another contraction, leading to yet another collapse, can be avoided. If successive collapses follow each other and collisions between stars form an essential part of each cycle, these collapses may be expected to change their character as the stars gradually die and the heaviest stars are degenerate dwarfs. Formation of a massive black hole could be the final stage in cluster evolution. Clearly there are many challenging problems still to be explored in following the evolutionary development of a cluster to its ultimate, perhaps even apocalyptic, conclusion!

Parts of this summary were prepared and discussed at the 1974 Theoretical Astrophysics Workshop at the Aspen Center for Physics.

References

Aarseth, S. J.: 1968, *Bull. Astron., Ser. 3* **3**, 105.

Aarseth, S. J.: 1975, 'Dynamical Evolution of Simulated Star Clusters. I: Isolated Systems' (preprint).

Aarseth, S. J. and Woolf, N. J.: 1972, *Astrophys. Letters* **12**, 159.

Albada, T. S. van: 1968, *Bull. Astron. Inst. Neth.* **19**, 479.

Ambartsumian, V. A.: 1938, *Ann. Leningrad State Univ.* No. 22 (Astron. Series, Issue 4).

Antonov, V. A.: 1962, *Bull. Leningrad State Univ.* (Ser. Math. Mech. Astr.), No. 7, 135.

Chandrasekhar, S.: 1942, *Principles of Stellar Dynamics*, University of Chicago Press, Chicago, Ch. II, V.

Colgate, S. A.: 1967, *Astrophys. J.* **150**, 163.

Heggie, D. C.: 1974, 'The Importance of Binaries in Cluster Dynamics' (preprint).

Hénon, M.: 1967, *Bull. Astron., Ser. 3* **2**, 91.

Hénon, M.: 1969, *Astron. Astrophys.* **2**, 151.

Hénon, M.: 1971, *Astrophys. Space Sci.* **13**, 284; **14**, 151.

Hénon, M.: 1973, in L. Martinet and M. Mayor (eds.), *Dynamical Structure and Evolution of Stellar Systems*, Swiss Society of Astronomy and Astrophysics Third Advanced Course, Geneva Observatory, p. 183.

Hoerner, S. von: 1960, *Z. Astrophys.* **50**, 184.

Hoerner, S. von: 1963, *Z. Astrophys.* **57**, 47.

Illingworth, G. and Freeman, K. C.: 1974, *Astrophys. J. Letters* **188**, L83.

Kamp, P. van de: 1971, *Ann. Rev. Astron. Astrophys.* **9**, 103.

King, I. R.: 1958a, *Astron. J.* **63**, 109.

King, I. R.: 1958b, *Astron. J.* **63**, 114.

King, I. R.: 1965, *Astron. J.* **70**, 376.

King, I. R.: 1966, *Astron. J.* **71**, 64.

Larson, R. B.: 1970, *Monthly Notices Roy. Astron. Soc.* **150**, 93.

Lynden-Bell, D. and Wood, R.: 1968, *Monthly Notices Roy. Astron. Soc.* **138**, 495.

Michie, R. W.: 1963, *Monthly Notices Roy. Astron. Soc.* **125**, 127.

Miller, R. H.: 1973, *Astrophys. J.* **180**, 759.

Ostriker, J. P., Spitzer, L., and Chevalier, R. A.: 1972, *Astrophys. J. Letters* **176**, L51.

Peterson, C. J. and King, I. R.: 1975, 'The Structure of Star Clusters. VI: Observed Radii and Structural Parameters in Globular Clusters' (preprint).

Prata, S.: 1971, *Astron. J.* **76**, 1017.

Rosenbluth, M. N., MacDonald, W. M., and Judd, D. L.: 1957, *Phys. Rev.* **107**, 351.

Sanders, R. H.: 1970, *Astrophys. J.* **162**, 791.

Spitzer, L.: 1940, *Monthly Notices Roy. Astron. Soc.* **100**, 396.

Spitzer, L.: 1969, *Astrophys. J. Letters* **158**, L139.

Spitzer, L. and Harm, R.: 1958, *Astrophys. J.* **127**, 544.

Spitzer, L. and Hart, M. H.: 1971a, *Astrophys. J.* **164**, 399 (Paper I).

Spitzer, L. and Hart, M. H.: 1971b, *Astrophys. J.* **166**, 483 (Paper II).

Spitzer, L. and Shapiro, S. L.: 1972, *Astrophys. J.* **173**, 529 (Paper III).

Spitzer, L. and Thuan, T. X.: 1972, *Astrophys. J.* **175**, 31 (Paper IV).

Spitzer, L. and Chevalier, R. A.: 1973, *Astrophys. J.* **183**, 565 (Paper V).

Spitzer, L. and Shull, J. M.: 1975, 'Random Gravitational Encounters and the Evolution of Spherical Systems. VII: Systems with Several Mass Groups' (preprint).

Wielen, R.: 1968, *Bull. Astron., Ser. 3* **3**, 127.

DISCUSSION

Lecar: What is the advantage of integrating the stellar orbits (in contrast to Hénon's method)?

Spitzer: The main advantage is that you can determine the escape rate. To evaluate this rate it is necessary to avoid going to the limit of large N at the beginning. When N is infinite, escape of stars becomes indistinguishable from accumulation in the far halo. For some purposes you don't really care which is which, in which case this large-N limit is appropriate to start with. But if you are interested in finding the actual escape rate, then you must keep N finite until you determine this rate, and then let N increase. As it turns out, the escape rate determined in this way is virtually independent of N.

Ipser: What is the mass to radius ratio of your core? How does it vary as the core contracts?

Spitzer: The ratio of mass to radius does not change very rapidly. This ratio varies with the mean square velocity, which goes up but rather slowly. So the ratio increases slightly with time.

King: In your first comparison of envelope profile with the $r^{-3.5}$ law it seemed to me that the slope was more like r^{-3}, which is more like what is observed. In particular, Hubble's law for elliptical galaxies, which has this slope, has always been an enigma.

Spitzer: For model F, you start with no halo at all, so it is quite difficult to build up a halo; as a result the envelope is rather curved on a log-log plot of ϱ against r. In Model D a conspicuous halo is produced in the initial collapse, and a well developed halo exists throughout the evolutionary life of the system; for this halo the $r^{-3.5}$ relationship fitted the density quite well. We tried to fit an inverse cube law to the density, but were unable to do so.

King: Can you say something about the economics of your calculations? How much does it cost to run a model, and how many man years have gone into the programming?

Spitzer: One typical evolutionary model on the Princeton IBM 360/91 costs us about $600. The computing program was written and applied by one graduate student during about half a year; in successive years three other graduate students have extended the program, again each devoting about half a year to this research effort.

HOMOLOGY IN THE EVOLUTION OF CLUSTER CORES

D. LYNDEN-BELL

Institute of Astronomy, The Observatories, Cambridge, U.K.

Abstract. It is pointed out that in advanced phases of evolution the cores of clusters ought to evolve homologously and reasons are sought as to why this evolution should be at constant core energy. It is pointed out that continued evolution after infinite core densities have been achieved is an important area for future research.

The aim of this paper is to re-emphasize certain simplifying features of cluster evolution which have not been used in the most powerful modern methods of tackling the problem. Their reintroduction might lead to a simplified theory and a greater-understanding. Research problems along these lines are re-emphasized.

At least for large clusters of equal mass stars in advanced stages of evolution the cluster cores have most of their mass at energies so well bound that Maxwell's distribution is a good approximation (Woolley, 1954; King, 1966; Spitzer *et al.*, 1972). The evolution proceeds through the changing temperature and density of this core. Now isothermal gas spheres have the same structure as one another in the sense that a scaling in radius and density suffices to bring their density profiles into the same standard shape. It is thus true that the central cores of clusters evolve almost homologously. This homology may extend even beyond the exactly isothermal energies, but it cannot extend to the whole cluster for reasons outlined below. Hénon's beautiful homological model of a whole cluster was only achieved at the cost of assuming an energy input at the centre (Hénon, 1961). It is important to discover in a neat form the homological structure of the evolving core and to predict the rate of core evolution. The discussion of the thermodynamics of isothermal spheres and their truncations given by Lynden-Bell and Wood (1968) shows that evolution of a well concentrated isothermal is not because of escape; rather it is because of the intrinsic gravothermal instability of the isothermal sphere. This gives one the hope that even the rate of the final dive to very great densities is not dependent on the outer parts of the cluster, but that the core evolves homologously and independently of the halo once it has developed sufficient central concentration.

The gravothermal instability occurs for a truncated isothermal sphere of equal mass stars when the dimensionless energy as measured from the centre is given by $u = \beta(\varepsilon + \psi_0) \sim 8.5$ where we have taken the Maxwellian at low energies to be

$$f \propto \exp(-\beta\varepsilon). \tag{1}$$

Here $\varepsilon = v^2/2 - \psi$ is the specific energy in the gravitational potential $\psi(r)$.

We shall find it useful in what follows to think of a star cluster as stratified in slices of different energy, rather as in stellar evolution a star is considered as stratified in shells of different mass (see Figure 1). For non-isothermal clusters it is useful to define

$\beta(\varepsilon)$ an inverse temperature or coolness at each energy by $\beta(\varepsilon)=-\mathrm{d}\log f(\varepsilon)/\mathrm{d}\varepsilon$. We shall denote the least value $\beta(-\psi_0)$ by β_0. A useful dimensionless variable proportional to the excess energy above the lowest value is

$$u=\beta_0(\varepsilon+\psi_0). \tag{2}$$

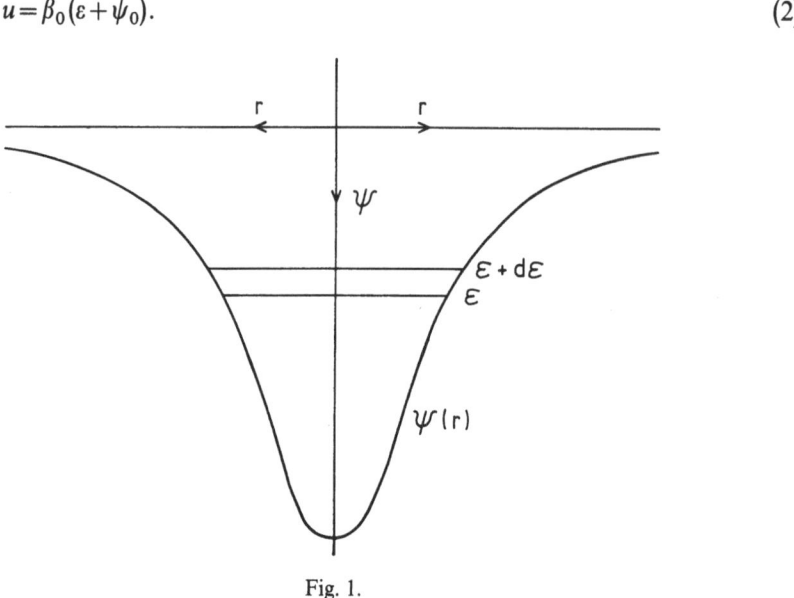

Fig. 1.

Since u is dimensionless it gives a scale invariant way of defining how far up the core a particular energy is. Since the core is evolving homologously the 'edge of the core' in energy will be defined by a particular value of u for all time. For definiteness we shall take this to be $u=8.5$ by analogy with the place at which the gravothermal instability occurs.

Let the central core density be ϱ_0, the total mass of the core be M_c all of which mass has $u<8.5$. Further let the internal energy of the core, that is the kinetic energy of these masses less their mutual potential energy, be E_c. Notice this definition has no contribution from the gravitational potential of the rest of the cluster. A good homological model would calculate $\beta_0(t)$, $\varrho(t)$, $M_c(t)$ and $E_c(t)$ from first principles as well as the distribution function $f(u)$. Let us first look at the dimensions of these quantities and the constant of gravity G.

$$\begin{aligned}
G&=[M^{-1}L^3T^{-2}]\\
\beta_0&=[L^{-2}T^2]\\
\varrho_0&=[ML^{-3}]\\
M_c&=[M]=[\beta_0^{-3/2}G^{-3/2}\varrho_0^{-1/2}]\\
E_c&=[ML^2T^{-2}]=[\beta_0^{-5/2}G^{-3/2}\varrho_0^{-1/2}]\\
u&=[1]
\end{aligned} \tag{3}$$

By the homology assumption the quantities $G^{3/2}\beta_0^{3/2}\varrho_0^{1/2}M_c$ and $G^{3/2}\beta_0^{5/2}\varrho_0E_c$ are dimensionless and will thus be time independent. Now as Hénon has pointed out

(Hénon, 1969) there is no escape from an isolated cluster if its evolution is governed solely by a diffusion equation. Thus if we tried to make our whole cluster homologous we would have E and M constant and we could therefore deduce β_0 and ϱ_0 were constant. However there is always evolution in the presence of temperature gradients and there is no finite mass isothermal sphere. Hence our homology assumption applied to an isolated cluster is wrong. If alternatively we try to build a homologous evolution with a tidal mass loss which confines the mean density $M(\frac{4}{3}\pi r_e^3)^{-1}$ within the tidal radius to be constant, then by homology ϱ_0 must be constant. Further during escape by diffusion

$$\dot{E} = M\dot{\varepsilon}_e = -\dot{M}\frac{GM}{r_e} = -\dot{M}GM^{2/3}\left(\frac{3\varrho_t}{4\pi}\right)^{1/3}. \tag{4}$$

Thus

$$E + \tfrac{3}{5}GM^{5/3}\left(\frac{3\varrho_t}{4\pi}\right)^{1/3} = E + \tfrac{3}{5}\frac{GM^2}{r_e} = constant = E_1. \tag{5}$$

The constant E_1 has the dimensions of energy, so $\beta^{5/2}G^{3/2}\varrho^{1/2}E_1$ must be constant by homology, which implies that β_0 is also constant unless E_1 is zero. If we define $\bar{r}$ by writing $E = (-GM^2/2\bar{r})$ the $E_1 = 0$ case gives $\bar{r} = \tfrac{5}{6}r_e$ which is so for a cluster of uniform density and can only be the case for a cluster with a weak central concentration. Homology therefore fails for complete clusters. However, let us return to homology as applied to the cores alone and consider escape from the core. We can write:

$$\dot{E}_c = \dot{M}_c\varepsilon_e - \dot{Q}, \tag{6}$$

where $\dot{Q}$ is the heat flux that comes out of the core and $\dot{M}\varepsilon_e$ is the energy change due to mass loss. We may write $\dot{Q} = \alpha\dot{M}_c\varepsilon_e$ where α is a constant by homology. Using definitions of r_e and $\bar{r}$ defined now for the core rather than the cluster we have

$$\dot{E}_c = \left(\frac{-GM_c}{r_e}\right)(1-\alpha) = \dot{M}_c\frac{E_c}{M_c}\frac{2\bar{r}}{r_e}(1-\alpha). \tag{7}$$

Thus

$$E_c \propto M_c^{\zeta} \quad \text{where} \quad \zeta = \frac{2\bar{r}}{r_e}(1-\alpha). \tag{8}$$

Homology alone will not give the value of ζ without further physical reasoning to give us the value of α, the ratio of the heat loss to the energy carried away by mass loss. There is a weak but not to my mind wholly convincing argument for taking $\alpha = 1$ and $\zeta = 0$. This argument demands that as each star leaves the core, the core must supply the wherewithal for its removal from core influence. Thus as each star leaves the core there must be just sufficient heat given out to ensure that the energy is available to free it completely from the gravity of the core. This implies.

$$\dot{M}_c\varepsilon_e = \dot{Q} \tag{9}$$

and therefore

$$\dot{E}_c = 0. \tag{10}$$

There is some evidence for the constancy of E_c from the detailed calculations of several investigators (Hénon, 1971, 1973; Spitzer *et al.*, 1972; Spitzer, 1973; Aarseth 1974; Larson, 1970) but I am not yet fully persuaded whether ζ is rather small as one would expect for $r_e \gg \bar{r}$ and $\alpha \sim \frac{5}{6}$ say, or whether ζ is really zero. If indeed E_c is accurately constant, surely there must be simple and fully convincing reason why it must be so. The problem of finding such an argument is an important challenge to theory.

It is perhaps worth recording the consequences of the $E_c =$ constant assumption although equivalent results have been given many times before.

(1) Since E_c and $G^{3/2} \beta_0^{5/2} \varrho_0^{1/2} E_c$ are both constant hence $\beta_0^5 \varrho_0$ is constant. Since $\beta_0^{3/2} \varrho_0^{1/2} M_c G^{3/2}$ is constant we have $M_c \propto \beta_0 \propto \varrho_0^{-1/5}$.

(2) Following a well-trodden path and introducing a relaxation time T we have by homology that

$$\frac{d\varrho_0}{dt} = c\frac{\varrho_0}{T}, \tag{11}$$

where c is a constant which depends on the precise definition of T. Now in homological evolution

$$T \propto \frac{v^3}{G^2 m\varrho \log N} \propto \frac{\beta_0^{-3/2}}{G^2 m\varrho_0 \log N}, \tag{12}$$

where m is the stellar mass and N is the number of stars in the cluster. Thus

$$\frac{1}{M_c}\frac{dM_c}{dt} = -\frac{1}{5}\frac{1}{\varrho_0}\frac{d\varrho_0}{dt} = -c_1 \beta_0^{3/2} \varrho_0 \log N = -c_2 M_c^{-7/2} \log N \tag{13}$$

if we take N to be the constant number of stars in the whole system or ignore the variation in $\log N$ (see Appendix) we have

$$dt \propto -M_c^{5/2}\, dM_c$$
$$t_0 - t \propto M_c^{7/2}. \tag{14}$$

Thus

$$M_c \propto (t_0 - t)^{2/7} \propto \beta_0$$
$$\varrho_0 \propto (t_0 - t)^{-10/7} \tag{15}$$
$$r_c \propto (t_0 - t)^{4/7}$$

and of course by assumption $E_c =$ constant. Notice that the temperature and density become infinite as $t_0 \to t$ but the mass of the core becomes zero leaving the energy constant.

Similar results can of course be written down for any particular value of ζ. In all cases the calculation of the distribution function and the value of the evolution is a harder task, but Hénon's vintage paper of 1961 shows us the way.

Since no mass is involved in the final dense core we should not stop here, but should be willing to attack the further evolution of clusters after infinite central density is formally achieved. Do they eventually create binaries at the centre and evolve into Hénon's homologous model with the otherwise mysterious energy source at the singularity, or do they continue with a non-homologous evolution? I should point out that Hénon's infinite density homologous model has a finite central temperature in contrast to the constant E_c assumption. At infinite density this can only be achieved with β_0 constant and $\zeta = 1$ and our formula for ζ makes this unreasonable.

What is certain is that the many body problem eventually gets replaced by the few body problem at its very centre. It is likely that in the end Aarseth's heavy central binary may form with a significant fraction of the cluster energy. It will be important to discuss the scenarios after that, bearing in mind that such heavy stars are short lived.

Appendix

Although the theory of the cut off in the calculation for the relaxation time in a cluster of variable density is not well established, it is probably more realistic not to take N constant but instead equal to the number of stars in the core, N_c. It is, I believe, an accident that this actually leads to a slow-down in cluster evolution just before infinite density is achieved because one gets

$$\frac{1}{N_c}\frac{dN_c}{dt} = -c_3 N_c^{-7/2} \log N_c$$

and so

$$t_0 - t \propto \int \frac{N_c^{5/2}\, dN_c}{\log N_c}$$

which may be expressed in terms of the exponential integral; for $N_c \gg 1$ a rough approximation is $\propto N_c^{7/2}/\log N_c$ which is no doubt marginally better then the neglect of the $\log N$ variation, but it is a price not worth paying for the ugly formulae that result. As the real changes occur when the core is reduced to the few body problem the theory is no good at this level anyhow.

References

Aarseth, S. J.: 1974, *Astron. Astrophys.* **35**, 237.
Hénon, M.: 1961, *Ann. Astrophys.* **24**, 369.
Hénon, M.: 1969, *Astron. Astrophys.* **2**, 151.
Hénon, M.: 1971, *Astrophys. Space Sci.* **13**, 284.
Hénon, M.: 1973, in L. Martinet and M. Mayor (eds.), *Dynamical Structure and Evolution of Stellar Systems*, Swiss Society of Astronomy and Astrophysics Third Advanced Course, Geneva Observatory.
King, I. R.: 1958, *Astron. J.* **63**, 114.
King, I. R.: 1966, *Astron. J.* **71**, 64.
Larson, R. B.: 1970, *Monthly Notices Roy. Astron. Soc.* **150**, 93.
Lynden-Bell, D. and Wood, R.: 1968, *Monthly Notices Roy. Astron. Soc.* **138**, 495.
Miller, R. H. and Parker, E. N.: 1964, *Astrophys. J.* **140**, 50.
Spitzer, L. and Thuan, T. X.: 1972, *Astrophys. J.* **175**, 31.
Spitzer, L.: 1973, *Astrophys. J.* **183**, 565.
Woolley, R. v. d. R.: 1954, *Monthly Notices Roy. Astron. Soc.* **114**, 191.

DISCUSSION

Miller: It is always surprising that one thinks that thermodynamics might work for self-gravitating systems. Could you say something about why you think thermodynamics might provide a valid description for a star cluster?

Lynden-Bell: I would consider the only systems to which thermodynamics are not applicable in equilibrium are those which have divergences in phase space. Here the frozen equilibrium concept is important for eliminating binaries and consideration of the core alone eliminates the divergence at infinity.

Feix: Validity of thermodynamics for physical systems is not only connected to infinities (divergences) for large negative energies (or zero energy) but is a more general question which arises for all systems. In fact it is a question of how the 'total information' ($6N$ 'data') can be reduced to a few characteristics numbers (how and which are these numbers). Thermodynamics of equilibrium systems tells us that density and temperature are enough but the question is when can we tell that a system is in equilibrium. Plasma and self gravitating gas imply certainly more sophisticated information, probably non local, maybe non Markovian (i.e. implying the past history of the system). It is interesting to notice that this question is both fundamental in statistical physics and very practical in computational physics where obviously the amount of returned information must be finite.

CORRELATION FUNCTIONS FOR THE
GRAVITATIONAL FORCE

L. COHEN*

Hunter College of The City University of New York, New York, N.Y., U.S.A.

Abstract. Spatial and time correlations of the force acting on a star are derived for finite gravitational systems. It is shown that their behavior is qualitatively different than that for infinite mediums.

The dissolution time for a binary system is considered. We explain why Chandrasekhar's dissolution time differs from that given by Ambartsumian and Oort in that it does not depend on the velocities of the field stars. We show that the difference lies in the definition of what constitutes 'relative change in velocity' of the two stars in the binary. Indeed, using the general approach of Chandrasekhar and von Neumann (appropriately modified) we derive a velocity dependent dissolution time.

1. Introduction

The total force on a star may be expressed as the sum of two forces which have different behavior. One force, due to the smoothed out distribution of the system as a whole, changes very slowly and can be expressed as the gradient of the smoothed out potential. The random part of the force is due to the rapid change of the positions of the nearby stars and its description must be stochastic. Chandrasekhar and Von Neumann (Chandrasekhar, 1941, 1944a, b, c; Chandrasekhar and Von Neumann, 1942, 1943) have developed a statistical theory of the random force under certain simplifying assumptions. They assumed that the field stars are not correlated with themselves or the test star, and that the field stars in the neighborhood of the test star are uniformly distributed. Furthermore, the number of field stars were taken to be infinite (i.e., an infinite medium) in such a manner as to keep the density constant.

Numerical experiments performed to test various aspects of the theory (Ahmad and Cohen, 1972, 1973, 1974) have shown excellent agreement with theory for the distribution of random force (Holtsmark distribution), the time rate of change of the random force and dynamical friction. The experiments to verify the two time autocorrelation function showed that the experimental curve decreased faster for large times than that predicted by theory.

Chandrasekhar's result for the time autocorrelation for force is that the decrease with time, for large times, is extremely slow, namely as $1/t$. Similarly, for large separation distance, the force correlation acting at two different points at the same time was shown to decrease only as its inverse.

We shall calculate below space and time autocorrelation functions for force for bounded gravitational systems and show that its behavior is qualitatively different from that for infinite systems. Also, we shall calculate the correlations of the forces acting at two different points at two different times and discuss its application to the problem of the stability of binaries.

* Work supported by a grant from The City University Faculty Research Award Program.

Hayli (ed.), Dynamics of Stellar Systems, 33–45. All Rights Reserved.

2. Force Correlations

We consider a finite spherical system of radius λ, density n, and assume that the field stars move in linear orbits. To keep the density constant, we shall introduce stars into the system at the same rate they are leaving it. To accomplish this, we distribute stars over all space with a density n, and with a velocity probability distribution, $\tau(\mathbf{v})$, as the stars in the real system. It is clear that as long as the velocity distribution is not a function of position, the density within λ will be constant. A field star will be 'counted' in any averaging only if at that time it is inside the sphere λ. If two times are involved, then it must be within the sphere at both times. That is, at time t a field star whose original position and velocity is $\mathbf{r}$, $\mathbf{v}$, must satisfy

$$|\mathbf{r}_i + \mathbf{v}_i t| < \lambda. \tag{1}$$

Also, we shall impose a lower limit, ε_i, restricting the distance with which a field star can approach the test star which is at position $\mathbf{r}_1$, say;

$$\varepsilon_i < |\mathbf{r}_i + \mathbf{v}_i t - \mathbf{r}_1|. \tag{2}$$

The lower limit ε_i, is a function of the relative velocity of the field and test star and is to be obtained by requiring the relative energy to be positive. Otherwise it would be a binary.

In the following we shall take all the masses to be equal and consider the case where the test stars are stationary. Further we shall not take the ε_i's as function of the relative velocities but take them to be constants and estimate them by

$$\varepsilon_i \equiv \varepsilon \sim \frac{2Gm}{\langle v^2 \rangle}. \tag{3}$$

The force correlation function at two points $\mathbf{r}_1, \mathbf{r}_2$, at two different times, t_1, t_2 is then

$$\langle F(\mathbf{r}_1, t_1) \cdot F(\mathbf{r}_2, t_2) \rangle = G^2 m^2 \sum_{\substack{ij \\ i \neq j}} \left\langle \frac{\mathbf{r}_i + \mathbf{v}_i t_1 - \mathbf{r}_1}{|\mathbf{r}_i + \mathbf{v}_i t_1 - \mathbf{r}_1|^3} \cdot \frac{\mathbf{r}_j + \mathbf{v}_j t_2 - \mathbf{r}_2}{|\mathbf{r}_j + \mathbf{v}_j t_2 - \mathbf{r}_2|^3} \right\rangle +$$

$$+ G^2 m^2 \sum_i \left\langle \frac{\mathbf{r}_i + \mathbf{v}_i t_1 - \mathbf{r}_1}{|\mathbf{r}_i + \mathbf{v}_i t_1 - \mathbf{r}_1|^3} \cdot \frac{\mathbf{r}_i + \mathbf{v}_i t_2 - \mathbf{r}_2}{|\mathbf{r}_i + \mathbf{v}_i t_2 - \mathbf{r}_2|^3} \right\rangle; \tag{4}$$

$$|\mathbf{r}_i + \mathbf{v}_i t_1| < \lambda$$
$$|\mathbf{r}_i + \mathbf{v}_i t_2| < \lambda$$
$$\varepsilon < |\mathbf{r}_i + \mathbf{v}_i t_1 - \mathbf{r}_1|$$
$$\varepsilon < |\mathbf{r}_i + \mathbf{v}_i t_2 - \mathbf{r}_2|.$$

Since the field stars are uncorrelated, the first part of (4) vanishes.

$$\langle F(\mathbf{r}_1, t_1) \cdot F(\mathbf{r}_2, t_2) \rangle = G^2 m^2 \sum \left\langle \frac{\mathbf{r}_i + \mathbf{v}_i t_1 - \mathbf{r}_1}{|\mathbf{r}_i + \mathbf{v}_i t_1 - \mathbf{r}_1|^3} \cdot \frac{\mathbf{r}_i + \mathbf{v}_i t_2 - \mathbf{r}_2}{|\mathbf{r}_i + \mathbf{v}_i t_2 - \mathbf{r}_2|^3} \right\rangle$$

$$= G^2 m^2 n \int \frac{\mathbf{r} + \mathbf{v}t_1 - \mathbf{r}_1}{|\mathbf{r} + \mathbf{v}t_1 - \mathbf{r}_1|^3} \cdot \frac{\mathbf{r} + \mathbf{v}t_2 - \mathbf{r}_2}{|\mathbf{r} + \mathbf{v}t_2 - \mathbf{r}_2|^3} \tau(\mathbf{v}) \, d\mathbf{r} \, d\mathbf{v} \quad (5)$$

$$|\mathbf{r} + \mathbf{v}t_1| < \lambda, \qquad |\mathbf{r} + \mathbf{v}t_2| < \lambda$$

$$\varepsilon < |\mathbf{r} + \mathbf{v}t_1 - \mathbf{r}_1|, \quad \varepsilon < |\mathbf{r} + \mathbf{v}t_2 - \mathbf{r}_2|.$$

Performing the transformation

$$\boldsymbol{\varrho} = \mathbf{r} + \mathbf{v}t_1; \qquad d\boldsymbol{\varrho} = d\mathbf{r} \tag{6}$$

$$\langle \mathbf{F}(\mathbf{r}_1, t_1) \cdot \mathbf{F}(\mathbf{r}_2, t_2) \rangle = G^2 m^2 n \int \frac{\boldsymbol{\varrho} - \mathbf{r}_1}{|\boldsymbol{\varrho} - \mathbf{r}_1|^3} \cdot \frac{\boldsymbol{\varrho} + \mathbf{v}t - \mathbf{r}_2}{|\boldsymbol{\varrho} + \mathbf{v}t - \mathbf{r}_2|^3} \tau(\mathbf{v}) \, d\boldsymbol{\varrho} \, d\mathbf{v} \tag{7}$$

$$|\boldsymbol{\varrho}| < \lambda, \ |\boldsymbol{\varrho} + \mathbf{v}t| < \lambda, \ |\boldsymbol{\varrho} - \mathbf{r}_1| > \varepsilon, \ |\boldsymbol{\varrho} + \mathbf{v}t - \mathbf{r}_2| > \varepsilon$$

where $t = t_2 - t_1$. Hence,

$$\langle \mathbf{F}(\mathbf{r}_1, t_1) \cdot \mathbf{F}(\mathbf{r}_2, t_2) \rangle = \langle \mathbf{F}(\mathbf{r}_1, 0) \cdot \mathbf{F}(\mathbf{r}_2, t_2 - t_1) \rangle. \tag{8}$$

Evaluation of (7) in its full generality is quite involved although one can find series expansions when a particular parameter (e.g., the separation $|\mathbf{r}_2 - \mathbf{r}_1|$) is small. This will be done in the last section where it will be applied to the problem of the stability of binaries. In the next two sections, we consider special cases of (7), namely the correlations at two different points at the same time and the correlation at the same point at two different times.

But in the case of infinite systems and $\varepsilon = 0$, (7) can be evaluated explicitly. Taking

$$\mathbf{r}_1 = -\mathbf{r}_2 = \mathbf{s}$$
$$\mathbf{a} = 2\mathbf{s} \tag{9}$$

we have

$$\langle \mathbf{F}(-\mathbf{s}, 0) \cdot \mathbf{F}(\mathbf{s}, t) \rangle = G^2 m^2 n \int \frac{\boldsymbol{\varrho} - \mathbf{s}}{|\boldsymbol{\varrho} - \mathbf{s}|^3} \cdot \frac{\boldsymbol{\varrho} + \mathbf{v}t + \mathbf{s}}{|\boldsymbol{\varrho} + \mathbf{v}t + \mathbf{s}|^3} \tau(\mathbf{v}) \, d\boldsymbol{\varrho} \, d\mathbf{v} =$$

$$= G^2 m^2 n \int \frac{\boldsymbol{\varrho}}{\varrho^3} \cdot \frac{\boldsymbol{\varrho} + \mathbf{v}t + \mathbf{a}}{|\boldsymbol{\varrho} + \mathbf{v}t + \mathbf{a}|^3} \tau(\mathbf{v}) \, d\boldsymbol{\varrho} \, d\mathbf{v}. \tag{10}$$

The spatial integration is straightforward

$$\int \frac{\boldsymbol{\varrho}}{\varrho^3} \cdot \frac{\boldsymbol{\varrho} + \mathbf{v}t + \mathbf{a}}{|\boldsymbol{\varrho} + \mathbf{v}t + \mathbf{a}|^3} \, d\boldsymbol{\varrho} = \frac{4\pi}{|\mathbf{v}t + \mathbf{a}|} \tag{11}$$

and accordingly

$$\langle \mathbf{F}(-\mathbf{s}, 0) \cdot \mathbf{F}(\mathbf{s}, t) \rangle = 4\pi G^2 m^2 n \int \frac{\tau(\mathbf{v})}{|\mathbf{v}t + \mathbf{a}|} \, d\mathbf{v}. \tag{12}$$

If we specialize to the Gaussian distribution

$$\tau(\mathbf{v}) = \frac{j^3}{\pi^{3/2}} \exp(-j^2 v^2) \tag{13}$$

and note that the angular integration gives

$$2\pi \int_{-1}^{1} \frac{d\mu}{(a^2 + avt\mu + v^2t^2)^{1/2}} = 4\pi \begin{cases} \dfrac{1}{a} & vt < a \\[2mm] \dfrac{1}{vt} & vt > a \end{cases} \tag{14}$$

we have after simplification

$$\langle \mathbf{F}(0,0) \cdot \mathbf{F}(\mathbf{a},t) \rangle = \frac{4\pi G^2 m^2 n}{a} \varphi\left(\frac{ja}{t}\right), \tag{15}$$

where $\varphi(x)$ is the error function

$$\varphi(x) = \frac{2}{\sqrt{\pi}} \int_{0}^{x} \exp(-x^2) \, dx. \tag{16}$$

Asymptotically the behavior of (15) is

$$\langle \mathbf{F}(0,0) \cdot \mathbf{F}(\mathbf{a},t) \rangle \sim \frac{8\sqrt{\pi}\, G^2 m^2 nj}{t} \qquad t \to \infty \tag{17}$$

$$\sim \frac{4\pi G^2 m^2 n}{a} \qquad a \to \infty. \tag{18}$$

We thus see that for infinite systems the behavior of the correlation function is inversely proportional to both the separation distance and the time when one of them approaches infinity.

2.1. CORRELATION AT TWO DIFFERENT TIMES AT THE SAME SPATIAL POINT

We now consider the correlation of the force, for finite systems, at two different times but at the same point. A detailed discussion of the two time autocorrelation has been given by Cohen and Ahmad (1974). We shall derive here the same result, for the case of a Gaussian distribution, in a more direct way.

Taking $r_1 = r_2 = 0$, (7) becomes

$$\langle \mathbf{F}(0) \cdot \mathbf{F}(t) \rangle = G^2 m^2 n \int \frac{\mathbf{\varrho}}{\varrho^3} \cdot \frac{\mathbf{\varrho} + \mathbf{v}t}{|\mathbf{\varrho} + \mathbf{v}t|^3} \tau(\mathbf{v}) \, d\mathbf{\varrho} \, d\mathbf{v} \qquad \begin{array}{l} \varepsilon < \varrho < \lambda \\ \varepsilon < |\mathbf{\varrho} + \mathbf{v}t| < \lambda. \end{array} \tag{19}$$

One could perform the ϱ integration first and thus keep the velocity distribution arbitrary (Cohen and Ahamd, 1974). This procedure is quite cumbersome and algebraically tedious. Specializing immediately to a Gaussian distribution and performing the transformation

$$\mathbf{x} = \mathbf{\varrho} + \mathbf{\eta}, \qquad d\mathbf{x} = d\mathbf{\eta}, \qquad \mathbf{\eta} \equiv \mathbf{v}t \tag{20}$$

we have

$$\langle \mathbf{F}(0) \cdot \mathbf{F}(t) \rangle = \frac{G^2 m^2 n \omega^3}{\pi^{3/2}} \int \frac{\mathbf{\varrho} \cdot \mathbf{x}}{\varrho^3 x^3} \exp\left[-\omega^2 (\mathbf{\varrho} - \mathbf{x})^2\right] d\mathbf{x} \qquad \begin{array}{l} \varepsilon < x < \lambda \\ \varepsilon < \varrho < \lambda. \end{array} \tag{21}$$

$$\omega \equiv j/t$$

Using the direction of $\mathbf{x}$ as the z axis and performing the angular integration of both $\mathbf{x}$ and $\boldsymbol{\varrho}$, we obtain

$$\langle F(0) \cdot F(t) \rangle = 8\sqrt{\pi}\, G^2 m^2 n \int_\varepsilon^\lambda \int_\varepsilon^\lambda \int_{-1}^1 \exp\left[-\omega^2 (\mathbf{x} - \boldsymbol{\varrho})^2\right] \mathrm{d}\,\mu\, \mathrm{d}x\, \mathrm{d}\varrho \qquad (22)$$

$$= 8\sqrt{\pi}\, G^2 m^2 n \left\{ \int_\varepsilon^\lambda \int_\varepsilon^\lambda \left(\frac{1}{2\omega^2 x\varrho} - \frac{1}{(2\omega^2 x\varrho)^2} \right) \times \right.$$

$$\times \exp\left[-\omega^2(x-\varrho)^2\right] \mathrm{d}x\, \mathrm{d}\varrho + \int_\varepsilon^\lambda \int_\varepsilon^\lambda \left(\frac{1}{2\omega^2 x\varrho} + \frac{1}{(2\omega^2 x\varrho)^2} \right) \times$$

$$\left. \times \exp\left[-\omega^2(x+\varrho)^2\right] \mathrm{d}x\, \mathrm{d}\varrho \right\}. \qquad (23)$$

To proceed further, we shall integrate by parts the second term in each of the integrals of (23). We note that

$$\int_\varepsilon^\lambda \frac{\exp\left[-\omega^2(x\pm\varrho)^2\right]}{\varrho^2} \mathrm{d}\varrho = \frac{1}{\varepsilon} \exp\left[-\omega^2(x\pm\varepsilon)^2\right] - \frac{1}{\lambda} \exp\left[-\omega^2(x\pm\lambda)^2\right] -$$

$$- 2\omega^2 \int_\varepsilon^\lambda \frac{\varrho\pm x}{\varrho} \exp\left[-\omega^2(\varrho\pm x)^2\right] \mathrm{d}\varrho \qquad (24)$$

and repeated use of (24) yields

$$\int_\varepsilon^\lambda \int_\varepsilon^\lambda \frac{\exp\left[-\omega^2(x+\varrho)^2\right]}{x^2\varrho^2} \mathrm{d}x\, \mathrm{d}\varrho = \frac{1}{\varepsilon^2} \exp\left[-\omega^2(2\varepsilon)^2\right] + \frac{1}{\lambda^2} \exp\left[-\omega^2(2\lambda)^2\right] -$$

$$- \frac{2}{\varepsilon\lambda} \exp\left[-\omega^2(\lambda+\varepsilon)^2\right] -$$

$$- \frac{2\omega}{\varepsilon} \sqrt{\pi}\left(\varphi(\omega(\lambda+\varepsilon)) - \varphi(2\omega\varepsilon)\right) +$$

$$+ \frac{2\omega}{\lambda} \sqrt{\pi}\left(\varphi(2\omega\lambda) - \varphi(\omega(\lambda+\varepsilon))\right) -$$

$$- 2\omega^2 \int_\varepsilon^\lambda \int_\varepsilon^\lambda \frac{\exp\left[-\omega^2(x+\varrho)^2\right]}{x\varrho} \mathrm{d}x\, \mathrm{d}\varrho \qquad (25)$$

$$\int \int \frac{\exp\left[-\omega^2(x-\varrho)^2\right]}{x^2\varrho^2} \mathrm{d}x\, \mathrm{d}\varrho = \frac{1}{\varepsilon^2} + \frac{1}{\lambda^2} - \frac{2}{\varepsilon\lambda} \exp\left[-\omega^2(\lambda-\varepsilon)^2\right] -$$

$$-\frac{2\omega}{\varepsilon}\sqrt{\pi}\,\phi(\omega(\lambda-\varepsilon))+$$

$$+\frac{2\omega}{\lambda}\sqrt{\pi}\,\varphi(\omega(\lambda-\varepsilon))+$$

$$+2\omega^2\int\limits_{\varepsilon}^{\lambda}\int\limits_{\varepsilon}^{\lambda}\frac{\exp[-\omega^2(\varrho-x)^2]}{x\varrho}\,d\varrho\,dx. \quad (26)$$

Substituting (25) and (26) into (23)

$$\langle \mathbf{F}(0)\cdot\mathbf{F}(t)\rangle = 16\sqrt{\pi}\,G^2 m^2 nj\left\{\tfrac{1}{2}\left[\frac{1}{t_1}S(t_1/t)+\frac{1}{t_4}S(t_4/t)\right]+\right.$$

$$\left.+\frac{1}{t_1 t_4}\left[t_2 S(t_2/t)-t_3 S(t_3/t)\right]-\tfrac{1}{2}t\left[\frac{1}{t_1^2}+\frac{1}{t_4^2}\right]\right\}, \quad (27)$$

where we have defined

$$t_1 = 2\varepsilon j; \qquad t_2 = (\lambda-\varepsilon)\,j; \qquad t_3 = (\lambda+\varepsilon)\,j; \qquad t_4 = 2\lambda j$$

$$S(x)=\sqrt{\pi}\,\varphi(x)+\frac{\exp(-x^2)}{x}. \quad (28)$$

Asymptotic expansions can be obtained

$$\langle \mathbf{F}(0)\cdot\mathbf{F}(t)\rangle = 4\pi G^2 m^2 n$$

$$\frac{1}{\varepsilon}-\frac{1}{\lambda}-\frac{t}{2\sqrt{\pi}j}\left(\frac{1}{\varepsilon^2}+\frac{1}{\lambda^2}\right) \qquad t\ll t_1 \qquad (29)$$

$$\frac{2j}{\sqrt{\pi}}\frac{1}{t}-\frac{1}{\lambda}-\frac{t}{2\sqrt{\pi}\,\lambda^2 j} \qquad t_1\ll t<t_4 \qquad (30)$$

$$\frac{2j^5}{3\sqrt{\pi}}\frac{(\lambda^2-\varepsilon^2)^2}{t^5} \qquad t_4\ll t. \qquad (31)$$

If λ is taken to be infinite, we recover the $1/t$ dependence obtained by Chandrasekhar (1944b) and Lee (1968). But as long as λ is kept finite, the dependence for large times is $1/t^5$.

The two time autocorrelation function affords a straightforward method of calculating the mean square velocity change. Under the assumption that the autocorrelation function is an even function of the difference in the two times, then the mean square change in the velocity within a time T can be obtained from

$$\langle(\Delta\mathbf{v})^2\rangle = 2\int\limits_{0}^{T}(T-t)\,\langle\mathbf{F}(0)\cdot\mathbf{F}(t)\rangle\,dt. \quad (32)$$

As these conditions are met in our case, we can substitute (27) into (32) and obtain, after some algebra

$$\langle(\Delta\mathbf{v})^2\rangle = 16\sqrt{\pi}\ G^2 m^2 nj \Big\{ t_1 H(t_1/T) + t_4 H(t_4/T) +$$

$$+ \frac{2}{t_1 t_4}\ [t_2^3 H(t_2/T) - t_3^3 H(t_3/T)] - \frac{T^3}{6}\Big(\frac{1}{t_1^2} + \frac{1}{t_4^2}\Big)\Big\}, \tag{33}$$

where

$$H(x) = \tfrac{1}{3}\frac{\exp(-x^2)}{x}\Big(\frac{1}{2x^2} - 1\Big) + \frac{\sqrt{\pi}}{3} + \frac{1}{2x}\ \mathscr{E}_i(x^2) + \sqrt{\pi}\ \varphi(x)\Big(\frac{1}{2x^2} - \tfrac{1}{3}\Big) \tag{34}$$

and $\mathscr{E}_i(x)$ is the exponential integral.

$$\mathscr{E}_i(x) = \int\limits_x^\infty \frac{\exp(-u)}{u}\ du. \tag{35}$$

Asymptotically

$$\langle(\Delta\mathbf{v})^2\rangle = 16\pi G^2 m^2 n$$

$$\Big(\frac{1}{\varepsilon} - \frac{1}{\lambda}\Big)\frac{T^2}{4}, \qquad T \ll t_1, \tag{36}$$

$$\frac{j}{\sqrt{\pi}}\ T \ln\frac{T}{2\varepsilon j}, \qquad t_1 \ll T \ll t_4 \tag{37}$$

$$-\frac{(\lambda-\varepsilon)^2}{3\lambda}\ j^2 + \frac{j}{\sqrt{\pi}}\ T\Big(\ln\frac{(\lambda+\varepsilon)^2}{4\varepsilon\lambda} + \frac{(\lambda-\varepsilon)^2}{2\varepsilon\lambda}\ \ln\frac{\lambda+\varepsilon}{\lambda-\varepsilon}\Big), \qquad t_4 \ll T. \tag{38}$$

The infinite system case can be obtained from (37) and we note that it has a $T \ln T$ dependence for all time. This has been derived by other methods by Hénon (1958), Ostriker and Davidson (1968), Lee (1968), and Prigogine and Severne (1960).

For finite systems $\langle(\Delta\mathbf{v})^2\rangle$ changes from a $T \ln T$ dependence for times up to t_4 into a T dependence for long times. This behavior has been discussed by Hénon (1958).

2.2. THE CORRELATION IN THE FORCE AT TWO POINTS AT THE SAME TIME

Taking $t = 0$ and $\mathbf{r}_1 = -\mathbf{r}_2 = \mathbf{s}$, we have from (7)

$$\langle \mathbf{F}(-\mathbf{s})\cdot\mathbf{F}(\mathbf{s})\rangle = G^2 m^2 n \qquad \int \frac{\boldsymbol{\varrho}-\mathbf{s}}{|\boldsymbol{\varrho}-\mathbf{s}|^3}\cdot\frac{\boldsymbol{\varrho}+\mathbf{s}}{|\boldsymbol{\varrho}+\mathbf{s}|^3}\ d\boldsymbol{\varrho} \qquad \begin{matrix} \varrho < \lambda \\ \varepsilon < |\boldsymbol{\varrho}-\mathbf{s}| \\ \varepsilon < |\boldsymbol{\varrho}+\mathbf{s}|. \end{matrix} \tag{39}$$

We shall not evaluate (39) but only give appropriate asymptotic expansions to illustrate the behavior for large separation distances in the case of finite systems.

$$\langle F(-s) \cdot F(s) \rangle = 4\pi G^2 m^2 n$$

$$\frac{1}{\varepsilon} - \frac{1}{\lambda} + \frac{1}{6}\left(\frac{1}{\lambda^3} - \frac{1}{\varepsilon^3}\right) a^2, \qquad a \ll \varepsilon \tag{40}$$

$$\frac{1}{a} - \frac{1}{\lambda} + \frac{1}{6}\frac{a^2}{\lambda^3}, \qquad \varepsilon \ll a \ll \lambda \tag{41}$$

$$-\frac{16}{3}\frac{\lambda^3}{a^4} + \frac{64}{5}\frac{\lambda^5}{a^6}, \qquad \lambda \ll a. \tag{42}$$

Again, for infinite λ we recover the $1/a$ dependence but as in the time correlation case the decrease is much faster when λ is kept finite.

3. Dissolution of Binary Systems

An application which Chandrasekhar (1944c) made of his theory is to the problem of the dissolution time of binary systems. The expression he obtained is fundamentally different from that obtained by Ambartsumian (1937) in that the dissolution time did not depend on the mean velocity of the field stars. Oort (1950) has also obtained an expression for the dissolution time and although it differs somewhat from Ambartsumian's, it does depend on the mean velocity of the field stars. Heggie (1974) has also considered the problem.

Cruz-Gonzalez and Poveda (1972) performed numerical experiments to test for agreement with theory. They found that none of the three expressions of the dissolution time was in conformity with experiment although they did find dependence on the velocity of the field stars. (But see note added in proof.)

We shall explain the reason as to why Chandrasekhar's expression is independent of the mean velocity. In particular, we shall show that the reason is not due to the statistical theory but in the definition of the 'relative velocity change' of the two components of the binary system. Indeed, we shall use the general approach of Chandrasekhar, as modified above, to obtain a velocity dependent dissolution time.

Essentially, in all three approaches the dissolution time is obtained by finding the relative absolute velocity change of the binary component within a time T and defining the dissolution time as the amount of time needed for the relative velocity change of the two components to be of the same order as the initial relative velocity. Or, equivalently, the square of the velocity change is equated to twice the mean kinetic energy of the binary. Chandrasekhar calculates the relative change, $\Delta v_1 - \Delta v_2 \equiv \Delta v_{12}$ between stars 1 and 2, constituting the binary, by considering the component of $\mathbf{F}_1 - \mathbf{F}_2$ in the direction of one of the two forces. This is appropriate since the average relative velocity change in the perpendicular direction is zero. The forces $\mathbf{F}_1$ and $\mathbf{F}_2$ are due to the field stars only.

$$\langle \Delta v_{12} \rangle = \int \left\langle \frac{(\mathbf{F}_1 - \mathbf{F}_2) \cdot \mathbf{F}_1}{|\mathbf{F}_1|} \right\rangle dt. \tag{43}$$

It is clear that if we assume field stars to be distributed over all space, $\langle (F_1 - F_2) \cdot F_1 \rangle / |F_1|$ will be independent of time and

$$\langle \Delta v_{12} \rangle = \left\langle \frac{(F_1 - F_2) \cdot F_1}{|F_1|} \right\rangle T. \tag{44}$$

In a previous paper, Chandrasekhar (1944b) obtained*, for small separation, a,

$$\left\langle \frac{(F_1 - F_2) \cdot F_1}{|F_1|} \right\rangle \sim 4\pi Gmna \tag{45}$$

where m is the average mass of a field star. Equating (45) to

$$\left(\frac{G(m_1 + m_2)}{a} \right)^{1/2}, \tag{46}$$

where m_1 and m_2 are the masses of the stars forming the binary, the dissolution time is then

$$\tau = \frac{(m_1 + m_2)^{1/2}}{4\pi G^{1/2} mna^{3/2}}. \tag{47}$$

On the other hand, Oort and Ambartsumian (using a theory developed by Bohr for the ionization of hydrogen) calculate $\langle (\Delta v_{12})^2 \rangle$, which, as will be clear from the considerations below, brings in the field velocities. We shall now proceed to calculate in the context of Chandrasekhar's Theory with the modifications described above. The introduction of ε, the cut off at small distances is essential; otherwise the integrals appearing would diverge.

We remark that we will keep the two stars forming the binary stationary. A more refined derivation would allow for the motion and take into account the interaction between them.

Also we shall assume that the separation distance, a, is much smaller than ε, in which case we can neglect r_1 and r_2 appearing in the constraints in (7). Placing the two components of the binary at positions $-s$ and s, $(a = 2s)$, we have for the mean square change in the relative velocity,

$$\langle (\Delta v_{12})^2 \rangle = \left\langle \left(\int_0^T F_1 \, dt - \int_0^T F_2 \, dt \right)^2 \right\rangle =$$

$$= \int_0^T \int_0^T \langle F(-s, t_1) \cdot F(-s, t_2) \rangle \, dt_1 \, dt_2 +$$

---

* It may be of interest to point out that $\langle (F_1 - F_2) \cdot F_1 / |F_1| \rangle$ can be estimated from the usual tidal force argument. If d is the distance to a field star, the difference in force on the two components in the direction of the field star is $(2Gm/d^3)\, a$. Inserting 4π to take into account averaging over the sphere of radius a and estimating d by the interparticle distance $d \sim n^{-1/3}$, we have

$$\langle (F_1 - F_2) \cdot F_1 / |F_1| \rangle \sim 8\pi Gmna.$$

$$+ \int \int \langle \mathbf{F}(\mathbf{s}, t_1) \cdot \mathbf{F}(\mathbf{s}, t_2) \rangle \, dt_1 \, dt_2 -$$

$$-2 \int \int \langle \mathbf{F}(-\mathbf{s}, t_1) \cdot \mathbf{F}(\mathbf{s}, t_2) \rangle \, dt_1 \, dt_2. \tag{48}$$

Using (32) and remembering that we will take $a \ll \varepsilon$

$$\langle (\Delta \mathbf{v}_{12})^2 \rangle = 4 \int_0^T \langle \Delta F_{12} \rangle (T-t) \, dt, \tag{49}$$

where, for convenience, we have defined

$$\langle \Delta F_{12} \rangle = \langle \mathbf{F}(\mathbf{s}, 0) \cdot \mathbf{F}(\mathbf{s}, t) \rangle - \langle \mathbf{F}(-\mathbf{s}, 0) \cdot \mathbf{F}(\mathbf{s}, t) \rangle \tag{50}$$

$$= \frac{G^2 m^2 n j^3}{\pi^{3/2}} \int \left(\frac{\mathbf{r}+\mathbf{s}}{|\mathbf{r}+\mathbf{s}|^3} - \frac{\mathbf{r}-\mathbf{s}}{|\mathbf{r}-\mathbf{s}|^3} \right) \cdot \frac{\mathbf{r}+\mathbf{s}+\mathbf{v}t}{|\mathbf{r}+\mathbf{s}+\mathbf{v}t|^3} \exp(-j^2 v^2) \, d\mathbf{r} \, d\mathbf{v}$$

$$\varepsilon < r < \lambda; \quad \varepsilon < |\mathbf{r}+\mathbf{v}t| < \lambda. \tag{51}$$

Following the same procedure as in Section 2.1., we have

$$\langle \Delta F_{12} \rangle = \frac{G^2 m^2 n \omega^3}{\pi^{3/2}} \int \left(\frac{\boldsymbol{\varrho}+\mathbf{s}}{|\boldsymbol{\varrho}-\mathbf{s}|^3} - \frac{\boldsymbol{\varrho}-\mathbf{s}}{|\boldsymbol{\varrho}-\mathbf{s}|^3} \right) \cdot \frac{\mathbf{x}+\mathbf{s}}{|\mathbf{x}+\mathbf{s}|^3} \times$$

$$\times \exp\left[-\omega^2 (\mathbf{x}-\boldsymbol{\varrho})^2 \right] d\mathbf{x} \, d\boldsymbol{\varrho} \tag{52}$$

$$\varepsilon < x < \lambda; \quad \varepsilon < \varrho < \lambda.$$

As s is small in comparison to ε, we can expand (52) as a power series in s.

$$\left(\frac{\boldsymbol{\varrho}+\mathbf{s}}{|\boldsymbol{\varrho}+\mathbf{s}|^3} - \frac{\boldsymbol{\varrho}-\mathbf{s}}{|\boldsymbol{\varrho}-\mathbf{s}|^3} \right) \cdot \frac{\mathbf{x}+\mathbf{s}}{|\mathbf{x}+\mathbf{s}|^3} \sim \frac{2\mu_1 - 6\mu\mu_2}{\varrho^3 x^2} s +$$

$$+ \frac{2 - 6\mu_2^2 - 6\mu_1^2 + 18\mu\mu_1\mu_2}{\varrho^3 x^3} s^2 \cdots, \tag{53}$$

where the cosine of the angles are defined as follows

$$\boldsymbol{\varrho} \cdot \mathbf{s} = \varrho s \mu_2$$
$$\mathbf{x} \cdot \mathbf{s} = x s \mu_1 \tag{54}$$
$$\mathbf{x} \cdot \boldsymbol{\varrho} = x \varrho \mu.$$

If we take a spherical coordinate system with the z axis in the direction of $\mathbf{x}$ and place $\mathbf{s}$ in the y, z plane, then

$$\mu_2 = \mu_1 \mu + \sqrt{1-\mu_1^2} \sqrt{1-\mu^2} \cos\varphi, \tag{55}$$

where φ is the azimuthal angle.

Consider the angular integrations

$$\int_0^{2\pi} \mu_2 \, d\varphi = 2\pi \mu_1 \mu \tag{56}$$

$$\int_0^{2\pi} \mu_2^2 \, d\varphi = 2\pi \mu_1^2 \mu^2 + \pi(1 - \mu_1^2)(1 - \mu^2) \tag{57}$$

$$\int_{-1}^{1} \mu_1 \, d\mu_1 = 0. \tag{58}$$

Because of (56) and (58), the s term of (53) is zero.

Integrating all the angles, except μ, in the s^2 term gives

$$\langle \Delta F_{12} \rangle = 16\sqrt{\pi} \, G^2 m^2 n \omega^3 s^2 \times$$

$$\times \int_\varepsilon^\lambda \int_\varepsilon^\lambda \int_{-1}^1 \frac{3\mu^2 - 1}{x\varrho} \exp[-\omega^2 (x - \varrho)^2] \, dx \, d\varrho \, d\mu. \tag{59}$$

Performing the μ integration we obtain

$$\langle \Delta F_{12} \rangle = -16\sqrt{\pi} \, G^2 m^2 n \omega^3 s^2 \times$$

$$\times \left\{ \int_\varepsilon^\lambda \int_\varepsilon^\lambda G(x, \varrho) \, dx \, d\varrho + \int_\varepsilon^\lambda \int_{-\varepsilon}^{-\lambda} G(x, \varrho) \, dx \, d\varrho \right\}, \tag{60}$$

where

$$G(x, \varrho) = \frac{1}{x\varrho} \left(\frac{1}{\omega^2 x\varrho} + \frac{3}{2} \frac{1}{\omega^4 x^2 \varrho^2} + \frac{3}{4} \frac{1}{\omega^6 x^3 \varrho^3} \right) \exp[-\omega^2 (x + \varrho)^2]. \tag{61}$$

The indefinite integral of $G(x, \varrho)$ can be obtained by integration by parts,

$$\int \int G(x, \varrho) \, d\varrho \, dx = \left\{ \frac{1}{12\omega^6 x^3 \varrho^3} - \frac{1}{6\omega^4 \varrho^3 x} - \frac{1}{6\omega^4 \varrho x^3} + \right.$$

$$\left. + \frac{1}{6\omega^4 \varrho^2 x^2} \right\} \exp[-\omega^2 (x + \varrho)^2] -$$

$$- \frac{1}{3\omega^3} \left(\frac{1}{\varrho^3} + \frac{1}{x^3} \right) \frac{\sqrt{\pi}}{2} [\omega(x + \varrho)]. \tag{62}$$

We shall consider here the case of $\lambda = \infty$, in which case

$$\langle \Delta F_{12} \rangle = 6\sqrt{\pi} G^2 m^2 n \omega^3 s^2 \left\{ \left(\frac{1}{12\omega^6 \varepsilon^6} - \frac{1}{6\omega^4 \varepsilon^4} \right) \{1 - \exp[-(2\omega\varepsilon)^2]\} - \right.$$

$$\left. - \frac{1}{3\omega^4 \varepsilon^4} + \frac{\sqrt{\pi}}{3\omega^3 \varepsilon^3} \phi(2\varepsilon\omega) \right\} \tag{63}$$

and inserting (63) into (49)

$$\langle (\Delta v_{12})^2 \rangle = 16\sqrt{\pi}\ G^2 m^2 n j^3 a^2 \Big\{ \tfrac{4}{15} \frac{T^5}{(2\varepsilon j)^6} \{ 1 - \exp[-(2\varepsilon j/T)^2] \} +$$

$$+ \tfrac{4}{3} \frac{T^3}{(2\varepsilon j)^4} \{ \tfrac{4}{5} \exp[-(2\varepsilon j/T)^2] - 1 \} +$$

$$+ \tfrac{8}{15} \frac{T}{(2\varepsilon j)^2} \exp[-(2\varepsilon j/T)^2] - \frac{8\sqrt{\pi}}{15(2\varepsilon j)} \left[1 - \varphi\left(\frac{2\varepsilon j}{T}\right) \right] +$$

$$+ \tfrac{4}{3} \frac{\sqrt{\pi}}{(2\varepsilon j)^3} T^2 \varphi(2\varepsilon j/T) \Big\}. \tag{64}$$

For long times (64) asymptotically approaches

$$\langle (\Delta v_{12})^2 \rangle = 16\sqrt{\pi}\ G^2 m^2 n j^3 a^2\ \frac{2T}{(2\varepsilon j)^2}, \qquad T \to \infty. \tag{65}$$

Equating (65) to the square of (46), and using

$$j^2 = \frac{3}{2\langle v^2 \rangle}$$

$$\varepsilon \sim \frac{2Gm}{\langle v^2 \rangle},$$

we have for the time of dissolution (for equal masses)

$$\tau = \sqrt{\frac{2}{3\pi}} \frac{Gm}{a^3 v^3 n}. \tag{66}$$

This agrees, in functional dependence, with the expression of Ambartsumian for the case of $a \ll 2Gm/\langle v^2 \rangle$.

Acknowledgement

The author gratefully acknowledges a number of criticisms and corrections made by Drs S. Aarseth and D. Heggie and for bringing to the authors attention the paper by M. Hénon.

Note added in proof. Dr Heggie has made an extensive study of the evolution of binary stars ('The Dynamical Evolution of Binary Stars', Thesis, Cambridge University Press, 1971). His result for hard binaries is that their disruption rate is exponentially small. Dr Heggie has pointed out (private communication) that the assumption of $a \ll \varepsilon$ implies (if the result of Equation (65) is to be applied to the binary problem) that the relative motion of the binary components is much faster than that of the field stars and hence the assumption of keeping s constant may be a poor one. The calculation can be modified by taking s to be time dependent and a function of the relative velocity of the binary components. This may change the final result significantly. But nonetheless, it is clear that the statistical theory of Chandrasekhar and Von Neumann will give a field star velocity dependence for the dissolution time if the autocorrelation function is modified as described in Section 1.

Regarding the numerical experiments of Cruz-Gonzalez and Poveda, M. Hénon (*Astron. Astrophys.* **19** (1972), 488), has shown that the method of simulating the field stars was incorrect and that when proper account is taken of this fact the numerical results yield a better agreement with Oort's formula for the dissolution of binaries.

References

Ahmad, A. and Cohen, L.: 1972, *Phys. Letters* **42A**, 243.
Ahmad, A. and Cohen, L.: 1973, *Astrophys. J.* **179**, 885.
Ahmad, A. and Cohen, L.: 1974, *Astrophys. J.* **188**, 469.
Ambartsumian, V. A.: 1937, *Astron. Zh.* **14**, 207.
Chandrasekhar, S.: 1941, *Astrophys. J.* **94**, 511.
Chandrasekhar, S.: 1944a, *Astrophys. J.* **97**, 47.
Chandrasekhar, S.: 1944b, *Astrophys. J.* **99**, 25.
Chandrasekhar, S.: 1944c, *Astrophys. J.* **99**, 54.
Chandrasekhar, S.: 1960, *Principles of Stellar Dynamics*, Dover Publications Inc., New-York.
Chandrasekhar, S. and von Neumann, J.: 1942, *Astrophys. J.* **95**, 489.
Chandrasekhar, S. and von Neumann, J.: 1944, *Astrophys. J.* **97**, 1.
Cohen, L. and Ahmad, A.: submitted to *Astrophys. J.*
Cruz-Gonzalez, C. and Poveda, A.: 1972, in M. Lecar (ed.), *Gravitational N-Body Problem*, D. Reidel Publ. Co., Dordrecht-Holland, p. 99.
Heggie, D. C.: 1975, this volume, p. 73.
Hénon, M.: 1958, *Ann. Astrophys.* **21**, 186.
Lee, E.: 1968, *Astrophys. J.* **151**, 687.
Oort, J. H.: 1950, *Bull. Astron. Inst. Neth.* **11**, 91.
Ostriker, J. P. and Davidson, A. F.: 1968, *Astrophys. J.* **151**, 679.
Prigogine, I. and Severne, G.: 1966, *Physica* **32**, 1376.

DISCUSSION

Severne: Is it consistent here to use simultaneously the approximation of straight line trajectories and finite system size?

Cohen: One could take other than linear orbits depending on the problem of hand. But for most situation linear orbits are a good approximation – and simple to work with.

Lynden-Bell: I would just like to get clear exactly what stars you consider the force from. Your calculation essentially considers only forces from the stars that move with the point considered.

Cohen: Yes. Also, stars are counted in the averaging only if they are within the system at *both* times.

VIBRATIONS OF NON-UNIFORM SPHERICAL SYSTEMS

G. SEVERNE

Fakulteit Wetenschappen, Vrije Universiteit Brussel, Brussels, Belgium

and

A. KUSZELL

Institute of Nuclear Research, Warsaw-Swirck, Poland

Abstract. Dispersion relations have been obtained and analyzed for a non-uniform, non-rotating spherical gravitational system. A restriction to short wavelengths makes it possible to consider a linearized form of the collisionless Boltzmann equation, differing from that for homogeneous systems by the appearance of a term expressing the effect of the mean self-gravitational field upon the motion. The mean field affects the radially directed wave perturbations, with a breaking of symmetry.

Inwardly and outwardly directed modes have quite different propagation characteristics, inward modes being preferentially propagated. Locally, the stability of the system is enhanced due to the effect of the mean field.

From the point of view of statistical mechanics, and more particularly with reference to the crucial problems of the approach towards a quasi equilibrium and the degree of applicability of a thermodynamic description, it must be recognized that there is still much to accomplish in the study of gravitational systems. A major difficulty is the existence of a strong field and of a large inhomogeneity. The motivation for the work now being reported (Severne and Kuszell, 1975, further denoted as Ref. I) was the isolation of a strong field effect, which is simple enough to be treated with reasonable rigour.

To this end, we have reconsidered the classical problem of the vibrations of a stellar system, as treated first by Lynden-Bell (1962). We however take a non-rotating spherical system, so that in the linearized collisionless Boltzmann equation

$$\frac{\partial f}{\partial t} + \mathbf{v} \cdot \frac{\partial f}{\partial \mathbf{r}} + \mathbf{F} \cdot \frac{\partial f^{(0)}}{\partial \mathbf{v}}(\mathbf{r}, \mathbf{v}) + \mathbf{F}^{(0)}(\mathbf{r}) \cdot \frac{\partial f}{\partial \mathbf{v}} = 0 \tag{1}$$

there now appears a mean field term, in $\mathbf{F}^{(0)}(\mathbf{r})$. Here $f(\mathbf{r}, \mathbf{v}, t)$ is the perturbed mass distribution in phase space. The acceleration field $\mathbf{F}(\mathbf{r}, t)$ is self-consistently determined by the Poisson equation

$$\operatorname{div} \mathbf{F}(\mathbf{r}, t) = -4\pi G \int d\mathbf{v}\, f(\mathbf{r}, \mathbf{v}, t). \tag{2}$$

The perturbations are with respect to the stationary but inhomogeneous, i.e. position dependent state: $f^0(\mathbf{r}, \mathbf{v})$, $F^0(\mathbf{r})$.

The analysis remains quite simple if we restrict attention to plane waves, of wavelength small on the scale of the inhomogeneity

$$\lambda \ll L_H \tag{3}$$

Hayli (ed.), Dynamics of Stellar Systems, 47–52. All Rights Reserved.

The scale L_H itself is determined by the unperturbed state $f^{(0)}$, $F^{(0)}$, and is thus of the order of the dimensions of the system. The restriction (3) to perturbations of small wavelength means that $f^{(0)}$ and $F^{(0)}$ can locally be taken as constant. One can then, in the usual manner, take advantage of a Fourier-Laplace transformation, according to

$$\phi_{\mathbf{k}}(z) = \int_0^\infty dt \int d\mathbf{r}\, e^{-i(\mathbf{k}\cdot\mathbf{r}-zt)}\, \phi(\mathbf{r}, t). \tag{4}$$

When one further integrates over the velocity components perpendicular to $\mathbf{k}$

$$\tilde{f}_{\mathbf{k}}(u, z) = \int d\mathbf{v}_\perp\, f_{\mathbf{k}}(\mathbf{v}, z),$$

$$u = \mathbf{k}\cdot\mathbf{v}/k, \qquad \mathbf{k}\cdot\mathbf{v}_\perp = 0, \tag{5}$$

the collisionless Boltzmann equation becomes

$$i(ku - z)\, \tilde{f}_{\mathbf{k}}(u, z) - h_{\mathbf{k}}(u) + F_{\mathbf{k}}(z)\frac{\partial \tilde{f}^{(0)}}{\partial u} = -\frac{F^{(0)}\cdot\mathbf{k}}{k}\frac{\partial \tilde{f}_{\mathbf{k}}(u, z)}{\partial u}, \tag{6}$$

with $F_{\mathbf{k}}(z)$ the Fourier-Laplace transform of the field, according to (4), and $h_{\mathbf{k}}(u)$ the perturbation at initial time:

$$h_{\mathbf{k}}(u) = \tilde{f}_{\mathbf{k}}(u, t=0). \tag{7}$$

Equation (6) differs from that for homogeneous systems by the appearance of a r.h.s. Thus, due to the mean field, $\tilde{f}_{\mathbf{k}}$ is determined by a differential equation instead of by an algebraic equation. The differential equation is however elementary, its solution being

$$\tilde{f}_{\mathbf{k}}(u, z) = e^{-i\alpha(u)} \int_{-\infty\,\mathrm{Sgn}\,F^{(0)}\cdot\mathbf{k}}^{u} du'\, e^{i\alpha(u')}\, g(u', z), \tag{8}$$

$$\alpha(u) = \frac{k}{F^{(0)}\cdot\mathbf{k}}(\tfrac{1}{2}ku^2 - uz), \tag{9}$$

$$g(u, z) = \left[h_{\mathbf{k}} - F_{\mathbf{k}}(z)\frac{\partial \tilde{f}^{(0)}}{\partial u}\right]\frac{k}{F^{(0)}\cdot\mathbf{k}}. \tag{10}$$

Due to the $\exp(i\alpha)$ factor of the integrand, one will expect to encounter complex error or related functions, whatever the velocity distribution.

Substitution of (8) into the transform of the Poisson Equation (2) yields for the perturbation to the mean field an expression of the form

$$F_{\mathbf{k}}(z) = \frac{1}{D(\mathbf{k}, z)} H_{\mathbf{k}}(z), \tag{11}$$

and the vibration modes are thus determined by the solutions of a dispersion relation

$$D(\mathbf{k}, z) = 0. \tag{12}$$

D plays the role of a dielectric function for our problem. $H_k(z)$ expresses transient effects from the initial perturbation (7).

To discuss the vibration modes, it is indispensable to specify the velocity distribution. For the Maxwellian,

$$f^{(0)}(\mathbf{r}, \mathbf{v}) = \varrho_0(\mathbf{r}) [\sqrt{2\pi}\,\sigma(\mathbf{r})]^{-3} \exp[-v^2/2\sigma(\mathbf{r})], \qquad (13)$$

and in terms of the parameters

$$\kappa = k/k_0, \qquad \varepsilon = |\mathbf{F}^{(0)} \cdot \mathbf{k}|/k\sigma\omega_0, \qquad (14)$$

$$\omega_0 = (4\pi G\varrho)^{1/2}, \qquad s = \mathrm{Sgn}(\mathbf{F}^{(0)} \cdot \mathbf{k}), \qquad (15)$$

the dispersion relation takes the explicit form:

$$D = 1 + \frac{dZ/d\zeta}{2(\kappa^2 + is\kappa\varepsilon)} = 0, \qquad (16)$$

$$Z = 2i \exp(-\zeta^2) \int_{-\infty}^{i\zeta} dt \exp(-t^2), \qquad \zeta = \frac{z/\omega_0}{\sqrt{2\kappa^2 + 2is\kappa\varepsilon}}. \qquad (17)$$

In Z one will recognize the plasma dispersion function. The standard results for a homogeneous unperturbed state and the Maxwellian distribution are recovered simply by annulling the field parameter ε.

The dispersion relation (16) allows for an infinity of vibration modes. A detailed discussion is to be found in Ref. I. Here we present only the main results.

Having separated the complex frequency z into real and imaginary parts

$$z = \omega - i\eta \qquad (18)$$

it is essential to distinguish *inward waves*, whose radial projection is directed along the field and for which $s\omega > 0$, from *outward waves*, directed against the field and such that $s\omega < 0$. As could *a priori* have been expected, a first effect due to the field is a breaking of symmetry: inward and outward waves have quite different propagation characteristics, while azimuthal waves, for which $\varepsilon = 0$, are unaffected by the field. Figures 1 and 2 give the computed curves for the reduced frequency ω/ω_0 and damping η/ω_0 for different modes (roman label) and values of ε (latin label). For inward modes (+index), the frequency increases with the field while the damping decreases (except for the central '0' mode). For outward modes the converse holds: there is thus preferential propagation of inwardly directed perturbations.

However, as in the field free calculation, the damping remains quite large. It is in fact questionable if these waves can be of physical significance. The relevant parameter is the relative damping η/ω, which on Figure 3 is given only for the preferentially propagated inward modes. While the field does have a large effect, for realistic conditions the field parameter is limited to $\varepsilon \lesssim 3$. Moreover the scaling condition (3) imposes a lower limit on κ, and one is led to conclude that the relative damping remains in excess of 0.3 (e-folding time of 0.5 period). This is far from small. Other velocity

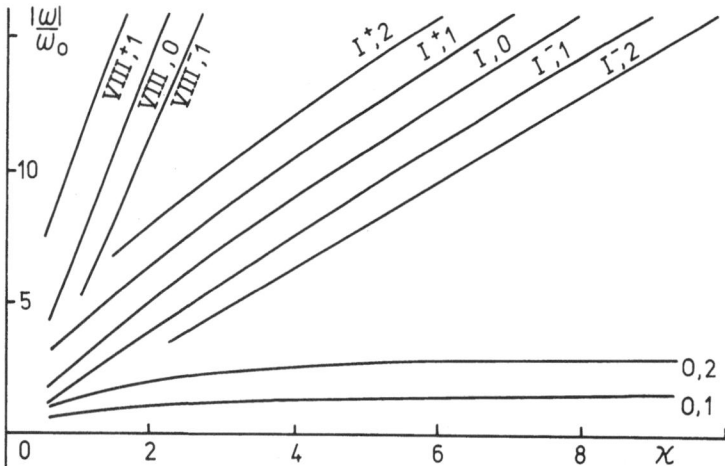

Fig. 1.   Dispersion curves for the maxwellian velocity distribution. The curves are labelled by their mode index (roman numeral), and by the value of the field parameter ε (latin numeral). The superscripts $\pm$ indicate inward $(+)$ or outward $(-)$ modes.

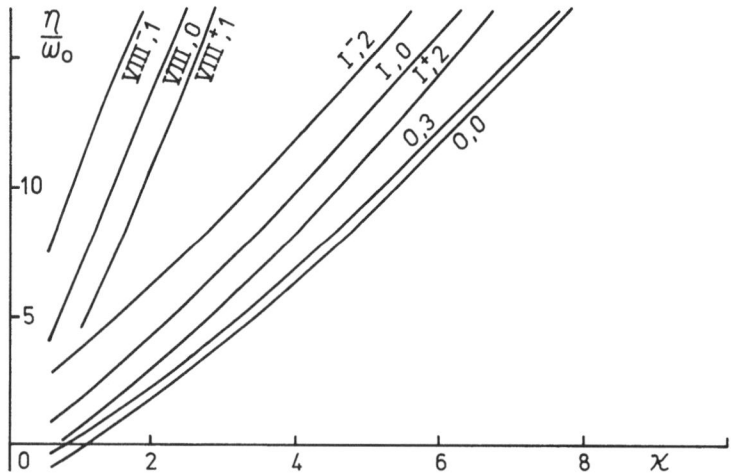

Fig. 2.   Reduced damping in function of the reduced wave vector κ. Conventions are as for Figure 1.

distributions do give more favorable results. By way of example, and also to indicate the sensitivity of the calculations to the choice of the velocity distribution, we have on Figure 3 also given some curves for the Lorentzian

$$f^{(0)}\,(\mathbf{r},\,\mathbf{v})=\varrho_0(\mathbf{r})\,\pi^{-2}\sigma(\mathbf{r})\,(v^2+\sigma^2)^{-2}. \tag{19}$$

The physical mechanism underlying the preferential propagation inwards appears directly when one analyses how the Landau damping of density waves is affected by the presence of the radial mean field. Landau damping results from the interaction of the density wave and the group of resonant stars, travelling at approximately the phase velocity of the wave: the faster particles give up energy to the wave, while the more numerous slower stars absorb energy, the net result being a damping of the wave. If in addition there is a field, the wave will also interact with the field, through the group of resonant stars. For simplicity, consider a radial wave. If the wave is directed inwards, the stars with which it is interacting are being accelerated by the field: the wave thus on average picks up energy from the field and its damping is reduced. If the wave is

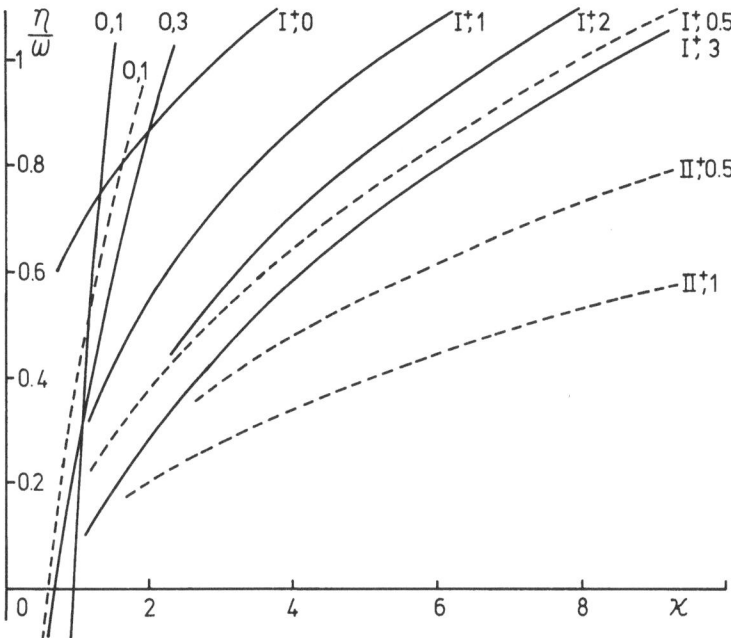

Fig. 3.   Relative damping. The dashed lines give the curves for a lorentzian distribution. Other conventions are as for Figure 1. No curves are given for the '−' modes, for which $\eta/\omega > 1$.

travelling outwards, the converse occurs: the field exerces a decelerating force on the resonant stars, and the damping of the wave is increased.

The field has also, and predictably, a favorable influence upon the stability of the system. The critical value of the wave vector κ_c is that for which the damping changes sign. It can be seen, from Figure 2 or Figure 3, that the '0' mode is here determinant, and that κ_c is reduced as ε increases. It should finally be remarked that κ_c, like all the parameters in this problem, is position dependent. (In fact, it is only for small r that one can extend the validity of the calculation down to $\kappa \simeq \kappa_c$ without violating the scaling condition (3)). The local character of the stability criterion makes it possible for an inwardly propagating perturbation to excite an instability only in the inner region of the system. Thus while in a purely local description the field enhances stability it may have a destabilizing influence on a larger scale. Insofar as our linear analysis remains valid, such an instability would remain restricted to the central part of the system, since all outwardly propagating vibrations are very strongly damped.

References

Lynden-Bell, D.: 1962, *Monthly Notices Roy. Astron. Soc.* **124**, 279.
Severne, G. and Kuszell, A.: 1975, *Astrophys. Space Sci.* (to appear, denoted as Ref. I).

DISCUSSION

Feix: The question can be raised about the possibility of treating an homogeneous self gravitating system with a steady state field since this self consistent steady field will be due to the inhomogeneity and the two terms are two aspects of the same phenomenon (finiteness of self gravitating systems). Taking only very small wavelength can be the answer but in that case the field effect is unimportant. Extending the results to the interesting wavelengths (of the system order of length) is of course forbidden.

Severne: Force and inhomogeneity effects can be separated because we have a third parameter at our disposal: the wavelength λ of the perturbation. For $\lambda \ll L_H$ the separation can be made. The problem is the meaning of the symbol '$\ll$'. For Plummer's model, $\gamma \ll L_H$ implies $\kappa \gg 0.37 \left(1 + (4r^2/r_{max}^2)\right)$, with r_{max} taken as the effective radius of the system and including 0.72 of the total mass.

TIDAL EFFECTS ON SPHERICAL STELLAR SYSTEMS

D. W. KEENAN and K. A. INNANEN

Physics Dept., York University, Downsview, Ontario, Canada

Many self-gravitating stellar systems are satellites of larger galaxies and must therefore be subjected to the tidal field of the parent system. Examples are the globular clusters and dwarf elliptical galaxies, which are satellites of our Galaxy. Most previous studies of tidal effects have been highly simplified, e.g. clusters in circular planar galactic orbits (Bok, 1934), or have assumed that the tidal field acts to limit the size of a star cluster without any effects on its internal structure or stability (Spitzer and Shapiro, 1972; Spitzer and Thuan, 1972).

To determine the effects of a realistic time varying tidal field on a stellar system where binary encounter processes are unimportant, we computed the orbits of a large and representative sample of test particles in the field of a spherically symmetric smooth model of a star cluster which, in turn, moved in its own orbit around a model galaxy. The effect of the tidal fields on the stellar orbits was investigated by observing the evolution of the stars in energy E and angular momentum h. E and h are, of course, conserved quantities for an isolated spherically symmetric field. Three systems were tested. Two involved a point mass galaxy with the cluster in an elliptical orbit about it. The third system used a rotationally symmetric model of a disk halo galaxy. The mass model was constructed to fit the mass distribution and kinematics of our Galaxy (Innanen, 1973). This system was an analogue for a 'typical' halo globular cluster in our Galaxy.

The three body equations of motion were integrated using an IBM 370/75 system and a fast double precision algorithm kindly supplied by Dr F. T. Krogh of J. P. L. (Krogh, 1969, 1971).

The main conclusion reached in these numerical experiments is that star clusters rotating in a retrograde sense compared with their galactic orbital motion, are much more stable in a tidal field than clusters with either direct rotation or no rotation. Observational support is given to this by measurement of the rotation of ω Centauri (Harding, 1965; Keenan *et al.*, 1973) and of M13 (Griffin, 1972a, b, 1974). The most stable stellar orbits are low-eccentricity, low inclination retrograde orbits. The least stable are similar direct orbits. For the case of a disk model galaxy, the stellar orbits inclined at large angles to the galactic plane are also highly perturbed due to the large field gradients across the disk in the vertical direction.

The galactic orbits of the stars which escaped from the model cluster due to the tidal interaction were also computed. The escapers were grouped into two 'clumps' in energy-angular momentum phase space. These clumps are separated from each other and from the point occupied by the parent cluster in orbital energy and angular momentum phase space. One 'clump' has lower energy and angular momentum than the cluster while the other has higher.

Hayli (ed.), Dynamics of Stellar Systems, 53–55. All Rights Reserved.

This clumping of the escapers' orbits may be explained by the existence of 'quasi-integrals' analogous to the Jacobi integral in the restricted problem of three-bodies (Szebehely, 1967), which restricts the motion of the body to certain regions of phase space.

A 'randomizing' action of the tidal field on the energies and angular momenta of the bound stars was also noted. This would be analogous to the 'shock induced relaxation' discussed by Spitzer and Chevalier (1973).

This work will be published in greater detail elsewhere (Keenan and Innanen, 1975).

Acknowledgements

We would like to thank Dr I. R. King for many useful discussions. This research was supported by the National Research Council of Canada.

References

Bok, B. J.: 1934, *Harvard College Obs. Circular* 384.
Harding, G. A.: 1965, *Roy. Obs. Greenwich-Cape Bull.* No. 99, 14S.
Griffin, R.: 1972a, *Observatory* **92**, 29.
Griffin, R.: 1972b, *Quart. J. Roy. Astron. Soc.* **13**, 442.
Griffin, R.: 1974, Private Communication.
Innanen, K. A.: 1973, *Astrophys. Space Sci.* **22**, 393.
Keenan, D. W., Innanen, K. A., and House, F. C.: 1973, *Astron. J.* **78**, 173.
Keenan, D. W. and Innanen, K. A.: 1975, *Astron. J.* (in press).
Krogh, F. T.: 1969, *T.U. Doc. No. CP-2308, J.P.L.,* Pasadena.
Krogh, F. T.: 1971, *T.U. Doc. No. CP-2586, J.P.L.,* Pasadena.
Spitzer, L. and Shapiro, S. L.: 1972, *Astrophys. J.* **173**, 529.
Spitzer, L. and Thuan, T. X.: 1972, *Astrophys. J.* **175**, 31.
Spitzer, L. and Chevalier, R. A.: 1973, *Astrophys. J.* **183**, 565.
Szebehely, V.: 1967, *Theory of Orbits*, Academic Press, New-York.

DISCUSSION

Freeman: Could you say what fraction of escapers are lost during passage through the galactic plane compared with other escapers.

Innanen: We have not looked at this question quantitatively, but we hope to do so.

Freeman: Your comment about a quasi-Jacobi-integral is very likely correct, because the time for a star to cross the cluster is short compared to the other time-scales involved.

Miller: Are there effects due to repeated perigalacticon passage – some stars that are set up on one passage that escape on next passage? Do some become more stable?

Innanen: Yes. Yes.

Baldwin: Are there any significant changes in the net angular momentum of the dwarf spheroidal system during the passages?

Innanen: Probably not.

Gott: ω Cen is the most massive globular cluster and also has the most elliptical isophotes. It would be your contention that it is in retrograde rotation?

Innanen: Yes. The clusters with predicted greater stability are those whose rotation is in the opposite sense to their sense of revolution.

Gott: Also would escape of the direct orbit stars be expected to leave all globular clusters with a net retrograde rotation?

Innanen: I don't know; the escaping stars would escape mainly from the outer parts of these systems.

The dense inner parts which are the ones that are really observed are not very strongly affected by these computations. Those parts which are interesting from the escaping or non escaping point of view are really well out in the halo and therefore are difficult to establish observationally, at least at the present time.

Lynden-Bell: I think ω Cen has a retrograde galactic orbit so we must get straight the meanings of retrograde and direct in this discussion.

Innanen: Same answer as given to Gott.

King: Is it not possible that the stability of retrograde orbits is due to the presence of a pseudo-integral? This seemed to be the case in the orbital studies of the restricted three-body problem by Hénon and by Jefferys. It appeared there that some bodies that were not bound were nevertheless bounded, by the existence of a pseudo-integral.

Innanen: No. We have not examined our results from the Hénon phase-space point of view.

Hénon: In numerical studies of the restricted three-body problem, one finds exactly the same result, namely that retrograde orbits are stable at larger distances than direct orbits. In celestial mechanics also this effect is well known: cf. for instance the fact that the four outer satellites of Jupiter are on retrograde orbits.

Kalnajs: Although the orbital periods of your clusters around the galaxy are much larger than the orbital periods of the stars in the cluster, the rate of change of the tidal field as seen by star crossing the galactic plane seems fast. How does the time-scale of this change compare to the periods of the pre-escapers?

Innanen: The time of application of the impulsive force when crossing the plane is, in effect, comparable to the orbital period of the star in the cluster. It does however depend considerably on the *inclination* of the cluster orbit to the galactic plane, which is sometimes high, and sometimes quite low.

N-BODY SIMULATIONS

S. J. AARSETH

Institute of Astronomy, Cambridge, U.K.

Abstract. Numerical results on escape and binary evolution in N-body systems are summarized and some results of a new calculation with initial binaries are described.

This brief review of the N-body problem concentrates on the two central aspects, escape and binary evolution. It is particularly important to examine these processes in relation to the fast Monte Carlo and Boltzmann moment methods which are based on simplifying assumptions. One task for N-body calculations is therefore to establish the region of validity of the approximations, since there is no *a priori* guarantee that such results can be applied even to relatively large systems.

Direct integrations of small particle numbers ($N=10$) indicate that multiple encounters contribute significantly to the escape rate, whereas theoretical considerations are based on two-body encounters. However, further calculations show that the latter do increase in relative importance for larger N. These simulations also yield more statistical information about the actual escape mechanisms. Two quantities are particularly useful for the analysis of this data. Let us denote the absolute value of the binding energy per unit mass before and after escape by β and α, respectively, both being expressed in terms of the initial mean kinetic energy, $\frac{1}{2}m v^2$. Typically, $\alpha + \beta \simeq 2$ for equal-mass systems with $N=250$, indicating that escape is still due to discrete events. Note that binaries are absent for most of the time. The distribution of escape energies, α, is very wide for unequal masses; in a cluster model with $N=500$ there are 18 escapers out of 46 with $\alpha > 1$. Some of these fast particles are produced by 'super-elastic' encounters with energetic binaries, but others are due to close two-body encounters which also dominate when $\alpha < 1$. These results have implications for the approximate methods which are based on distant two-body encounters. Thus according to theory, the close encounters (i.e. separations $r \leqslant 2G\bar{m}/\bar{v}^2$) should only contribute in the ratio $1/5 \ln(0.4\,N)$ with respect to the distant encounters. Because of the numerical results, the neglect of escape by close encounters in large systems ($N \sim 10^5$) may only be justified if the theoretical N-dependence is too weak.

In order to discuss the particle evaporation, we introduce the relative escape rate per crossing time, $L = \Delta N t_{\mathrm{cr}}/Nt$. A total of four equal-mass models with $N=250$ have been studied (including one by Dr Wielen), giving $L \simeq 5 - 11 \times 10^{-4}$ for $t/t_{\mathrm{cr}} \simeq 30 - 40$. The introduction of a mass spectrum leads to an increased escape rate, i.e. $L \simeq 4 \times 10^{-3}$, whereas $L \simeq 3 \times 10^{-3}$ for a similar model with $N=500$ and $t=30 t_{\mathrm{cr}}$. In the latter case, only about 10% of the escapers have $\beta < 1$, although there has been ample time for such orbits to return from the halo for further interactions. Thus it appears that the increased relaxation time of elongated halo orbits is not compensated by the smaller binding energy. This feature is also consistent with the absence of preferential escape among all the light particles. Furthermore, well bound particles can only acquire positive energy by large transitions and the presence of such events demonstrates the importance of close encounters.

The N-body calculations show that binaries play a crucial role in small stellar systems. In the first place, a close binary inevitably forms as the end product of the core evolution. Mass segregation favours the combination of heavy particles but, where appropriate, the subsequent evolution proceeds further in the direction of increased mass by capture or exchange. At the same time, the binding energy rapidly exceeds 50%; even for $N = 500$ the corresponding time-scale is only about $12\,t_{cr}$. This energy sink behaviour can be understood in terms of an asymmetry between incoming and ejected particles since only the latter may exceed the escape velocity. Capture, which is the opposite process to escape, may be viewed as a short-lived event, giving much the same end result. Most of the energy absorbed by the central binary is due to the ejection of strongly bound particles into the halo, whereas escape contributes less to the core evolution. Consequently, the decreased core density leads to a slower evolution rate. At some stage, the core expansion is halted and gives way to a secondary contraction, culminating in a stable hierarchical triple system. Subsequent disruption by external perturbations often produces significant recoil kinetic energy which is transferred to other core members by two-body encounters. The equivalent development is somewhat delayed in equal-mass systems, but the higher number density in the core compensates to some extent for the lack of heavy particles.

It is not yet clear whether binary effects may be neglected in the larger systems simulated by fast methods. Although the presence of only one binary could at most influence the inner core, multiple binary formation during the collapse phase appears likely on theoretical grounds. Another possibility is that binaries are present initially in significant numbers. N-body calculations have recently been made in order to study this effect and some preliminary results are available.

For simplicity we adopt an equal-mass system with 100 particles of mass m_1 and replace 20 of these particles by 10 close binaries, each with a binding energy $E_b = 5\,m_1 \bar{v}_1^2$. This choice of energy is based on theoretical considerations of maximum efficiency discussed elsewhere in this volume. In order to set up a self-consistent dynamical structure, the cluster is exposed to violent relaxation in the form of a moderate initial contraction. The next phase is characterized by mass segregation; hence a modest initial binary population may eventually dominate the central region if their combined mass exceeds the average mass of the single particles. At this stage $(t \simeq 5\,t_{cr})$, the number of halo-type orbits is very similar in the comparison system where 10 single particles of mass $m_2 = 2\,m_1$ replace the binaries. Although one of the close binaries is destroyed in a two-binary collision, their total internal energy is slightly increased. Further favourable interactions occur during the subsequent evolution, until finally at $t \simeq 24\,t_{cr}$ about 75% of the total energy is contained in bound pairs compared to 50% initially. Hence treating the binaries as single particles implies that the cluster members are less bound by a factor of 2 and the dynamical time-scale is increased by $2\sqrt{2}$. The binaries are also less bound to the cluster centre by a similar factor, with the most energetic binary $(E_b \simeq 20\, \bar{m}_1 \bar{v}_1^2)$ in an elongated halo orbit.

Of the nine surviving binaries, three remained unperturbed, even preserving their eccentricity to two decimal places, and two more retained their identity with somewhat increased binding energy. A further three suffered exchange of one companion and one binary actually lost both of its original members. Significantly, the binaries with new companions are also the most energetic. The modest escape rate, i.e. $L \simeq 1 \times 10^{-3}$, reflects the core expansion effect. In the comparison system one particle escapes during ten crossing times, also giving $L \simeq 1 \times 10^{-3}$. The heavy particles are now much more strongly bound to the centre, with two of them combined into a close binary of energy $E_b \simeq 4\, m_1 \bar{v}_1^2$. Considering the future evolution of the first system, it is likely that the binaries will eventually be destroyed by further collisions or be ejected altogether, in both cases leaving behind a more loosely bound cluster. Finally, we note that the number of centrally concentrated binaries was only three or four on average. This small population illustrates the effectiveness of superelastic encounters between the binaries, a process which acts in the opposite sense to the mass segregation.

A more extensive survey of the N-body problem will be published jointly with Dr M. Lecar in *Annual Review of Astronomy and Astrophysics*, Volume 13.

Some of the ideas discussed above were formulated at the 1974 Theoretical Astrophysics Workshop held at the Aspen Center for Physics.

DISCUSSION

King: The disagreement of the number of close to distant encounters is because you have applied the theory where it was not meant to apply. The theoretical escape formula depends on assuming that $\langle v_e^2 \rangle / \langle v^2 \rangle = 4$, which is far from true in the region from which escape takes place in your systems, and the resulting escape rate is very sensitive to this number. The formula is more applicable if there is a tidal limit, and I think that you would find that escape by diffusion was relatively much more important there. Correspondingly, the existence in your systems of Hénon's paradox (that stars close to escape have vanishingly small diffusion) is also due to the lack of a tidal limit. In fact, Hénon's paradox is asymptotically true for real systems, as the tidal limit becomes infinitely large; but for a typical rich cluster, with a typical tidal limit, escape should be predominantly by diffusion.

Aarseth: The above discussion is only concerned with isolated systems, as are the Monte-Carlo calculations. Furthermore, I am primarily questioning the escape mechanism rather than the escape rate and the former depends only weakly on the velocity ratio.

Spitzer: I would like to point out that the difference between close encounters and distant encounters becomes rather small if N is small, and even for N equal to 250, it may be difficult to tell these two apart. For example, I compute that for a cluster with $N = 250$ the mean change of energy for a star of zero energy making one traverse through the core and experiencing many so-called 'distant encounters' is about $\frac{3}{5}$ of the mean kinetic energy for all cluster stars; this 'step size' for energy changes is evidently, much larger than one would commonly associate with a diffusion picture.

Miller: In your discussion for $\Delta = (\frac{1}{2} v_\infty^2 - E') / \frac{1}{2} \overline{mv^2}$, the energy change leading to escape, what happens with encounters that lead to large energy changes but do not lead to escapes? Have they been as carefully studied? Are there collisions, starting from lower E' which have just as large a Δ but do not lead to escape? What is known of the distribution of Δ, as it depends on E' (but independent of whether process leads to escape)?

Aarseth: There is no detailed analysis of energy changes which do not lead to escape. It is quite conceivable that there are such large energy changes but the number of small changes will in any case be much greater.

Severne: Can you indicate what value the ratio of close to distant encounters should attain to give a value $\bar{\Delta} \simeq 2$?

Aarseth: The ratio should probably be somewhat greater than one.

THE SIMULATION OF NON-ISOLATED EXPANDING
GRAVITATIONAL SYSTEMS

G. JANIN

European Space Operations Centre, Darmstadt, F.R.G.

and

M. J. HAGGERTY

Centre for Statistical Mechanics and Thermodynamics,
The University of Texas at Austin, Tex., U.S.A.

Abstract. Numerical experiments on the gravitational N-body problem are reviewed for the case of non-isolated systems. The effect of an external field on expanding cubical systems is discussed.

1. Numerical Experiments on Non-Isolated Spherical Systems

Very few cases of non-isolated gravitational systems have been studied by the method of numerical experiments. Usually the gravitational system is defined in a universe where everything else is excluded. This is, of course, a non-realistic approximation.

Among the attempts to take into account an external gravitational field, one should first mention the successful work of Hayli (1970) concerning the tidal effect on galactic cluster, where the mechanism of disruption of the cluster is analysed in full detail.

Another external effect on galactic clusters, the disruption of clusters through passing clouds of interstellar matter, was investigated by Bouvier and Janin (1970). The disruption time observed in these numerical experiments was found to be roughly two times faster than the theoretical value estimated by Spitzer (1958) corrected by taking into account the close encounters of the clouds with the cluster (Bouvier, 1971).

In the numerical model of Bouvier and Janin, a finite number of idealized clouds of interstellar matter are moving inside a sphere of radius R. Whenever one of the clouds hits the surface of the sphere, the direction of its velocity is randomly redistributed so as to send it back into the sphere. This procedure is supposed to reproduce correctly the mean density of the galactic cloud population [a slight mishandling in the procedure as described in the paper of Bouvier and Janin is mentioned by Hénon (1972)]. The star cluster is initially centered at the centre of the sphere. But as soon as a star leaves the central region of the sphere, it undergoes a force towards the centre due to the uniform distribution of clouds.

It would be feasible to compensate for this force by introducing an artificial gravitational field. This field should induce a centrifugal force, whose intensity depends on the mean cloud mass inside a concentric sphere of radius equal to the distance of the star to the centre.

But this field, abruptly cut off at radius R, creates a discontinuity at the surface of the sphere of clouds and reduces the credibility of this model.

No really satisfactory model has yet been found.

Hayli (ed.), Dynamics of Stellar Systems, 61–63. All Rights Reserved.

2. Cubic Systems for Representing Large Expanding Clusters

Numerical experiments on a large uniform expanding gravitational system (Janin and Haggerty, 1974) with regard to an application to cosmological systems (Haggerty and Janin, 1974), have only been possible thanks to two crude assumptions, made in order to limit the number of particles involved in the numerical integration:
  – reduction of strongly bound subclusters to single particles at their centres of mass;
  – division of the system in a net of subsystems of cubical shape having similar global evolution.

The first assumption is reasonably valid for expanding systems, where the probability of a disintegration of a bound subcluster by a passing particle is small.

The second assumption is made in order to restrict the study of the whole system, if homogeneous, to that of one of the cubical subsystem, having the freedom to associate several subsystems side by side.

Studying only a part of a system gives birth to the problem of taking into account the exterior of the subsystem. An external field is introduced for this purpose. This field is equivalent to having a fictitious uniform negative mass density in the region containing particles. A central attractive field that is linear in the displacement from the centre is added. The spurious effect due to the absence of spherical symmetry is therefore counter-balanced. As the system is expanding, so the additional field is time dependent.

As in the spherical system described in the preceding section, the delicate point is the boundary effects. A systematic study has been made to measure the effect of the cube sides on near-boundary particles.

A general conclusion is that the motion of the particles near the boundary is severely affected by the presence or absence of discrete particles, as distinct from either a uniform fluid or vacuum, on the other side of the boundary. However, the nature of the evolution of small regions is not greatly affected by changes in the mass distribution within other small regions at a great distance.

Among the deviations of parameters observed under the effect of the external field, the most significant encountered is that of the total energy (constant for isolated systems) of the simulated system. In some cases, the magnitude of the energy variation is such that the virial ratio, initially equal to 2, decreased to 1 when the system expanded by a factor 4 in radius. Later, the energy variation slows down abruptly and can even reverse itself.

It was expected to observe in this representation an energy exchange between the exterior and the interior of the system because of the non-symmetric role played by these two parts: the interior undergoes an evolution while the exterior keeps being a continuous distribution of matter. The initial direction of the energy exchange is shown by the escapers from the interior to the exterior. But the magnitude of the energy stream was a surprise. No completely satisfactory explanation has yet been found.

References

Bouvier, P.: 1971, *Astron. Astrophys.* **14**, 341.
Bouvier, P. and Janin, G.: 1970, *Astron. Astrophys.* **9**, 461.
Haggerty, M. J. and Janin, G.: 1974, *Astron. Astrophys.* **36**, 415.
Hayli, A.: 1970, *Astron. Astrophys.* **7**, 17.
Hénon, M.: 1972, *Astron. Astrophys.* **19**, 488.
Janin, G. and Haggerty, M. J.: 1974, *J. Comput. Phys.* **16**, 76.
Spitzer, L.: 1958, *Astrophys. J.* **127**, 17.

DISCUSSION

Bouvier: In connection to our former investigation about the influence of clouds on star clusters, I wish to point out that we had been led at a disruption time which appeared shorter than the Spitzer disruption time of 1958, by a factor 3 or 4. The Spitzer scheme did not take into account either the inner evolution cluster since it was based on the so called 'impulsion approximation' or the closest encounters of the cluster with the clouds, and it was only after having improved the Spitzer deduction that a fair agreement was reached between the disruption times completed numerically and estimated theoretically, in spite of the neglect of the steady-mean field of the clouds in the numerical experiment.

Wielen: It seems to me absolutely necessary to compensate somehow the unrealistic effect of the mean field of the clouds in numerical experiments. As I pointed out to Dr Janin in 1972, I fear that the results obtained by Bouvier and Janin (1970) are severely affected by the unrealistic simulation of the clouds. The unrealistic mean gravitational field towards the center of the cluster, due to the average density of clouds inside a sphere of radius r around the cluster, is equal to the gravitational attraction by the cluster at $r=8$ pc, and is 30 times as large as that of the cluster for $r \geqslant 26$ pc. Hence nearly no star could escape from the system. Most of the observed change in the total energy is probably due to a few stars in the outermost regions absorbing much energy from the clouds instead of escaping. In this case, the dissolution time of the cluster as derived from monitoring the total energy of all the stars, is certainly too short. A model in which the mean field of the clouds is just compensated by an appropriate opposite field, should provide more reliable results.

Hohl: Could many of the difficulties associated with the external field be eliminated by using Fouries transform techniques to obtain the field for periodic systems? The field for the central system would still be for the isolated system.

Janin: The use of Fourier transform techniques for estimating the field for periodic gravitational systems is certainly of high interest for our problem, providing a satisfactory handling of the three-dimensional case is available.

Lynden-Bell: In your last experiment, do you start with everything within the cube expanding?

Janin: Yes. We give initial peculiar radial velocities to the particles within the cube which is also expanding at the same rate, more or less like a fluid.

Lynden-Bell: Is it a bound system?

Janin: Yes.

Lynden-Bell: If you have a bound system and you give it a radial pulsation that converts kinetic energy into potential energy the virial ratio goes down.

NUMERICAL EXPERIMENTS ON THE TENSOR VIRIAL

IN N-BODY SYSTEMS

R. H. MILLER

Dept. of Astronomy, University of Chicago, Chicago, Ill., U.S.A.

Abstract. Tensor generalizations of the virial theorem were checked in a 100-body integration. The virial theorem was remarkably well satisfied, and the calculation confirmed the generalized Lagrange-Jacobi identities. The potential energy tensor, the kinetic energy tensor, and the virial tensor showed surprisingly long correlation times of about $\frac{1}{3}$ of a crossing time.

1. Introduction

The tensor generalization of the Lagrange-Jacobi identities for self-gravitating n-body systems (Chandrasekhar, 1964) can provide an independent check on n-body calculations beyond that provided by the usual first integrals of motion, even though it leads to no new constants of motion. These equations read:

$$\frac{1}{2}\frac{d^2 I_{ij}}{dt^2} = 2\,T_{ij} + W_{ij} = V_{ij}, \tag{1}$$

where I_{ij} is usually called the 'moment of inertia tensor' (it is not the same as the object called by that name in the mechanics books; cf. Goldstein, 1960), T_{ij} is the kinetic energy tensor, and W_{ij} is the potential energy tensor. The tensor that appears on the right hand side of Equation (1) is called V_{ij} for convenience. These tensors are defined by

$$I_{ij} = \sum_{\alpha=1}^{n} m_\alpha x_i^{(\alpha)} x_j^{(\alpha)}, \tag{2}$$

$$T_{ij} = \frac{1}{2} \sum_{\alpha=1}^{n} m_\alpha \dot{x}_i^{(\alpha)} \dot{x}_j^{(\alpha)}, \tag{3}$$

and

$$W_{ij} = -G \sum_{\alpha=1}^{n-1} m_\alpha \sum_{\beta=\alpha+1}^{n} m_\beta \frac{\left(x_i^{(\alpha)} - x_i^{(\beta)}\right)\left(x_j^{(\alpha)} - x_j^{(\beta)}\right)}{r_{\alpha\beta}^3}. \tag{4}$$

Here, α and β are indices that identify particles ($\alpha, \beta = 1, 2, \ldots, n$), $x_i^{(\alpha)}$ is the i-component of the position vector of particle α, $\dot{x}_i^{(\alpha)}$ is its velocity, m_α is its mass, and $r_{\alpha\beta}$ is the (scalar) distance between particles α and β. Each of these tensors is manifestly symmetrical; the traces of T_{ij} and of W_{ij} are the usual (total) kinetic and potential energy of the cluster and the trace of Equation (1) is the usual Lagrange-Jacobi identity (Chandrasekhar, 1960).

Several remarks are in order concerning this system of equations. First, Equation (1) is a set of six independent ordinary differential equations concerning integral properties of the system which follow directly from the equations of motion as an identity; they are not an approximation. The more familiar 'virial theorem', which follows from the assertion that the time-average of the left-hand side of Equation (1) should approach zero if the system is in a 'dynamically steady state', is, by contrast, a severe approximation. (For an illuminating discussion of the approximation from a mathematical point of view, see Pollard, 1966). Some comments on the apparent validity of these approximations appear later in this note.

Second, the fact that Equation (1) is a set of ordinary differential equations means that it does not lend itself to a check on n-body calculations as directly as do the usual ten first integrals (total energy, total angular momentum, and so on) because it does not lead to conserved quantities that lend themselves to simple checking or to use as controls on the integration (Miller, 1971; Nacozy, 1971). Rather, they require either another numerical integration or a numerical differentiation. Either of these procedures introduces its own difficulties into the overall numerical integration of the n-body system. Similarly, the "virial theorem" (and its tensor generalizations) yields, at best, a crude *a posteriori* check on the integration because the right-hand side of Equation (1) can depart substantially from zero at any instant.

Third, computation of W_{ij} entails summation over pairs of particles, and thus tends to be expensive in calculations with reasonable numbers of particles. In addition, the quantities W_{ij} and T_{ij} vary substantially on the time scale of encounters between individual pairs of particles, so integration of the system (1) requires time-steps as fine as the shortest time-steps in use for any star of the cluster. This feature, coupled with the cost of determining W_{ij}, militates against using the tensor generalization of the Lagrange-Jacobi identities as a check or control on the integration.

Nonetheless, the 'tensor virial' equations can provide some insights into the behavior of star clusters, and thus are interesting on their own merits quite apart from any possible utility in providing an additional check on the integration. For example, they promise to provide a useful tool in the study of strongly rotating systems, which are expected to depart significantly from spherical symmetry. The studies reported here were not based on a strongly rotating system; the most interesting results relate to the correlation times that emerged in attempts to verify that the system obeys Equation (1).

2. The N-Body Integration

These experiments made use of a data file generated for use as a classroom exercise for a course in stellar dynamics; while this file was a convenient source of data, it was somewhat restrictive in that it was not convenient to reconstruct features not already present in the file and because the file was designed for other purposes. The n-body calculation was patterned after that of Wielen (1967); a 100-body system was integrated for about 8 crossing-times. All particle masses were equal $(G=m=1)$, and a low value of total energy was chosen to scale the time-steps conveniently $(E=-26.4$, so T_{cr} is about 260; energy was conserved to about 0.3% over the entire run). The initial condition was generated by a Monte Carlo process to simulate a Plummer model. It had nonvanishing, but small, angular momentum; merely that left over because the Monte Carlo stopped after loading 100 particles instead of going on forever. The systematic rotation represented about 0.25% of the total kinetic energy at the beginning of the calculation. It was purposely left nonzero for the classroom exercises for which this run was designed. The data file consists of a set of 42 'snapshots' each giving the instantaneous coordinates and velocities for each of the 100 particles. These snapshots were made at equal intervals of time, $\Delta t = 50$, or about five per crossing-time, covering the interval $t = 0$ to $t = 2050$.

The classroom exercises included (a) verification of the usual first integrals as checks on the calculation, (b) plots of some particle trajectories, (c) follow the changes in radial density distribution (watch for development of a 'core-halo' structure), (d) checks for formation of long-lived binaries, (e) checks for escaping particles, (f) determination of scatter in cluster mass estimates by the usual scalar virial theorem, making use of only the radial velocities and projected positions as seen with real clusters, and (g) some simple checks on the tensor virial relations. No strongly bound long-lived binaries were found, and two stars escaped from the cluster, but had not gone more than 3 cluster radii away by the end of the calculation at 8 crossing-times.

The results reported here were obtained in a more thorough study than was attempted in the student exercises. For each snapshot, the tensors T_{ij}, W_{ij}, I_{ij}, and V_{ij} of Equations (2)–(4) were computed; the tensor elements were also averaged over the set of snapshots. Further, the second differences of elements of the tensor I_{ij} (from successive snapshots) were computed as numerical approximations to the second time derivatives that appear in Equation (1).

3. The Tensors

The tensors T_{ij} and W_{ij}, evaluated at each snapshot, were nearly diagonal and the diagonal elements of each tensor were nearly equal. There was considerable scatter in the individual values for the elements of each tensor. The diagonal elements of V_{ij} were usually small – noticeably smaller than typical elements of either $2T_{ij}$ or of W_{ij}, and not significantly larger than the off-diagonal elements. Although V_{ij} was diagonalized at each snapshot, the directions of its principal axes jumped round so much that no pattern emerged. The elements of I_{ij} grew to larger magnitude during the run; the diagonal elements were typically about 10 times as large as the off-diagonal elements.

The arithmetic-mean tensors averaged over all 42 snapshots showed a little more structure. Again, T_{ij} and W_{ij} were nearly diagonal (off-diagonal elements about $\frac{1}{30}$ of the diagonal elements), with the diagonal elements nearly equal. The scatter of each of these tensor elements, as evaluated for each snapshot, can be taken as some indication of the statistical accuracy with which the mean values are determined. On this basis, the diagonal elements of T_{ij} and W_{ij} were nearly equal (within 2 standard deviations); in both, the 33-component was about 10–13% larger than the 11 or 22 elements. Similarly, the mean off-diagonal elements are about 1.5–2 standard deviations away from zero. In view of the expected irregularity of these tensor elements, these tensors cannot convincingly be argued to be other than multiples of an identity matrix.

The arithmetic mean of the virial tensor, V_{ij}, yielded no element that was significantly nonzero by the same kind of statistical test. The fairly rapid convergence of the arithmetic-mean of the virial tensor to zero provides experimental evidence for the operation of a 'tensor virial theorem'; the left hand side of Equation (1) can be replaced by zero to a remarkably good approximation.

It is unfortunate that there does not seem to be a natural way to estimate the covariances of these tensor elements by appeal to higher order tensors constructed along the lines of T_{ij}, W_{ij}, or I_{ij}. A theory indicating how this might be done, and how the magnitudes of the elements might be estimated from integral cluster properties, relating the whole picture to first principles, should be interesting.

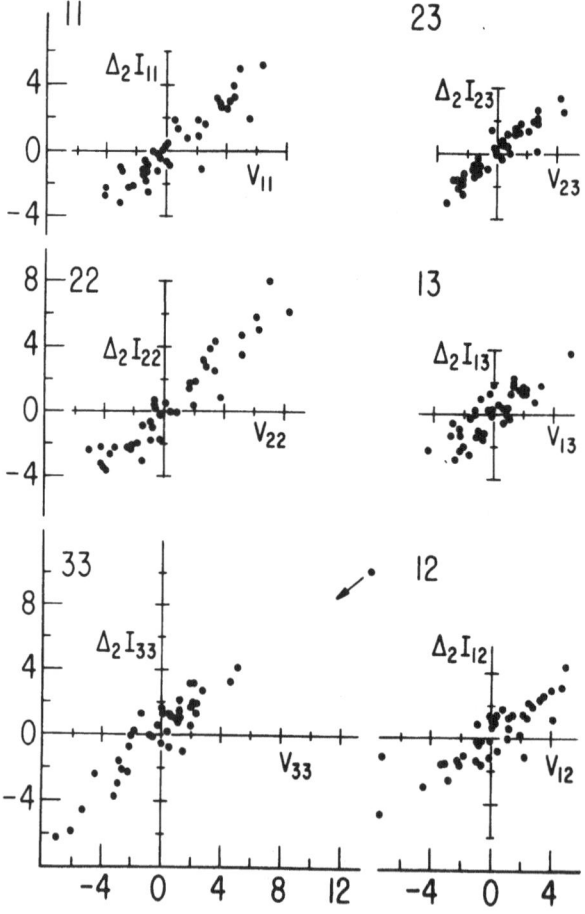

Fig. 1.   Experimental checks on the identities of Equation (1). Each point represents one snapshot of the evolution of the 100-body system, with ordinate $(I_{ij}(s+1)+I_{ij}(s-1)-2I_{ij}(s))/5000$, and abscissa $V_{ij}(s)$, where s is the 'snapshot number', $(s=2, 3, ..., 41)$. All points should lie on a straight line of unit slope through the origin. The scatter results from infrequent sampling, so the basic picture is confirmed.

4. The Identities

As argued earlier, direct integration of Equation (1) is impractical. Both T_{ij} and W_{ij} vary on the time scale of individual encounters, so accurate integration would require use of time-steps as small as the shortest in use anywhere in the system. This would require very frequent evaluation of W_{ij}, which is expensive because it involves sums over pairs of particles. Samplings at $\frac{1}{5}$ of a crossing-time are nowhere nearly frequent enough to permit reliable integration of Equation (1).

The identities may be investigated as a differential relation, rather than by integration. The most convenient way to do this numerically is to compare values of the second differences of I_{ij} with values of V_{ij}. A plot of half the second difference of I_{ij} at time t (suitably scaled to allow for the sampling interval) against V_{ij} for the same time t, should yield a set of points that lie on a straight line through the origin with unit slope. Such a plot was made for each of the 6 independent tensor elements; but the points did not all lie on the line (Figure 1) rather, they scattered about the line. The underlying line is clearly discernable, but the scatter of points is appreciable. Evidently the sampled value of V_{ij} need not accurately represent the average value over the sampling interval; similar changes can occur in I_{ij}. The second difference is notoriously difficult to obtain numerically in any case.

The scatter of points indicates that the sampling interval is not close enough to permit precise testing of Equation (1), but the appearance of the line confirms the basic picture. This is not simply a matter of everything being correlated with everything – a plot of the second difference of I_{22} vs V_{33}, for example, gives a fairly complete scatter diagram. We conclude that the sampling interval used in these experiments is adequate to confirm the basic trend, but insufficient to show exact agreement. The evidence of time-scales is perhaps the most interesting result of these experiments; a fuller discussion of this point follows.

5. Correlation Time

The arithmetic means used in Section 3 are easily evaluated and thus represent the most natural experimental approach to the study of the 'virial theorem'. But they are neither a proper time-average nor a proper ensemble-average:

$$\langle T_{ij} \rangle = \tfrac{1}{42} \sum_{s=1}^{42} T_{ij}(s\Delta t) \neq \frac{1}{42\Delta t} \int_0^{42\Delta t} d\tau \, T_{ij}(\tau) \neq \lim_{\mathbb{T} \to \infty} \frac{1}{\mathbb{T}} \int_0^{\mathbb{T}} d\tau \, T_{ij}(\tau). \qquad (5)$$

The key to this question is the correlation time for typical changes in the relevant quantities. If the correlation time is at least on the order of Δt (but significantly shorter than $42\,\Delta t$ – the duration of the experiment), then the arithmetic mean approximates a time-average (over the restricted time available to the experiment; the 'significantly shorter than $42\,\Delta t$' is required to approximate the arbitrarily long time interval in the definition of a time-average). On the other hand, if the correlation-time is much shorter than Δt, successive samples are statistically independent and the arithmetic mean approximates an ensemble average (again, with the proviso that 42 samples is hardly the very large number imagined in the definition of an ensemble average). A fundamental difficulty is that there is no *a priori* way to estimate the correlation-time for these quantities from first principles. Some attempts have been made along these lines for typical statistical-mechanical systems by Lebowitz *et al.* (1967) and by Zwanzig and Ailawadi (1969); however even for these simpler systems, general rules for computing correlation times for arbitrary quantities are not available.

The best recourse again seems to be experimental. Correlation times were studied through lagged products of the tensor elements, thus forming lagged autocorrelations. In all cases, the autocorrelation at lag 1 was 30–40% of that at lag 0 (the variance), and that at lag 2 or greater was essentially zero. Thus, experimentally, the correlation time for elements of V_{ij} and for off-diagonal elements of T_{ij} and W_{ij} is about $\Delta t = 50$, or $\frac{1}{5}$ of a crossing-time. Changes in the autocorrelation were masked by the large mean values for the diagonal elements of T_{ij} and of W_{ij}.

It was rather surprising that so long a correlation-time appeared experimentally in view of the pathological variations in kinetic and potential energies noted in most integrations of *n*-body systems (see, for example, Wielen, 1967). Both $|T_{ij}|$ and $|W_{ij}|$ increase sharply as a pair of stars enters a close encounter. Because of the large number of close encounters that can occur, each of these quantities can vary quite strongly and quite rapidly. The strong encounters are of short duration, and a cluster looked at at some arbitrary time is not likely to be undergoing a particularly strong encounter. They also involve only two particles, typically, leaving the slower evolution of the entire system that dominates the longer time structure. This allows the weaker long-lag autocorrelation to appear, but loses a significant short-lag contribution.

The comparison of second differences of I_{ij} with V_{ij} discussed in Section 4 involves similar considerations of correlation time. If there were rapid variations in either, the comparison should not work out with this kind of sampling. The scatter of points away from the line $\frac{1}{2}\ddot{I}_{ij} = V_{ij}$ indicates that some detail has been lost. But some information survives even this infrequent sampling, indicating that the correlation time for the second difference of I_{ij} and for V_{ij} is on the order of the sampling interval. Lagged plots, like those of Figure 1 in which the points are located by the second difference of I_{ij} at snapshot number s and by V_{ij} at snapshot number $(s+1)$, show barely discernable traces of the line. This, too, implies that the correlation time for values of these tensor elements is about one sampling interval, but does not extend over twice the sampling interval.

6. Discussion and Conclusions

The elements of the virial tensor, V_{ij}, fluctuated rapidly with mean zero and standard deviation about 2% of the cluster energy for this 100-body system. Presumably larger fractional variations would occur with fewer bodies. The tensor generalization of the virial theorem was surprisingly well confirmed for this system.

Correlation times for rapidly varying quantities, with the kinetic energy tensor, the potential energy tensor, and the virial tensor as prototypes, were about $\frac{1}{5}$ of a crossing-time for this system. The correlation times describe the dominant time-scale of fluctuations of these quantities, and probably give some measure of the time-scale of phase-mixing.

The identities of Equation (1) do not provide a useful check for n-body integrations because they are expensive to evaluate and difficult to usc. Howcver, it has been confirmed that the integration generates a system that obeys these identities.

The interpretations given in this note are subject to the usual difficulties that attend numerical experiments with self-gravitating n-body systems. These arise from the errors of numerical integration and from a tendency to overinterpret results because one knows too much about what is going on in the system. Both sets of problems are best handled by using several samples of data – calculations run from different sets of initial conditions to sample the parameter space of initial conditions. That recourse was not available for this experiment for economic reasons, so other arguments are required.

The numerical integration errors can be disposed of because the time scales of the quantities studied in this experiment are not long compared to the e-folding time for error growth; however, those time scales might be of the same order of magnitude. The dangers of over-interpreting one experiment are not troublesome here because the experiment was undertaken to check on some calculation details in a sample integration. As more general statements are made, the position becomes more precarious. The conclusions about correlation time may be affected by the fact that there were no very close encounters in this system. But a close encounter in which the potential energy of one star pair is equal to the potential energy of all the rest of the cluster must occur much less frequently with 100 bodies than with 16. This system may be special, but there is no strong reason to believe that it is; at any rate, this is a danger that affects all experimentation. We conclude that the results presented here withstand scrutiny on both counts.

References

Aarseth, S. J.: 1971, *Astrophys. Space Sci.* **14**, 118.
Chandrasekhar, S.: 1960, *Principles of Stellar Dynamics*, Dover Publications Inc., New-York.
Chandrasekhar, S.: 1964, in W. E. Brittin (ed.), *Lectures in Theoretical Physics*, University of Colorado Press, Boulder, Colo., Vol. 6, p. 1.
Goldstein, H.: 1960, *Classical Mechanics*, Addison-Wesley, Cambridge, Mass.
Hoerner, S. von: 1960, *Z. Astrophys.* **50**, 184.
Lebowitz, J. L., Percus, J. K., and Verlet, L.: 1967, *Phys. Rev.* **153**, 250.
Miller, R. H.: 1971, *J. Comput. Phys.* **8**, 464.
Nacozy, P. E.: 1971, *Astrophys. Space Sci.* **14**, 40.
Pollard, H.: 1966, *Mathematical Introduction to Celestial Mechanics*, Prentice Hall Inc., Englewood Cliffs, N.J.
Wielen, R.: 1967, *Veröffentl. Astron. Rechen-Inst. Heidelberg*, Nr. 19.
Zwanzig, R. and Ailawadi, N. K.: 1969, *Phys. Rev.* **182**, 280.

R. H. MILLER

DISCUSSION

King: Your mention of ensemble averaging raises a serious general problem: how do we extract meaningful distribution functions from a small number of bodies in a simulation? Clearly the answer is in the averaging of snapshots, as you have suggested; but I think that the general statistical problem is one on which we need serious new work.

Miller: One of the usual pitfalls is a tendency to take snapshots at a fixed number of integration steps, which makes it more likely to take the snapshot during a close encounter. This leads to a strong bias in sampled quantities, which is, however, well-known to those who do such n-body integrations.

THE DYNAMICAL EVOLUTION OF BINARIES IN CLUSTERS

D. C. HEGGIE

Trinity College, and Institute of Astronomy, Cambridge, U.K.

Abstract. Using information on the rates at which binaries suffer encounters in a stellar system (Heggie, 1974a), we here study the effects of such processes on the evolution of the system itself. First considering systems with no binaries initially, we show that low-energy pairs attain a quasi-equilibrium distribution comparatively quickly. Their effect on the evolution of the cluster is negligible compared with that of two-body relaxation. In small systems energetic pairs may form sufficiently quickly to exercise a substantial effect on its development and on the escape rate, but in large systems their appearance is delayed until the evolution of the core is well advanced. In that case they appear to be responsible for arresting the collapse of the core at some stage.

Binaries of low energy, even if present initially in large numbers, are likely to have at most only a temporary effect on the evolution of the system. High-energy pairs are not easily destroyed, and so, if present initially, their effect is persistent. It competes with two-body relaxation especially when the fraction of such pairs and the total number-density are high, as in the core, where, in addition, binaries tend to congregate by mass segregation. When encounters with binaries become important, being mostly 'superelastic' they enhance escape and lead to ejection of mass from the core into the halo, thus accelerating the rate at which mass is lost by tidal forces. It is difficult to decide observationally whether globular clusters possess sufficiently large numbers of binaries for these effects to be important.

1. Introduction

Results from explicit computations of small N-body systems with up to 500 members (von Hoerner, 1963; van Albada, 1968; Aarseth, 1971a; Heggie, 1974a) have consistently revealed that phenomena involving the dynamical formation and development of binary stars play some role in the evolution of clusters of stars. On the other hand, the study of much larger collisional systems is at present practicable only if use is made of rapid approximate methods based on the Fokker-Planck equation (Larson, 1970a, b; Hénon, 1971; Spitzer and Hart, 1971a, etc.). Since such methods implicitly ignore three-body encounters, which are essential for the binary phenomena whose effects in explicit N-body calculations are so striking, it has become a matter of some importance to establish whether or not the binaries themselves have a significant influence on the overall evolution of the cluster, or whether, on the contrary, they form and develop automatically as a by-product in response to the evolution of the cluster by the usual relaxation processes in which binaries do not essentially participate.

Although it seems almost paradoxical, the simplified methods, which ignore three-body interactions, agree in predicting that three-body phenomena become important; for, as Hénon (1971) noted, the effect of collisional relaxation, whereby an outward flux of mass accompanies an inward flux of binding energy, leads to conditions favourable for the formation of one or more energetic binaries. Therefore he supposed that the binaries observed to form in direct N-body calculations were a symptom of this phenomenon, and thus a consequence of evolution of the cluster rather than a cause. Spitzer and Hart (1971a) improved the argument by outlining some

Hayli (ed.), Dynamics of Stellar Systems, 73–94. All Rights Reserved.
Copyright © 1975 by the IAU.

facts on the formation and evolution of binaries in clusters. If we define 'soft' binaries to be those with binding energies x satisfying $\beta x < 1$, where $\frac{3}{2}\beta^{-1}$ is the local mean kinetic energy of the single stars, they remarked that the effect of such pairs was negligible because their total binding energy must be small and, furthermore, they must exhibit a tendency to be disrupted. For 'hard' binaries, which by definition satisfy the inequality $\beta x > 1$, the argument differs, for these may absorb binding energy, but Spitzer and Hart showed, by a calculation to orders of magnitude, that the number of such pairs forming per unit relaxation time must vary approximately as N^{-1}, where N is the total number of particles in the system. Since the evolution of the cluster will be over after at most a hundred relaxation times, the conclusion is that the formation of hard pairs can be entirely ignored, provided that the system is of sufficient size.

Since considerable theoretical effort has been applied to a detailed, independent investigation of dynamical processes involving binaries, the rate of their occurrence can now be stated with some accuracy in most cases (Heggie, 1974a). In this paper these newly available results are applied to a re-examination of the role played by binaries in the evolution of clusters, with especial regard to the questions discussed by Spitzer and Hart. In the next two sections, devoted respectively to soft and hard pairs, the arguments will be seen to differ in some respects from those quoted above, although the broad conclusions of Spitzer and Hart are preserved. In particular it is confirmed that in a large system there is insufficient time for the formation of enough binaries to affect significantly the evolution of the system. In the fourth section we consider how these conclusions must be modified if we do not wait for the binaries to form by these inefficient dynamical processes: supposing that there may already exist numbers of binaries when the star cluster has formed, we enquire how abundant and how energetic they need to be in order to exert a significant influence on the dynamical evolution of the whole cluster. Throughout most of the paper, except where otherwise stated, the discussion is restricted to systems whose members all have the same mass, m, and so the final section indicates certain avenues which require exploration before the results may be applied to real systems, with especial regard to the effects of a mass-spectrum. At this point some suggestions are made for contact with observations.

First, however, we remark on the assumptions and approximations made in order to arrive at the 'reaction rates' which we shall quote, though full details will be found elsewhere (Heggie, 1974a). The results are expressed in terms of rate functions $Q(x, y)$, analogous to those of atomic physics, defined thus:

Let n be the number-density of single stars near a binary having 'internal' binding energy $x > 0$. Then the probability that, during an encounter with a single star, the binary suffers a change in its binding energy lying in the range $(y, y + dy)$ during the time interval $(t, t + dt)$ is defined to be $nQ(x, y)\, dy\, dt$.

This definition is easily extended to accommodate the case in which stars with different masses are present, but we note that the encounter responsible for a certain change in energy is idealised to occur instantaneously. Since only the *local* number-density enters, we consider solely binaries with semi-major axes much smaller than the local density scale height in the cluster. For the same reason the results are restricted to encounters with stars at impact parameters much less than this scale height; however, since the gravitational perturbation of a distant star on the internal motion of a binary is tidal, the total effect of the neglected very distant encounters is normally weak and negligible. It is assumed that, prior to the encounter, the distributions of the binary and the single star participating in it are uncorrelated, and that the distributions of the velocity of the single star and of the velocity of the mass-centre of the binary are Maxwellian. The initial energy of the binary is fixed at x, and the orientation and the phase of its orbit are random. The distribution of its eccentricity, e, is assumed to be $f(e)=2e$; there is substantial theoretical justification for this and it is well confirmed numerically (Heggie, 1974a; also examples in van Albada, 1968; and Aarseth and Hills, 1972). The results of other numerical experiments suggest that the analytically derived rates are correct within a factor of two, and often better than this.

Another useful preliminary is an expression for the relaxation time, for which we adopt the expression

$$t_{rh} = \frac{0.0600}{\log(0.4N)} \left(\frac{NR_h^3}{Gm} \right)^{1/2} \tag{1}$$

(Spitzer and Hart, 1971a), where logarithms to the base of ten are intended, and R_h is the radius containing the innermost half of the mass. From the same source we have

$$\frac{GNm^2}{R_h} \simeq 7.5 \langle \beta^{-1} \rangle, \tag{2}$$

where the average value of any local quantity, f, is defined to be

$$\langle f \rangle \equiv N^{-1} \int fn \, d^3\mathbf{r}, \tag{3}$$

the integral being taken over the volume of the cluster. These authors also define a dynamical time, closely related to the crossing time, by the equation

$$t_{dh} = \frac{1.58 \, R_h^{3/2}}{(GNm)^{1/2}},$$

and, finally, we introduce the quantity n_h, defined to be the mean number-density within a distance R_h of the centre of the cluster, whence

$$n_h \equiv \frac{3N}{8\pi R_h^3}.$$

D. C. HEGGIE

2. Soft Binaries

It can be shown by several methods (e.g. Gurevich and Levin, 1950) that, if pairs of stars in a system are uncorrelated, the space density of pairs with binding energies in the range $(x, x+dx)$ is approximately $n_0(x)\,dx$, where

$$n_0(x)=\tfrac{1}{2}\pi^{3/2}n^2\beta^{3/2}G^3m^6x^{-5/2},\tag{4}$$

provided that single stars have a Maxwellian distribution of velocities and

$$N^{-1}\ll\beta x\ll 1.\tag{5}$$

This means that we consider only very soft pairs with separations much less than the typical density scale height, whence once again knowledge of only the local number-density will suffice. A rather descriptive argument (Heggie, 1974a) may be used to show that (4) will generally be the distribution of binding energies observed by soft pairs with energies in the range $6N^{-1}\ll\beta x\ll 3N^{-1/2}$, but the following more thorough discussion is applicable also to the range (5).

Now we may remark that $n_0(x)$ is approximately an equilibrium distribution in the range (5), provided that the cluster has evolved to an extent sufficient to ensure that the distribution of velocities is approximately Maxwellian. To demonstrate this we need the kinetic equation for $n(x)$, the absence of the subscript zero implying that we here do not necessarily suppose that the number-density of binaries has the form (4). It is

$$\frac{1}{n}\frac{\partial n(x)}{\partial t}=n^2Q(x)-n(x)\,Q(x,-\infty)+$$

$$+\int_0^\infty dx'\,\{n(x')\,Q(x',x-x')-n(x)\,Q(x,x'-x)\},\tag{6}$$

where the new quantities $Q(x)$, $Q(x,-\infty)$ are defined to be, respectively, the formation and destruction rates for pairs of energy x, in a manner akin to $Q(x, y)$.

It may be shown by general arguments that the rate functions Q must satisfy the 'detailed balance' conditions (cf. Ross *et al.*, 1969)

$$n^2Q(x)=n_0(x)\,e^{\beta x}Q(x,-\infty)$$

and

$$n_0(x)\,e^{\beta x}Q(x,x'-x)=n_0(x')\,e^{\beta x'}Q(x',x-x').$$

Indeed, by consideration of the dynamics of individual encounters (Heggie, 1974a) one obtains approximate expressions for the rate functions, and these are actually found to satisfy the quoted relations within the accuracy of the approximations. One implication is that the distribution

$$n(x)=n_0(x)\,e^{\beta x}\tag{7}$$

is an equilibrium solution of (6), provided that n and β are constants.

This distribution is physically unrealistic for $\beta x \to \infty$, but in the range specified by (5) it is approximately $n_0(x)$. Hence the distribution (4) is approximately an equilibrium distribution, provided that n and β are constants. Generally speaking, this will also be true approximately even if (7) is not satisfied for $\beta x \gtrsim 1$, provided that changes in energy between values within the range (5) occur with far greater frequency than changes from energies within this range to energies in the range $\beta x \gtrsim 1$. That this is the case can be seen by examining the appropriate form (Heggie, 1974a)

$$Q(x, y) \propto G^2 m^{7/2} \beta^{1/2} y^{-2} \left(4 \frac{x}{y} + 7 \right) \left(\frac{x}{x+y} \right)^{5/2} \quad \text{when} \quad (\beta x)^{3/2} \ll \beta y,$$

for the rate function: if $\beta x \ll 1$ the rate for energy changes y satisfying $\beta y \simeq 1$ is much less than that for changes satisfying the inequality $\beta y \ll 1$. Hence we conclude that (4) is approximately an equilibrium distribution of binding energies in the range (5), if n and β are constants.

The initial conditions in most computational studies of N-body systems are such that there is no pair correlation to begin with, but in general binaries may not be present initially with the distribution (4) and, even if they are, the quantities n and β will change with time as a result of the familiar relaxation processes, and so the distribution will not remain in equilibrium for long. We therefore investigate whether deviations from equilibrium, however generated, tend to grow or to decline. Setting $n(x) \equiv n_0(x) + \delta n(x)$, we see that the deviation $\delta n(x)$ satisfies the same equation as $n(x)$, i.e. (6), except that the first ('creation') term on the right-hand side is absent. Now soft binaries are normally destroyed at a rate given by

$$Q(x, -\infty) = \tfrac{20}{3} \sqrt{\frac{\pi}{3}} G^2 m^{7/2} \beta^{1/2} x^{-1} \tag{8}$$

approximately, while the rate at which the energy of a soft pair changes as a result of encounters which do not destroy it is

$$\dot{x} \simeq nG^2 m^{7/2} \beta^{1/2} \left(-\frac{\pi}{2} + 4 \sqrt{\frac{\pi}{3}} \left\{ \tfrac{5}{3} - \ln \frac{4}{\beta x} \right\} \right) \tag{9}$$

(Heggie, 1974a). When binaries are distributed according to (4), the formation of new pairs approximately balances the destruction of those present. In the absence of a creation mechanism, the number of soft pairs would diminish because of their destruction, and by (9) the mean energy of those surviving would decrease, thus accelerating the destruction process. Gurevich and Levin (1950) and Spitzer and Hart (1971a) showed that this tendency of soft pairs to be disrupted by encounters could be understood using an equipartition argument. We conclude that δn tends to zero, i.e. the distribution of soft pairs tends to approximately $n_0(x)$ given by (4), in consequence of three-body encounters.

The time scale for the approach of the distribution of soft pairs to equilibrium may be estimated from (1) and (8) as

$$t_s = \frac{1}{nQ(x, -\infty)},$$ (10)

$$\simeq 1.00 \log(0.4N)\, \beta x (n^{-1} n_h)\, (\beta \langle \beta^{-1} \rangle)^{-3/2} t_{rh}.$$

As Gurevich and Levin in effect pointed out, a simple argument shows that t_s must be much less than t_{rh} for very soft pairs, for the latter is the time required on average for changes in energy of order β^{-1}, while the disruption of a soft pair requires only a much smaller change, of order x. By comparison, Larson's estimate for the time required for the complete evolution of a dense central core is at least $16 t_{rh}$ (cf. Spitzer and Hart, 1971b). We therefore expect that, even if soft pairs are not initially distributed according to (4), their distribution will take approximately this form after a time which is small compared with the time taken for what Spitzer and Hart term 'complete collapse' of the core. Then, as the parameters n and β vary because of the evolution of the system, soft pairs will remain approximately in equilibrium provided that such changes take place on a time scale which is long compared with t_s. If the time scale is of the order of t_{rh}, this will certainly be true for very soft pairs, and their distribution will change on a time scale which is entirely governed by that of the evolution of the single-particle distribution function. From now on we shall assume that soft pairs satisfying the inequality (5) will be distributed in accordance with (4).

Using this distribution, the total number and the total binding energy of such binaries are found to be of the order of

$$90\, N^{1/2} \langle n\beta^3 \rangle \langle \beta^{-1} \rangle^3 n_h^{-1}$$ (11)

and

$$200\, N^{-3/2} E \langle n\beta^2 \rangle \langle \beta^{-1} \rangle^2 n_h^{-1}$$ (12)

respectively, where E is the total energy of the system. The total number of such pairs and, even more so, the total binding energy, both become relatively negligibly small for a sufficiently large system. For example, in the case of Plummer's model, $\langle n\beta^3 \rangle \langle \beta^{-1} \rangle^3 n_h^{-1} \simeq 0.36$, although (4), (11) and (12) all require some adjustment of coefficients for this model (Heggie, 1974a) because of the non-Maxwellian distribution of velocities.

There is another reason why the existence of very soft pairs poses no threat to the validity of fast methods for studying the evolution of clusters on the basis of two-body encounters. In such systems pairs of stars are uncorrelated except during hyperbolic encounters, and we recall that the equilibrium distribution of bound soft pairs (4) corresponds to an absence of correlation in the two-body distribution function. Therefore these same fast methods correctly, if implicitly, describe the equilibrium distribution to which we expect very soft pairs will adhere throughout most of the lifetime of the system. Finally, since the soft pairs generally exist in the absence of correlations, there is no justification for regarding them as donating a separate contribution to the binding energy of the system.

3. Hard Binaries

The destruction of very hard pairs of energy x takes place on a time scale t_h which is defined by analogy with t_s in (10) and has the approximate value

$$t_h \simeq 0.7 \log(0.4N)\,(n^{-1}n_h)\,(\beta\langle\beta^{-1}\rangle)^{-3/2}(\beta x)^2\,e^{\beta x}t_{rh}. \tag{13}$$

It will be noted that the values quoted for t_s and t_h are very comparable when $\beta x \simeq 1$, as indeed they should be, and that t_h is approximately equal to the quoted estimate for the time of complete core evolution for a certain value of βx which is fairly insensitive to the values of n and β. In case $N = 10^5$ and $\beta\langle\beta^{-1}\rangle = 1$, some representative values are given in Table I. For substantially more energetic pairs, the proba-

TABLE I

Values of βx at which $t_h \simeq 16 t_{rh}$

n/n_h	βx
1	1.2
10	2.3
100	3.6

bility of direct disruption may be ignored but, as in the case of soft pairs, we must enquire whether 'cascades' (i.e. successions of changes in binding energy less than that required for disruption) act so as to increase or to decrease the lifetime of a hard pair.

For this and other purposes we require the following approximate expressions for the rate functions for hard pairs (Heggie, 1974a): from 'close' encounters

$$Q(x,y) \simeq \begin{cases} 45G^2m^{7/2}\beta^{1/2}x^{-2}e^{\beta y} & (y<0) \tag{14} \\ 45G^2m^{7/2}\beta^{1/2}x^{5/2}(x+y)^{-9/2}, & (y>0) \tag{15} \end{cases}$$

and from 'distant' encounters

$$Q(x,y) \simeq \begin{cases} 4.9G^2m^{7/2}\beta^{1/2}x^{-1}|y|^{-1}\left|\ln\dfrac{|y|}{x}\right|^{-1/3}e^{\beta y} \\ \qquad\qquad (-\beta x \ll \beta y \ll -1) \tag{16} \\[2ex] 2.4G^2m^{7/2}\beta^{1/2}x^{-1}|y|^{-1}\left|\ln\dfrac{|y|}{x}\right|^{-1/3} \\ \qquad\qquad (\beta y_0 \ll |\beta y| \ll 1) \tag{17} \\[2ex] 4.9G^2m^{7/2}\beta^{1/2}x^{-1}y^{-1}\left|\ln\dfrac{y}{x}\right|^{-1/3}, \\ \qquad\qquad (1 \ll \beta y \ll \beta x) \tag{18} \end{cases}$$

where βy_0 is a certain very small positive number. In order to evaluate the average rate of change, the ranges over which the latter set of three expressions are held to be valid must be extended slightly so as to be contiguous. Thus the boundary between (16) and (17) is taken to lie at $\beta y = -1$, and (16) and (18) are taken to hold only when

$$|\beta y| < A\beta x, \tag{19}$$

where A is some number less than unity. Hence from close encounters we obtain

$$\dot{x} \simeq 0.8\,(nn_h^{-1})\,(\beta\langle\beta^{-1}\rangle)^{3/2}\,\frac{1}{\log(0.4N)}\,\beta^{-1}t_{rh}^{-1} \qquad (\beta x \gg 1). \tag{20}$$

From distant encounters the contribution is of the same functional form, but the co-efficient is about 1.0 when $A = 1$ and roughly proportional to A. The result that the rate of change of binding energy is independent of x may seem surprising, but it can be understood easily from a simple argument (Heggie, 1974b).

The result (20), together with the similar result for distant encounters, implies that hard pairs which escape disruption rapidly become harder, thus strongly reducing the probability of disruption. In fact from (20) we may obtain an upper limit for the time scale over which a hard pair increases its energy by an amount of order β^{-1}, and Table I implies that during this time the disruption probability decreases by a factor of about ten. Hence a more accurate estimate of the binding energy at which disruption becomes unimportant may be obtained by equating this time scale with t_h, whence the critical value of the binding energy is given by $\beta x \simeq 0.9$. Many of the quoted expressions are inaccurate for such low values of βx, but we are probably safe in neglecting the disruption of pairs with energies exceeding $B\beta^{-1}$, where B is some number around two.

The total rate at which such pairs form, including those created by cascade from the population of soft binaries distributed according to (4), is approximately

$$\frac{300}{\log(0.4N)}\,\frac{\langle\beta^{9/2}n^2\rangle}{\langle\beta^{-1}\rangle^{-9/2}n_h^2}\,B^{-7/2}N^{-1}t_{rh}^{-1}, \tag{21}$$

where again we have been obliged to extend the ranges of validity of some of our formulae; for example it is assumed that (4) holds for $N^{-1} \ll \beta x \lesssim 1$. The average energy of these new binaries is about

$$\tfrac{7}{5}B\langle\beta^{-1}\rangle\,\frac{\langle\beta^{7/2}n^2\rangle}{\langle\beta^{-1}\rangle\,\langle\beta^{9/2}n^2\rangle}.$$

From (21), using the values of the averages for a uniform sphere, we expect to obtain approximately

$$\frac{4.8\times 10^3}{N\log(0.4N)}\,B^{-7/2}$$

hard pairs after the lapse of about $16t_{rh}$. In the case of Plummer's model the corresponding constant is about 1.5×10^3. Hence, taking $B \simeq 2$ in the case of a uniform sphere, at least one hard pair can form and survive within the time required for collapse of the core only if the total number of particles is less than about 200, although such results are rather sensitive to the chosen value of B. At any rate, in such a case a hard binary could, according to (20), absorb only about 2% of the total energy of the system, which is $\frac{3}{2}N\langle\beta^{-1}\rangle$, in the lifetime of the core, and the initial energy of the pair would be typically rather modest, i.e. $\simeq 3\langle\beta^{-1}\rangle$ on average.

For smaller systems the number of pairs expected, and the fraction of the energy of the system which they might be expected to absorb, should both increase. On the other hand, the dependence of the formation rate (21) on N^{-1}, which was obtained by Spitzer and Hart (1971a) using a much simpler argument, ensures that hard pairs will not normally have time to form in systems containing more than several hundred members. Furthermore, even if a hard pair were to form, the energy which it could absorb within the time required for complete collapse of the core would not be a significant fraction of the total energy of the system. To sufficiently large systems, therefore, fast methods of studying evolution which implicitly ignore processes involving binary stars may, in this respect at least, be applied with confidence. Direct comparison between the Monte Carlo method and explicit N-body integrations (Hénon, 1971; Aarseth et al., 1974) provides complementary empirical support for the conclusion that this assumption made in the former method is approximately valid, certainly as far as early evolution of the cluster is concerned.

One qualification remains to be raised, for the computational evidence on systems with $N=250$ or 500 and a spectrum of masses seems at first sight to be in contradiction with the predictions given above, in that a large fraction of the total energy comes to reside in hard pairs in a time comparable with that required for complete collapse of the core (Aarseth, 1971a). Admittedly the evidence from the equal-mass models is less clear-cut (Aarseth, 1974) and it is for this case that the above theory has been developed, but in any event there is one aspect of the theory which we have ignored so far.

The factor $\langle\beta^{9/2}n^2\rangle$ in (21) is changing with time as a result of the usual relaxation mechanisms, which have nothing to do with the evolution of binaries, and possibly in addition as a result of the development of binaries. Late in the evolution of the core of a stellar system the number-density at a small distance, r, from its centre takes a form which has been quoted for small N as

$$n \propto r^{-12/5} \tag{22}$$

(von Hoerner, 1968), while Larson (1970b) obtains a steeper profile, with the logarithmic density gradient around -3 or -4, in a region which extends ever closer to the centre as time proceeds. The variation of β with r is much less steep, and so the quantity $\langle\beta^{9/2}n^2\rangle$ is increasing with time as evolution of the core draws towards its close, and by (21) the rate at which hard binaries form should increase proportionately. Alternatively one might regard the core crudely as an independently evolving N-body system with a decreasing number of members; when its membership is sufficiently small, (21) shows that hard binaries will form and evolve at a rate large enough to be important.

At this stage it becomes of interest to consider how the number of binaries formed during the whole collapse of the core depends on N. For this purpose we shall assume that the central number-density and mean kinetic energy vary asymptotically with time approximately as

$$n_c = n_0 \left(\frac{\tau}{\tau_0}\right)^m, \qquad \beta_c^{-1} = \beta_0^{-1}\left(\frac{\tau}{\tau_0}\right)^n, \qquad (23)$$

where m and n are constants whose values, drawn from several sources, are given in Table II, $-\tau$ is the time measured in such a way that collapse of the core occurs as $\tau \to 0$, and a subscript zero denotes the initial value. Calculations show (Larson, 1970a, b) that, close to the centre of a cluster with a highly evolved core, there is a region of approximately uniform density, and we shall estimate its radius by treating this region as a cluster to which (2) is applicable with central values of density and 'temperature'.* Hence we find the number of stars within the central region to vary as

$$N_c = N_0 \left(\frac{\tau}{\tau_0}\right)^{-(m/2)+(3n/2)},$$

and its kinetic energy as

$$E_c = E_0 \left(\frac{\tau}{\tau_0}\right)^{-(m/2)+(5n/2)}. \qquad (24)$$

Taking $\tau_0 = 16t_{rh}$, we obtain from (21) the result that the total number of hard persistent pairs formed within the central region during the time $-\tau_0 < -\tau < -\tau_f$ is approximately

$$\frac{9.6 \times 10^3 \, B^{-7/2}}{|6n-3m-2| \log(0.4N)} N_0 N^{-1} (n_0 n_h^{-1})^2 (\beta_0 \langle \beta^{-1}\rangle)^{9/2} \times$$

$$\times \left\{\frac{N_0}{2N}\left(N\frac{N_0}{2N}\right)^{(3n-2m-2)/(3n-m)} - \frac{1}{N}\right\}, \qquad (25)$$

where $-\tau_f$ is taken to be the time at which the number of core members reaches the minimum value consistent with finite binding energy, i.e. two (cf. von Hoerner, 1968). When the density profile is as steep as, say, that given by (22), the number of binaries formed outside the core is at most comparable with the number formed inside it, which is given by (25).

The prediction of the actual numbers of binaries formed requires a careful choice of the initial conditions β_0, n_0 and N_0, which in turn depend on N, but, if we assume that such ratios as N/N_0 are independent of N, then the only factors which will vary

---

* The resulting exponents agree to within 0.04 of those obtained for the radius at which the density falls to 10^{-2} of its central value, as obtained from the small-scale figures in Larson's papers. The differences are systematic, however, in that the decrease of radius with time is slightly steeper when taken from the figures than when obtained via (2).

with N in the above formula are the logarithmic term in the first denominator, and the term within the bracket. Now we require values of the indices m and n, and a list of those which have been obtained from theoretical or computational studies is given in Table II. The conclusion from the theoretical models is clear enough: from

TABLE II

Exponents for the evolution of the core

Source	m	n	$\dfrac{3n-2m-2}{3n-m}$	$m-\tfrac{3}{2}n+1$	$-\dfrac{5n-m}{3n-m}$
von Hoerner (1958)	-1.56	-0.37	0.02	-0.01	0.64
Miller and Parker (1964) $\left.\rule{0pt}{3.2ex}\right\}$ Spitzer and Saslaw (1966)	$-10/7$	$-2/7$	0	0	0
von Hoerner (1968)	$-4/3$	$-2/9$	0	0	$-1/3$
Larson (1970a)	-1.34	-0.225	0.01	-0.00	-0.32
Larson (1970b)[a]	-1.53	-0.30	0.25	-0.08	-0.05
Larson (1970b)	-1.54	-0.30	0.28	-0.09	-0.06
Larson (1970b)	-1.67	-0.29	0.59	-0.24	-0.28
Larson (1970b)	-1.47	-0.30	0.07	-0.02	0.05

[a] Larson notes that the behaviour of n_c and of β_c^{-1} is not precisely represented by a power law, the indices changing slightly with τ. The behaviour found by King (1958) is qualitatively similar in this respect, although τ cannot be defined in quite the same way because the particle number does not quite vanish after any finite time in this theory. However, these two effects are not related, for in King's theory it arises because of the variation with N of the logarithmic term in the expression for the relaxation time, and in Larson's calculations this term is taken as a constant.

the third column of figures we see that the number of persistent hard binaries formed should be approximately independent of the number of particles. For some of Larson's models, however, many more pairs will form in larger systems as the core collapses than in small ones.

When such a binary forms in the core, it does so with an energy which is a few times the current value of the mean kinetic energy per particle. The binary then hardens at a rate given by (20), while β_c is varying in accordance with (23). If $\beta_c x$ decreases with time, the binary effectively becomes softer, and it will ultimately be disrupted. On the other hand, if $\beta_c x$ increases with time, then the typical changes which its binding energy exhibits as a result of encounters with single stars will eventually become so much larger than β_c^{-1}, which must be of the order of the energy of escape from the core, that the binary will be ejected from the core. Only if $\beta_c x$ is approximately constant will the binary persist in the core as a hard pair. From (20) and (23) we readily find that

$$\beta_c x \sim \left(\frac{\tau}{\tau_0}\right)^{m-(3n/2)+1}$$

as $\tau \downarrow 0$, and the exponent here is given in the second last column of Table II. The prediction is, then, that in those cases where only a few binaries are expected to form, they will remain as hard pairs in the core; whereas if conditions are such that many pairs will form in large systems, then they will be continually ejected from the core, if not from the cluster. Such loss of mass by the core naturally tends to inhibit its further development if the flux is large enough.

It is unclear what will be the total energy of the binaries. If the energy of the core is taken as an upper limit, we see from (24) and the kind of argument used to arrive at (25) that the final energy of the core is roughly

$$E_0 (N_0/2)^{-(5n-m)/(3n-m)}.$$

The exponent here is given in the final column of Table II, from which we remark only that it is uncertain whether the fraction of energy absorbed by hard binaries increases or decreases as we proceed to a consideration of larger systems. For this and several other purposes it would be very helpful to possess a theory of core evolution by which the differences in the exponents found by Larson could be understood, as well as their relationship with the exponents predicted from previous theories.

At any rate, the conclusion must be that, in large systems, binaries evolve as a consequence of the development of the system by other processes, and it is only towards the very end of its evolution that binaries will appear. Therefore, fast methods which ignore the formation of binaries should be applicable throughout all except the very last stages of core evolution, and estimates of the time scale for the complete collapse of the core derived by such methods should be reliable to this extent.

Some confirmation of these conclusions emerges from a study of the results of published N-body experiments, if we compare the time scale over which hard binaries evolve with that for complete collapse of the core. In a small system the two should be comparable, while in a large system the evolution of binaries should occur rather suddenly after a lengthy period in which the system evolves in the absence of hard pairs. Somewhat striking evidence that this is so is provided by graphs showing the binding energies of hard pairs present at different times. When $N = 10$ (van Albada, 1968), the energy of the hardest pair present increases fairly steadily, in marked contrast with the rather abrupt appearance and development of a hard pair in a much larger system ($N = 500$: Aarseth, 1971a), although one cannot easily rule out the effects of a different choice of mass-spectrum.

At this stage it is pertinent to remark on the mass-dependence of binary processes, a point which will be taken up again in the final section. First, the formation rate for pairs of a certain energy (Heggie, 1974a) increases strongly with increasing masses of the components, a fact that can be understood also by a phase-space volume argument (Miller, 1975). This is even true when the mass distribution is as steeply weighted against large masses as that used by Aarseth (1971b), viz. $f(m) \propto m^{-2}$. Second, the number-density of the largest masses is especially high where binary formation is most rapid, i.e. at the centre of the cluster, because of the phenomenon of mass segregation, a fact which further favours the formation of binaries with components of high mass.

Once a binary has formed, its components may change identity only through 'exchange' or 'resonance'. It has been noted (Heggie, 1974a) that it is unlikely for either component of a massive pair to be replaced by a particle of much smaller mass by exchange, while there is substantial numerical evidence (Anosova, 1969; Szebehely, 1972; Valtonen, 1974) that the least massive component of a bound triple system has much the highest probability of escape. Aarseth (1971a) has emphasised, in fact, that the final components of energetic pairs tend to be drawn from the two or three most massive particles even in the largest N-body systems studied.

Comparatively little attention has been paid either theoretically or computationally to the evolution of clusters after the development of the core has reached completion. Clearly the formation of new pairs will continue in the old halo, though at a rate much lower than that during the collapse of the core, and their energies are unlikely to be as great as those of the binaries which comprise the residue of the core. If there are many such pairs they may interact among themselves, leaving a few very energetic binaries, and thus resulting in a situation not unlike that which would have arisen had the evolution of the core led to the creation of only a small number of pairs in the first place.

Encounters with a hard pair of energy x have several important effects on the single stars, and one of these we now discuss. If the encounter is sufficiently distant, the energy of the binary is almost unchanged and the encounter is 'elastic'. It merely contributes to two-body relaxation, the binary behaving like a single star whose mass equals the combined mass of its components, and the amount of energy exchanged is at most of order β^{-1}. If, however, the single star approaches to a distance sufficiently close that the change, y, in the binding energy of the binary considerably exceeds β^{-1}, the encounter is essentially 'superelastic', since $y > 0$ in general. If ε is the binding energy of the star to the cluster, then ε suffers a change $\Delta\varepsilon \simeq -\frac{2}{3}y$. Since y can be of order x, typically, a close encounter between a hard binary and a single star can lead to the escape of the star with high energy, and possibly also to the ejection of the binary. In fact there is direct evidence from computational studies of systems with $N \leqslant 500$ (van Albada, 1968; Aarseth, 1971a, 1974; Heggie, 1974a) linking hard binaries with escape at high energy.

In the presence of one hard pair of energy x, the escape rate is

$$\mathcal{R} = n \int_{3\varepsilon/2}^{\infty} Q(x, y) \, dy.$$

Using (15) and (18) we readily obtain the result

$$\mathcal{R} \simeq 4.1 \frac{n}{n_h} \langle \beta^{-1} \rangle^{3/2} \beta^{1/2} \varepsilon^{-1} N^{-1} f\left(\frac{\varepsilon}{x}\right) t_{dh}^{-1} \qquad (1 \lesssim \beta\varepsilon), \qquad (26)$$

where f is a function represented in Figure 1 by the appropriately labelled continuous line, and t_{dh} is the dynamical time defined by Spitzer and Hart (1971a). Here we have taken $A = e^{-1}$ in (19) for convenience, while the dotted extension to the left shows the form that f would take were distant encounters ignored entirely. Likewise the energy flux is

$$\mathscr{E} \simeq 4.1 \frac{n}{n_h} \langle \beta^{-1} \rangle^{3/2} \beta^{1/2} N^{-1} g \left(\frac{\varepsilon}{x} \right) t_{dh}^{-1} \qquad (1 \lesssim \beta \varepsilon), \qquad (27)$$

where the function g is plotted also in Figure 1, the two curves having the same

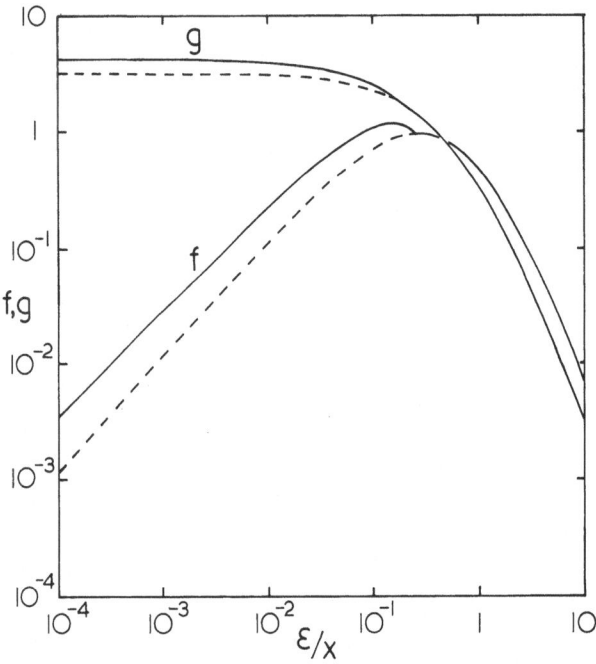

Fig. 1. Graphs of the functions appearing in Equations (26), (27) and (31), the argument being ε/x. The distinction between the continuous and dashed curves is explained in the text.

meaning as those for f. In this case, however, numerical integration is needed to obtain the results displayed. It is helpful to record Hénon's two-body escape rate (Hénon, 1969) in comparable notation, when it yields

$$\mathscr{R} = 0.011 \, t_{dh}^{-1} \qquad (28)$$

and

$$\mathscr{E} = 0.0050 \langle \beta^{-1} \rangle \, t_{dh}^{-1}.$$

In small systems containing at most a few hundred members it is clear that the mechanism of escape by encounters with a binary is competitive whenever a suitable binary is present. In large systems it is likely to be of importance only as soon as the core has evolved to the extent that the central density is high enough, although probably this also enhances the rate of two-body escape processes. Throughout the remainder of the evolution of the core the importance of the three-body process will persist, although the total mass lost during this phase need not be large because this part of the evolution is very brief. Furthermore, a single change in the internal binding energy of a pair which is sufficiently large to lead to the escape of a star from the cluster is probably also large enough to remove the binary at least into the halo, thus reducing the rate of further encounters with members of the core.

4. Initial Binaries

Too little is known about the process of star formation to rule out the possibility that a substantial fraction of stars are born as binaries, and yet enough is known to suggest strongly that it is likely (Larson, 1972). In this section we shall estimate the minimum abundance in which such 'initial binaries' would have to be present in order to exert a substantial influence on the evolution of the cluster.

Considering first soft binaries, we denote by $n(x) \, dx$ the initial number-density within the energy interval $(x, x+dx)$ in excess of what one would expect from the equilibrium distribution (4). These binaries would be destroyed on a time scale given by (10), and the rate at which their energy would be released is given by

$$\dot{x} \simeq \frac{-1.2\left(0.6 + \ln \dfrac{1}{\beta x}\right)}{\log(0.4N)} \frac{n}{n_h} (\beta \langle \beta^{-1} \rangle)^{3/2} \beta^{-1} t_{rh}^{-1}$$

(Heggie, 1974a), which is different from (9) because we here include those encounters which lead to the destruction of a binary. Hence the local rate per unit volume at which the binding energy of soft pairs would change is

$$-\frac{1.2}{\log(0.4N)} \int \left(0.6 + \ln \frac{1}{\beta x}\right) \frac{n(x) \, dx}{n} \frac{n}{n_h} (\beta \langle \beta^{-1} \rangle)^{3/2} n \beta^{-1} t_{rh}^{-1}.$$

The result has been cast in this form to facilitate comparison with the rate at which energy is exchanged by collisional relaxation. This is of order $\frac{3}{2} n \beta^{-1} t_r^{-1}$, where t_r is the local relaxation time, and using Chandrasekhar's expression (Chandrasekhar, 1942, p. 201) we obtain the result

$$1.1 (\beta \langle \beta^{-1} \rangle)^{3/2} \frac{n}{n_h} n \beta^{-1} t_{rh}^{-1}.$$

From these formulae we see that the release of energy from an initial excess of soft pairs would have an influence on the evolution of the cluster comparable with that due to collisional relaxation only if their numbers were comparable with those of the single stars. Very soft pairs would be the most efficient, but the effect is a weak logarithmic one.

The release of binding energy by soft pairs would lead to a contraction of the cluster, but the mechanism would only persist for a time of order t_s, given by (10). Therefore the total effect of the binary process could be small, even though it might dominate for a short time, if most of the pairs were very soft and hence rapidly destroyed. An estimate of the total effect of their destruction can be obtained by computing their binding energy per unit volume, which is

$$\int \beta x \, \frac{n(x) \, dx}{n} \, n\beta^{-1}.$$

Only if this were comparable with the local density of kinetic energy, i.e. $\frac{3}{2}n\beta^{-1}$, would the cumulative effect of initial soft pairs be important. The presence of the factor βx implies that very soft pairs would be relatively inefficient.

The fact that the destruction of hard pairs may be neglected implies that the effects of possible initial hard binaries would not be confined to a brief early period within the first relaxation time. If, as a result of an encounter between a binary and a single star, the internal binding energy of the binary were to change by an amount y, then the binding energy of the single star with respect to the cluster would change by an amount $\Delta \varepsilon \simeq -\frac{2}{3}y$ (cf. the discussion of escape rates in the previous section). For a typical single star the rate at which such encounters would occur is

$$\frac{3}{2} \int n(x) \, Q\left(x, -\frac{3}{2}\Delta\varepsilon\right) dx, \tag{29}$$

and so the mean rate at which the energy of single stars would change as a result of close encounters with hard pairs would be

$$-\frac{0.50}{\log(0.4N)} \int \frac{n(x) \, dx}{n} \, \frac{n}{n_h} \, (\beta \langle \beta^{-1} \rangle)^{3/2} \, n\beta^{-1} t_{rh}^{-1} \tag{30}$$

per unit volume, using (20). If $N = 10^5$, this would be comparable with the effect of collisional relaxation only if the total number of hard pairs were as much as about ten times the number of single stars. We have neglected a sizable contribution to (30) from distant encounters, however.

If the number of binaries were really so large, other effects would require consideration. So far we have only considered 'superelastic' encounters between single stars and binaries, but relaxation would also occur as a result of 'elastic' encounters, in which binaries may be treated as point masses with twice the mass of the single stars. The presence of such a species would considerably decrease the two-body relaxation time, and a system consisting almost entirely of hard binaries would relax four times more quickly than one whose members were mostly single stars (cf. Chandrasekhar, 1942, p. 201). Finally, if the proportion of binaries were large, the effect of non-elastic encounters between pairs would probably dominate that of superelastic encounters between pairs and single stars.

Since the effect of hard binaries would be to cause a decrease in the binding energy of the single stars, their existence initially in sufficiently large numbers would not only inhibit the formation in the system of a dense core of particles with high binding energy, but would also lead to enhanced loss of mass, as we shall see later. Furthermore, the consequent overall expansion of the system would much increase the efficiency of mass loss by tidal forces (Hayli, 1971), although the rate of occurrence of other dynamical processes would decrease as the mass-density decreased.

There are two respects in which (30) is misleading, in addition to that pointed out by Dr. Hénon in the discussion of this paper. It implies, first, that the efficiency of hard pairs would be independent of their binding energy, x. This statement ignores the stochastic nature of encounters, for encounters with very hard pairs would typically yield very large changes of energy but these would occur very infrequently. Second, (30) obscures the fact that changes in energy due to encounters would not occur primarily as a succcssion of numerous small changes, but discretely. Thus only one change of order $-\varepsilon$ to the energy of each star, during the evolution of the system towards high central densities, would be sufficient to prevent formation of a core, even though the average rate of energy exchange by binaries could then be considerably smaller than that by collisional relaxation. To complement (30) on all three criticisms, therefore, we shall calculate the expected number of times, ν, that the energy of a single star suffers changes satisfying the inequality $\Delta\varepsilon < -\varepsilon_0$, where ε_0 is a positive constant, during a time interval τt_{rh}, where τ is a measure of the interval in units of t_{rh}. From (15) and (18), using the sorts of approximations employed in deriving the escape rate in Section 3, we obtain

$$\nu \simeq \frac{0.16\tau}{\log(0.4N)} (\beta \langle \beta^{-1} \rangle)^{3/2} \frac{n}{n_h} \frac{1}{\beta\varepsilon_0} \int \frac{n(x)}{n} f\left(\frac{\varepsilon_0}{x}\right) dx \qquad (\beta\varepsilon_0 \gtrsim 1), \qquad (31)$$

where f is the function plotted in Figure 1.

Now we see that binaries with energies slightly larger than the required energy change are the most efficient. To estimate the rate of escape of stars from the system, we take for ε_0 the average change in energy required for escape, viz. $\frac{9}{2}\langle \beta^{-1} \rangle$ (Chandrasekhar, 1942, p. 206). Taking a uniform sphere with $N = 10^5$ and $\tau = 16$ we find that the coefficient of the integral is about 0.12, and so it is again clear that only very large populations of hard binaries, now confined to a restricted range of energies between, say, $2\beta^{-1}$ and $20\beta^{-1}$, could lead to complete loss of single stars from the system during the time that would normally be required for complete collapse of the core. Clearly, considerably smaller populations could still be of importance in influencing the development of the core, especially when regard is paid to the facts that the logarithmic factor in (31) should really depend on the structure of the cluster as well as its total membership, and, a fortiori, that (31) depends on the spatial density, n.

From the discussion of Equation (29) it will be recalled that the population of hard pairs was required to be about ten times that of single stars in order that they should redistribute kinetic energy as efficiently as two-body relaxation. At first sight, therefore, it is curious that the rates of escape due to the two mechanisms are comparable when the number-density of hard binaries, with energies in the range of greatest efficiency, is as low as $0.02n$, which we see by comparing (28) with the discussion of (31). The explanation is that the importance of small energy changes relative to large changes is much greater for two-body relaxation than for the binary mechanism;

therefore a given population of binaries is much more efficient at producing escapers, to which process possibly only large changes in energy contribute, than at exchanging energy. Likewise, energy exchange by hard binaries is considerably enhanced relative to that by single stars if we consider the mean square energy change rather than the mean. Hence the minimum population of hard binaries needed to exert a significant influence on the evolution of the system would depend very much on the aspect under consideration.

5. Outlook

In the absence of a simple analytical theory for the evolution of stellar systems it is difficult to be more precise about the magnitude of the effect on the evolution of a system which the presence of a specified initial number of hard binaries would exert. Numerical experiments are needed in order to study this question in detail, and in this connexion the above calculation implies that pairs with a binding energy of about five or ten times the mean kinetic energy would be the most effective if included in an N-body computation or a Monte Carlo scheme.

Such studies would also establish the possible importance of three other effects which have a bearing on these questions. First, as mentioned above, from the point of view of two-body relaxation a hard binary may be regarded as a point mass with twice the mass of the single stars. Hence the process of mass segregation (e.g. Aarseth, 1973) might enhance the relative density of binaries near the centre of the system by over an order of magnitude, and so binaries may influence the evolution of the core even if their initial abundance is much less than the previous estimates of the minimum number required to have a significant effect on the development of the whole cluster. A second mechanism of a slightly different nature, but having a similar effect, comes into play as soon as energetic encounters between single stars and binaries begin to occur with significant frequency. If the internal energy of the binary changes by an amount y, then for the single star the change in binding energy relative to the cluster is $\Delta\varepsilon \simeq -\frac{2}{3}y$, and for the binary $\Delta\varepsilon \simeq -\frac{1}{3}y$. Since it is the binding energy per unit mass which discriminates core and halo members, we see that encounters in which significant changes occur in the internal energy of the binary are much more efficient at removing single stars from the core than binaries. Finally, in considering an abundance of binaries comparable with that of single stars, we repeat that the frequency of encounters between binaries may rival that of other types of encounter, and furthermore the cross-section for energetic encounters between two binaries almost certainly exceeds that for encounters between binaries and single stars.

The immediate difficulty in applying these results to real clusters is our ignorance of the dependence on the masses of the rate functions Q in some important cases, although the following order-of-magnitude calculation suggests that the presence of different masses may not alter some of our conclusions. We consider encounters between a single star of mass m_3 and hard binaries of energy x, whose components have masses m_1, m_2. Let $n(x)$ be the number-density of such pairs. Then ε, the binding energy of the single star, changes as a result of (a) 'elastic' encounters, and (b) 'superelastic' encounters. The effect of the first mechanism is to increase ε at a rate of order

$$\dot{\varepsilon} \sim n(x)\, m_3 M_{12} M_{123} \frac{G^2}{\langle v \rangle} \qquad (32)$$

(Chandrasekhar, 1943), where $M_{12} \equiv m_1 + m_2$, $M_{123} \equiv M_{12} + m_3$, and $\langle v \rangle$ is the mean particle speed in the system. Now let us consider the second mechanism. Taking a Keplerian approximation to the relative motion of the binary and the third body, the impact parameter required for an approach to a distance comparable with the semi-major axis of the binary is of order

$$\frac{G}{\langle v \rangle} \left(\frac{m_1 m_2 M_{123}}{x} \right)^{1/2} .$$

If $m_3 \ll m_1, m_2$ it is reasonable to suppose that the change in the velocity of the single star relative to the binary is approximately independent of m_3 and typically comparable with the relative velocity of its components, and so the average change in ε is of order $(m_3 M_{12}^3 / m_1 m_2 M_{123}^2)\, x$. Hence the average rate of change due to inelastic encounters is of order

$$\dot{\varepsilon} \sim -n(x) \frac{m_3 M_{12}^3}{M_{123}} \frac{G^2}{\langle v \rangle},$$

which has almost the same mass-dependence as (32) if m_3 is very small. The conclusion would then be that the relative importance of the two relaxation processes does not change significantly in the presence of different masses.

The observational position is also complicated by the existence of a spectrum of masses. Here the boundary between hard and soft pairs may be taken to lie at the energy of the most energetic single member of the cluster. If this is a star of mass $\mathcal{M}_{\max}$ say, and a velocity of 0.5 km s^{-1}, typical of open clusters, the corresponding semi-major axis is

$$a \simeq 4 \times 10^3 \frac{\mathcal{M}_1 \mathcal{M}_2}{\mathcal{M}_{\max}} \text{ AU}, \qquad (33)$$

where $\mathcal{M}_1$, $\mathcal{M}_2$ are the masses of the components, all masses being expressed in solar units. With components exceeding the sun in mass, such binaries should be resolved visually at distances even as great as 1 kpc. In globular clusters, for which the coefficient in the corresponding expression would be smaller by a factor of about 100, such binaries could not be detected by visual inspection.

Following the results of Section 2, binaries present initially with semi-major axes exceeding these limits would not be expected to persist for more than a relaxation time, at the most. In globulars, they would be difficult to detect, and should not have been expected to survive in the relaxed central regions of these objects. They should be sought in young open clusters less than about 10^7 years old, and since our only interest is in an excess of very soft pairs over the numbers corresponding to an uncorrelated pair distribution, they could be detected in relatively nearby clusters by considering the statistics of pair separations among cluster members, or by dividing the area of the cluster into squares and comparing the star counts with a Poisson distribution.

A search for hard pairs, especially those with semi-major axes between about 0.05 and 0.5 times the value in (33), could usefully be conducted in the central regions of open clusters of all ages. They would be detectable by direct visual techniques only in the most nearby systems, but in the remainder only interferometric methods and, in special cases, occultation photometry would be feasible. Spectroscopy does not afford much prospect of success except in globular clusters.

The other source of information on the abundance of binaries is the solar neighbourhood, although the extent of its relevance to the stellar population in clusters is unknown. Considering the twelve known pairs within 5 pc (van de Kamp, 1971) having semi-major axes between 1 and 100 AU, we find that $n^{-1} \int n(x) \, dx \simeq 0.4$. The masses cover a wide range exceeding a decade, but the average is about 0.6 $\mathcal{M}_\odot$, and so these binaries, were they situated in an open or globular cluster, would roughly include those values of the binding energy which are most effective in influencing the evolution of the system. Furthermore, there is no guarantee from the conclusions of Section 4 that an abundance of this order would not have a significant effect on the development of a core, and it would certainly enhance substantially the rate of escape from the cluster as a whole.

Acknowledgements

I am grateful to many people for numerous discussions on binary problems, but especially to Dr S. J. Aarseth, who suggested to me some years ago that the topic was ready for detailed study, and who has followed the progress of the work with deep interest. I am happy to acknowledge in addition the financial support of the U.K. Science Research Council and of Trinity College, Cambridge, as well as the hospitality of the former Institute of Theoretical Astronomy, and now of the Institute of Astronomy, Cambridge.

References

Aarseth, S. J.: 1971a, *Astrophys. Space Sci.* **13**, 324.
Aarseth, S. J.: 1971b, *Astrophys. Space Sci.* **14**, 20.
Aarseth, S. J.: 1973, *Vistas in Astronomy* **15**, 13.
Aarseth, S. J.: 1974, *Astron. Astrophys.* **35**, 237.
Aarseth, S. J., Hénon, M., and Wielen, R.: 1974, *Astron. Astrophys.* **37**, 183.
Aarseth, S. J. and Hills, J. G.: 1972, *Astron. Astrophys.* **21**, 255.
Anosova, Zh. P.: 1969, *Astrophysics* **5**, 81.
Chandrasekhar, S.: 1942, *Principles of Stellar Dynamics*, University of Chicago Press, Chicago, Ill.
Chandrasekhar, S.: 1943, *Astrophys. J.* **97**, 255.
Gurevich, L. E. and Levin, B. Yu.: 1950, *Astron. Zh.* **27**, 273, also NASA TT F-11,541 (1968).
Hayli, A.: 1971, *Astrophys. Space Sci.* **13**, 309.
Heggie, D. C.: 1974a, 'Binary Evolution in Stellar Dynamics', to be submitted.
Heggie, D. C.: 1974b, in Y. Kozai (ed.), 'The Stability of the Solar System and of Small Stellar Systems', *IAU Symp.* **62**, 225.

Hénon, M.: 1969, *Astron. Astrophys.* **2**, 151.
Hénon, M.: 1971, *Astrophys. Space Sci.* **13**, 284.
King, I. R.: 1958, *Astron. J.* **63**, 114.
Larson, R. B.: 1970a, *Monthly Notices Roy. Astron. Soc.* **147**, 323.
Larson, R. B.: 1970b, *Monthly Notices Roy. Astron. Soc.* **150**, 93.
Larson, R. B.: 1972, *Monthly Notices Roy. Astron. Soc.* **156**, 437.
Miller, R. H.: 1975, this volume, p. 65.
Miller, R. H. and Parker, E. N.: 1964, *Astrophys. J.* **140**, 50.
Ross, J., Light, J. C., and Schuler, K. E.: 1969, in A. R. Hochstim (ed.), *Kinetic Processes in Gases and Plasmas*, Academic Press, New York and London, p. 281.
Spitzer, L., Jr and Hart, M. H.: 1971a, *Astrophys. J.* **164**, 399.
Spitzer, L., Jr and Hart, M. H.: 1971b, *Astrophys. J.* **166**, 483.
Spitzer, L., Jr and Saslaw, W. C.: 1966, *Astrophys. J.* **143**, 400.
Szebehely, V.: 1972, *Celes. Mech.* **6**, 84.
Valtonen, M. J.: 1974, in Y. Kozai (ed.), 'The Stability of the Solar System and of Small Stellar Systems', *IAU Symp.* **62**, 211.
van Albada, T. S.: 1968, *Bull. Astron. Inst. Neth.* **19**, 479.
van de Kamp, P.: 1971, *Ann. Rev. Astron. Astrophys.* **9**, 103.
von Hoerner, S.: 1958, *Z. Astrophys.* **44**, 221.
von Hoerner, S.: 1963, *Z. Astrophys.* **57**, 47.
von Hoerner, S.: 1968, *Bull. Astron., 3ème Série* **3**, 147.

DISCUSSION

Lynden-Bell: I would like to ask a strange question: is the average total mass of a binary pair greater or less than the average mass of single stars? The point is that if a gas cloud either forms a pair or a single star there might be no reason why binaries should be heavier and should sink to the middle of a cluster.

Heggie: The general question here is how does the abundance of binaries depend on the masses of the components? Your remark would imply that it is small for the highest masses, but observational evidence should also be considered. Prof. Spitzer has pointed out privately that since stellar evolution removes the largest masses first, after a time the masses of binary components could be as large as those of single stars even if your idea is correct.

Hayli: How much are binaries responsible for the disruption of soft binaries?

Heggie: Hard binaries act on soft binaries approximately as single stars with a mass equal to the combined mass of their components. Heavy stars are more effective at disruption than light ones, but the relative effect of binaries on the disruption of soft pairs is chiefly determined by the abundance of hard binaries relative to single stars.

Hénon: Is not it true that a large fraction of the energy given away by the hard binaries is carried out of the cluster by the escaping stars? In this case, the evolution of the cluster would be less affected than your computation indicates.

Heggie: It is true that mass lost from the system has a smaller effect on the evolution of the core than mass which is transferred from the core to the halo. This is another reason why it is better to calculate the number of energy changes rather than the rate at which energy is transferred. Then if we consider the binaries whose energies lie in the range of greatest efficiency, we see from the expression for v that the number of energy changes leading to escape is considerably less than that which leads to transfer from core to halo. This is not true for extremely hard pairs, and so the effect of your remark is, again, to exclude such pairs from consideration.

Lecar: What happens to binaries with $\beta x \approx 1$?

Heggie: Obviously there is some value of βx, about one, where $\langle \dot{x} \rangle = 0$, but I have not calculated what it is. As shown in the paper, binaries which are only slightly more energetic than this are very unlikely to be disrupted. It is more difficult to discuss with any precision the behaviour of soft pairs in some range less than this value, which leads, for example, to considerable uncertainty in the rate at which hard pairs form from soft ones. More work is required here.

Lecar: Why do 'soft' binaries become softer and 'hard' binaries become harder?

Heggie: Gurevich and Levin remarked that the internal kinetic energy of a very soft pair is typically much less than the kinetic energy in the relative motion of the binary and a third body, whence equipartition leads to increased internal kinetic energy in the binary, i.e. to disruption. If a third body closely approaches a very hard binary of energy x, such that the energy of relative motion is of order $\beta^{-1} \ll x$, two things may happen: either x increases, when the third body can escape again; or else x decreases. In this case, since the relative velocity of the binary and the third body at the time of encounter is in general almost independent of its initial value, the change in energy must be of order $-x$, thus binding the third body to the binary. When the resulting triple system disrupts, the third body will escape with energy of order x, and so in this case also the net change in the binding energy of the binary is positive.

Spitzer: It would certainly be interesting to incorporate these binary star effects in the Monte Carlo calculations. An important process is likely to be collision between pairs of binaries, since as you have suggested the number density of binaries may exceed that of single stars, at least in the cluster core. Will your paper include cross-sections and similar data for these encounters between two binary systems?

Heggie: No, although it ought to be possible to estimate such results, from the data on encounters between binaries and single stars, if the energies of the two pairs are very different. Numerical computations are probably necessary when this condition is not met.

Lecar: What is the distribution of eccentricities of hard binaries?

Heggie: Numerical evidence shows that the distribution is well relaxed, i.e. $f(e) = 2e$. Analysis of the effects of encounters indeed indicates that eccentricity relaxes faster than binding energy as a result of "distant" encounters.

BINARY EVOLUTION IN STELLAR SYSTEMS*

R. H. MILLER

Dept. of Astronomy, University of Chicago, Chicago, Ill., U.S.A.

Abstract. Aarseth has shown by means of n-body calculations that, in star systems with a range of particle masses, the most massive stars quickly form a binary which soon takes up a large fraction of the total binding energy of the cluster. Similar effects appear in other kinds of physical systems as well; mesic atoms behave in much the same way. The phase volumes of two otherwise equivalent stellar systems, each dominated by a tightly bound binary, favor exchange to incorporate the more massive star in the binary by a factor equal to the cube of the ratio of masses.

Self-gravitating n-body systems with a few hundred particles of different masses have been shown to form a tight binary rather quickly, with most of the total binding energy in the binary (Aarseth, 1971). The binary is usually formed of the most massive particles. This is similar to the state that yields the greatest phase volume in the microcanonical ensemble (Miller, 1973, 1974). It is remarkable, however, that the two most massive stars find their way into the binary so rapidly, and that the state dominated by a single binary is reached so quickly.

The situation is reminiscent of atomic processes when negative μ-mesons are stopped in condensed matter. The muons rapidly find their way to the ground state of an atom formed with the heaviest nucleus or with the nucleus having the greatest charge. A particularly interesting case occurs when muons are stopped in liquid hydrogen. The muon finds its way to a deuteron (even in hydrogen with 10^{-4} atomic contamination by deuterium), where it can cause a fusion $p + d \rightarrow He^3$. The muon is forcefully ejected after the fusion and is ready to start the process again. Several spectacular bubble chamber photographs show two or three fusion reactions during the 2.2 μs mean lifetime of a muon. The only difference between $(p\mu d)^+$ and $(ped)^+$ ions is the closer spacing caused by the much greater mass of the muon (207 electron masses); fusion can occur because the HD molecule is bound by the muon with a very small internuclear distance. A discussion of this process, together with some photographs of triple fusion reactions is given by Doede (1964).

There are essential differences between the muon-hydrogen system and stellar systems. While the μ-mesic atom is a quantum mechanical system, and, once in the ground state, can only become more tightly bound by transferring to a more massive or a higher-Z nucleus, a binary star can become more tightly bound without exchanging members. Still the analogy is suggestive, and illustrates how rapidly and effectively systems can make transitions toward states that afford a larger phase volume. In particular, the analogy suggests that a significantly larger phase volume must be available to states in which more massive particles are bound into a binary.

In a stellar system, the phase volume available within the energy interval from E to

---

* This paper appeared at the meeting as a comment to D. C. Heggie's talk.

Hayli (ed.), Dynamics of Stellar Systems, 95–97. All Rights Reserved.

$E + dE$, with different particle masses, is the sum of contributions due to each particle-pair (Miller, 1973). It is

$$\sigma_{\beta\gamma}(E) = D \left(G \sum_{\alpha=1}^{N} m_\alpha \right)^{3N-6} [Gm_\beta m_\gamma]^3 \left\{ \frac{\prod_{\alpha=1}^{N} m_\alpha}{\sum_{\alpha=1}^{N} m_\alpha} \right\}^{3/2} (-E)^{(1-3N)/2}, \qquad (1)$$

for the pair of particles with indices β, γ, where D is a (large) numerical constant. The curly bracket comes from the integral over the $3N$-dimensional momentum space. The remaining coefficients come from the integral over the $3N$-dimensional configuration space, which is kept finite by placing a cutoff radius far beyond the present limits of the star cluster. The cutoff radius appears as a constant multiplying the virial theorem cluster dimension; the constant is absorbed into D. The configuration volume is the cartesian product of the configuration volume accessible to the star-pair (β, γ) as a binary (square bracket) and of the configuration volume available to the rest of the particles independently (round bracket). The total phase volume, $\sigma(E)$, is

$$\sigma(E) = \sum_{\text{pairs}} s_{\beta\gamma}(E), \qquad (2)$$

and is dominated by states in which two particles are bound into a binary with all the binding energy of the cluster while the remaining particles have low velocities at large distance. The total angular momentum can be set to the required value with a small velocity assigned to particles at great distances, leaving the entire energy in the binary.

The phase volume available with a certain pair of particles is proportional to $(m_\beta m_\gamma)^3$, if all other features of the system are the same. In particular, the ratio of phase volumes accessible to otherwise equivalent systems in which the binary is made out of particles α, β to that in which it is made out of α, γ is $(m_\beta/m_\gamma)^3$. This is the increase of phase volume accessible to the system by substitution of particle β for particle γ in a binary with particle α. It is independent of the number of particles in the system. The cutoff distance cancels in the ratio.

In Aarseth's (1971) 250-body Case I, the phase volume with the two most massive particles (mass 25) together is 125 times as great as that with one massive particle and one particle from the second most massive set (mass 5). The two most massive particles found each other to form a binary after only two crossing times and quickly absorbed about 80% of the energy in the cluster.

Aarseth's (1971) 500-body Case V has a mass spectrum such that the ratio of phase volumes does not favor the two most massive particles nearly as strongly. Relative to the phase volume accessible to a system dominated by a binary formed of particles 1 and 2 (particles are numbered in the order of decreasing mass), a system dominated by the pair 1-3 has 0.84, 1-4 has 0.72, and 1-5 has 0.62. The pair 2-3 has 0.70, 2-4 has 0.60, and so on. Experimentally, after a sequence of particle exchanges, particles 1-3 formed a tight binary at about 12 crossing times, and soon about half the cluster energy was in that binary.

The general picture afforded by the comparisons of systems accords with Aarseth's results. Typically, systems with all masses equal do not form tight binaries nearly as rapidly – indeed, it is difficult to argue that binary formation plays much of a role in such systems. We stress that phase-volume considerations, while taking the $1/r^2$-force law explicitly into account, may indicate a preponderance of states that can only be reached after very long times when actual phase-space trajectories are taken into account.

Experimental results from n-body calculations dramatically confirm the picture provided by accessible phase-volumes in the microcanonical ensemble – that the final state of an n-body system is a tight binary with all other particles removed to infinity. It has long been suspected that this is the only stable final state, but the microcanonical ensemble arguments strengthen the conjecture substantially. The remarkable feature of the experimental results is the rapidity with which the system tends toward that state.

References

Aarseth, S. J.: 1971, *Astrophys. Space Sci.* **13**, 324.
Doede, John H.: 1964, *Phys. Rev.* **132**, 1782.
Miller, R. H.: 1973, *Astrophys. J.* **180**, 759.
Miller, R. H.: 1974, *Advances in Chemical Physics* **26**, 107.

DISCUSSION

Wielen: I have studied recently the Ursa Major star cluster. It is a gravitationally bound, sparse open cluster, which has probably lost most of its members. The binary ζ UMa AB = Mizar may well be identified with the central binary or triple system (together with 80 UMa = Alcor) which is formed dynamically in n-body simulations.

THE STRUCTURE OF ROUND STELLAR SYSTEMS:
OBSERVATION AND THEORY

IVAN R. KING

Dept. of Astronomy, University of California, Berkeley, Calif. 94720, U.S.A.

Abstract. In addition to its traditional roles of checking theories and applying them, observation also plays an important role in guiding theory and indicating the form that it should take. This paper reviews types of observation, types of stellar system (omitting the highly flattened ones), and the relations between them. Observational limitations are indicated, as well as places where observations of particular sorts are needed, either for guidance of theories or for their application.

1. Introduction

This review will attempt to assess the role of observations of spherical and moderately flattened stellar systems, from the point of view of theory. The main concern will be with two questions: what *can* observations do for theory, and what *do* observations say about theories? Some discussion will also be given of how observations can be used to apply theory, in order to derive significant astronomical results.

In principle, observations can assist the theoretician in several different ways. First, they can act as a general guide for the theory, indicating the directions in which it should go. The fortunate cases are very few in which a theory can start completely *a priori*, idealizing the problem in a reasonable way and proceeding to a correct theoretical picture. More often, the theoretical problem lacks clear definition, and we must first look at the Universe to see what the significant conditions and parameters are. An example of this situation is the dynamics of globular clusters, where the observation of a tidal limit, in the very first star count made on a Palomar Schmidt plate, showed that the tidal cutoff played a vital role in the dynamical problem (King, 1962). The second major way in which observations can contribute to theory is to discriminate between various theoretical possibilities, each of which predicts different observational consequences. We shall see an example of this below, in the attempt to distinguish between various possible cutoffs in a velocity distribution. Later on, of course, there is the task of testing a fully developed theory, to see whether its predictions are in agreement with observation. Finally, when a theory is accepted as being valid, observational data are needed in order to apply it, to reach concrete conclusions about the Universe.

In principle, we can observe almost everything about the state of a stellar system – at least at the present moment of time – but in practice there are many difficulties and limitations. We can observe two of the three position coordinates, lacking only the line-of-sight component, and in principle we can observe all three components of the velocity of each star, by combining radial velocities and proper motions. In practice, however, it is almost never possible to derive such detailed information, at least with accuracy that has any practical significance.

Hayli (ed.), Dynamics of Stellar Systems, 99–117. All Rights Reserved.

Accuracy is of course a serious problem in any set of observations, but this fact is not always as obvious as it should be. There are sets of observations in the literature that are afflicted with serious errors. This is not to say that all observations should be rejected, or regarded as suspicious; but one must always have an eye to the quantitative reliability of the observations that are in hand.

Even when the accuracy is completely adequate, observations of the structure of stellar systems are limited by three fundamental problems. The first problem is faintness: we need to observe the number and distribution of faint stars, the brightness distribution of the integrated light in faint envelopes, and radial velocities from faint spectra. In every case the faintness problem limits both the quantity and the quality of the observations, which often stop tantalizingly short of the interesting level. The second problem is background. The dynamical understanding of stellar systems often depends heavily on knowing the density distribution in their outer envelopes, and here the background is nearly always a limitation. Statistical fluctuations in the number of background stars limit the distance to which we can detect the outer envelopes of globular clusters, and this interference makes it difficult to study the envelopes of open clusters at all. For clusters of galaxies the problem is even worse, because of clustering of the background galaxies. Finally, the outer envelopes of elliptical galaxies are lost in the light of the night sky. The third problem is small numbers. Sometimes this is merely an economic problem: for example, where the limitation of observing time has allowed velocities of only a dozen objects to be measured. Sometimes, however, it is an essential problem: for example, how does one discuss the dependence of distribution on stellar mass, in a cluster that has a total of only 50 stars?

This last question raises another fundamental problem which pervades the interface between observation and theory. Just how are the two to be compared when the observation always provides only a subset of the quantities that exist in a theoretical model? The answer, I believe, is clear and unequivocal – and very important to follow in practice. The observations, limited as they are in accuracy, should never be transformed into a different set of quantities, for comparison with theoretical predictions. It is rather for the theoretical results to be transformed into the observational domain, for direct comparison with unadulterated observations. Quite aside from the question of whether the theory is correct or completely wrong, it can be carried out to as many significant figures as one desires; hence it does not suffer from statistical errors that are magnified in the transformation. The observations, however, have accidental errors that are nearly always increased when they are transformed to another domain. (Orthogonal transformations, such as the Fourier transform, are the one exception.)

A good example is density distributions, where the observations are made in a two-dimensional domain. Transformation from a projected to a spatial density distribution requires, whatever the method may be, a numerical differentiation of observed densities, which greatly increases the accidental errors. A theoretical distribution in three dimensions, however, can easily be projected into a two-dimensional distribution without any loss of accuracy that matters. Particularly insidious is the observational use of strip counts, in which perfectly good two-dimensional densities are degraded into a one-dimensional marginal distribution before being processed further.

2. Types of Observation

For the stellar systems considered here, the observations consist mainly of density distributions. Velocity distributions, valuable though they are, are much more difficult to obtain, and consequently much less velocity information is available.

Density distributions may be determined by counting individual stars, or by surface photometry of the integrated light. The data may be registered photographically, photoelectrically, or electrographically. Each technique has its advantages and its disadvantages. The great advantage of direct photography is that a single photograph can take in a very wide area. Not only is this an economic advantage, but it also avoids the sometimes costly errors that are involved in converting separate observations onto the same basis. Its disadvantages are its relatively low photometric accuracy and the difficulty of converting the non-linear response of the photograph into a correct intensity scale.

For accuracy and linearity, photoelectric observations are far superior. (Not all the photoelectric observations in the literature, however, have as high an accuracy as one might hope.) But observations made with conventional photoelectric photometers have two severe disadvantages. First, they are slow and uneconomical, requiring the point-by-point scanning of large areas. Second, they are unable to work at high spatial resolving power. The effects of time-changes in seeing require photoelectric observers to use an aperture that is far larger than the resolution element of a photograph taken under the same conditions. Both these disadvantages are avoided by panoramic detectors, which are coming into increasing use. In fact, because of the inherently higher quantum efficiency of photoelectric surfaces, panoramic detectors are much more efficient than conventional photographs – over the limited area that they cover. For large areas, however, photographs are unmatched, giving as many as 10^8 picture elements, compared with 10^5 for a large panoramic detector.

Electrography is a technique that shares some of the advantages of both houses. It combines the accuracy, linearity, and speed of its photoelectric detector with the resolving power and capacity of its photographic registration. Although the photosurfaces themselves are similar in size to those of panoramic detectors, the fine-grain photographic emulsions that are used in electrography give numbers of picture elements far superior to those of panoramic photoelectric detectors. Electrography has its disadvantages, however. Only a small number of electrographic devices have worked successfully, and the plates introduce special measuring problems, with their sharp resolving power and very high density range.

Each type of stellar system poses its own observational problems and suggests different observational techniques. For open clusters the overriding problem is the small number of stars. Surface-brightness techniques are useless; one must study the individual stars. Background numbers are a serious problem, since nearly all open clusters lie in the rich star fields of low galactic latitudes. Fortunately, proper motions are available for individual stars in a number of clusters; with these, the field stars can be weeded out with a high degree of probability. Even so, the star numbers are small. Nevertheless, systems with a small number of stars pose, as we know, a dynamical problem that is essentially different from that of systems with large N; and open clusters are the place where we can study such systems observationally.

Globular clusters can be studied by star counts or by surface photometry. The latter technique suffers somewhat from statistical uncertainties, since the integrated light of each region comes from a relatively small number of luminous stars (King, 1966b). In the center of a globular cluster, however, we must resort almost completely to surface photometry, because the individual stars are too crowded to count. Crowding is an even more serious problem than a glance at a photograph would suggest. Comparison of large-scale with small-scale photographs shows that crowding begins to affect star counts as soon as there is one star per 50 or 100 picture elements.

Again here, electrography comes closest to having the best of both worlds. It can reach faint individual stars at the edges of clusters, and it can measure accurate surface brightnesses in the centers. Most important of all, electrographic studies can disentangle crowded and even overlapping star images, because of the high resolution, the linearity, and the large dynamic range.

Even so, globular clusters present a frustrating problem of incomplete observations: we have the distribution of bright stars only at the center and the distribution of faint stars only at the edge. At the center of a typical globular cluster, crowding makes it impossible to study any except the brightest stars. (Surface photometry pushes the effective magnitude limit only a little fainter, since nearly all the light is contributed by the most luminous stars.) Outside the crowded central region, where the fainter stars can be studied, the distribution of bright stars is poorly determined, just because their numbers are so small. Thus no single group can be observed from center to edge, and the observations give only very limited information about one of the most interesting dynamical questions, the relative distribution of stars of different mass. To make this situation even worse, nearly all of the magnitude range that is observed in a globular cluster is inhabited by stars that have evolved from a small part of the original main sequence, around the present-day turnoff; hence these stars all have nearly the same mass. The observations end at a magnitude only a little below the main-sequence turnoff, just where mass is beginning to vary significantly with magnitude.

For elliptical galaxies, where the stars are too faint to detect individually, distributions can be studied only by surface photometry. Near the center, where the brightness is ample, the only serious problem is angular resolving power; at the distances of even nearby ellipticals, core radii are often of the order of a second of arc. Farther out, the eventual limitation is the light of the night sky – or rather its uncertainties and fluctuations; galaxy profiles begin to lose accuracy seriously when the light of the galaxy adds only a few percent to the night sky.

Density distributions in clusters of galaxies have problems similar to those of open star clusters: first, the numbers are generally small, and second, the envelope of the cluster is too easily lost in the background. Here again, it is extremely valuable to use motion criteria to separate members from background galaxies; in this case the criterion is radial velocity.

Velocities, which are much less available than positional distributions, can be applied in various ways to the study of stellar systems. As already indicated, they can be used to eliminate non-members. Their sizes – more specifically, the velocity dispersion – can be used to determine the mass of the system. The velocity dispersion can be determined from radial velocities of individual objects (or, for a few of the nearest open clusters, from proper motions), and in the high-surface-brightness centers of globular clusters and elliptical galaxies it can be determined from the broadening of lines in the integrated spectrum. Both individual velocities and integrated spectra can be used to study the rotation of a stellar system, whose relation to the velocity dispersion is of great dynamical significance.

The paramount problem in velocity observations is the faintness of the light. It restricts the observations to low spectroscopic dispersion, with a consequent low accuracy. For individual velocities, only luminous stars can be reached, and their numbers are few. For the integrated light, spectra are available only in a small central region of a stellar system. These restrictions leave us with a very limited picture of velocity distributions. Where observations exist at all, they usually provide only a value for the overall velocity dispersion. In a few cases, central angular velocities of rotation are available. Detailed velocity distributions are simply too much to hope for, but it seems possible at least to determine the gross features of some rotation curves, as well as the change of velocity dispersion as a function of radius.

3. Globular Clusters

Stellar systems pose theoretical problems of several different sorts, connected respectively with the initial rapid changes, with the nearly-steady condition into which the system then settles, and with the evolutionary processes that change that condition. Although observations do shed indirect light on evolutionary processes, we will concentrate here on the steady phase and its characteristics, because this is the dynamical problem that the observations illuminate most directly. Within this context, it is natural to devote our attention to globular clusters. Not only are they rich enough systems that their density distributions can be determined with a high degree of significance; their dynamical time scales are also short enough that all types of dynamical processes should be important.

On the crudest level, we can ask what kind of density distributions globular clusters have. Observationally, the answer is very simple; a cluster is described by three parameters: the core radius, the limiting radius, and a number factor such as the integrated magnitude. Since the core radius is set independently by the cluster's gravitational binding energy, the limiting radius by the tidal field of the Milky Way, and the number factor by its total mass, the absence of further differences shows that globular clusters are as similar as they can possibly be. The dynamical reason is easy enough to see: clusters are more than a relaxation time old, and relaxation has made them what they are.

More generally, the density distribution in a stellar system is our clue to its velocity distribution, which is in turn our clue to its internal dynamics. The velocity and density distributions are rigidly connected by the time-independent Liouville equation (along with Poisson's equation to describe the gravitational field), so that a velocity distribution corresponds to a unique density distribution. Roughly speaking, the low-velocity center of the velocity distribution corresponds to the core of the density distribution, and the high-velocity tail corresponds to the spatial envelope.

Mixing processes – either the initial violent relaxation or the subsequent relaxation through stellar encounters – make the velocity distribution in a stellar system tend toward a Gaussian form. Because the position space of the system is infinite in extent, however, relaxation can never be complete. Deviations from relaxation, and from the corresponding Gaussian velocity distribution, will thus show themselves in the envelope of the system and correspondingly in the tail of the velocity distribution. This means that stellar systems should all have similar cores, while the envelopes can be expected to show the nature of the relaxation process and the way in which it has affected the cluster.

In globular clusters the relaxation comes from stellar encounters; and after the initial settling down, its chief effect is to drive stars across the tidally set boundary of the cluster. Thus the form of the envelope is determined simply by the location of the boundary. Dynamically, however, the limitation is set not by the spatial location of the boundary but rather by the energy that a star needs in order to reach it. Since relaxation drives a steady flow outward through velocity space, the velocity distribution drops smoothly to zero at the cutoff velocity (Spitzer and Härm, 1958); and the corresponding spatial distribution fits the envelopes of globular clusters quite well (King, 1966a). Figures 1 and 2 show examples. If the shape of the cutoff were different, the profile of the cluster's envelope would also be different, as shown in Figure 3. (For extensive collections of globular-cluster data, see King et al., 1968, and Peterson and King, 1975.)

The close agreement shown in Figures 1 and 2 does not mean that we understand the envelope of globular clusters completely, however. The models that produced the curves in Figure 3 were all based on velocity distributions that are everywhere isotropic; yet it is quite possible that the velocities in the outer parts of a stellar system may be quite anisotropic. The theoretical case can be argued either way. On the one hand, the initial formation of a stellar system probably involves a collapse that endows stars with radial motions; furthermore, the ejection of stars from the core into the envelope, by encounters, leaves them with motions that are largely radial. On the other hand, stellar encounters should randomize the angular momenta in the cluster core, and tidal forces should randomize them in the envelope; randomized angular momenta correspond to an isotropic velocity distribution.

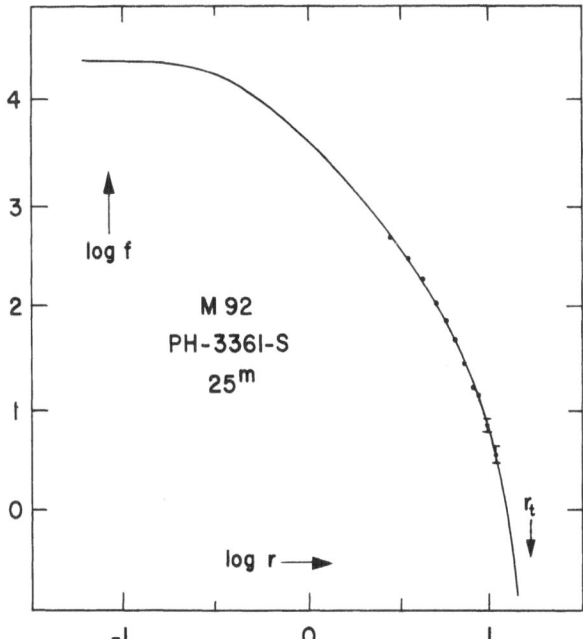

Fig. 1. Fit of observed surface densities in a high-concentration globular cluster. Points represent surface densities, f, determined from star counts in rings; vertical bars are statistical uncertainties. Central region was too crowded to count. Curve represents projected densities from a theoretical model with tidal limit at r_t.

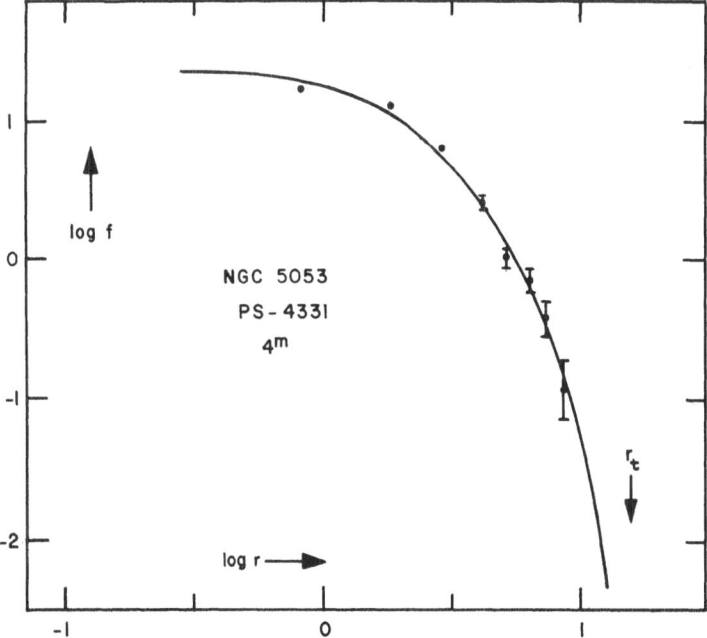

Fig. 2. Fit of observed surface densities in a low-concentration globular cluster. Representation as in Figure 1.

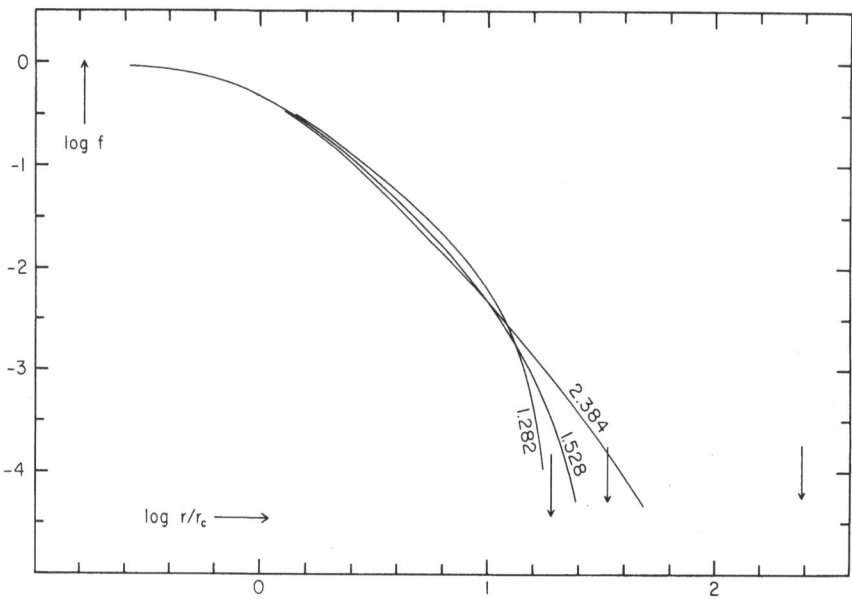

Fig. 3. Projected density vs radius for theoretical models with different velocity cutoffs. Label on each curve is $\log r_t/r_c$; all have same r_c. In order of increasing label, models have their velocity distribution truncated at escape velocity v_e, going to zero linearly as $(v - v_e)$, and going to zero quadratically as $(v - v_e)^2$.

Anisotropy in a velocity distribution has its effect on the spatial density distribution, and we may ask whether such effects are observable. Figure 4 shows an example of the effect of anisotropy; for an identical core and inner envelope, the outer envelope of the anisotropic model has a much smoother gradient and a much greater extent. The differences resemble, in an unfortunate way, the effects of changing the shape of the velocity cutoff, and it is thus difficult to distinguish between these two types of phenomenon. There seem to be two ways in which the distinction can in principle be made. First, Figure 4 makes it clear that an anisotropic-velocity envelope has a larger limiting radius than that of the corresponding isotropic model; anisotropic models can, in fact, be designed to have envelopes whose extent is infinite (Oort and van Herk, 1959; Michie and Bodenheimer, 1963). Since we know, from calculations of the galactic tidal field, an upper limit to the distance to which the envelope of any given cluster can extend, this allows us to set a corresponding upper limit on the anisotropy of its velocity distribution.

Probably the best way to study the question of anisotropy would be through measurements of the variation of the line-of-sight velocity dispersion with distance from the center. If the velocities in the envelope are largely radial in direction, they will have only a small component along the line of sight, and the velocity dispersion will drop sharply in the envelope. To interpret such an effect, however, it is essential to use a complete and self-consistent dynamical model of the cluster, since the mere existence of a finite limiting radius will cause the velocity dispersion to fall off radially even when the velocity distribution is isotropic (King, 1966a, Table I; King, 1972,

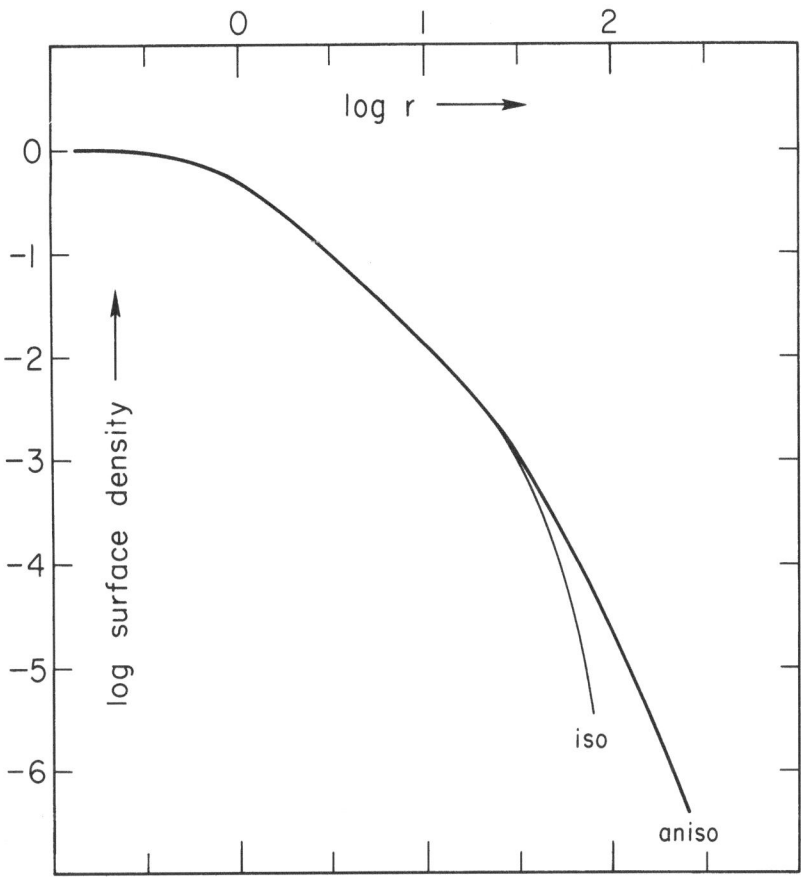

Fig. 4. Projected density vs radius, in two models with same core, to illustrate effect of anisotropy in the velocity distribution.

Table 2). Observations that would settle this question appear to exist (Gunn and Griffin, 1971; Griffin, 1972); when these data are made available to the astronomical community, a long-standing problem will be removed.

Determination of limiting radii of star clusters have other applications. Peterson (1974) has used them to study the shapes of globular-cluster orbits in the Milky Way, arguing that the limiting radius of each cluster shows how close its orbit takes it to the galactic center. Hodge (1966) has made similar calculations for neighboring dwarf elliptical galaxies.

The foregoing discussion has touched on one level of the dynamics of star clusters, but it has not yet distinguished between stars of different mass. Relaxation through stellar encounters is a mechanism that does make a distinction between stellar masses, and systems that have undergone such relaxation should show differences in distribution between stars of high and low mass. Not so for systems that have undergone only an initial violent relaxation, however, since violent relaxation treats all masses equally. Thus star clusters, with their relatively short relaxation times, should show mass segregation, while elliptical galaxies should not.

Differences in distribution, for stars of different mass, are perversely difficult to determine in globular clusters. As previously indicated, the bright stars can be studied in the core but are too few to give good statistics in the envelope, while the faint stars are well observed in the envelope but cannot be resolved in the core. Only in the globular clusters of lowest central concentration can we observe faint stars in the center, but here our dynamical study is frustrated again. In low-concentration clusters the density distributions are determined much more by the cutoff than by the circumstances in the core, and the distributions of the different stellar types thus look very similar. It is in the theoretical models of highly concentrated clusters that segregation effects are most striking, but there we are unable to observe enough of the center to detect the effects that we are looking for.

This situation is illustrated by Figures 5 and 6, which were plotted from theoretical models in which the stellar masses and the luminosity function are chosen to represent those in a real globular cluster. The first of the figures shows that in a low-concentration cluster the distributions of bright and faint stars are practically indistinguishable. The second figure shows how the differences predicted in a high-concentration cluster elude observation.

There is one situation, however, in which a study of mass segregation has the

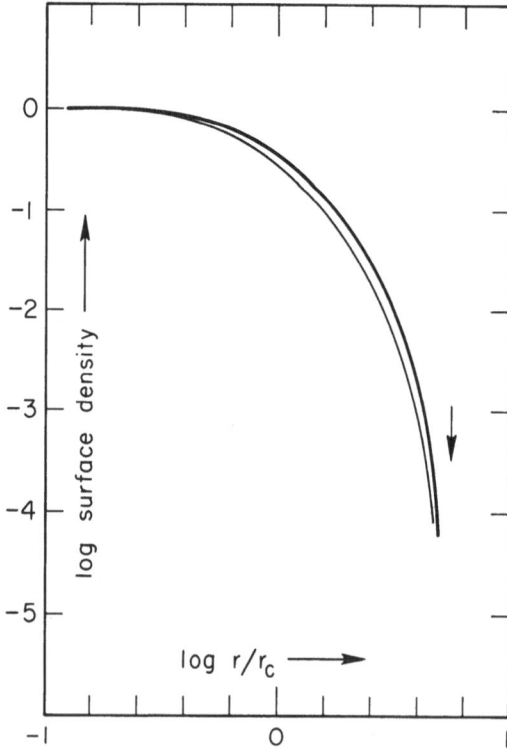

Fig. 5.   Surface density vs radius in a low-concentration globular-cluster model. Lower and upper curves refer to stars down to limits $M_V = +2$ and $+10$, respectively. Central densities have been normalized to same value.

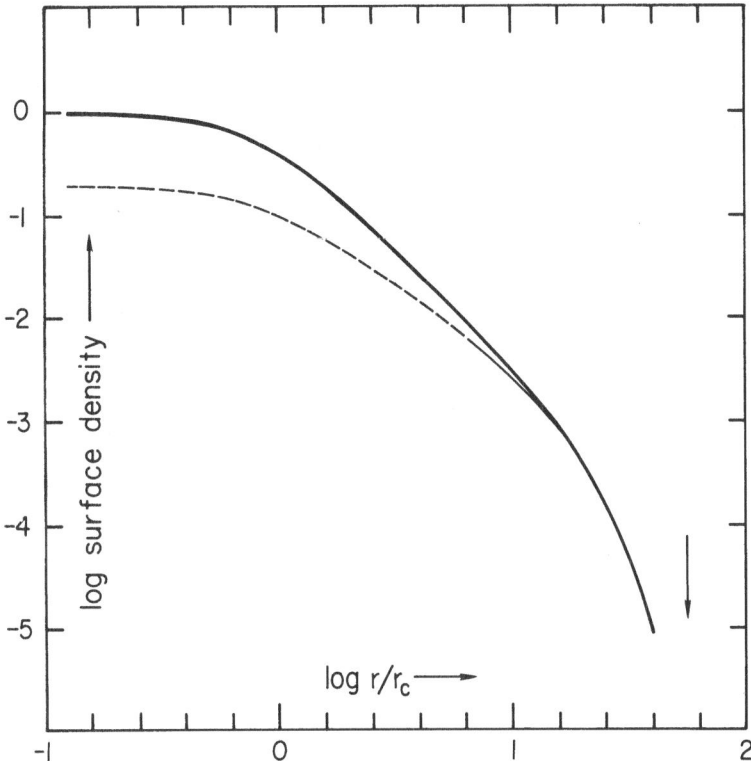

Fig. 6. Surface density vs radius in a high-concentration globular cluster. Upper and lower curves refer to stars down to limits $M_V = +2$ and $+10$, respectively, but dashed part is unobservable because of crowding. Outer densities have been normalized to same values.

possibility of being successful. If a group among the brighter, evolved stars has lost mass and then relaxed to a higher velocity dispersion, their broader spatial distribution might be observable. The number of stars is small, but these brighter stars can be observed throughout the entire cluster, on short-exposure plates that are bothered very little by crowding effects.

Although such studies can be carried out in principle, in practice I do not know of any observational data that are adequate. I once tried to do a study of this sort, using published counts of stars of various types in concentric annuli. The analysis consisted of comparing the observed numbers with numbers computed for a model in which stars of different types were distributed in accordance with their masses. For each stellar type the low-velocity part of the velocity distribution corresponded to the appropriate Maxwellian distribution, but the high velocities of course had a cutoff of the sort that has already been described. The technique was to use a chi-squared test of goodness of fit, comparing an observed distribution with the computed distribution of stars of each mass. The resulting curve of chi-squared against mass would show a minimum at the mass to which the observed group fitted best. At first the results looked very encouraging: the minimum value of chi-squared corresponded to a good probability level, and in many cases a change of only 20% in the mass would make the fit quite bad. I abandoned the study, however, when it turned out that data from different, but equally credible, sources gave different results. An example is shown in Figure 7. I believe that the difficulty of determining good stellar distributions in the centers of clusters has been underestimated. Perhaps this is another task for electrography.

A related problem, in which mass segregation plays a role and must be allowed for, is the determination of luminosity functions in globular clusters. While the brighter stars are counted throughout the cluster, the fainter stars can be counted only in an outer annulus, and their total number involves an extrapolation into the center, which can be made correctly only by use of a dynamical model of the cluster. Numbers in an actual case are illustrated in Figure 8. When model fitting is used in this way (Wilson and King, 1975), the previously reported differences between luminosity

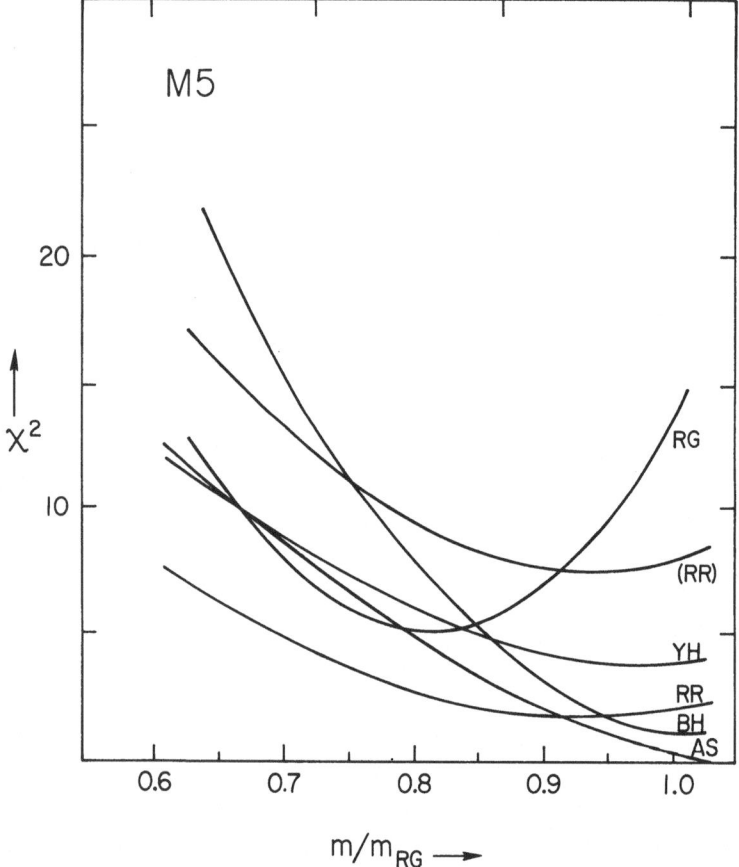

Fig. 7.   Goodness-of-fit for observed distributions of stars of various types in M5, when compared with theoretical distributions of stars of various mass. Abbreviations stand for red giant, asymptotic branch, yellow horizontal branch, blue horizontal branch, and RR Lyrae. Data from Simoda and Tanikawa (1970, Table 3), except for (RR), which is from Oort and van Herk (1959, Table 15). Latter has 3 degrees of freedom; other distributions have 2. Red-giant star-count distribution disagrees with model, which was fitted to a surface-brightness distribution that is dominated by red giants. The two versions of the RR Lyrae distribution disagree with each other.

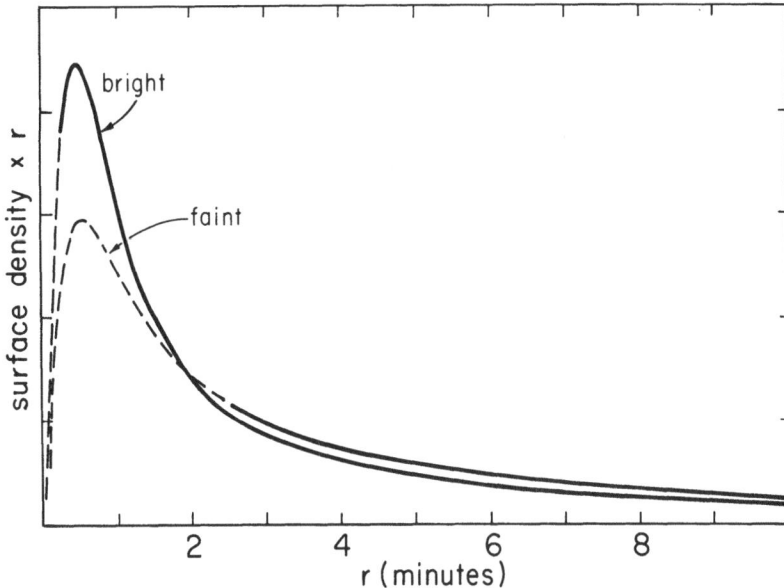

Fig. 8.   Star numbers in M3, plotted in such a way that area under the curves is proportional to total star numbers. Dashed portions are unobservable, with extrapolation provided by dynamical model.

functions of different globular clusters disappear. It is interesting to note that the largest remaining uncertainty is due to not knowing the degree of anisotropy that should be put into the velocity distribution.

Another important application of cluster models is in determining the total masses of clusters from their observed internal velocity dispersions (Illingworth, 1975). It turns out, fortunately, that the observations allow a fairly direct determination of the central density in a cluster (see the discussion of elliptical galaxies, below); but to extrapolate from the central value to a total requires a model in which the stars of different mass have the correct relative distribution. It is interesting to note that observations of this sort give us our only information about the number of low-mass stars in globular clusters. The stars that we can see, which contribute practically all of the light, go only a little below the main-sequence turnoff; the less-massive stars make themselves known only by their gravitation.

Along with the ways in which observation works hand in hand with theory, we should also take note of cases in which star clusters have not been observed to behave in the way predicted by theory. The most striking of these is the case of the runaway centers. Numerical simulations have repeatedly shown cluster cores that become denser and denser, producing a central singularity at a finite point in time. Many clusters are old enough, dynamically, that they should already have reached this stage of central catastrophe, yet nowhere do we find either a central collapse going on, or else evidence that it has happened at some time in the past. Globular clusters have smooth central distributions, typically with a core radius of the order of a parsec, and there is no sign of additional central peaks. If as much as 100 solar masses of additional material were placed at the center of a high-concentration cluster of 10^5 solar masses, its gravitation would perturb the distribution of the observed stars in a noticeable way. I know of only one cluster in which such a central excess might possibly exist. The central region of M15 shows a small excess of brightness over the profile that an isothermal core would give. The excess may or may not be observationally significant, but it certainly deserves closer examination.

One might imagine that open clusters are an even more likely place to look for central runaways, since their relaxation times are even shorter. For these poorer systems, however, it may be that the evolution of the small number of stars in the core soon becomes dominated by the massive central binary that always seems to form in the N-body calculations. The 'runaway' then consists merely of a continued transfer of energy between the central binary and its surroundings. If we think naively about the time scales, then nearly every open cluster should already have developed a massive central binary. A more realistic picture, however, includes stellar evolution; since the most massive stars have the shortest lifetimes, will they last long enough to play their assigned dynamical roles – and if not, how do the dynamic predictions change? Apparently, more theoretical work is needed on this problem. On the observational side, however, I do not know of any study at all in which open clusters are systematically searched for the present of a central, massive binary.

The presence of less spectacular binaries in clusters also poses an interesting observational problem. In a globular cluster, for instance, where the most massive stars have masses characteristic of the middle main sequence, W Ursae Majoris binaries should be the most massive stars in the cluster, and they should therefore concentrate strongly to the cluster center. Since they are far from being the brightest stars, these binaries will be very hard to detect; only in globular clusters of the lowest concentration can we hope to see individual stars of such low luminosity in the central regions. Unfortunately, the clusters of low concentration tend to be poor systems, with few stars of all kinds; and only a very few rich clusters have centers that are open enough for the search to be made. I do not know of any case where it has been done.

4. Stellar Systems of Other Types

Open clusters suffer, relative to globular clusters, from a smaller star number and a richer stellar background. Nevertheless they offer us several unique opportunities in supporting our dynamical theories with observational facts. Because they are younger in years than the globulars, open clusters still contain stars with a much wider range in mass. Furthermore, in many open clusters we can observe the distribution of the stars down to a lower limit of absolute magnitude than we can reach in any globular cluster. In clusters where we can use proper motions to separate faint cluster stars from the field, the magnitude range is especially large. Thus the nearby Hyades, although poor in stars, is in a sense the best-studied of all clusters (Pels *et al.*, 1975).

Although younger in years, open clusters are older dynamically than are globular clusters, because their relaxation times are so much shorter. Most of the open clusters

that were ever born are already dead, for dynamical reasons (Oort, 1958); and a
statistical study of the ages and sizes of existing clusters gives us a check on their
dynamical mortality (Wielen, 1971).

But from the observational point of view, perhaps the greatest value of open clusters is their very poorness in stars. These systems fall in the range of N where our
theoretical dynamics is very weak. For stellar encounters, these are the systems where
large velocity changes are of comparable overall importance to small changes; for
overall dynamics, these are the systems where the relaxation time is hardly longer
than the crossing time. For both these situations (which are of course dynamically
related), we lack a good theoretical treatment. Perhaps this is an area in which observations can guide a fledgling theory. As a starting point, one extensive set of star
counts already exists (van den Bergh and Sher, 1960).

Elliptical galaxies lie at the opposite extreme, dynamically, from open clusters. They
have so many stars that there is no doubt about the validity of large-N representations,
but they are less than one relaxation time old. What dynamical process determines
their structure? The first step in answering this question is to examine the profiles
of some actual systems; two of these are shown in Figures 9 and 10. In each case the
curve that is fitted, more or less well, is that of a relaxed star-cluster model. Since the
model has a nearly-Gaussian distribution of stellar velocities, the fit shows that
elliptical galaxies have an internal velocity distribution that is nearly Gaussian. This
should be a consequence, of course, of the violent relaxation (Lynden-Bell, 1967)
that took place at the time when the galaxy formed.

Violent relaxation leads to a result that differs from that of stellar encounters in one

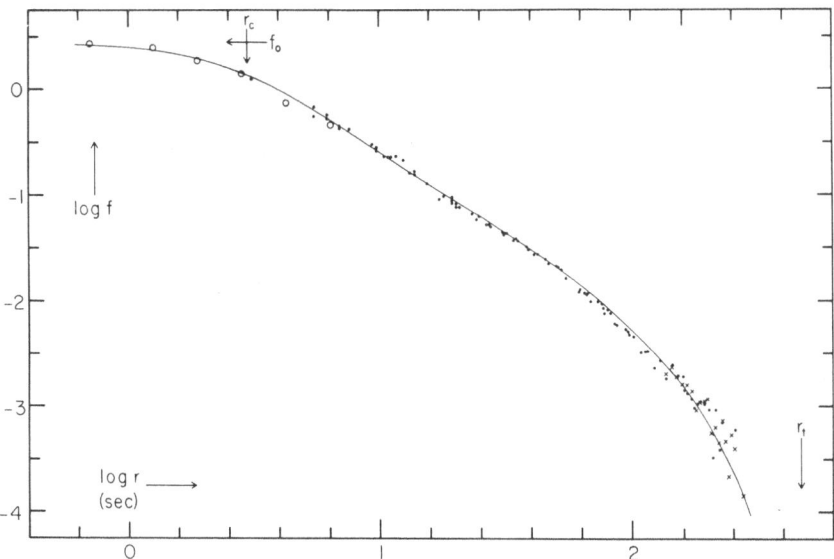

Fig. 9. Surface brightnesses in the elliptical galaxy NGC 3379, from data by Miller and Prendergast
(1962). Curve is a star-cluster model.

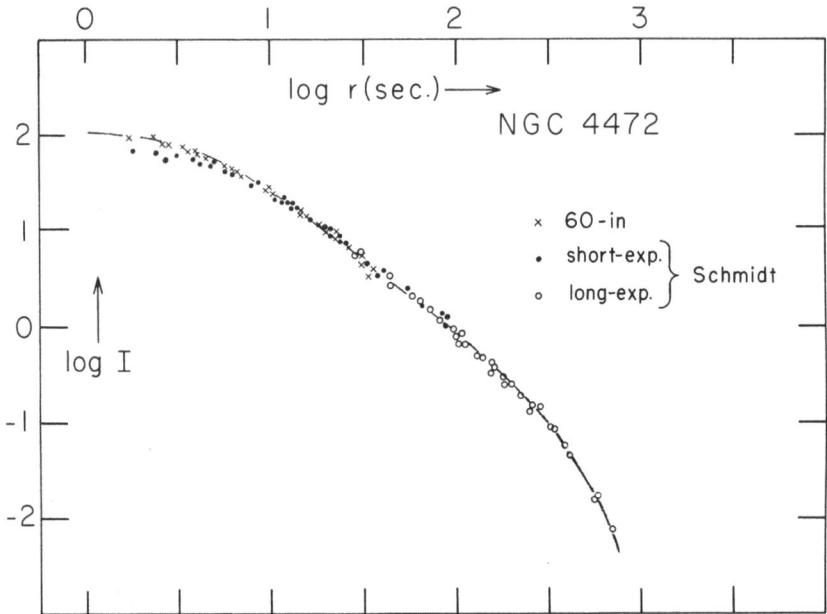

Fig. 10.   Photographic surface brightnesses in the elliptical galaxy NGC 4472, determined by the author.
Same curve as used in Figure 9.

important way: it does not produce equipartition between stars of different mass. It is easily verified that equipartition does not exist in elliptical galaxies; if it did, then the envelope would consist almost completely of red dwarf stars, and its color would be a deep and unmistakable red.

The profile of elliptical-galaxy envelopes poses a problem, however, which should not be obscured by the apparent ease of fitting star-cluster models to them. The theoretical curve used in fitting the galaxies in Figures 9 and 10 has an envelope shape that is determined by the particular ratio of the tidal cutoff radius to the core radius of the system, but this cannot possibly be the mechanism that molds the envelopes of elliptical galaxies. As Hubble showed long ago (Hubble, 1930), all elliptical galaxies have profiles that resemble these, with the surface density falling approximately as r^{-2} (space density as r^{-3}). It is inconceivable that every elliptical galaxy could have a tidal cutoff distance that is always the same multiple of its core radius.

What then has determined the profile of the envelopes of elliptical galaxies? Since relaxation has not changed these systems, we must be seeing a result of the process that originally formed them. Specifically, the profile of the envelope is determined by the particular way in which the high-energy tail of the velocity distribution drops to zero, and this is what we should examine.

There is more to elliptical galaxies than a mere radial profile; they are, after all, elliptical rather than spherical. Rotation obviously plays an important role in determining their forms. Theoretical models have been developed by Wilson (1975a, b); they fit the observed isophote shapes to a large extent, but there are some isophotal

profiles that Wilson's models are unable to fit. The greatest observational lack is
detailed information about the rotations of elliptical galaxies. Do the systems with
different isophotal profiles also have different patterns of rotation? We would like
very much to know.

An application for which it is important to have a good dynamical picture of elliptical galaxies is the determination of their masses. From an observation of the internal
velocity dispersion and the relative distribution of density, it is possible in principle
to determine the mass of the system. One might imagine naively that all that is needed
is to apply the virial theorem. Such an approach can unfortunately produce very bad
results, since it invariably involves serious and uncertain extrapolations. (For a
detailed discussion, see Rood *et al.*, 1972, Section VIII.)

In fact, the quantity that should be determined from the observations is the quantity
that really does follow directly from them: the central density. In practice, the only
place at which we know the velocity dispersion in an elliptical galaxy is in its bright
central region; hence this is the only place in which we can determine masses. Since
observations give us the central distribution of light, we can also derive a central
M/L, which is one of our intrinsic objectives in any case. But all else is extrapolation.

It should be no surprise that clusters of galaxies have density profiles like those
inside of single elliptical galaxies. They also have relaxation times longer than their
ages but were subjected to violent relaxation at the time of their formation. Figure 11
shows the distribution of galaxies in the Coma cluster, fitted with the same curve
that was used for the elliptical galaxies in Figures 9 and 10. The detailed problems
may turn out to be very different, however. Even though we still have relatively little

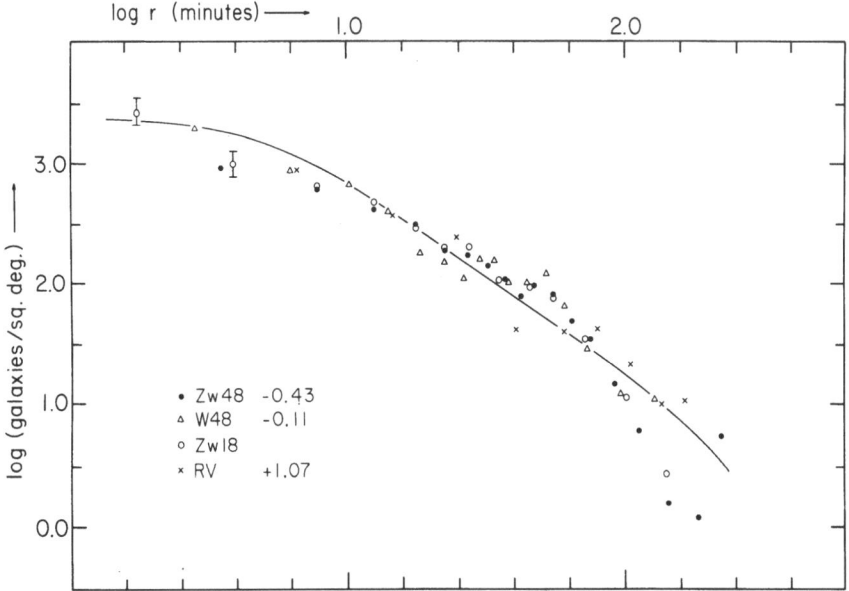

Fig. 11.   Surface densities of galaxies in the Coma cluster, as collected from various sources and fitted
together by Rood *et al.* (1972). Same curve as used in Figures 9 and 10.

information about clusters of galaxies, it is clear that new problems arise – for example, the clusters whose centers contain a single, dominant supergiant galaxy. This is usually a cD galaxy, whose profile appears to be different from that of a 'normal' elliptical. What are these galaxies? It is likely that the first steps toward answering such questions will be observational.

5. Conclusion

In this review I have tried to emphasize the importance that observations should have for a right-thinking theoretician. The nature of the Universe may eventually be rationalized by theoretical understanding, but it will not be discovered in the first place by pure thought. Any theory rests upon idealizations, and we should be sure that they are the correct ones. This admonition applies equally to simulations, where there is an equal danger of solving the wrong problem.

At the same time, I have tried to indicate directions in which theory indicates that observation should go. Our future progress will depend on a well-chosen combination of our various techniques.

Acknowledgements

The preparation of this paper was supported by NSF Grant GP-40483X.

References

Griffin, R. F.: 1972, *Observatory* **92**, 28.
Gunn, J. E. and Griffin, R. F.: 1971, in *Ann. Report Dir. Hale Obs.*, p. 411.
Hodge, P. W.: 1966, *Ann. Rev. Astron. Astrophys.* **9**, 35.
Hubble, E.: 1930, *Astrophys. J.* **71**, 231.
Illingworth, G. D.: 1975, this volume, p. 151.
King, I. R.: 1962, *Astron. J.* **67**, 471.
King, I. R.: 1966a, *Astron. J.* **71**, 64.
King, I. R.: 1966b, *Astron. J.* **71**, 276.
King, I. R.: 1972, *Astrophys. J.* **174**, L123.
King, I. R., Hedemann, E., Hodge, S. M., and White, R. E.: 1968, *Astron. J.* **73**, 456.
Lynden-Bell, D.: 1967, *Monthly Notices Roy. Astron. Soc.* **136**, 101.
Michie, R. W. and Bodenheimer, P. H.: 1963, *Monthly Notices Roy. Astron. Soc.* **126**, 269.
Miller, R. H. and Prendergast, K. H.: 1962, *Astrophys. J.* **136**, 713.
Oort, J. H.: 1958, in D. J. K. O'Connell (ed.), *Stellar Populations*, North Holland Press, Amsterdam, p. 510.
Oort, J. H. and van Herk, G.: 1959, *Bull. Astron. Inst. Neth.* **14**, 299 (No. 491).
Pels, G., Oort, J. H., and Pels-Kluyver, H. A.: 1975, this volume, p. 159.
Peterson, C. J.: 1974, *Astrophys. J.* **190**, L17.
Peterson, C. J. and King, I. R.: 1975, *Astron. J.* **80** (in press).
Rood, H. J., Page, T. L., Kintner, E. C., and King, I. R.: 1972, *Astrophys. J.* **175**, 627.
Simoda, M. and Tanikawa, K.: 1970, *Publ. Astron. Soc. Japan* **72**, 143.
Spitzer, L. and Härm, R.: 1958, *Astrophys. J.* **127**, 544.
van den Bergh, S., and Sher, D.: 1960, *Publ. David Dunlap Obs.*, Vol. II, No. 7.
Wielen, R.: 1971, *Astron. Astrophys.* **13**, 309.
Wielen, R.: 1975, this volume, p. 119.
Wilson, C. P.: 1975a, *Astron. J.* **80**, 175.
Wilson, C. P.: 1975b, this volume, p. 207.
Wilson, C. P. and King, I. R.: 1975, *Astron. J.* **80** (in press).

DISCUSSION

Miller: A warning to the theoreticians here – although King has stressed the extent to which observations delimit theories, there is still a good deal of prejudice in the analytic forms used. This is common to all such attempts (if anything, King is more candid than most in admitting to such prejudices). If the problem is turned around, and regarded as a problem in statistical inference on some parameters, it usually turns out that the parameters are remarkably poorly determined. With allowance for backgrounds, King's models allow 4 adjustable parameters, and should surely give a good fit to any reasonable set of observations. I've tried some cluster fits to a Plummer model plus background, and gotten remarkable good fits (3 parameters), but then tried with another analytic form (also 3 parameters) and found an equally good fit. In fact, the fits were *too* good.

King: I admit the prejudice freely. It is a question of philosophy: I am quite happy to accept an agreement of observations with a theoretical picture that is *a priori* highly plausible. As for backgrounds, I do not think that it is at all correct to call the background a fourth parameter. The background level is indeed determined observationally, but these observations are quite separate from the ones used to determine the other three parameters. Typically we counted an external surrounding region larger than the cluster itself, so that the statistical error in the background determination is negligibly small. It is very unlikely that the background levels have systematic errors either; we checked for this by counting entire fields that did not have a cluster in then, and then testing for uniformity. Finally, I suspect that your success in fitting cluster data with a Plummer model was due to using data with large statistical uncertainties. I don't think that you could fit good-quality globular-cluster data with a Plummer model at all.

Brosche: If three parameters are sufficient for the description of globular clusters and elliptical galaxies, are they necessary as well, in other words, does any relation exist between those parameters?

King: Peterson has looked at all sorts of correlations between quantities, and the only one that seems to hold up consistently is the correlation between observed limiting radius and calculated limiting radius. Otherwise there seem to be no significant correlations.

Lecar: How do you determine the tidal radius as related to the orbit of the globular cluster? Do you use current position or peri-galacticon?

King: In the diagram that I showed, the observed radii were determined from observation, by fitting standard curves. The 'calculated' values were for the cluster's present position. Thus the ratio of the two should be an indication of the ratio between the cluster's present distance from the Galactic Center and its perigalactic distance.

Kalnajs: I presume that color variations across clusters do not give any useful information about mass-segregation. Can you tell us why not?

King: The color changes to be expected are quite small – a couple of hundredths of a magnitude in $B-V$. When you look at the statistical uncertainties in the colors of the regions that have to be measured, because of the small number of stars that contribute most of the light, the uncertainties are as large as the quantities that you are trying to measure.

DYNAMICS OF STAR CLUSTERS:
COMPARISON OF THEORY WITH OBSERVATIONS
AND SIMULATIONS

R. WIELEN

Astronomisches Rechen-Institut, Heidelberg,

and

Hamburger Sternwarte, Hamburg-Bergedorf, Germany

Abstract. Stellar dynamical theories for collisional systems are tested by numerical N-body simulations of isolated star clusters with N up to 500. For the dynamical evolution of the density distribution in star clusters, good agreement is found between Monte Carlo results and N-body models. There remains a discrepancy between theory and simulations in the rate of stellar escape from clusters and/or in the mechanism by which escapers are produced. Simulations of non-isolated star clusters are compared with observations of open clusters. The observed age distribution of open clusters can be explained by the dynamical dissolution of clusters. Stars of low mass do not escape relatively more frequently than stars of average mass.

1. Introduction

In stellar dynamics, we have essentially three categories of knowledge: observations, simulations and theories. Observations deal with stellar systems actually in the sky. Theories aim at a mathematical description of stellar systems on the basis of plausible physical and statistical assumptions. Simulations are numerical experiments in which we use directly the basic law of gravitation without any other a priori assumption. When we try to compare the results of these three approaches in order to understand stellar systems, it is easiest to compare theories and simulations, because we have rather complete control over the objects under investigation. For such a comparison of theory versus simulation, models of stellar systems should be used which are as simple as possible in order to test the basic assumptions of a theory. For example, isolated systems with stars of equal mass can be investigated first. Only if the theory has successfully passed such a test, should more realistic situations be studied. In a comparison of numerical simulations with astronomical observations, the models must be as realistic as possible. We can then test whether we have included in our simulations all the relevant physical phenomena (external forces, mass loss of stars etc.) and whether we have used realistic values for the parameters of the model (e.g. for its total mass). A direct comparison of a simplified version of a theory with observations is usually more difficult to judge in its implications. This is especially true if such a theory is in conflict with the results of simulations, but nevertheless is able to reproduce some observed properties of actual stellar systems.

In this review, we shall consider only stellar systems in which encounters among the stars are important ('collisional systems'). Open star clusters are typical examples of such systems. Since a number of review papers on the dynamical evolution of

Hayli (ed.), Dynamics of Stellar Systems, 119–131. All Rights Reserved.

collisional systems have been published recently (Aarseth, 1973; Hénon, 1973; King, 1974; Wielen, 1974), we shall concentrate here on recent results. Comparisons between theory and simulation and between observations and simulations will be presented. For a comparison between theory and observations, we refer to King (1975) for globular clusters and to Prata (1971) for the rich open cluster M67.

2. Comparison of Theory with Simulations

In this section, we discuss isolated star clusters, containing stars of equal mass or stars with a realistic mass spectrum. First, we compare the results of theory and simulations for the evolution of the spatial structure of a cluster. Second, the results on the escape of stars from a cluster will be compared.

In the Figures 1 and 2 (Aarseth *et al.*, 1974), we compare results for the evolution of the spatial structure of a cluster by plotting the radii of spheres which contain 10%, 50% and 90% of the total mass as a function of time. The unit of time corresponds to about 11 relaxation times t_{rh}, defined by Spitzer and Hart (1971). In all the calculations, the Plummer model has been adopted as a common initial state of the cluster.

Figure 1 shows the results for the equal-mass case. In the N-body simulations by Aarseth (1974) and by Wielen (1974), the dynamical evolution of the cluster has been followed by integrating the complete equations of motion of the N-body problem for $N = 250$ or 100 stars. The Monte Carlo results obtained by Hénon (1973) and by Shull and Spitzer (1974) rely on the classical theory of relaxation. The cumulative effect of weak two-body encounters is described by the Fokker-Planck equation which is solved by a numerical Monte Carlo procedure. The fluid-dynamical approach of Larson (1970) is also based on the Fokker-Planck equation, but further approximations are introduced in order to simplify the numerical treatment. The general shape of the theoretical and experimental curves is quite similar. A slight discrepancy in the time scale of evolution is indicated: The Monte Carlo models evolve faster than the N-body models by a factor of about 1.5, and the fluid-dynamical model by a factor of 2 or 3. After about 1.5 units of time, the central density becomes infinite according to the theoretical models. At about the same time, a close central binary forms near the center in the N-body models.

In Figure 2, we present results for the case of a realistic distribution of stellar masses listed in Table III. Six N-body models (Wielen, 1974) with N up to 500 stars are compared with three new Monte Carlo models obtained by Hénon. Again, we find a nice qualitative agreement between theory and simulations. In the case of unequal masses, the Monte Carlo models evolve faster than the N-body models by a factor of at least 2. Also the central collapse of the Monte Carlo models occurs earlier ($t \sim 0.15$) than the formation of a close central binary in the N-body models ($t \sim 0.5$). A preliminary comparison of the segregation of stars of different masses suggests good agreement between the Monte Carlo computations and the N-body models in this respect.

From the comparisons shown in Figures 1 and 2, we conclude that the classical theory of relaxation, if used in a realistic form such as the Monte Carlo scheme, is

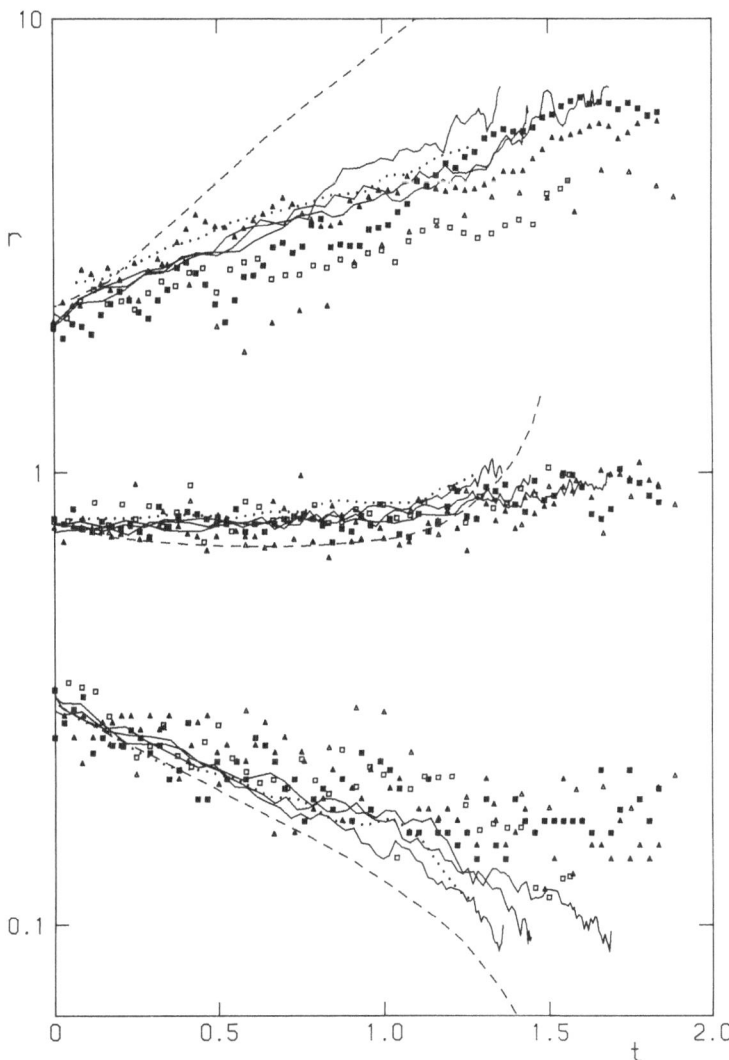

Fig. 1. Radii r of spheres containing 10%, 50% and 90% of the total mass, as a function of time t, for a cluster containing stars of equal masses. N-body simulations by Aarseth ($N=250$, filled symbols) and by Wielen ($N=250$, open squares; $N=100$, open triangles). Monte Carlo models by Hénon (solid lines) and by Shull and Spitzer (dotted lines). Fluid-dynamical results by Larson (dashed lines).

able to predict rather successfully the evolution of the spatial structure of star clusters observed in N-body simulations. This confirms the basic assumption of the theory, namely that the diffusion effect of independent, weak two-body encounters is the main source of relaxation in star clusters. The discrepancy between theory and simulations regarding the time scale of the evolution can be removed by a proper modification of the theory (Hénon, 1975). The comparison indicates furthermore that the range of validity of N-body simulations and Monte Carlo computations

R. WIELEN

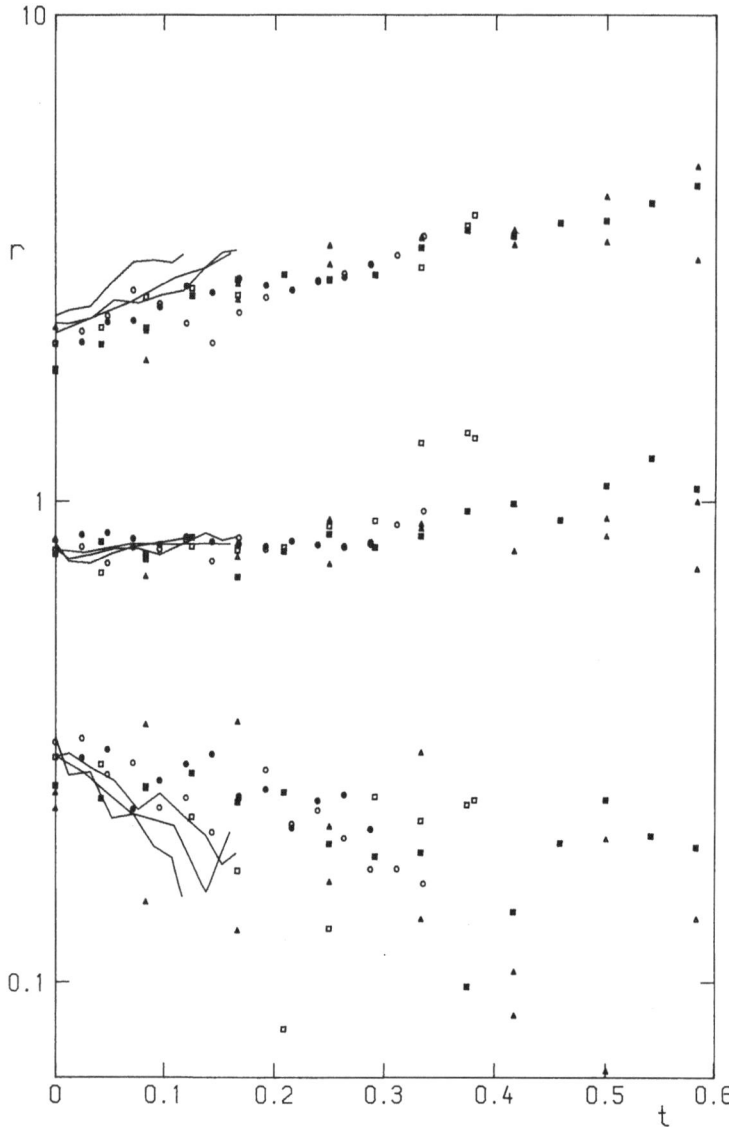

Fig. 2.   The same quantities as shown in Figure 1, but for a cluster containing stars with a realistic mass spectrum. N-body simulations by Wielen: $N=100$ (triangles), $N=250$ (squares), $N=500$ (circles). Monte Carlo models by Hénon (solid lines).

overlaps for $N \sim 100$ to 500 stars. Hence the N-body results for $N \leqslant 500$ may now be extrapolated to much higher values of N with some confidence.

Beside the positive conclusions, there remain some problems. It is unclear whether the dominant central binary usually formed in N-body integrations can be essentially identified with the infinite central density predicted by the theory. Even within the frame of the theory itself, the central collapse can be interpreted in different ways: Spitzer advocates that a finite fraction ($\sim 10\%$) of the total mass of a cluster is involved in the collapse; in Hénon's picture, the mass of the collapsing core becomes zero when the central density becomes infinite. It is not an easy task to continue the Monte Carlo scheme beyond the time of the central collapse. Recently, Hénon (1975) has proposed a way of extending the Monte Carlo computations past the collapse time by introducing an artificial 'well' which absorbs the negative energy flowing towards the center. As we shall discuss below, the Monte Carlo method does not seem to reproduce correctly the escape of stars. Both difficulties, the handling of the collapsed core and the escape of stars, cast some doubts on whether the theory in the present form is able to describe accurately the long-term evolution and the final dissolution of star clusters.

The N-body simulations give rather detailed information on the mechanism by which stars can gain enough energy to escape from the cluster. The classical theories predict a slow diffusion of stars towards positive total energy due to distant encounters. Hénon (1960, 1969), however, has advocated that in isolated clusters escapers are mainly produced by single, close two-body encounters. The N-body models show that the total energy of an escaping star always jumps suddenly to a positive value, and that typical escapers are strongly bound to the cluster before they gain the energy for escape (see e.g. Figures 4 and 5 in Wielen, 1974). The basic mechanism for this sudden change of the energy is in most cases a close encounter of the star which escapes afterwards, either with another single star or with a binary or a small group of stars in the core of the cluster. A detailed description and statistics on such 'escape events' have been given by Aarseth (1974). Hence the N-body models favor Hénon's theory of escape and contradict the classical picture of diffusion as a mechanism for escape of stars. It may be argued that the diffusion process becomes more important for rich star clusters, because the theoretically predicted ratio between the frequency of distant and close encounters is proportional to $\log N$. However, even if 10% of the escapers were caused by diffusion for $N = 500$ stars, the expected contribution of distant encounters would remain small also for globular clusters with $N \sim 2 \times 10^5$, namely about 20% (Wielen, 1974).

The rate of escape, $\dot{N} T_{cr}$, defined as the average number of stars which escape in one crossing time T_{cr}, is shown in Table I for stars of equal mass, and in Table II for a realistic spectrum of stellar masses. The numbers quoted for the classical theories refer to $N = 100$ stars, but increase only very slowly with N. The number quoted for Spitzer and Thuan (1972) refers to the average escape rate of their model E1. A comparison of the theoretical predictions with the experimental results in Table I shows that all the old classical theories overestimate the escape rate in the case of equal masses. The escape rate derived from the Monte Carlo calculations of Spitzer and Thuan (1972) is in rather good agreement with the results of N-body simulations (Table I). This coincidence may, however, be regarded as fortuitous as long as the difference in the basic mechanism of producing escapers remains unexplained. The converse situation occurs if we compare Hénon's theory of escape with the N-body models. While this theory predicts correctly the escape mechanism, it fails to reproduce the escape rates. The discrepancy is even stronger for equal masses (Table I) than for unequal masses (Table II). The escape rate predicted by Hénon may be too

TABLE I

Escape rate for isolated clusters (Stars of equal masses)

Number of escapers per crossing time $\dot{N}\, T_{\mathrm{cr}}$ [stars]	Source	
0.70	Spitzer (1940)	
0.55	Chandrasekhar (1942)	
1.18	Chandrasekhar (1943)	
1.58	White (1949)	
0.98	Spitzer and Härm (1958)	
0.31	King (1958)	
0.4	Larson (1970)	
0.14	Spitzer and Thuan (1972)	
0.027	Hénon (1960)	
0.08	Wielen Model HE	$N = 50$
0.06	Wielen Model E	$N = 100$
0.13	Wielen Model DE	$N = 250$
0.17	Aarseth Model III	$N = 250$
0.28	Aarseth Model VI	$N = 250$
0.22	Aarseth Model VII	$N = 250$

TABLE II

Escape rate for isolated clusters
(Realistic spectrum of stellar masses)

Number of escapers per crossing time $\dot{N}\, T_{\mathrm{cr}}$ [stars]	Source	
0.48	Hénon (1969)	
1.0	Wielen Model HP	$N = 50$
0.8	Wielen Models P + P2 + R	$N = 100$
1.1	Wielen Models DP + DP2	$N = 250$
1.1	Aarseth Model IV	$N = 250$
0.6	Wielen Models FP + FP2	$N = 500$
1.6	Aarseth Model V	$N = 500$

small for various reasons: The theory neglects multiple encounters, binaries, core formation and mass segregation. All these phenomena enhance the escape rate in actual systems.

In Table III, we compare the relative escape rates of stars of different mass. The last three columns give results of N-body calculations (Wielen, 1974). The models DG3 and FG3 are non-isolated clusters. The N-body integrations carried out by Aarseth (1973, 1974) are in good agreement with our results. For isolated systems, the theories predict a strong increase in the relative escape rate with decreasing stellar mass m, in contradiction to the N-body models. This discrepancy may be due to the strong mass segregation and the formation of massive binaries, which occur

TABLE III

Relative rate of escape for stars of different masses

$$(\Delta N_m/N_m)/(\Delta N/N)$$

$\dfrac{m}{\langle m \rangle}$	$\dfrac{N_m}{N}$	Chandrasekhar (1942)	Spitzer and Härm (1958)	Hénon (1969)	Isolated models $N=50\text{–}500$	Model DG3 $N=250$	Model FG3 $N=500$
4.4	4%	1×10^{-8}	~0	0.05	0.4:	0.3:	0:
2.2	10%	2×10^{-4}	~0	0.25	0.6:	0.4:	0:
1.1	24%	0.27	0.23	0.71	1.0	1.1	0.8
0.55	62%	1.51	1.52	1.30	1.1	1.1	1.3

in the N-body models but are neglected in the theories. We shall mention here that even for the more realistic, non-isolated N-body models (such as DG3 or FG3 in Table III), the relative escape rate does not depend significantly on the stellar mass for masses below the average mass. Only more massive stars escape less frequently than indicated by the overall escape rate. The implications of N-body results for the relative escape rates for the luminosity function of the Hyades have been discussed by Aarseth and Woolf (1972).

3. Comparison of Observations with Simulations

What astronomical observations give relevant information about the dynamical evolution of star clusters? This question has been discussed during this Symposium by King (1975) for the case of globular clusters. Hence, we shall concentrate here on open clusters.

The density distribution and the segregation of stars of different masses, although observed in projection only, must contain important information on the dynamics of clusters. But these observations hardly provide any significant test for a theory at present, since essentially all the dynamical theories (old and new ones) claim a rather perfect agreement with such observations. The luminosity function as well as the frequency of binaries in a cluster may only slightly reflect the dynamical evolution of a cluster. The relative scarcity of faint dwarfs in open clusters, compared with field stars, is probably due to special physical conditions during the formation of a cluster. Most binaries must have already existed as proto-binaries. The radius of a cluster, its overall velocity dispersion, its total mass and membership have to be considered in most cases as scaling parameters, not as independent test quantities. Of course, the knowledge of such parameters is indispensable for any reliable comparison of theories or simulations with actual clusters.

For star clusters, there exists an observable quantity which is often even more accurately known than the total mass, namely the age of a cluster. The observed distribution of the ages of open clusters contains rather direct information on the total lifetimes and hence on the time scale of the dynamical evolution of these clusters. From the observed age distribution, we can deduce that 50% of newly born open clusters disintegrate within 2×10^8 yr, 10% live longer than 5×10^8 yr and only 2% survive over 1×10^9 yr (Wielen, 1971). Hence the typical lifetime of an open cluster is rather short compared to the age of the Galaxy, but there exists a wide spread in the individual lifetimes. These total lifetimes of open clusters can serve as a powerful observational test of theories and simulations of the dynamical evolution of star clusters.

The dynamical evolution of open star clusters is mainly caused by the following effects: (1) relaxation by encounters among the cluster stars, (2) the gravitational tidal field of the Galaxy, (3) fluctuating gravitational fields of passing interstellar clouds, (4) mass loss of evolved massive stars in late phases of their internal evolution. The effects (1), (2) and partly (4) have been incorporated in the N-body simulations of open clusters carried out by Aarseth and by Wielen. Details on these N-body models can be found in the reviews (Aarseth, 1973; Wielen, 1974). In addition, we present here some results on a new N-body model, named FG3, which is similar to the models G3 and DG3 but contains $N = 500$ stars. The range of N, up to 500, now available covers most open clusters. Therefore, a rather direct comparison of the N-body models with open clusters is possible.

The total lifetimes of open star clusters as predicted from N-body simulations by Wielen, are shown in Figure 3. Since the escape rate $\dot{N}$ is rather constant during the dynamical evolution, the evaporation time $T_{ev} = N/\dot{N}$ is nearly identical with the total lifetime, if we use the initial parameters such as N_0, $\mathfrak{M}_0$ and R_0 of a cluster in the applications. In Figure 3 we have plotted the evaporation time of a cluster, T_{ev}, as a function of its median radius in projection R, for clusters containing $N = 100$ stars (total mass $\mathfrak{M} = 50 \ \mathfrak{M}_\odot$), 250 stars (125 $\mathfrak{M}_\odot$) and 500 stars (250 $\mathfrak{M}_\odot$). A detailed description of the construction of Figure 3 from the results of N-body simulations can be found in Wielen's review (1974). The lifetimes predicted by Aarseth (1973) are about twice as long as those obtained by Wielen. This discrepancy is due to slightly different procedures used for identifying escapers from non-isolated clusters. The difference may disappear, if the passing HI clouds are properly incorporated into the N-body models, since they should rapidly remove the outermost halo of a cluster to which most of the doubtful escapers belong.

Up to now, passing interstellar clouds have been neglected in the N-body simulations carried out by Aarseth and Wielen. The N-body experiments of Bouvier and Janin (1970) were done for very small systems ($N = 25$), without considering the tidal field of the Galaxy, and their results may be strongly biased by the unrealistic mean force field of the clouds present in their calculations (see my remark after Janin's paper (1975) during this Symposium). Hence, at present the effect of interstellar clouds can only be inferred from theoretical studies, especially from the Monte Carlo models of Spitzer and Chevalier. In Figure 4, the dissolution time $T_{dis} = T_{sh}/2$ of a cluster, due to the gravitational shocks of passing clouds alone (Spitzer and Chevalier, 1973), is compared with the N-body results of Figure 3 for $N = 500$ stars and $\mathfrak{M} = 250$ $\mathfrak{M}_\odot$ as a function of the median radius R of the cluster. Since the properties of interstellar clouds, e.g. their typical masses, are still very uncertain, we show in Figure 4

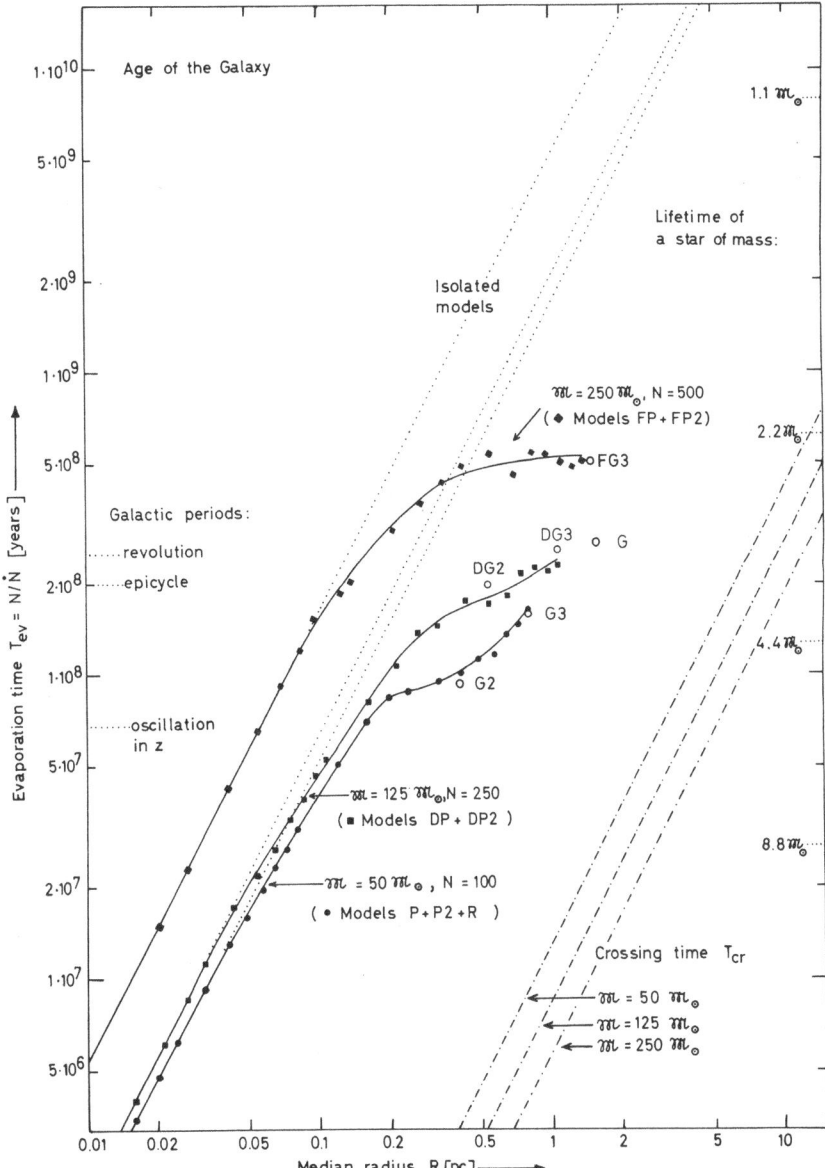

Fig. 3. Evaporation times of *N*-body models for open clusters as a function of the median radius *R* of the projected cluster for various total masses 𝔐, shown by the symbols and the fitted solid lines.

two cases: For the lower line, we have adopted the shock time T_{sh} according to the cloud data proposed by Spitzer and Chevalier (1973), while the upper line corresponds to the 'standard clouds' described by Spitzer (1968). The dashed line which connects the *N*-body results with the theoretical line for $T_{sh}/2$, is derived by adding the corresponding escape rates, i.e. by taking the harmonic mean of T_{ev} and $T_{sh}/2$. Figure 4 seems to indicate that the dynamical dissolution of an open cluster is sig-

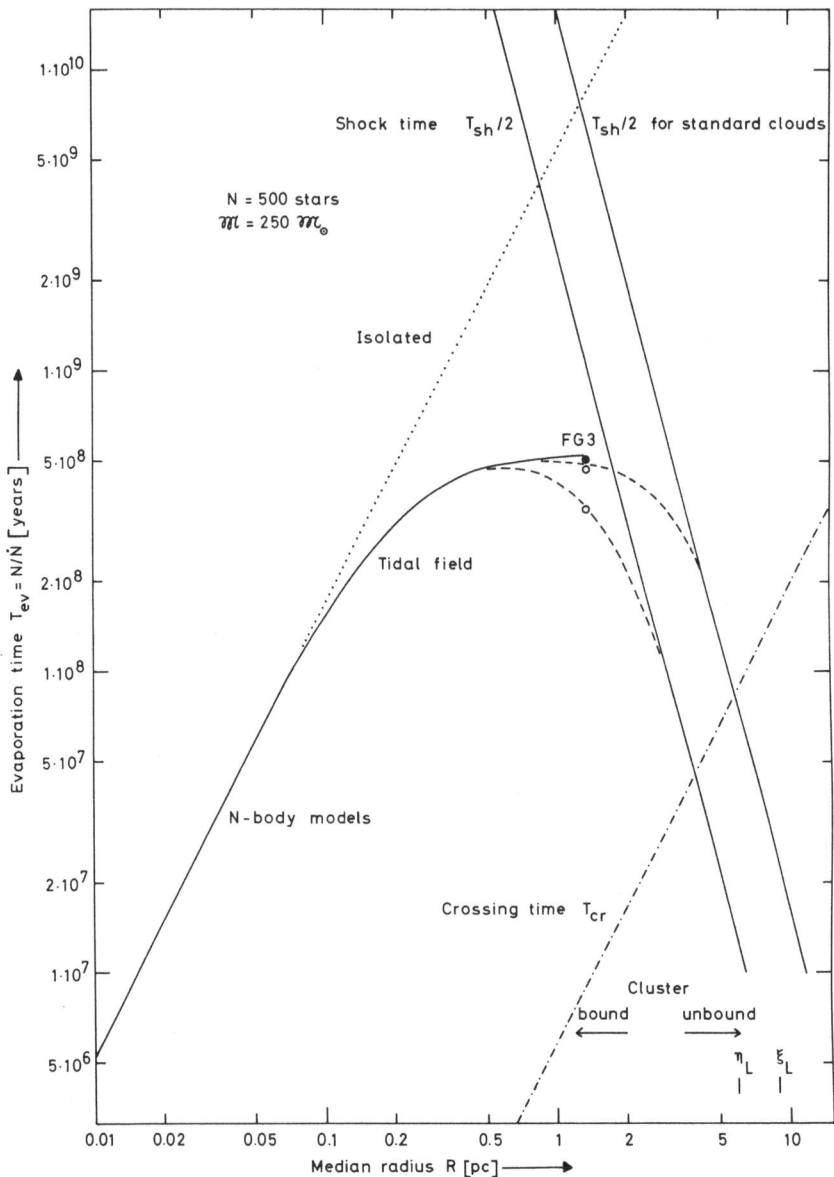

Fig. 4. Evaporation time of *N*-body models (solid curve, circles) and dissolution time by clouds, $T_{sh}/2$ (solid lines), predicted from Monte Carlo models of Spitzer and Chevalier, for a total mass of $\mathfrak{M} = 250\ \mathfrak{M}_\odot$ as a function of the median radius *R* of a cluster.

nificantly accelerated by passing interstellar clouds only if the cluster is already either weakly bound or even unstable because of the tidal field of the Galaxy (i.e. if $R \gtrsim 2$ pc for $\mathfrak{M} = 250\ \mathfrak{M}_\odot$).

The shape of the outermost regions of an open cluster may help to decide observationally whether gravitational shocks due to interstellar clouds are important. According to the N-body simulations without clouds, the tidal field of the Galaxy produces a significant flattening (up to $1:2$) of the halo of a cluster perpendicular to the galactic plane. If the clouds are dynamically important for an open cluster, the halo may be rounder and smaller in size, because it will not fill any more the critical volume (see Figure 6 of Wielen, 1974) corresponding to the tidal radius of the cluster, since the outermost stars are rapidly stripped off from the cluster by the gravitational interaction with the clouds. Unfortunately, the presently available observations of the sparsely populated halos of open clusters, e.g. of the Hyades, do not allow any accurate statement on the degree of a probable flattening.

In order to apply the results of N-body simulations, we need information on the initial total masses $\mathfrak{M}$ and median radii R of open clusters. From young clusters, we find that most clusters have median radii in projection between 0.5 pc and 3 pc; $R \sim 1$ pc is quite typical. It is much more difficult to estimate the total mass of a cluster. Only for a very few open clusters can the total mass be determined by applying the virial theorem. Total masses derived from extrapolating the luminosity function in a cluster from that of field stars, are very probably too high, because of the relative underabundance of low-mass stars in open clusters (e.g. van den Bergh and Sher, 1960). Total masses of open cluster may range from less than 50 $\mathfrak{M}_\odot$ to more than 10^3 $\mathfrak{M}_\odot$, but most clusters seem to have rather small values of $\mathfrak{M}$.

The observed typical total lifetime of open clusters, 2×10^8 yr, can be explained by the N-body models, if we accept $R \sim 1$ pc and $\mathfrak{M} \sim 100$ $\mathfrak{M}_\odot$ $(N \sim 200$ stars) as being typical for an open cluster of zero age. According to Figures 3 and 4 the lifetime is not very sensitive to variations of R for $R \sim 1$ pc, but increases roughly proportional to N for a fixed value of the mean stellar mass $\mathfrak{M}/N$. Since a typical value of $\mathfrak{M} \sim 100$ $\mathfrak{M}_\odot$ is not implausible, we may claim agreement between the observed and predicted typical lifetimes of open clusters for the time being. The observed range in the lifetimes of open clusters, roughly from 10^8 to 10^{10} yr, can be explained as due to the variety of total masses and median radii of open clusters. The lifetimes of the oldest open clusters, e.g. NGC 188 or M67, require an initial total mass of about 5000 $\mathfrak{M}_\odot$ for $R \sim 2$ pc.

4. Final Remarks

The discussion in Section 2 shows that a comparison of theory with simulations gives significant results which improve our theoretical understanding of stellar systems. In order to facilitate such comparisons, both the investigators working in the fields of theory and numerical simulations should cooperate as closely as possible in order to provide quantities which can be directly compared.

The comparison of observations with theory or simulations is still rather disappointing, because we have so few points of contact between the two kinds of investigation. In order to facilitate comparison of observations with theoretical results, the observers should be asked to publish also their raw data as far as possible and not to mix up their observations immediately with special theoretical conceptions. Much observational work, which should be especially designed for a meaning-ful comparison with theoretical work, remains to be done for star clusters.

References

Aarseth, S. J.: 1973, *Vistas in Astronomy* **15**, 13.
Aarseth, S. J.: 1974, *Astron. Astrophys.* **35**, 237.
Aarseth, S. J. and Woolf, N. J.: 1972, *Astrophys. Letters* **12**, 159.
Aarseth, S. J., Hénon, M., and Wielen, R.: 1974, *Astron. Astrophys.* **37**, 183.
Bergh, S. van den and Sher, D.: 1960, *Publ. David Dunlap Obs.* **2**, 203.
Bouvier, P. and Janin, G.: 1970, *Astron. Astrophys.* **9**, 461.
Chandrasekhar, S.: 1942, *Principles of Stellar Dynamics*, Chicago University Press, Chicago.
Chandrasekhar, S.: 1943, *Astrophys. J.* **98**, 54.
Hénon, M.: 1960, *Ann. Astrophys.* **23**, 668.
Hénon, M.: 1969, *Astron. Astrophys.* **2**, 151.
Hénon, M.: 1973, in L. Martinet and M. Mayor (eds.), *Dynamical Structure and Evolution of Stellar Sys-
    tems*, Swiss Society of Astronomy and Astrophysics Third Advanced Course, Geneva Observatory,
    p. 183.
Hénon, M.: 1975, this volume, p. 133.
Janin, G. and Haggerty, M. J.: 1975, this volume, p. 61.
King, I. R.: 1958, *Astron. J.* **63**, 109.
King, I. R.: 1974, *Celes. Mech.* **9**, 349.
King, I. R.: 1975, this volume, p. 99.
Larson, R. B.: 1970, *Monthly Notices Roy. Astron. Soc.* **150**, 93.
Prata, S. W.: 1971, *Astron. J.* **76**, 1017.
Shull, J. M. and Spitzer, L.: 1974, private communication.
Spitzer, L.: 1940, *Monthly Notices Roy. Astron. Soc.* **100**, 396.
Spitzer, L.: 1968, *Diffuse Matter in Space*, Interscience Publishers (Wiley), New York-London-Sydney-
    Toronto, p. 85.
Spitzer, L. and Chevalier, R. A.: 1973, *Astrophys. J.* **183**, 565.
Spitzer, L. and Härm, R.: 1958, *Astrophys. J.* **127**, 544.
Spitzer, L. and Hart, M. H.: 1971, *Astrophys. J.* **164**, 399.
Spitzer, L. and Thuan, T. X.: 1972, *Astrophys. J.* **175**, 31.
White, M. L.: 1949, *Astrophys. J.* **109**, 159.
Wielen, R.: 1971, *Astron. Astrophys.* **13**, 309.
Wielen, R.: 1974, in L. N. Mavridis (ed.), *Proceedings of the First European Astronomical Meeting (Athens
    1972)*, Springer Verlag, Berlin-Heidelberg-New York, Vol. 2, p. 326.

DISCUSSION

Larson: I was intrigued to note in your first diagram that the evolution time seems to increase system-atically as you go from my model which treats a cluster as a smooth fluid, to the Monte-Carlo models which represent the system as a collection of a few thousand spherical shells, and finally to the N-body simulations with only a few hundred discrete mass points. Thus there seems to be a trend of increasing evolution time with increasing 'graininess' of the model simulation. Do you think that there are any effects such as the formation of binaries that might tend to retard the evolution of a system with a more grainy structure (e.g. one with smaller N)?

Wielen: The N-body results do not indicate any systematic trend with N for N ranging from 100 to 500.

Spitzer: In connection with your last slide, I would question your use of standard clouds in computing tidal shocks. Clouds of all sizes are clearly present as is seen from the statistics of observed color excesses as well as from direct photographs of cloudy regions; and it is the larger clouds that make the greatest contribution to shock heating. I believe that the consideration of standard clouds only will give too small a heating rate.

Wielen: It would be very helpful for our understanding of the dynamics of open clusters to have more reliable data on the properties of interstellar clouds.

Pişmiş: I am interested to know a bit more about the rates of escape: Although you have shown, on one of your slides, rates of escape for numbers of mass points ranging from 10 to 500, the results are based on different models as well as probably on different computing procedures. How sensitive are these results to the adopted models and also on the different computing procedures? How would the rate of escape vary with the varying number of mass points if computations were based on the same model and on the same computing procedure?

Wielen: The escape rate measured in terms of escaping stars per crossing time seems to be rather constant in the range of N from 50 to 500. This is in agreement with theoretical predictions.

TWO RECENT DEVELOPMENTS CONCERNING
THE MONTE CARLO METHOD

M. HÉNON

Observatoire de Nice, Nice, France

Abstract. This paper consists of two independent parts.

(1) The Monte Carlo method for computing the evolution of spherical stellar systems has been modified so that the computation can be continued after the time of formation of the central singularity. Results are presented for systems with equal and unequal star masses. The initial core-halo formation is followed by a general expansion of the cluster, while the central singularity absorbs a growing fraction of the total negative energy.

(2) Theoretical expressions of the 'diffusion coefficients', which describe the effect of encounters in a stellar system, contain a factor $\ln(\gamma N)$ where N is the number of stars and γ is a constant usually taken to be of the order of 0.4. A reconsideration of the 'non-dominant terms' leads to a substantially lower value, of the order of 0.15 for equal masses and 0.075 for unequal masses with a typical distribution. This correction improves the agreement between N-body and Monte Carlo simulations of spherical systems.

1. Extension of Monte Carlo Models Beyond the Singularity

All methods used so far to study the dynamical evolution of a spherical cluster under the effect of encounters indicate that the core of the system contracts, and that the contraction ends in a singular event at a finite time. When the system is described by a continuous distribution function, the central density becomes infinite at the critical time (Hénon, 1961; von Hoerner, 1968; Larson, 1970). In N-body simulations, a close binary forms near the centre (see reviews by Aarseth, 1973, 1975; Wielen, 1974, 1975). In Hénon's Monte Carlo models, the innermost shell collapses (Hénon, 1971). In the models of Spitzer and associates, the computation does not quite reach the critical time, but a small fraction of the mass near the centre appears to be headed for collapse (Spitzer and Thuan, 1972; Spitzer, 1975).

A theoretical explanation of this phenomenon has been given by Lynden-Bell and Wood (1968), who treated stars as molecules and applied thermodynamical concepts: the core of the system is 'hotter' than the outer regions, so that there is an outward flux of heat; the core loses energy, contracts, and becomes hotter as a consequence of the virial theorem. This process accelerates until the central singularity is formed.

What happens after the critical time? Unfortunately many methods fail at or before that time, because of technical difficulties related to the appearance of the singularity. The N-body simulations, however, are able to survive the event. They indicate that the central binary progressively shrinks and absorbs a growing fraction of the total negative energy of the system. An early analytical model (Hénon, 1961) also indicated that after the central singularity has formed, it will emit a continuous flow of energy.

Again the explanation of this behaviour is given by Lynden-Bell and Wood's

Hayli (ed.), Dynamics of Stellar Systems, 133–149. All Rights Reserved.

mechanism, which does not stop when the singularity is formed: the central parts are still hotter than the outside, and there is still an outward heat flux. However, the system has exhausted its first source of energy, which was the contraction of the core. Therefore the energy must come now from the singularity itself. This will appear less mysterious if events are considered in more detail. Shortly before the critical time, only a small fraction of the core is rapidly contracting, and therefore feeding the energy flow; the rest of the system is in quasi-equilibrium on this short time scale and is only traversed by the energy flow, without contributing to it. The active region becomes smaller and smaller as the critical time is approached, and finally shrinks to a point. This is most clearly seen in Larson's (1970) figures, and was also predicted earlier (Hénon, 1961). Thus, at the critical time, the energy flux originates from the centre itself. Since no further contraction is possible, after the critical time the energy must continue to come from the central singularity.

In the N-body simulations with point masses, extracting energy from the singularity is no problem; the central binary can supply any amount of energy by shrinking. In Hénon's Monte Carlo method, as recently described (1973), the innermost shell begins to shrink at the critical time; it can also supply an unlimited amount of energy in this way, and thus it plays very much the same role as the central binary in the N-body simulations. Unfortunately, the results then become unreliable, for technical reasons: the method assumes that the shells which represent the system have closely spaced radii, approximating a continuous distribution; it does not function properly if a singular shell has a radius much smaller than the others. For this reason, the Monte Carlo computations have until now been stopped at the critical time.

On the other hand, simple estimates indicate that many real globular clusters, if not most, have already passed their critical time (Spitzer, 1975). Since the Monte Carlo method is especially designed for the case of globular clusters, it seems to be of prime importance to be able to extend it to the post-critical phase of evolution. We may note also that a recent comparison with N-body simulations for the case of unequal masses (Aarseth et al., 1975; also Wielen, 1975, Figure 2) was impeded by the early demise of the Monte Carlo models.

We present here an attempt to extend the Monte Carlo models past the critical time. The problem is to introduce at the centre a mechanism which can represent the singularity and act as an energy source, in a way which is physically reasonable and technically trouble-free. As already said, if the system is left to its own devices, it will use one shell for the purpose, and this creates technical difficulties. Therefore it was felt preferable to introduce at the centre an *artificial energy source*, entirely distinct from the shells. This source has no mass; its only role is to supply energy as required by the evolution of the cluster.

The detailed mechanism for the transfer of energy from the source to the rest of the system must now be specified. We take the view that *the rate of flow of the energy is controlled by the system as a whole, not by the singularity*. To justify this assumption, consider the near-real situation in N-body simulations, with mass points and a central binary. At any given time, the structure of the system corresponds to some definite temperature difference between the inner and outer parts, and therefore dictates a definite outward flux of energy. Suppose that the central singularity delivers too much energy with respect to this requirement. The excess energy will not be transmitted, but stored in the innermost fraction of the core, which will expand. This reduces the interaction between the central binary and the core, and the energy flux from the binary is brought down. Conversely, if the energy delivered is insufficient, the core contracts and the interaction with the binary increases. This regulating mechanism automatically adjusts the energy flux supplied by the binary to the amount required by the system. The situation is the same as in stellar interiors, where the rate of production of nuclear energy in the central region is automatically adjusted to the value required to maintain the overall equilibrium.

This has two consequences for our model. First, the physical nature of the energy source does not matter, and we shall in fact leave it unspecified. Second, the mechanism for the transfer of energy must be self-regulatory, as in the real case: the flux should increase when the innermost core contracts, and conversely, so as to bring about a stable state of affairs.

Let us define the *self-energy* of a shell as the energy which it would have in the absence of all other shells, i.e. (cf. Hénon, 1971):

$$\tfrac{1}{2}Km(v_r^2 + v_t^2) - \tfrac{1}{2}\frac{GK^2m^2}{r}, \tag{1}$$

where r is the radius of the shell; m, v_r, v_t are the mass, radial velocity and transverse velocity of each individual star in the shell, and K is the number of stars in the shell. For a typical shell, the first term in (1) is of order T/n, where T is the kinetic energy of the whole cluster and n is the number of shells; the second term is of order $|W|/n^2$, where W is the potential energy of the cluster. Thus the first term is usually much larger than the second, and the self-energy (1) is normally positive. On the other hand, experience with the previous models showed that when the innermost shell starts collapsing, its self-energy becomes negative. Physically, this means that the shell becomes an independent self-gravitating system, just as the binary in the N-body simulations.

Therefore the following simple procedure was introduced. The self-energy of the innermost shell is constantly monitored. Whenever it is found to be negative, it is brought back to a zero value; this is done by increasing the radial velocity v_r. The energy used in this operation is considered to have been given by the central energy source.

This simple device was found to work quite satisfactorily. Figure 1 represents three 'old' Monte Carlo computations (thin lines), already shown by Wielen (1975, Figure 1), and a new computation (thick line), in which a central energy source has been introduced as explained above. The initial conditions are of course identical in all cases. All stars have the same mass. The old models stop at about $t = 1.5$ because of the collapse of the innermost shell. In the new model, this collapse does not happen any more, and the computation can be continued without difficulty past the critical

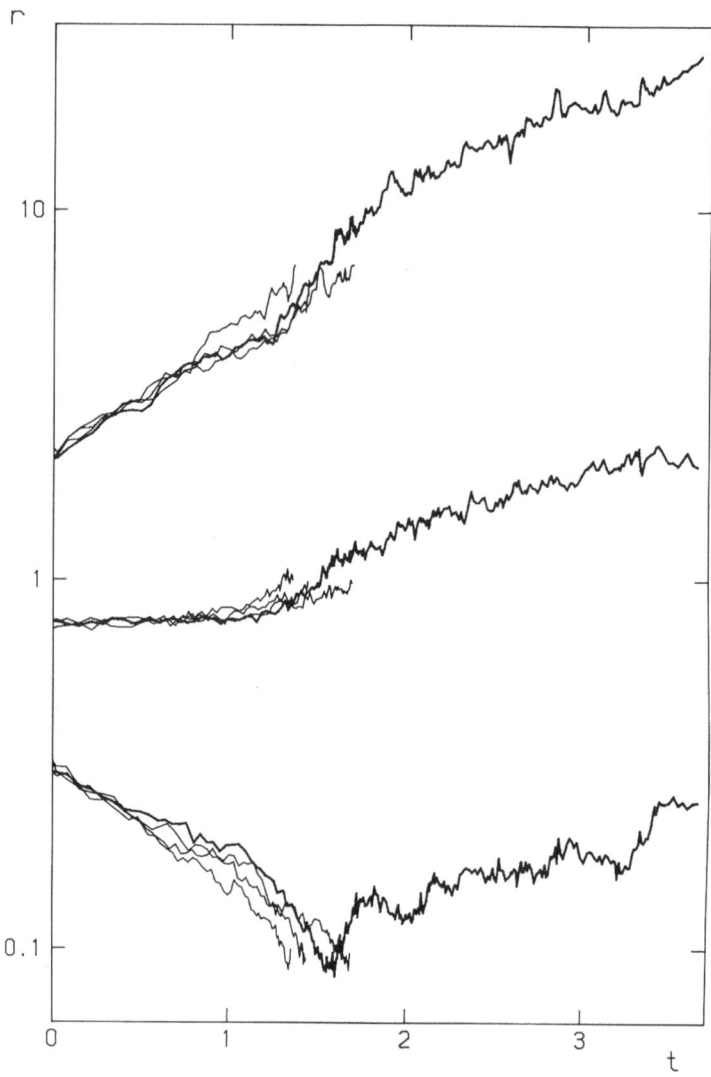

Fig. 1. Comparison of three old Monte Carlo models (thin lines) and one new model (thick line) which goes beyond the singularity. The curves represent the radii containing 10%, 50%, 90% of the mass, vs time. All stars have the same mass. The initial state is Plummer's model with isotropic velocity distribution.

time and for as long as desired. Figure 2 represents the total energy supplied by the central source since the beginning, as a function of time; the slope of the curve is the energy flux. It should be noted that the energy source is not just turned on at the critical time, but is present from the beginning. However, Figure 2 shows that the energy supplied by the source before the critical time is negligible. Therefore the source has no effect on the evolution during the first phase, and indeed Figure 1 shows that there is no significant difference between the new and the old models.

As the critical time is approached, however, the energy flux begins to increase sharply (Figure 2), while the contraction of the core is halted and even reversed (Figure 1).

Thus, the evolution appears to consist of two rather different phases. The first phase, which lasts until the critical time at about $t = 1.5$, is characterized by the usual core-halo formation. In the second phase, after the critical time, we observe a general

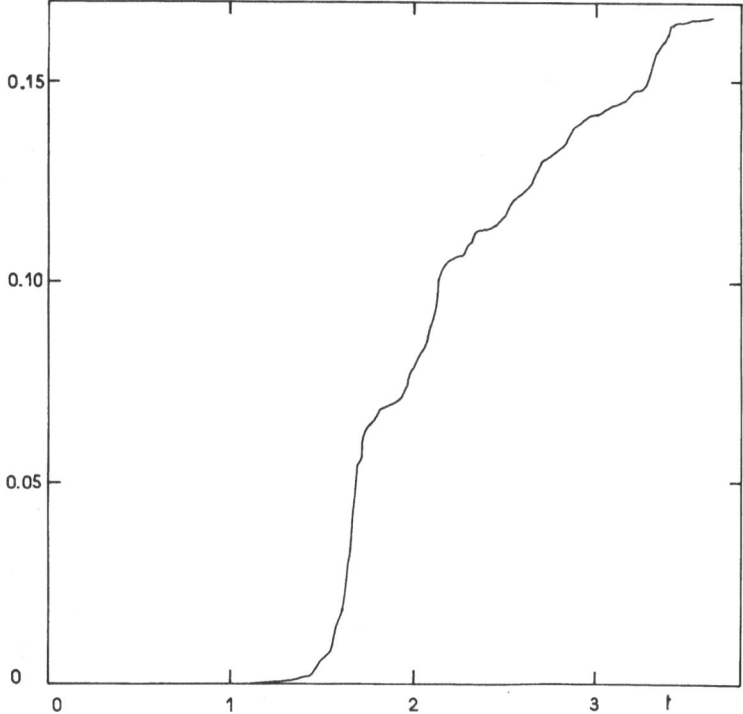

Fig. 2.   Energy delivered by the central source, as a function of time, for the new model of Figure 1.

expansion of the system; this is of course made possible by the fact that the system now receives energy from the central source. The initial total energy of the system is -0.25 in our units; Figure 2 shows that at the end of the computation, a negative energy equal to about -0.165 has been absorbed in the central singularity, so that the rest of the system is left with about $\frac{1}{3}$ of its initial negative energy. Further evolution will in all probability follow the same trend, with the cluster expanding indefinitely while more and more of the total negative energy is absorbed by the central source. The evolution slows down with time, because the relaxation time increases as the cluster expands. Therefore the curve of Figure 2 should asymptotically approach the limiting value 0.25.

This description of the final evolution applies only to the present idealized case of an isolated cluster. In real clusters, tidal effects will certainly affect the evolution, and probably bring about a full dissolution of the system after a finite time (Wielen, 1971).

Figure 1 shows also that the halo expands faster than the core. This, unfortunately, seems to rule out the use of a simple homological model for describing the final phase of the evolution.

Figure 3 again represents three old Monte Carlo models (thin lines), already shown by Wielen (1975, Figure 2), and a new model (thick line), for a case with un-

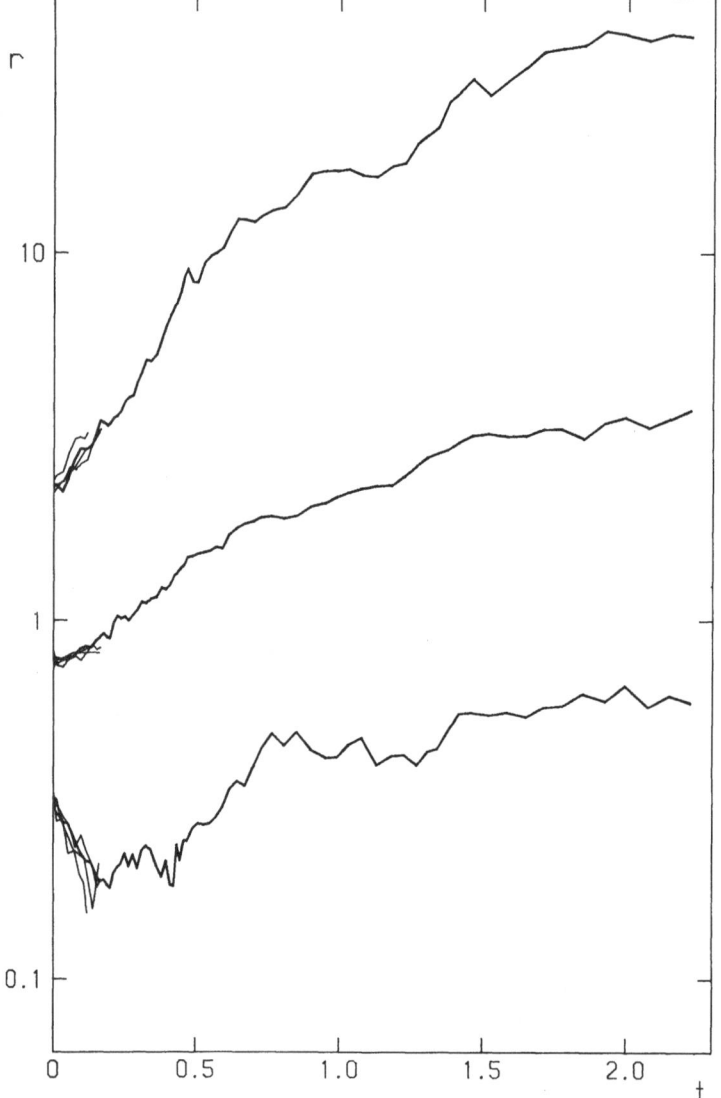

Fig. 3. Comparison of three old Monte Carlo models (thin lines) and one new model (thick line) which goes beyond the singularity. The curves have the same meaning as in Figure 1. The masses of the stars are distributed according to Wielen's law. The initial state is the same as in Figure 1.

equal masses. Here the progress is even more apparent: the old models are stopped in the vicinity of $t = 0.15$, while the new model has been computed beyond $t = 2$ and could be easily extended still further. Figure 4 represents the growth of the total energy supplied by the central source. These two figures are qualitatively similar to

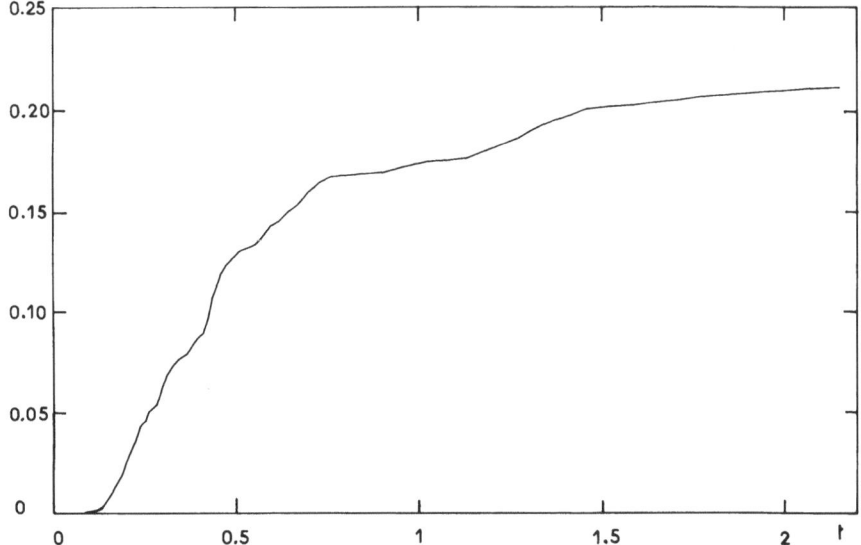

Fig. 4.   Energy delivered by the central source, as a function of time, for the new model of Figure 3.

Figures 1 and 2: again there is a first phase of core-halo formation, terminating at $t \approx 0.15$, and a second phase of general expansion; and the central source absorbs a growing fraction of the total negative energy. At the end of the computation, this fraction is about 85%. The asymptotic approach of the limiting value 0.25 is apparent on Figure 4. This figure is also in qualitative agreement with the results obtained by Aarseth (1971) from N-body simulations.

2. Non-Dominant Terms in the Diffusion Coefficients

The Monte Carlo method (Hénon, 1971, 1973) relies on the classical theory of relaxation due to two-body encounters. This theory involves a number of approximations. Until recently, there was no pressing need to introduce refinements, since no accurate data were available against which the theoretical results could be tested; observational evidence has been particularly disappointing in that respect (cf. King, 1975). Today, however, the situation is changing; N-body simulations provide a wealth of very detailed information, against which the theory can be matched (Wielen, 1975). There is also the hope that new observational techniques will improve our knowledge of real clusters, for instance by allowing a reliable determination of individual velocities. Therefore it may be worthwhile to reconsider some of the approximations in the theory.

One of the standard approximations consists in neglecting the 'non-dominant terms' (Chandrasekhar, 1941), i.e. the terms which are of order $1/\ln N$ with respect to the dominant terms. N is the number of stars in the system. This approximation is clearly justified in the limit of N very large. For N of the order of 100 to 500, however, the factor $1/\ln N$ is not so small and non-dominant terms could have a significant effect.

Consider a test star with mass m_1 and velocity V_1. Its change of velocity as a result of two-body encounters is governed by the 'diffusion coefficients', i.e. the components of $\langle \Delta V_1 \rangle$, $\langle \Delta V_1 \Delta V_1 \rangle$, etc. For simplicity, we shall consider here only one scalar coefficient: $\langle (\Delta V_1)^2 \rangle$, the mean square of the change of the velocity vector (not to be confused with the change in absolute velocity). This quantity may be considered as typical since it incorporates both radial and transversal diffusion. It is given by a comparatively simple expression (cf. Hénon, 1973):

$$\langle (\Delta V_1)^2 \rangle = 4\pi G^2 \Delta t \iiint f(V_2, m_2)\, m_2^2 V^{-1} \times$$

$$\times \ln\left\{ 1 + \left[\frac{l_{max} V^2}{G(m_1 + m_2)} \right]^2 \right\} dV_2\, dm_2. \qquad (2)$$

Here Δt is the interval of time considered; $f(V_2, m_2)$ is the distribution function, defined as number of points per unit volume in a seven-dimensional phase space (r_2, V_2, m_2); $V = |V_2 - V_1|$ is the relative velocity of the two encountering stars; l_{max} is the maximum impact distance. Formula (2) is exact, within the frame of the two-body encounter theory: no approximations have been made so far.

l_{max} is of the order of the dimensions of the system; we shall adopt here the value recommended by Spitzer and Hart (1971):

$$l_{max} = R_h, \qquad (3)$$

where R_h is the radius containing half the mass, itself given in good approximation by (Spitzer, 1969):

$$R_h = 0.4 GM/\langle V_1^2 \rangle, \qquad (4)$$

where M is the mass of the system and $\langle V_1^2 \rangle$ is the mean square velocity.

The argument of the logarithm in (2) is large: if we replace V^2 by its mean value, $2\langle V_1^2 \rangle$ (since V is the relative velocity between two stars), and also m_1 and m_2 by their mean value, $\langle m \rangle = M/N$, this argument becomes:

$$1 + (0.4N)^2. \qquad (5)$$

Therefore the term 1 can clearly be neglected.

From this point, the classical treatment continues by noting that since this large quantity is the argument of a logarithm, its exact value does not matter very much; therefore it is actually replaced by $(0.4N)^2$. Then the logarithm becomes a constant which can conveniently be taken out of the integrations. Here, we shall instead keep the exact expression (2), omitting only the constant 1. After substitution of (3) and (4), we write it in the form

$$\langle (\Delta V_1)^2 \rangle = 8\pi G^2 \Delta t \iiint f(V_2, m_2)\, m_2^2 V^{-1} \times$$

$$\times \{ \ln(0.4N) + \ln(V^2/2\langle V_1^2 \rangle) + \ln[2\langle m \rangle / m_1 + m_2)] \}\, dV_2\, dm_2 \qquad (6)$$

which clearly separates the dominant term (first term in braces) from the non-dominant terms (second and third terms). The classical approximation consists in omitting these non-dominant terms.

The diffusion coefficient (6) depends on the velocity V_1 and mass m_1 of the test star. In order to obtain a mean value for the effect of the non-dominant terms, we consider now the average of (6) over all test stars. In so doing, it seems physically most reasonable to weigh each star by its mass, so that we compute in effect

$$\langle\langle(\varDelta V_1)^2\rangle\rangle = \frac{\iiiint \langle(\varDelta V_1)^2\rangle f(V_1, m_1)\, dV_1 m_1\, dm_1}{\iiiint f(V_1, m_1)\, dV_1 m_1\, dm_1}. \tag{7}$$

The denominator is then simply the local density ϱ. We have

$$\langle\langle(\varDelta V_1)^2\rangle\rangle = 8\pi G^2 \varrho^{-1} \varDelta t [I_1 \ln(0.4N) + I_2 + I_3], \tag{8}$$

with

$$I_1 = \int_8 f_1 f_2 m_1 m_2^2 V^{-1}\, dV_1\, dV_2\, dm_1\, dm_2,$$

$$I_2 = \int_8 f_1 f_2 m_1 m_2^2 V^{-1} \ln(V^2/2\langle V_1^2\rangle)\, dV_1\, dV_2\, dm_1\, dm_2, \tag{9}$$

$$I_3 = \int_8 f_1 f_2 m_1 m_2^2 V^{-1} \ln[2\langle m\rangle/(m_1 + m_2)]\, dV_1\, dV_2\, dm_1\, dm_2,$$

$$f_1 = f(V_1, m_1), \qquad f_2 = f(V_2, m_2).$$

(8) can also be written

$$\langle\langle(\varDelta V_1)^2\rangle\rangle = 8\pi G^2 \varrho^{-1} \varDelta t I_1 \ln(\gamma N), \tag{10}$$

with

$$\ln\gamma = \ln(0.4) + I_2/I_1 + I_3/I_1. \tag{11}$$

Thus the effect of the non-dominant terms will be simply to modify the dimensionless constant γ in (10).

We consider first the case of equal masses:

$$f(V_1, m_1) = g(V_1)\, \delta(m_1 - m_0). \tag{12}$$

Then $I_3 = 0$, and

$$\frac{I_2}{I_1} = \frac{\int_6 g(V_1) g(V_2) V^{-1} \ln(V^2/2\langle V_1^2\rangle)\, dV_1\, dV_2}{\int_6 g(V_1) g(V_2) V^{-1}\, dV_1\, dV_2}. \tag{13}$$

We restrict ourselves now to isotropic velocity distributions, i.e. $g(V_1)$ will depend only on the velocity modulus V_1. The integrals in (13) can be evaluated exactly for some particular forms of g. For a Maxwellian distribution,

$$g(V_1) = a \exp(-j^2 V_1^2), \tag{14}$$

with a, j constants, the result is

$$I_2/I_1 = \ln(\tfrac{2}{3}) - C = -0.9827\ldots, \qquad \gamma = 0.1497\ldots \tag{15}$$

where $C = 0.5772\ldots$ is Euler's constant. For a truncated power law,

$$g(V_1) = aV_1^n \quad \text{for} \quad V_1 < V_0, \qquad g(V_1) = 0 \quad \text{for} \quad V_1 > V_0, \tag{16}$$

with a, V_0 constants and n an integer, we find for n even, $n \geqslant 0$:

$$\frac{I_2}{I_1} = \ln\frac{2(n+5)}{n+3} - 2 - \frac{2}{n+3} - \frac{2}{2n+5} +$$
$$+ \frac{2}{n+2}\left(1 + \tfrac{1}{3} + \tfrac{1}{5} + \cdots + \frac{1}{n+1}\right), \tag{17}$$

and for n odd, $n \geqslant -1$:

$$\frac{I_2}{I_1} = \ln\frac{2(n+5)}{n+3} - 2 - \frac{2}{n+3} - \frac{2}{2n+5} +$$
$$+ \frac{2}{n+2}\left(\ln 2 + \tfrac{1}{2} + \tfrac{1}{4} + \tfrac{1}{6} + \cdots + \frac{1}{n+1}\right). \tag{18}$$

These expressions are evaluated for a few values of n in Table I. The last line, $n = \infty$,

TABLE I

Values of γ for a velocity distribution of
the form (16)

n	I_2/I_1	γ
-1	-0.8941	0.1636
0	-0.8627	0.1688
1	-0.8917	0.1640
2	-0.9259	0.1585
3	-0.9571	0.1536
4	-0.9840	0.1495
$\ldots$		
∞	-1.3069	0.1083

corresponds to a δ-distribution where all stars have the same velocity modulus: $g \propto \delta(V_1 - V_0)$.

We note that I_2/I_1 is always negative. This can be explained by the presence of a factor V^{-1} in the integrals, which gives more weight to low values of V. In physical terms: encounters with a low relative velocity are more effective. As a result, the true mean value of the logarithm is lower than the value derived in the classical treatment, where V^2 is simply replaced by its average.

We note also from (15) and Table I that, if one excepts the rather unphysical case of equal velocity moduli ($n = \infty$ in Table I), the value of γ appears to be not very sensitive to the shape of the distribution function. This fortunate fact suggests that one can, with very little error, adopt a standard value for γ. We shall adopt here the value (15) corresponding to a Maxwellian distribution of velocities, rounded to

$$\gamma = 0.15. \tag{19}$$

Our results should be compared with an earlier computation of the effect of the non-dominant terms by Chandrasekhar (1941). A direct comparison is not possible, because Chandrasekhar considers a different quantity: $\langle(\Delta E)^2\rangle$, where ΔE is the energy exchanged between the two stars during an encounter. In our notations, there is

$$\Delta E = \Delta\left(\tfrac{1}{2}m_1\mathbf{V}_1^2\right) = m_1\left[\mathbf{V}_1\Delta\mathbf{V}_1 + \tfrac{1}{2}(\Delta\mathbf{V}_1)^2\right], \tag{20}$$

so that $\langle(\Delta E)^2\rangle$ actually involves diffusion coefficients up to the fourth order. One can, however, compute for this quantity the correction for non-dominant terms exactly as we did for $\langle(\Delta\mathbf{V}_1)^2\rangle$. We consider again the equal-mass case. For field stars with a maxwellian distribution of velocities, and for a test star of given velocity, $\langle(\Delta E)^2\rangle$ is given by Equation (81) in Chandrasekhar (1941). We define a mean value $\langle\langle(\Delta E)^2\rangle\rangle$ over all test stars. Here again it is found to be of the form

$$\langle\langle(\Delta E)^2\rangle\rangle \propto I_1 \ln(0.4\,N) + I_2 = I_1 \ln(\gamma N), \tag{21}$$

as in (8) and (10). The integrals I_1 and I_2 can be evaluated numerically, using Equation (81) and Tables 2 and 4 in Chandrasekhar (1941). The values of g for $x_0 < 0.6$, not given by Chandrasekhar, have been taken equal to zero. The maximum impact distance, l_{max}, is erroneously identified by Chandrasekhar with the average distance between neighbour stars, D_0; therefore D_0 should be replaced in his formulas by l_{max}, itself given by (3) and (4). The result is then

$$I_2/I_1 = -1.0046, \qquad \gamma = 0.1465, \tag{22}$$

which agrees rather well with (19). This indicates that γ is not very sensitive to the particular diffusion coefficient, or combination of them, which is considered.

Liboff (1959) computed the transport coefficients in a fully ionized plasma, with a Maxwellian distribution of velocities. Although his emphasis was on a more correct treatment of the long-distance cutoff, he also computed exactly the effect of the non-dominant terms for short distances. Our quantity $\langle\langle(\Delta\mathbf{V}_1)^2\rangle\rangle$ can be expressed in terms of Liboff's integral $\Omega_{1,c}^{(1)}$, and our result (15) can be shown to agree with his Equation (4.28).

Figure 5 shows again the four Monte Carlo models of Figure 1 (full lines), together with the results of N-body simulations by Aarseth (1974) and Wielen (1974) (filled and open symbols). The value $\gamma = 0.15$ is used. This figure should be compared with Aarseth et al. (1975, Figure 1) or Wielen (1975, Figure 1) where the same data are represented, but the classical value $\gamma = 0.4$ is used. It will be seen that the agreement is significantly improved with the new value of γ. There is, in fact, no noticeable deviation any more between the results of the two methods in Figure 5, except perhaps in the lower right, where the radius corresponding to 10% of the mass appears

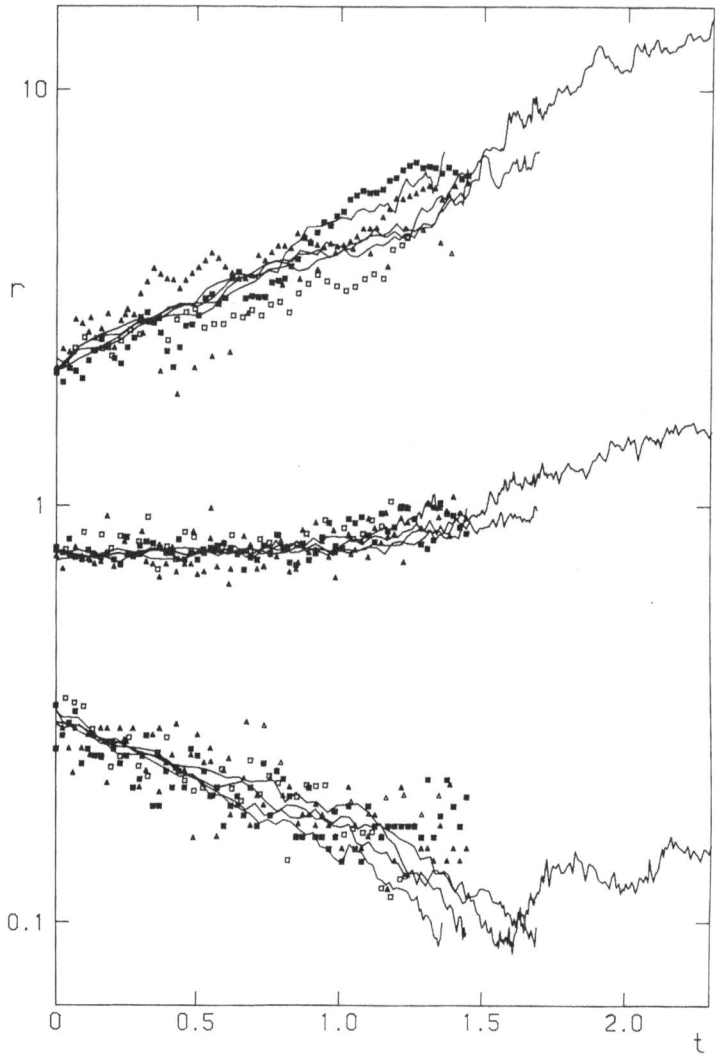

Fig. 5. Comparison of four Monte Carlo models (full lines) with four N-body simulations (symbols) for the equal-mass case, using the new value $\gamma = 0.15$. The data represent the radii containing 10%, 50%, 90% of the mass, vs time. The Monte Carlo models are the same as in Figure 1. Filled triangles and squares represent two N-body simulations by Aarseth, with $N = 250$. Open triangles and squares represent two N-body simulations by Wielen, with $N = 100$ and $N = 250$ respectively. The initial state is Plummer's model with isotropic velocity distribution.

to reach a minimum at a somewhat earlier time and at a higher value in the N-body simulations than in the Monte Carlo models.

A change in the value of γ affects not merely the comparison between Monte Carlo and N-body results, but also the internal comparison between N-body simulations for different values of N; for, in order to do such a comparison, the simulations must be reduced to a common evolutionary scale, and this can be done only by borrowing from the theory the law of dependence of the evolution rate on N. In the present paper, the unit of time for each model was taken as (Aarseth *et al.*, 1974):

$$t_0 = GM^{5/2}(-4\mathscr{E})^{-3/2} N/\ln(\gamma N), \tag{23}$$

where $\mathscr{E}$ is the total energy of the system; t_0 is the theoretical relaxation time, apart from a dimensionless multiplicative constant. As a result, when γ is replaced by a different value γ', the apparent rate of evolution of a N-body simulation is multiplied by

$$\ln(\gamma N)/\ln(\gamma' N). \tag{24}$$

This factor depends on N: for $\gamma = 0.4$ and $\gamma' = 0.15$, its value is 1.3622 and 1.2706 respectively for $N = 100$ and 250, which are the values used in the N-body simulations of Figure 5. On the other hand, the Monte Carlo results correspond to the limit $N \to \infty$, and are not affected by a change in γ: the factor (24) reduces to 1 in this limit.

We consider now the case of unequal masses. We shall assume for simplicity that the distribution function is separable, i.e. there is no mass segregation:

$$f(\mathbf{V}_1, m_1) = g(\mathbf{V}_1) h(m_1). \tag{25}$$

Then I_2/I_1 has the same expression (13) as before, depending only on the velocity distribution g. On the other hand, we must now compute the third term in (11), which is no longer zero:

$$\frac{I_3}{I_1} = \frac{\iint h(m_1) h(m_2) m_1 m_2^2 \ln[2\langle m\rangle/(m_1 + m_2)] \, dm_1 \, dm_2}{\int h(m_1) m_1 \, dm_1 \int h(m_2) m_2^2 \, dm_2}. \tag{26}$$

$\langle m \rangle$ is given by

$$\langle m \rangle = M/N = \int h(m) m \, dm \bigg/ \int h(m) \, dm. \tag{27}$$

The corrective term (26) depends only on the mass distribution h. We evaluate it first for a continuous distribution:

$$h(m) = am^{-2} \quad \text{for} \quad m_0 < m < Qm_0, \qquad h(m) = 0 \quad \text{elsewhere}, \tag{28}$$

with a, m_0, Q constants. We obtain

$$\frac{I_3}{I_1} = \ln\left(\frac{2Q \ln Q}{Q-1}\right) - \frac{Q \ln Q}{Q-1} - $$
$$- \frac{D(1+Q) + QD(1+1/Q) + (1+Q)\ln[4Q/(1+Q)^2] + (1+Q)\pi^2/12}{(Q-1)\ln Q}, \tag{29}$$

M. HÉNON

where D is the dilogarithm function (Abramowitz and Stegun, 1965):

$$D(x) = -\int_1^x \frac{\ln t}{t-1} dt. \tag{30}$$

The expression (29) has been tabulated for a few values of Q in Table II.

TABLE II

Values of the corrective term
I_3/I_1 for a mass distribution
of the form (28)

Q	I_3/I_1
2	−0.0498
4	−0.1964
8	−0.4322
16	−0.7458
32	−1.1246
64	−1.5562

We consider next the case of a discrete distribution:

$$h(m) = a \sum_{i=1}^{k} n_i \delta(m - m_i) \tag{31}$$

and the frequently used law for the masses m_i and the corresponding numbers n_i:

$$m_i = 2^{i-1}, \qquad n_i = 2^{k-i}. \tag{32}$$

The integrals in (26) are replaced by simple sums. Results are given in Table III.

TABLE III

Values of the corrective term
I_3/I_1 for a mass distribution
of the form (31, 32)

k	I_3/I_1
2	−0.1461
3	−0.3811
4	−0.6938
5	−1.0717
6	−1.5028

These results correspond closely to those of Table II for $Q = 2^k$, since (32) is the discrete equivalent of (28).

Finally, for Wielen's (1967) discrete distribution, which corresponds to $k = 4$;

$m_i = 1, 2, 4, 8$; $n_i = 62, 24, 10, 4$, we find

$$I_3/I_1 = -0.7014. \tag{33}$$

It can be seen that I_3/I_1 too is always negative. This can be explained by the presence of the factor m_2^2 in (26): heavy field stars have much more effect in encounters. Since $m_1 + m_2$ appears in the denominator of the argument of the logarithm, the true mean value of the logarithm is less than the classical value, obtained by replacing m_1 and m_2 by the average $\langle m \rangle$.

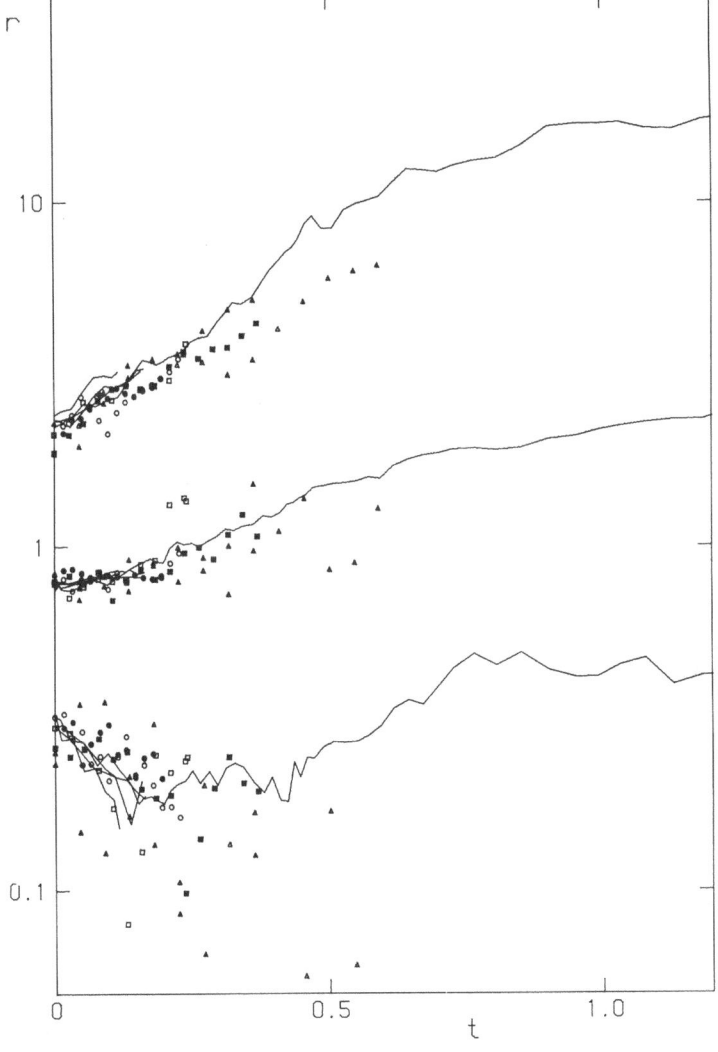

Fig. 6. Comparison of four Monte Carlo models (full lines) with six N-body simulations by Wielen (symbols) for the case of Wielen's mass distribution, using the new value $\gamma = 0.075$. The Monte Carlo models are the same as in Figure 3. Triangles, squares and circles correspond to $N = 100, 250, 500$ respectively. The initial state is the same as before.

We note also that the corrective term I_3/I_1 becomes more and more important as the dispersion of the masses increases. Therefore one cannot meaningfully quote a mean value here. In view of our previous result (19), the value of γ to be used should in general be

$$\gamma = 0.15 \exp(I_3/I_1), \tag{34}$$

where I_3/I_1 depends on the particular mass distribution, and should be evaluated from (26) or from one of our Tables.

For Wielen's distribution, we derive from (33) and (34) the rounded value

$$\gamma = 0.075. \tag{35}$$

Figure 6 shows again the four Monte Carlo models of Figure 3 (full lines), together with the results of six N-body simulations by Wielen (1974) (filled and open symbols). The value $\gamma = 0.075$ is used. This figure should be compared with Aarseth et al. (1975, Figure 2) or Wielen (1975, Figure 2), where the classical value $\gamma = 0.4$ was used. Here again the agreement is much better with the new value, although the Monte Carlo models still appear to run somewhat too fast. The corrective factor (24), with $\gamma = 0.4$ and $\gamma' = 0.075$, is 1.8308, 1.5711, 1.4619 respectively for $N = 100$, 250, 500, the values used in the N-body simulations of Figure 6. It is apparent here that the effect of the non-dominant terms is far from negligible.

The agreement found here, particularly in the case of equal masses (Figure 5), may be somewhat accidental, since there are many other approximations in the theory which we have not considered. For instance, we know that the largest impact distance should be of the order of the dimensions of the system; but its particular identification with the radius containing half the mass, in (3), is no more than a guess. If this maximum impact distance were halved or doubled, the constant γ would also be halved or doubled.

Nevertheless, our results show at least that there exists no significant disagreement between the results of N-body simulations and Monte Carlo models. The previously noted differences (Aarseth et al., 1975) can be entirely accounted for by the approximations in the theory. Thus the classical theory, founded on the assumption that evolution is due to the cumulative effect of two-body encounters, appears to be confirmed by the results of N-body simulations.

Acknowledgement

We thank Dr Aarseth for helpful comments on this paper.

References

Aarseth, S. J.: 1971, *Astrophys. Space Sci.* **13**, 324.
Aarseth, S. J.: 1973, *Vistas in Astronomy* **15**, 13.
Aarseth, S. J.: 1974, *Astron. Astrophys.* **35**, 237.

Aarseth, S. J.: 1975, this volume, p. 57.

Aarseth, S. J., Hénon, M., and Wielen, R.: 1975, *Astron. Astrophys.* **37**, 183.

Abramowitz, M. and Stegun, I. A.: 1965, *Handbook of Mathematical Functions*, Dover, New York, p. 1004.

Chandrasekhar, S.: 1941, *Astrophys. J.* **93**, 285.

Chandrasekhar, S.: 1942, *Principles of Stellar Dynamics*, Chicago University Press, Chicago.

Hénon, M.: 1961, *Ann. Astrophys.* **24**, 369.

Hénon, M.: 1971, *Astrophys. Space Sci.* **13**, 284 and **14**, 151.

Hénon, M.: 1973, in L. Martinet and M. Mayor (eds.), *Dynamical Structure and Evolution of Stellar Systems*, Swiss Society of Astronomy and Astrophysics Third Advanced Course, Geneva Observatory, p. 224.

King, I.: 1975, this volume, p. 99.

Larson, R. B.: 1970, *Monthly Notices Roy. Astron. Soc.* **147**, 323; **150**, 93.

Liboff, R. L.: 1959, *Phys. Fluids* **2**, 40.

Lynden-Bell, D. and Wood, R.: 1968, *Monthly Notices Roy. Astron. Soc.* **138**, 495.

Spitzer, L., Jr.: 1969, *Astrophys. J.* **158**, L139.

Spitzer, L., Jr.: 1975, this volume, p. 3.

Spitzer, L., Jr. and Hart, M. H.: 1971, *Astrophys. J.* **164**, 399.

Spitzer, L., Jr. and Thuan, T. X.: 1972, *Astrophys. J.* **175**, 31.

Von Hoerner, S.: 1968, *Bull. Astron., Sér. 3*, **3**, 147.

Wielen, R.: 1967, *Veröffentl. Astron. Rechen-Inst. Heidelberg*, Nr. 19.

Wielen, R.: 1971, *Astrophys. Space Sci.* **13**, 300.

Wielen, R.: 1974, in L. N. Mavridis (ed.), *Proceedings of the First European Astronomical Meeting (Athens 1972)*, Springer-Verlag, Berlin-Heidelberg-New York, Vol. 2, p. 326.

Wielen, R.: 1975, this volume, p. 119.

DISCUSSION

Larson: Do I understand correctly that the time scale correction that you have derived applies equally to any theory based on the Fokker-Planck equation?

Hénon: Yes. In particular, this correction will also improve the agreement between your fluid-dynamical models and the *N*-body simulations.

DYNAMICAL MASSES AND MASS-TO-LIGHT RATIOS
FOR GLOBULAR CLUSTERS

G. ILLINGWORTH

*Kitt Peak National Observatory, Tucson, Arizona, U.S.A.**

Abstract. Dynamical masses have been determined for 10 globular clusters. Comparison of the dynamically determined M/L values with those calculated from extrapolations of the observed upper main sequence/giant branch luminosity functions for M3 and M5 indicate that low mass stars ($\sim 0.2\ M_\odot$) comprise a large fraction of the total number.

The best means at present for obtaining data on the form of the luminosity function (and hence mass function) in globular clusters for the faint, unobservable low mass stars is from dynamical determinations of the cluster masses. The importance of the mass function is due to (i) the dependence of the dynamical evolution of the cluster upon it (see e.g. Prata, 1971; Spitzer and Hart, 1971) and (ii) the possibility of determining the initial mass function by the use of representative evolutionary models. A summary is given here of the results of an observational program to determine masses for 10 globular clusters. This is followed by a comparison of the dynamically determined M/L values with those calculated from extrapolations to the observed luminosity functions of M3 and M5.

1. Masses and M/L Ratios

Central velocity dispersions have been measured for 10 concentrated globular clusters from high dispersion spectra of the integrated light. These velocity dispersions are given in Table I. They were determined from the spectra by (i) comparing the

TABLE I

Observed data

NGC	$\langle V_r^2 \rangle^{1/2}$ km s^{-1}	$\log \dfrac{r_t}{r_c}$	r_c pc	D kpc	M_V
104	10.5 ± 0.4	2.03	0.48	4.1 ± 0.4	-9.2 ± 0.2
362	7.5 0.9	1.70	0.58	9.7 1.4	-8.5 0.2
1851	7.9 0.7	1.83	0.37	10.5 1.9	-8.3 0.3
2808	14.2 1.3	1.75	0.73	10.0 1.6	-9.7 0.2
6093	12.5 2.5	1.88	0.33	9.1 1.8	-8.1 0.2
6266	13.7 1.1	1.63	0.56	7.9 1.2	-9.3 0.3
6388	18.9 0.8	1.75	0.50	11.6 2.3	-9.6 0.4
6441	17.6 0.8	1.70	0.41	9.3 1.9	-8.9 0.4
6715	14.2 1.0	1.83	0.71	22.1 3.1	-9.6 0.3
6864	10.3 1.5	1.82	0.53	20.0 2.8	-8.5 0.3

* Operated by the Association of Universities for Research in Astronomy, Inc., under contract with the National Science Foundation.

globular cluster spectra with artificially broadened spectra of suitable comparison stars and (ii) comparison in the Fourier domain of the power spectra of the globular cluster spectra and comparison star spectra. Determination of the velocity dispersion with high accuracy was possible using the latter method. The parameters r_c (the core radius) and r_t/r_c (the ratio of the tidal and core radii) required for use of the self-consistent models of King (1966) were obtained from photoelectric surface photo-metry and star counts. r_c and $\log(r_t/r_c)$ are tabulated in Table I. In addition the distances and integrated magnitudes used are also given in Table I. The masses and M/L ratios calculated from the models using the data in Table I are given in Table II. More

TABLE II

Masses and mass-to-light ratios

NGC	Mass $(M_\odot \times 10^{-6})$	$(M/L_v)_\odot$
104	0.54 ± 0.07	1.4 ± 0.2
362	0.19 0.06	0.9 0.3
1851	0.16 0.05	0.9 0.3
2808	0.92 0.24	1.4 0.4
6093	0.39 0.18	2.8 1.3
6266	0.55 0.14	1.3 0.4
6388	1.12 0.26	2.0 0.5
6441	0.74 0.18	2.6 0.6
6715	1.03 0.23	1.8 0.4
6864	0.40 0.13	1.9 0.6

extensive data, details of the methods used and of the sources of error are given in Illingworth and Freeman (1974), Illingworth (1975), and Illingworth and Illingworth (1975).

2. Luminosity Functions

The mean M/L value for the clusters in Table II is

$$\langle M/L_v \rangle_\odot = 1.7.$$

This is a convenient dynamical value for comparison with M/L values derived from observed luminosity functions extrapolated to the unobservable faint low mass stars. The use of this mean M/L is not to imply that all these clusters are expected to have the same M/L. Differences in the dynamical evolution between clusters will quite likely result in different present M/L values even if the initial mass function was the same for all clusters.

The observed luminosity functions of M3 (Sandage, 1957) and M5 (Simoda and Tanikawa, 1972) have been chosen as representative of globular cluster upper main sequence/giant branch luminosity functions. The differences between these two functions probably stem from the methods used to extrapolate the counted numbers of stars in the outer regions to total numbers for the whole cluster. The need to use a dynamical model for this extrapolation has been pointed out by King and Wilson

(1972) since stars of different mass are distributed differently in the cluster. As luminosity functions properly corrected for this effect are not yet available those for M3 and M5 will be used as they represent the present extremes in the ratio of main sequence/giant branch star numbers (see Figure 2 of Simoda and Tanikawa).

The first extrapolation used was to fit the solar neighborhood luminosity function recently described by Wielen (1974). This is shown in Figures 1a and 1b. The apparent distance moduli used were:

$$\text{M3:} \quad (m-M)_{\text{app},V} = 14.83 \quad \text{Sandage (1970)}$$
$$\text{M5:} \quad (m-M)_{\text{app},V} = 14.39 \quad \text{Arp (1962)}.$$

The solar neighborhood curve has been normalized to the luminosity functions at $M_V \sim +6$. The Wielen function is similar to the combined van Rhijn (1936), Luyten (1939), and Kuiper (1942) luminosity function used by Sandage (1957) as an extension to the observed M3 luminosity function. The large, disk red dwarf population discussed by Weistrop (1972) is not included in the Wielen function. For M5 the Wielen function was fitted to the data compensated for the effects of equipartition, as is shown by a dotted line. The uncompensated curve is shown by a broken line. The peaking of the compensated curve at $M_V \sim +6$ indicates that the compensation was probably too extreme but still an improvement over the uncompensated function.

To determine the probable number of white dwarf members the Sandage (1957) modified Salpeter (1955) initial luminosity function has been fitted above the turnoff point. All stars having a mass greater than the current turnoff/red giant mass were

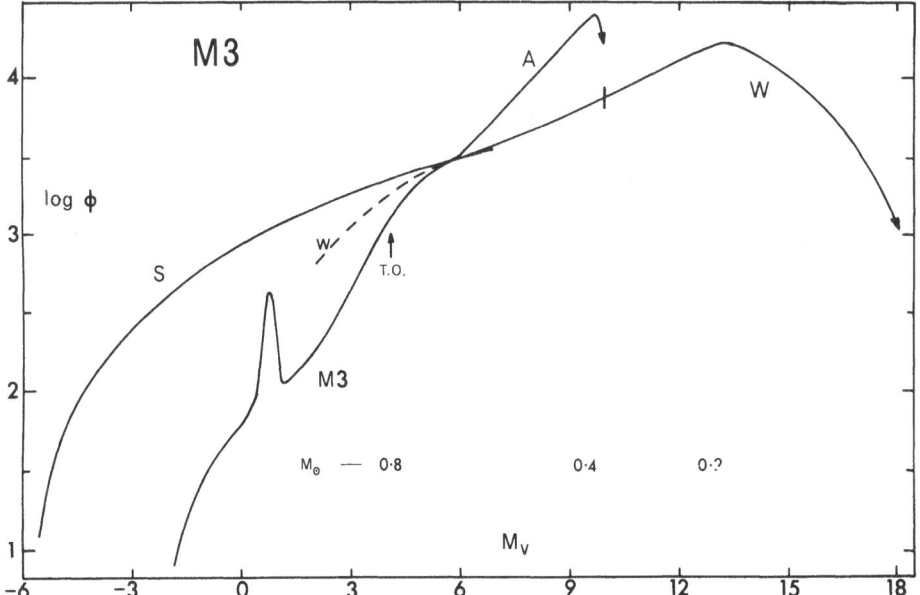

Fig. 1a.   The observed luminosity functions of M3 and M5 fitted with (W) the Wielen luminosity function for the solar neighborhood, (S) the Sandage modified Salpeter initial luminosity function, (A) an arbitrarily increased function with mass cutoff at $M \sim 0.4\,M_\odot$.

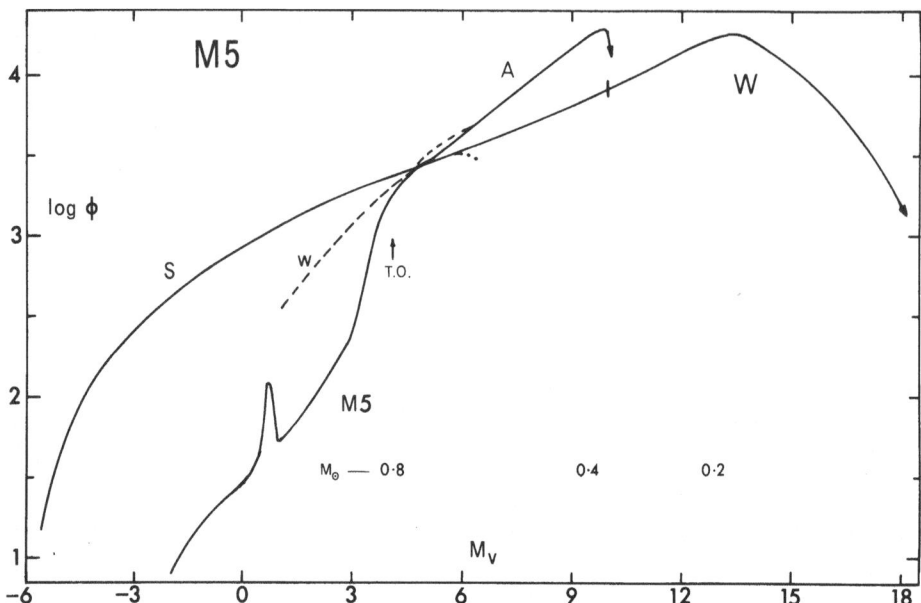

Fig. 1b. ϕ is the number of stars between $M_v + 0.1$ and $M_v - 0.1$. T.O. is the main sequence turnoff point. Further details are given in the text.

considered to have become white dwarfs. Prata (1971) found that the most likely interpretation of his results from an evolutionary study of M67 was that the initial mass function (for M67) was not a Salpeter function. The likely form was a power law with smaller slope (i.e. less low mass stars) at the low mass end but with a sharp increase in slope around 1.5–4 $M_\odot$. Thus the derived M/L values below are given both with and without the white dwarfs to show the effect of a sharply decreased white dwarf component.

The next step was to assign masses to the white dwarfs, the giant branch and the main sequence stars. The mass of the white dwarfs was taken from the recent results of Wickramasinghe and Strittmatter (1972): $\langle M \rangle \sim 0.6\ M_\odot$, Shipman (1972): $\langle M \rangle \sim 0.5$ $M_\odot$ and Trimble and Greenstein (1972): $\sim 0.65\ M_\odot < \langle M \rangle < \sim 0.87\ M_\odot$. The adopted mass was $M_{WD} = 0.6\ M_\odot$.

The results from color magnitude diagrams down to the main sequence for M3, M13, M15 and M92 (Sandage, 1970) combined with the main sequence/giant branch star models of Iben and Rood (1970a) indicate that the red giant mass $M_{RG} \sim 0.8\ M_\odot$ (see also Iben, 1971). Böhm-Vitense and Szkody (1973) using the main sequence data for M15 and M92 from the same paper by Sandage find $M_{RG} \sim 0.85\ M_\odot$. $M_{RG} = 0.8$ $M_\odot$ was used here.

For the horizontal branch stars $M_{HB} = 0.6\ M_\odot$ was adopted (Iben and Rood, 1970b; Iben, 1971).

For the upper main sequence, where the bolometric corrections are small, assigning masses was fairly easy. The brightening of the main sequence was considered. The lower main sequence masses are more uncertain. This is due to the lack of a well

established mass-luminosity relation and the uncertainty in the bolometric corrections. For the bolometric corrections on the main sequence the data considered was
from Harris (1963), Lamla (1965), Copeland *et al.* (1970), and Johnson (1966) with the
spectral type $(B-V)$ relation from Blaauw (1963). A mean relation was adopted.
Similarly a mean relation was adopted from the M_{bol}-mass relations given by Harris
et al. (1963), Eggen (1967), Copeland *et al.* (1970), and McCluskey and Kondo (1972).
The mass of the stars at the peak in the luminosity distribution at $M_V \sim 13.5$ was
$\sim 0.2\ M_\odot$.

If some degree of equipartition of energy occurs in globular clusters then this,
combined with the galactic tidal force and the effect of gravitational shocks, may well
have led to the loss of most of the low mass stars in the cluster (see Ostriker *et al.*, 1972).
The effect of this on the M/L values was estimated by arbitrarily cutting the Wielen
function at $M \sim 0.4\ M_\odot$. The M/L values [Table III; (b)] determined from this were
somewhat lower than the mean dynamical value.

A further example using the same cutoff mass but having a higher M/L value is
shown on Figures 1a and 1b. This arbitrarily increased function was the result of an
attempt to retain the low mass cutoff with an M/L nearer the mean dynamical value.

TABLE III

Mass-to-light ratios

	No white dwarfs		With white dwarfs	
	$(M/L_V)_\odot$	$\langle M \rangle$	$(M/L_V)_\odot$	$\langle M \rangle$
M3(a)	1.0	0.29	1.2	0.32
(b)	0.5	0.57	0.7	0.58
(c)	0.9	0.52	1.2	0.53
M5(a)	1.4	0.29	1.7	0.32
(b)	0.7	0.57	1.0	0.58
(c)	1.1	0.53	1.4	0.55

The derived $(M/L_V)_\odot$ values are given in Table III for M3 and M5 with the extrapolations:

(a) Wielen luminosity function,
(b) Wielen luminosity function cutoff at $M \sim 0.4\ M_\odot$,
(c) raised luminosity function with the same cutoff.

Results both with and without the white dwarfs are presented. The mean mass $\langle M \rangle$
(in $M_\odot$) is also given.

3. Discussion

Before the M/L values derived from extrapolation of the observed luminosity functions are compared with the mean dynamical M/L value some consideration needs to
be given to uncertainties in the dynamical value. The major uncertainty arises from

the use of a single mass model in the mass derivation. The likely result of using models (again with an isotropic velocity distribution function like that of the single mass models) covering a range of stellar masses will be to increase the mass somewhat but probably by a factor less than 2. Given this, the mean $\langle M/L_V \rangle_\odot = 1.7$ is then underestimated. Thus it is likely that case (b) in Table III (Wielen luminosity function cutoff at 0.4 $M_\odot$) is incorrect. Case (c), the raised luminosity function cutoff at 0.4 $M_\odot$ appears unlikely as well since the disagreement with the observed function around $M_V = +6$ for both M3 and M5 is large and will get worse for higher M/L values – although the luminosity functions are rather uncertain at this point due to the effects of mass differentiation in the cluster. In the case of M3 inclusion of these effects will be such as to worsen the disagreement (note the compensated and uncompensated curves for M5 in Figure 1b).

Thus from the luminosity functions considered here the most likely extrapolation to the observed luminosity functions appears to be one similar to the solar neighborhood dwarf luminosity function. This means that stars of low mass (i.e. around 0.2 $M_\odot$) still comprise a large fraction of the total number. A low mass cutoff from preferential depletion of low mass stars due to dynamical evolution is still possible but it will be closer to 0.2 $M_\odot$ (or less) than 0.4 $M_\odot$. Unless the number of white dwarfs is considerably larger than expected from the Salpeter initial mass function, they will not have an appreciable effect on these conclusions.

Definitive resolution, however, of the form of the mass function for low mass stars and of the number and mass of the white dwarfs will require a very detailed observational study (plus use of models covering a range of masses) of one or two clusters.

Acknowledgements

The observations reported here were carried out at Mt. Stromlo Observatory. I would like to thank the staff, in particular Ken Freeman, for their assistance. I am grateful for the use of the facilities at Mt. Stromlo and Siding Spring and for the assistance of an Australian National University Scholarship.

References

Arp, H. C.: 1962, *Astrophys. J.* **135**, 311.
Blaauw, A.: 1963, *Stars and Stellar Systems* **3**, 383.
Böhm-Vitense, E. and Szkody, P.: 1973, *Astrophys. J.* **184**, 211.
Copeland, H., Jensen, J. O., and Jørgensen, H. E.: 1970, *Astron. Astrophys.* **5**, 12.
Eggen, O. J.: 1967, *Ann. Rev. Astron. Astrophys.* **5**, 105.
Harris, D. L., III.: 1963, *Stars and Stellar Systems* **3**, 263.
Harris, D. L., III., Strand, K. Aa., and Worley, C. E.: 1963, *Stars and Stellar Systems* **3**, 273.
Iben, I., Jr.: 1971, *Publ. Astron. Soc. Pacific* **83**, 697.
Iben, I., Jr. and Rood, R. T.: 1970a, *Astrophys. J.* **159**, 605.
Iben, I., Jr. and Rood, R. T.: 1970b, *Astrophys. J.* **161**, 587.
Illingworth, G.: 1975, 'The Masses of Globular Clusters. II: Velocity Dispersions and Mass-to-Light Ratios', *Astrophys. J.* (in press).
Illingworth, G. and Freeman, K. C.: 1974, *Astrophys. J. Letters* **188**, L83.

Illingworth, G. and Illingworth, W.: 1975, 'The Masses of Globular Clusters. I: Surface Brightness Distributions and Star Counts', *Astrophys. J. Suppl.* (in press).

Johnson, H. L.: 1966, *Ann. Rev. Astron. Astrophys.* **4**, 193.

King, I. R.: 1966, *Astron. J.* **71**, 64.

King, I. R. and Wilson, C. P.: 1972, in A. G. Davis Philip (ed.), *The Evolution of Population II Stars* (Dudley Observatory Report, No. 4), p. 29.

Kuiper, G. P.: 1942, *Astrophys. J.* **95**, 201.

Lamla, E.: 1965, in H. H. Voigt (ed.), *Landolt-Bornstein*, Springer-Verlag, Berlin, **1**, p. 369.

Luyten, W.: 1939, *Publ. Minnesota Obs.* **2**, No. 7, 123.

McCluskey, G. E., Jr. and Kondo, Y.: 1972, *Astrophys. Space Sci.* **17**, 134.

Ostriker, J. P., Spitzer, L., Jr., and Chevalier, R. A.: 1972, *Astrophys. J. Letters* **176**, L51.

Prata, S. W.: 1971, *Astron. J.* **76**, 1017.

Rhijn, P. J., van: 1936, *Groningen Publ.* No. 47.

Salpeter, E. E.: 1955, *Astrophys. J.* **121**, 161.

Sandage, A.: 1957, *Astrophys. J.* **125**, 422.

Sandage, A.: 1970, *Astrophys. J.* **162**, 841.

Shipman, H. L.: 1972, *Astrophys. J.* **177**, 723.

Simoda, M. and Tanikawa, K.: 1972, *Publ. Astron. Soc. Japan* **24**, 1.

Spitzer, L., Jr. and Hart, M. H.: 1971, *Astrophys. J.* **166**, 483.

Trimble, V. and Greenstein, J. L.: 1972, *Astrophys. J.* **177**, 441.

Weistrop, D.: 1972, *Astron. J.* **77**, 849.

Wielen, R.: 1974, in G. Contopoulos (ed.), *Highlights of Astronomy*, Vol. 3, D. Reidel Publ. Co., Dordrecht-Holland, p. 395.

Wickramasinghe, D. T. and Strittmatter, P. A.: 1972, *Monthly Notices Roy. Astron. Soc.* **160**, 421.

DISCUSSION

King: This is a beautiful example of a new observational technique. As for the conclusions, I should like to emphasize that the main weight of the results goes into the number of low-mass stars which are too faint to observe in any other way.

STRUCTURE OF THE HYADES CLUSTER

G. PELS, J. H. OORT, and H. A. PELS-KLUYVER

Sterrewacht, Leiden, The Netherlands

H. A. Pels and J. H. Oort have discussed the results of a search for fainter Hyades that had been made by G. Pels. The search extended over a region of about 15° radius around the Hyades centre. It was made from proper motions determined from a comparison of Astrographic Catalogue positions with those on plates taken with the 34-cm Leiden refractor. The data extend to about 13^m. In part of the field comparisons were also made with plates taken for the Astrographic *Charts*, in order to extend the luminosity distribution to somewhat fainter magnitudes.

The survey has roughly doubled the number of known Hyades. A combination of the present survey with that of Van Bueren gave a regular distribution on the sky with no observable deviation from circular symmetry.

Using a plausible extension of the luminosity function for fainter members an approximate mass density distribution was computed. Using the corresponding gravitational field it was shown that the observed space distribution of the cluster members agrees with that expected for a cluster in dynamical equilibrium, having a Maxwellian velocity distribution with a cut-off corresponding with the outer limit of the cluster. The average velocity in one coordinate is about ± 0.2 km s^{-1}. The age of the cluster is indeed such that it should have reached approximate equilibrium.

There is a distinct segregation of stars of different mass.

The problem of outlying members is still under investigation.

A full account has been submitted for publication in *Astronomy and Astrophysics*.

DYNAMICAL FRICTION IN THE COMA CLUSTER

M. LECAR

Center for Astrophysics, Harvard College Observatory and Smithsonian Astrophysical Observatory,
Cambridge, Mass., U.S.A.

Abstract. If the galaxies composing the Coma Cluster were initially 'Ostriker-Peebles' galaxies (i.e., iso-thermal spheres extending to 250–500 kpc), the hidden mass is likely to be tidally torn-off halos. Owing to dynamical friction, the remnant cores tend to spiral in to the center of the cluster. This happens as long as the hidden mass, whatever its origin, consists of objects much less massive than galaxies.

Ostriker and Peebles (1973) suggested that, in order to be stable against bar formation (see also Bardeen, 1974), galaxies should have most of their mass in extended halos. An isothermal sphere density profile would also explain the observed flat rotation curves (Roberts, 1974), which requires that their masses increase linearly with their radii. For illustration, I will assume that an 'Ostriker-Peebles' galaxy can be characterized by the relation

$$M = 10^{10} M_\odot \times r(\text{kpc}). \tag{1}$$

For an isothermal sphere, $M \cong (2c^2/3G) \times r$, as r becomes large. The above relation (1) results from taking the internal dispersion velocity, $c = 254$ km s^{-1}.

Studies by Omer *et al.* (1965), Peebles (1970), Rood *et al.* (1972), Bahcall (1972), and Oemler (1973) provide a relatively consistent (to within a factor of 2) set of parameters for the Coma Cluster. Their results (adjusted to $H = 50$ km s^{-1} Mpc^{-1}) can be modeled by assuming Coma to be distributed like an isothermal sphere with

$$
\begin{aligned}
& M = 2 \times 10^{15} M_\odot, \\
& R = 4 \text{ Mpc}, \\
& c = 1800 \text{ km s}^{-1}, \\
& r_c = 0.25 \text{ Mpc}, \\
& \varrho_0 = 3 \times 10^{-3} M_\odot \text{ pc}^{-3}.
\end{aligned}
\tag{2}
$$

The core radius, r_c, is defined as that radius where the projected density falls to half its central value, and is equal to three isothermal structure lengths. The isothermal structure length, a, is related to the dispersion velocity, c, and the central density, ϱ_0, by

$$a^2 = \frac{c^2}{12\pi G\varrho_0} = \left(\frac{r_c}{3}\right)^2. \tag{3}$$

I will assume that the galaxies initially had masses of 2.5×10^{12} $M_\odot$ and radii of 250 kpc. If each galaxy in the cluster occupies a volume $4\pi r^3/3$, then the mean value of r is

$$\bar{r} = N^{-1/3} R. \tag{4}$$

By taking $N = 800$ and $R = 4$ Mpc, $\bar{r} = 430$ kpc. At 1 Mpc from the center, $r = 240$

Hayli (ed.), Dynamics of Stellar Systems, 161–166. All Rights Reserved.

kpc, decreasing to 60 kpc at the center. Thus, interior to 1 Mpc, the initial galaxies overlap, and it seems likely that they would lose their halos through tidal interactions with their neighbors. Alladin (1965, 1974), Alladin *et al.* (1974), Sastry and Alladin (1970), Gallagher and Ostriker (1972), Lauberts (1974), and Biermann (1974) have studied penetrating collisions between spherical galaxies. Biermann found, for conditions similar to what I have assumed for the Coma Cluster (relative velocities between the galaxies about 7 times their internal dispersion velocities), that for encounter distances $\frac{1}{4}$ of the radius of the galaxy, as much as $\frac{1}{4}$ of the mass could be dispersed. However, as the binding energy of the galaxy (characterized by $(254 \text{ km s}^{-1})^2$ per unit mass) is less than 3% of its kinetic energy of center of mass motion (characterized by $(1800 \text{ km s}^{-1})^2$ per unit mass), the galaxies should hardly slow down as they shed their halos.

I now turn to the orbits of the remnant cores swimming in a sea of halo material. If the sea consists of objects (e.g., M dwarfs) whose mass is much less than that of a galaxy, Spitzer (1969) and Saslaw and de Young (1971) have shown that the galaxies will tend to settle to the center of the cluster. Locally, the galaxies try to reach equipartition with the stars, and since the galaxies have too much kinetic energy, they slow down. But, in a gravitational field, to slow down means to fall, and to fall means to speed up. The galaxies can not attain equipartition, but they keep trying. A less picturesque description is given by 'dynamical friction'. The following formula for the rate of change of kinetic energy per unit mass is adapted from Hénon (1973),

$$\frac{\text{d}}{\text{d}t}(\tfrac{1}{2}V^2) = -\frac{2\pi G\varrho \cdot GM \ln \Lambda}{V}, \tag{5}$$

where V is the velocity of the galaxy, M is its mass, ϱ is the density of field stars, and Λ can be taken as $M_{\text{cluster}}/M = 800$.

I have assumed, in the above, that the galaxies have the same velocity dispersion as the field stars and that the fraction with rms velocity less than the mean is $\frac{1}{2}$. As only those field stars with velocities less than that of the galaxy contribute to dynamical friction, the process quenches quickly if the field stars heat up relative to the galaxies. The process becomes 10 times less efficient if the field stars are 3 times faster (rms) than the galaxies.

It is convenient to adopt the dimensionless variables appropriate for an isothermal sphere. If we let

$$V = c \cdot v$$
$$\varrho = \varrho_0 \cdot \delta \tag{6}$$
$$t = \omega^{-1} \cdot \tau, \qquad \omega = c/a = \sqrt{(12\pi G\varrho_0)},$$

Equation (5) can be written

$$\frac{\text{d}}{\text{d}\tau}(\tfrac{1}{2}v^2) = -\left\{\tfrac{1}{6}\left[\frac{(GM/c^2)}{a}\right] \ln \Lambda\right\} \cdot \frac{\delta}{v}. \tag{7}$$

As the galaxy loses kinetic energy, it becomes more tightly bound to the cluster. Consider a galaxy in a circular orbit with velocity V. Its orbit is described by

$$V^2 = \frac{GM(r)}{r},$$

$$E = \tfrac{1}{2}V^2 + \int_0^r \frac{GM(r)}{r^2}\,dr = \tfrac{1}{2}\frac{GM(r)}{r} + \int_0^r \frac{GM(r)}{r^2}\,dr, \tag{8}$$

$$\frac{dE}{dr} = \tfrac{1}{2}\left[\frac{GM(r)}{r^2} + 4\pi G\varrho r\right].$$

In isothermal units,

$$r = ax,$$

$$M(x) = 4\pi\varrho_0 a^3 \mu(x), \qquad \mu(x) = \int_0^x \delta(x)\,x^2\,dx,$$

$$c^2 = 12\pi G\varrho_0 a^2,$$
$$E = c^2\varepsilon,$$

and Equation (8) is written

$$v^2 = \tfrac{1}{3}\frac{\mu(x)}{x},$$

$$\varepsilon = \tfrac{1}{6}\frac{\mu(x)}{x} + \tfrac{1}{3}\int_0^x \frac{\mu(x)}{x^2}\,dx, \tag{9}$$

$$\frac{d\varepsilon}{dx} = \tfrac{1}{6}\left[x\delta(x) + \frac{\mu(x)}{x}\right].$$

Combining Equations (7) and (9) yields

$$\frac{dx}{d\tau} = \frac{dx}{d\varepsilon}\cdot\frac{d\varepsilon}{d\tau} = \frac{-K}{x\sqrt{\mu(x)/x}\,\{1 + [\mu(x)/x^3\delta(x)]\}} \equiv \frac{-K}{x\Phi(x)}, \tag{10}$$

where $K = \sqrt{3}\,[(GM/c^2)/a]\,\ln\Lambda$. The quantity $K^{-1} = 65 \times a(\text{Mpc})$ and the unit of time is $a/c = 5.4 \times 10^8$ yr $\times a(\text{Mpc})$. As $x \to 0$, $\delta \to 1$, $\mu \to x^3/3$, and $\Phi \to 4x/3\sqrt{3}$.
    In this limit, Equation (10) integrates to

$$\left(\frac{x}{x_0}\right)^3 = 1 - \frac{9\sqrt{3}}{4}\frac{K}{x_0^3}\tau,$$

and $(x/x_0) = \tfrac{1}{2}$ when $\tau = (7/18\sqrt{3})(x_0^3/K)$. Since $t = a/c$, $t_{1/2} = 8 \times 10^9$ yr $\times r^3(\text{Mpc})/a(\text{Mpc})$. As $x \to \infty$, $\delta \to 2/x^2$, $\mu \to 2x$, and $\Phi \to 2\sqrt{2}$.

In this limit, Equation (10) integrates to

$$\left(\frac{x}{x_0}\right)^2 = 1 - \frac{K}{\sqrt{2}} \frac{\tau}{x_0^2},$$

and $(x/x_0) = \frac{1}{2}$ when $\tau = (3/2\sqrt{2})(x_0^2/K)$. The quantity $t_{1/2} = 3.7 \times 10^{10}$ yr $\times r^2$ (Mpc), independent of a.

It turns out to be sufficiently accurate to use the small-x limit for $x \leqslant 3$ and the large-x limit for $x \geqslant 9$. In the overlap region, $3 \leqslant x \leqslant 9$, results can be obtained that are $\pm 50\%$ accurate by using whichever formula gives the smallest time. However, I have integrated numerically the equation

$$\int_{1/2x}^{x} dx\, x\Phi(x) \equiv x^2 F(x) = K\tau. \tag{11}$$

The results, accurate to $\pm 10\%$, are given in the appendix, and have been used to generate Table I.

The table indicates significant contraction interior to two core radii. However, there are two effects that would tend to increase the times. First, I have assumed circular orbits, while the typical orbits are probably highly elliptical. I have not, as yet, worked out the numbers for elliptical orbits, but I think including this effect would tend to shorten the times interior to the core and lengthen them exterior to the core. Second, I have assumed a galactic mass of 10^{12} $M_\odot$, which implies a radius (for an isothermal galaxy) of 100 kpc. As, in the center of the cluster, there is not room for

TABLE I

Times to reduce the radius to $\frac{1}{2}$ its initial value for two choices of the core radius of the Coma Cluster

R (arcmin)	R (Mpc)[a]	$t_{1/2}$ ($r_c = 0.25$ Mpc)	$t_{1/2}$ ($r_c = 0.50$ Mpc)
2.5	0.1	9.0×10^7 yr	4.8×10^7 yr
5.0	0.2	6.8×10^8	3.6×10^8
7.5	0.3	2.1×10^9	1.2×10^9
10.0	0.4	4.5×10^9	2.7×10^9
12.5	0.5	8.0×10^9	5.1×10^9
15.0	0.6	1.3×10^{10}	8.4×10^9
20.0	0.8	2.5×10^{10}	1.8×10^{10}
25.0	1.0	4.2×10^{10}	3.2×10^{10}
30.0	1.2	6.3×10^{10}	5.0×10^{10}
40.0	1.6	1.1×10^{11}	1.0×10^{11}
50.0	2.0	1.8×10^{11}	1.7×10^{11}

[a] Assumes 1 arcmin $= 40$ kpc.

100-kpc galaxies, the mass there should be reduced by at least a factor of 2, which doubles the lifetimes.

Dynamical friction leads to mass segregation, since time scales inversely with mass. Bahcall (1973) found, by use of galaxy counts, core radii for Coma of 0.25 Mpc ($\pm 20\%$) for three limiting galaxy brightness's that differed by about 3 magnitudes. This result implies, assuming constant M/L, only modest mass segregation.

Oemler (1973) found the curious result that the space density had a hollow (a dip followed by a rise followed by a steady decrease) at about 1.2 Mpc. Could it be that the galaxies interior to 1.2 Mpc spiralled in faster than those exterior to 1.2 Mpc?

I remind you that the spiralling in by dynamical friction does not rest on the more speculative scenario of torn-off halos. It is just another way of estimating the time scale for mass segregation in a self-gravitating system with a spectrum of masses. As Bahcall (1973) suggested, core radii of clusters of galaxies could be useful cosmological distance indicators. It is therefore of some importance to determine whether they evolve significantly in cosmological times.

I hope that I have convinced you that the parameters for the Coma Cluster justify an N-body simulation of this interesting object.

Appendix

The functions $\Phi(x)$ (defined by Equation (10)) and $F(x)$ (defined by Equation (11)) are presented below. $\Phi(x)$ has been calculated from tables of the isothermal sphere by Chandrasekhar and Wares (1949) and is accurate to the last digit. $F(x)$ was obtained by a three-point Simpson's Rule integration.

x	$\Phi(x)$	$F(x)$	x	$\Phi(x)$	$F(x)$
1	0.7458	0.22	14	3.4047	1.23
2	1.3770	0.42	15	3.4157	1.24
3	1.8673	0.58			
4	2.2424	0.71	16	3.4202	1.25
5	2.5321	0.82	17	3.4194	1.26
			18	3.4144	1.27
6	2.7575	0.91	19	3.4061	1.27
7	2.9331	0.98	20	3.3953	1.27
8	3.0693	1.04			
9	3.1740	1.08	21	3.3825	1.27
10	3.2534	1.13	22	3.3682	1.27
			23	3.3528	1.27
11	3.3126	1.16	24	3.3366	1.27
12	3.3554	1.19	25	3.3199	1.27
13	3.3853	1.21			

References

Alladin, S. M.: 1965, *Astrophys. J.* **141**, 768.
Alladin, S. M., Sastry, K. S., and Ballabh, G. M.: 1974, in B. Barbanis and J. Hadjidemetriou (eds.), *Galaxies and relativistic Astrophysics*, Springer-Verlag, Berlin, p. 129.
Alladin, S. M., Potdar, A., and Sastry, K. S.: 1974, this volume, p. 167.

Bahcall, N. A.: 1973, *Astrophys. J.* **183**, 783.

Bardeen, J. M.: 1974, this volume, p. 297.

Biermann, P.: 1974, 'Report to the Faculty of Mathematical Sciences', George August University, Gott-
ingen.

Chandrasekhar, S. and Wares, G. W.: 1949, *Astrophys. J.* **109**, 551.

Gallagher, J. S. III and Ostriker, J. P.: 1972, *Astron. J.* **77**, 288.

Hénon, M.: 1973, in L. Martinet and M. Mayor (eds.), *Dynamical Structure and Evolution of Stellar Systems*, Geneva Observatory, p. 214.

Lauberts, A.: 1974, *Astron. Astrophys.* **33**, 231.

Oemler, A., Jr.: 1973, 'The Systematic Properties of Clusters of Galaxies', California Institute of Technology (Ph.D. Thesis).

Omer, G. C., Jr., Page, T. L., and Wilson, A. G.: 1965, *Astron. J.* **70**, 440.

Ostriker, J. P. and Peebles, P. J. E.: 1973, *Astrophys. J.* **186**, 467.

Peebles, P. J. E.: 1970, *Astron. J.* **75**, 13.

Roberts, M. S.: this volume, p. 331.

Rood, H. J., Page, T. L., Kintner, E. C., and King, I. R.: 1972, *Astrophys. J.* **175**, 627.

Saslaw, W. C. and de Young, D. S.: 1971, *Astrophys. J.* **170**, 423.

Sastry, K. S. and Alladin, S. M.: 1970, *Astrophys. Space Sci.* **9**, 261.

Spitzer, L., Jr.: 1969, *Astrophys. J.* **158**, L139.

DISCUSSION

Miller: If you were to observe the cluster after this process, and use the virial theorem to estimate the mass, what would you get?

Lecar: The virial theorem would underestimate the mass of the cluster by the usual amount, because in this model, most of the mass of the cluster is in low luminosity halo material.

Hénon: It seems to me that this last argument of your students amounts in fact to considering only encounters between the galaxies and neglecting encounters between the galaxies and the hidden mass. So this might explain why you don't have dynamical friction and spiraling anymore. You still have encounters between galaxies in order to achieve hydrostatic equilibrium.

Lecar: In the problem considered by my students, the test galaxies were assumed to have negligible mass compared to the unseen (unspecified) material providing the gravitational field. This, admittedly, is an unrealistic dynamical model, but serves to illustrate the variety of solutions that are consistent with Poisson's equation and the equation of hydrostatic equilibrium (without, however, specifying how hydrostatic equilibrium is attained).

Gott: The rate at which a galaxy of mass M_g spirals in due to dynamical friction is similar to the 2-body relaxation time calculated imagining the cluster mass to be composed completely of galaxies of mass M_g. Thus as in two body relaxation one finds that the most massive galaxies spiral toward the center most rapidly.

Lecar: Not quite. If the field galaxies and the test galaxy had the same mass, on the average, the test galaxy would not lose energy (although it's direction in space would be randomized). In the model I'm considering, the test galaxy is much more massive than the field stars and loses energy almost in every encounter.

Pişmiş: I like to make a comment: on observational grounds, it is shown independently by Page and Poveda that the percentage of the missing mass in systems of stars, or galaxies, is related to the total mass of the system in the sense that the larger the total mass of the system the larger is the percentage of the missing mass. I suspect your treatment would not take care of this property.

Saslaw: Do you think this process could account for the mass discrepancy claimed for small groups of galaxies?

Lecar: I don't think this (that is ripping off halos) would work for small groups of galaxies.

FORMATION OF DOUBLE GALAXIES BY TIDAL CAPTURE

S. M. ALLADIN, A. POTDAR, and K. S. SASTRY

Centre of Advanced Study in Astronomy, Osmania University, Hyderabad, India

Abstract. The conditions under which double galaxies may be formed by tidal capture are considered. Estimates for the increase in the internal energy of colliding galaxies due to tidal effects are used to determine the magnitudes V_{cap} and V_{dis} of the maximum relative velocities at infinite separation required for tidal capture and tidal disruption respectively. A double galaxy will be formed by tidal capture without tidal disruption of a component if $V_{cap} > V_i$ and $V_{cap} > V_{dis}$ where V_i is the initial relative speed of the two galaxies at infinite separation. If the two galaxies are of the same dimension, formation of double galaxies by tidal capture is possible in a close collision either if the two galaxies do not differ much in mass and density distribution or if the more massive galaxy is less centrally concentrated than the other. If we assume, as statistics suggest, that the mass of a galaxy is proportional to the square of its radius, it follows that the probability of the formation of double galaxies by tidal capture increases with the increase in mass of the galaxies and tidal disruption does not occur in a single collision for any distance of closest approach of the two galaxies.

We know from classical dynamics that two stars approaching each other from a distance cannot form a binary system without the aid of a third perturbing body. An encounter between two galaxies is, however, an inelastic collision and may lead to the formation of a double galaxy under certain circumstances.

Consider a galaxy of mass M_1 approaching a galaxy of mass M_2 from a great distance. Let U be the total internal energy of a galaxy (i.e. the sum of the potential energy of the galaxy and the kinetic energy of the stars constituting the galaxy) and let E be the total external energy of the galaxies (i.e. the sum of the potential energy due to the mutual gravitational attraction of the two galaxies and the kinetic energy due to the orbital motion of the two galaxies). The total energy,

$$\mathscr{E} = E + U_1 + U_2 \tag{1}$$

is conserved.

The tidal effects of the encounter result in increase in the velocities of the stars in the two galaxies, so that the total internal energy, $U_1 + U_2$ of the galaxies increases at the expense of the external energy E. Let ΔU_1 and ΔU_2 be the increments in the internal energies of the two galaxies due to the encounter, then,

$$E_i = E_f + \Delta U_1 + \Delta U_2, \tag{2}$$

where E_i and E_f are the initial and final values of the external energy.

Following Rood (1965) we define the capture velocity, V_{cap}, as the initial relative velocity at infinite separation for which $E_f = 0$, so that,

$$\frac{1}{2} \frac{M_1 M_2}{M_1 + M_2} V_{cap}^2 = \Delta U_1 + \Delta U_2. \tag{3}$$

We have obtained ΔU_1 and ΔU_2 at various distances of closest approach of the two

Hayli (ed.), Dynamics of Stellar Systems, 167–175. All Rights Reserved.

galaxies by selecting 48 stars as representative of the test galaxy and calculating at each instant the tidal force exerted on them by the field galaxy. The tidal force, integrated over all time, gives the change in velocity of the representative stars during the collision, from which the change in the internal energy of the entire galaxy during the collision is inferred.

The following simplifying assumptions have been made in the determination of ΔU:

(i) The galaxies are spherically symmetric configurations whose density distribution does not change during the encounter. With this assumption it is convenient to represent the density distribution in galaxies by that of a polytrope and to derive the relevant forces from the theory of polytropes as explained in Alladin (1965), Sastry and Alladin (1970), and Potdar and Ballabh (1974). Rood (1965) indicated that the polytrope of index $n=4$ closely represents the density distribution of a typical globular galaxy. Unless otherwise stated this model of density distribution is adopted for the galaxies.

(ii) The relative motion of the two galaxies is considered as uniform rectilinear motion with the distance, p, and velocity, V_p, at closest approach, same as those for the actual relative orbit of the galaxies. Since the tidal forces due to the collisions are of impulsive nature and much of the gain in kinetic energy of the galaxies occurs when the galaxies are close to each other, this assumption does not lead to much loss of accuracy.

(iii) The motion of the stars in the galaxies is neglected in comparison with the orbital motion of the galaxies during the encounter. The adequacy of this impulsive approximation is discussed in Sastry and Alladin (1970).

With these assumptions and the use of the theory of polytropes, we get:

$$\Delta U_2 = \tfrac{1}{2} \frac{G^2 M_1^2 M_2}{V_p^2 R_2^2} \, J(R_1/R_2, \, p/R_2, \, n_1, \, n_2), \tag{4}$$

where G is the gravitational constant, R_1 and R_2 are the radii of the two galaxies represented as polytropes of indices n_1 and n_2. J has been obtained numerically as a function of the distance of closest approach, p, for various values of n_1 and n_2 for $R_1 = R_2$ by Sastry (1973) and for various values of R_1/R_2 for $n_1 = n_2 = 4$ by Potdar (1974).

If the two galaxies do not interpenetrate each other during the encounter J varies as $(R_2/p)^4$ and Equation (4) reduces to the same form as that obtained by Spitzer (1958) for the change in the internal energy of a galactic cluster due to the tidal effects of a passing interstellar cloud.

Under certain circumstances the tidal forces may be so strong that one of the two galaxies may be disrupted during the collision. The condition $\Delta U/|U| = 1$ is a convenient order of magnitude discriminant of the disruptive effects of the tidal forces. We shall regard a galaxy to be disrupted if $\Delta U/|U| \geqslant 1$. It should be noted, however, that $\Delta U/|U| = 1$ does not necessarily imply total disruption of the galaxy; a significant fraction of its mass may still remain bound in this case. We define the disruption velocity, V_{dis}, as that relative velocity at infinite separation before the collision for which $\varDelta U$ for one of the galaxies is of the order of its initial internal energy $|U|$. $|U|$ may be conveniently obtained by Virial theorem from the potential energy of a polytrope.

In order that a double galaxy may be formed by tidal capture without one of the galaxies getting disrupted we require that,

$$V_{cap} > V_i, \qquad V_{cap} > V_{dis}, \tag{5}$$

where V_i is the initial relative velocity of the galaxies at infinite separation.

To obtain velocities at infinite separation from the corresponding velocities at closest approach, we make use of the following relation which expresses the conservation of energy:

$$E_i = E_p + (\varDelta U_1)_p + (\varDelta U_2)_p \tag{6}$$

The subscript p denotes the value of the quantity at the distance of closest approach. In writing (6) we have assumed that the tidal effects of the collision are symmetrical about the distance of closest approach so that $(\varDelta U_1)_p + (\varDelta U_2)_p = \frac{1}{2}(\varDelta U_1 + \varDelta U_2)$. Strictly speaking $(\varDelta U_1)_p + (\varDelta U_2)_p < \frac{1}{2}(\varDelta U_1 + \varDelta U_2)$ since the relative velocity of the galaxies is smaller in the second half of the encounter.

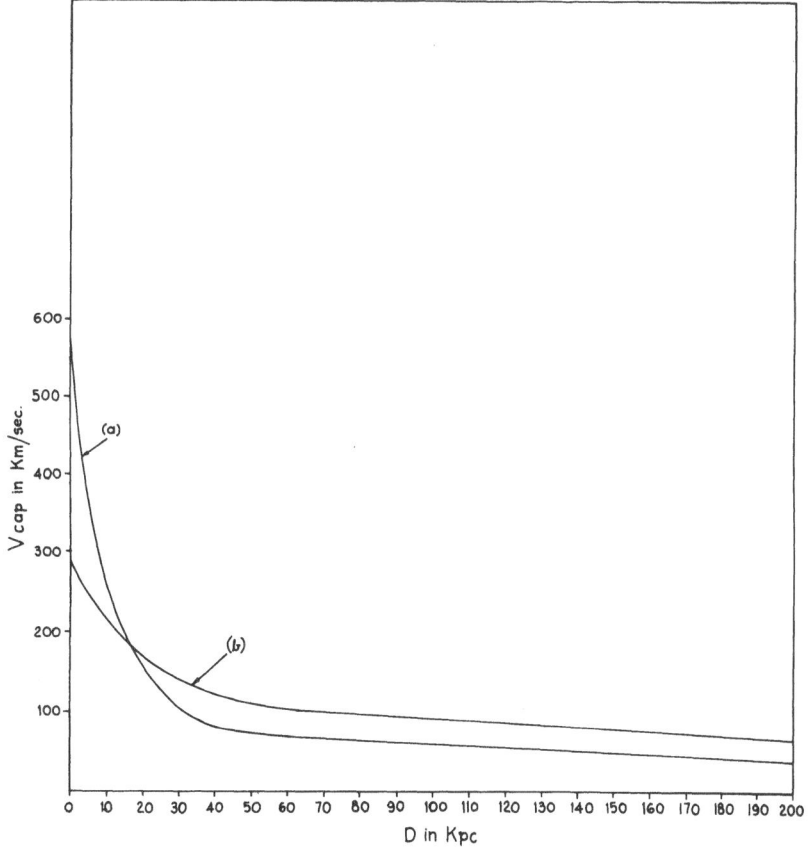

Fig. 1.   The run of V_{cap} with D for cases (a) and (b).

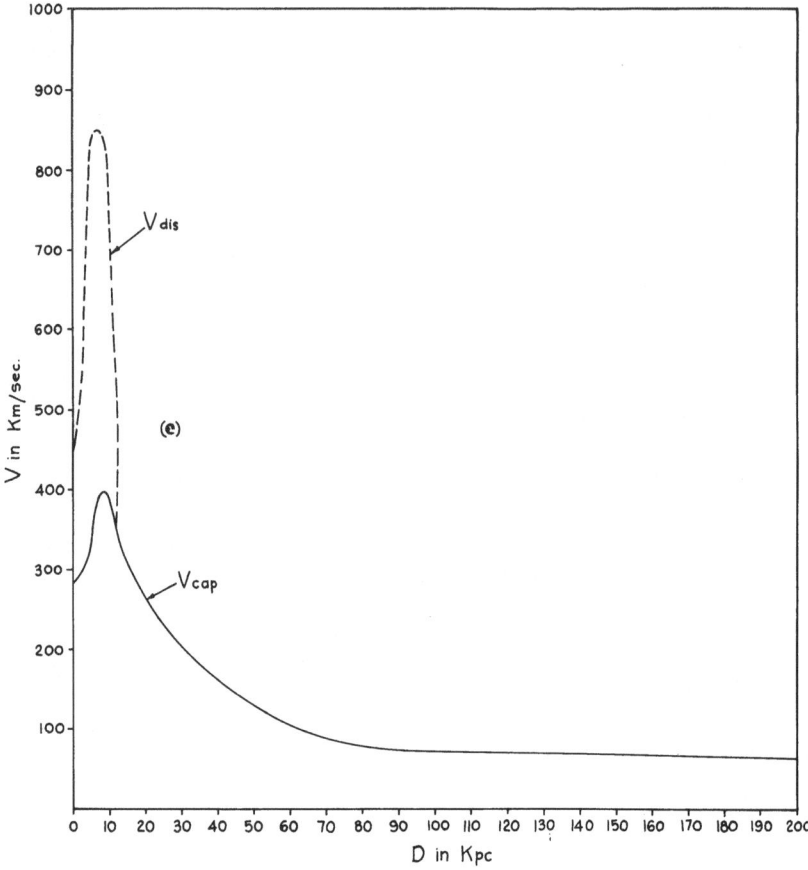

Fig. 2.   The run of V_{cap} and V_{dis} with D for case (c).

TABLE I

Impact parameters D_c and D_d in kpc corresponding to V_{cap} and V_{dis}
as a function of p (kpc)

Case	p	0	2	6	10	20
(a)	D_c	0	4.5	36	60	250
(b)	D_c	0	3.9	13	27	60
(c)	D_c	0	3.6	8.3	15	80
	D_d	0	2.7	6.8	12	–
(d)	D_c	0	4.4	21	57	250
	D_d	0	2.2	–	–	–
(e)	D_c	0	3.7	12	26	60
	D_d	0	2.1	6.4	11	30
(f)	D_c	0	4.9	11	20	110
(g)	D_c	0	19	130	H [a]	H
(h)	D_c	0	26	H	H	H

[a] H denotes values higher than 300.

The impact parameter, D, has been obtained by assuming that the stars in the galaxies have spherically symmetric distribution of velocities so that from the law of conservation of angular momentum we have:

$$DV_i = pV_p. \tag{7}$$

The capture velocities are given as a function of the impact parameter for encounters between galaxies of same dimension in Figures 1 to 5. The disruption velocities are also indicated whenever $V_{dis} > V_{cap}$. The values of the impact parameter corresponding to various values of p for capture and disruption cases are given in Table I. D_c and D_d denote the values of D for collisions with velocities V_{cap} and V_{dis} respectively.

The following cases are considered:

(a) $R_1 = R_2 = 10$ kpc, $M_1 = M_2 = 10^{11} M_\odot$, $n_1 = n_2 = 4$
(b) $R_1 = R_2 = 10$ kpc, $M_1 = M_2 = 10^{11} M_\odot$, $n_1 = n_2 = 0$
(c) $R_1 = R_2 = 10$ kpc, $M_1 = M_2 = 10^{11} M_\odot$, $n_1 = 4,$ $n_2 = 0$
(d) $R_1 = R_2 = 10$ kpc, $M_1 = 10^{12} M_\odot$, $M_2 = 10^{11} M_\odot$, $n_1 = n_2 = 4$
(e) $R_1 = R_2 = 10$ kpc, $M_1 = 10^{12} M_\odot$, $M_2 = 10^{11} M_\odot$, $n_1 = n_2 = 0$
(f) $R_1 = R_2 = 10$ kpc, $M_1 = 10^{12} M_\odot$, $M_2 = 10^{11} M_\odot$, $n_1 = 0, n_2 = 4$.

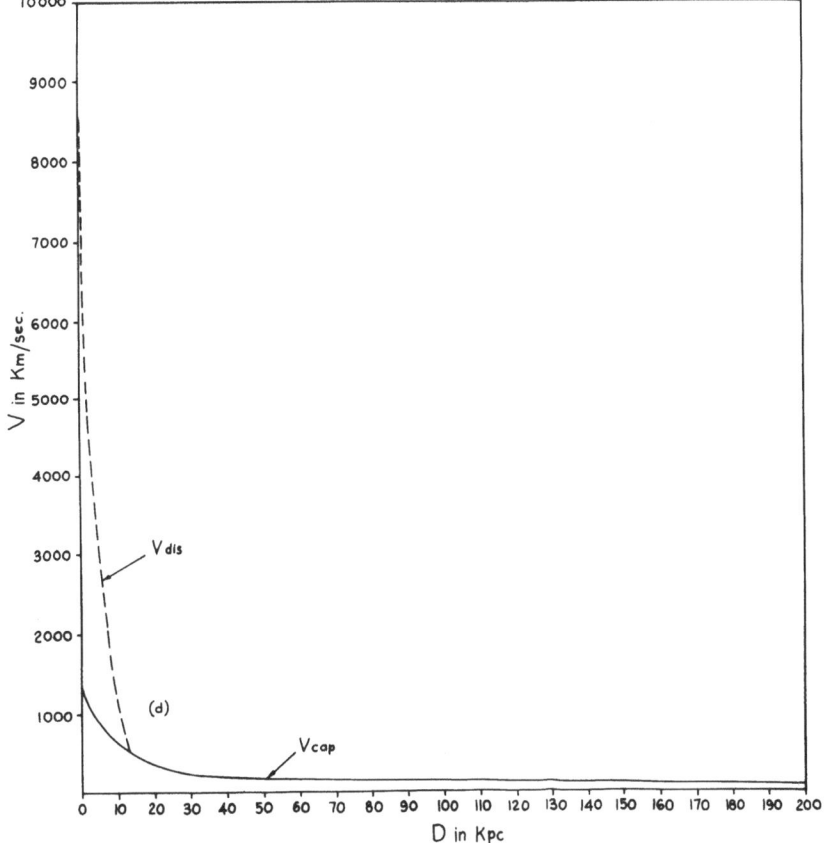

Fig. 3. The run of V_{cap} and V_{dis} with D for case (d).

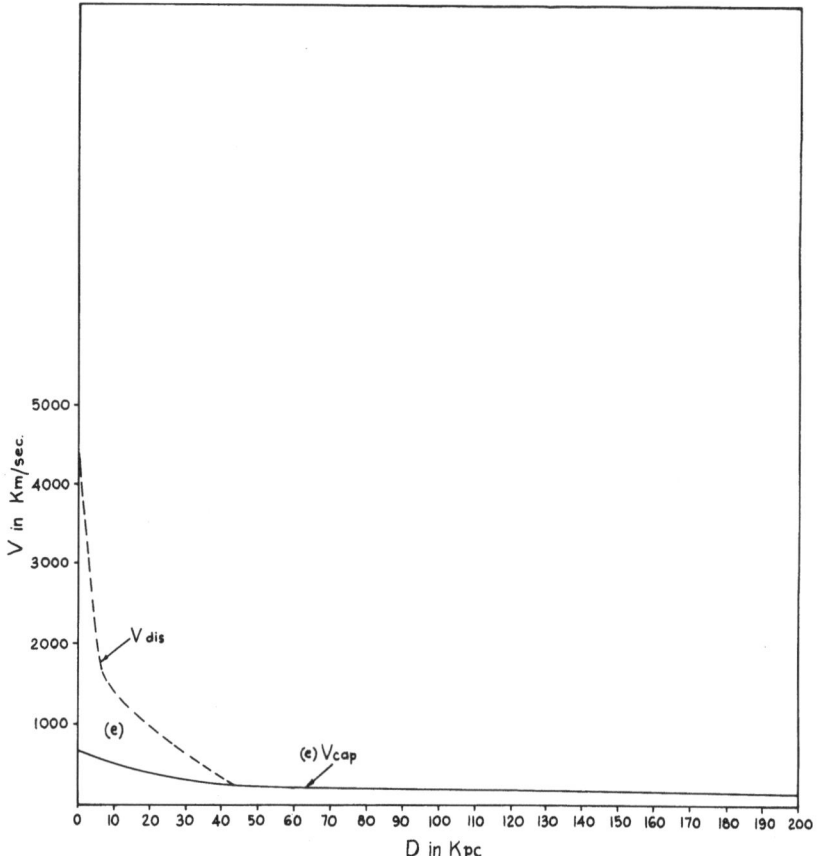

Fig. 4.   The run of V_{cap} and V_{dis} with D for case (e).

We conclude from the figures that formation of double galaxies by tidal capture without disruption of one of the galaxies is possible in a close collision between two galaxies of same dimension either if the two galaxies do not differ much in mass and density distribution or if the more massive galaxy is less centrally concentrated than the other. In cases (c), (d), and (e), M_2 will disrupt in close collisions for $V_i < V_{dis}$. Disruption does not generally occur in distant collisions.

Cox and Toomre have studied the case of a head-on collision between two equal and approximately spherical galaxies presumed to have started from rest at infinity. Their results indicate that the tidal friction is so considerable that at least 80% of the total mass of the two galaxies soon tumbles into a single heap. (Toomre, 1974).

In order to study the effects of change in the dimensions of the galaxies on V_{cap} we shall make use of the empirical statistical relationship between the potential energy and the mass of an elliptical galaxy obtained by Fish (1964):

$$|\Omega| = 9.6 \times 10^{-8} \, M^{3/2} \qquad \text{(cgs units)} \qquad (8)$$

Comparing this with the potential energy of a polytrope of index $n = 4$ (Chandrasekhar,

1939), we have,

$$M = 0.23\,R^2 \qquad \text{(cgs units)} \qquad (9)$$

We shall assume this relationship approximately and consider a giant galaxy to be of mass $10^{11}\,M_\odot$ and radius 10 kpc, and a dwarf galaxy to be of mass $10^9\,M_\odot$ and radius 1 kpc.

The curves (a), (g), and (h) in Figure 6 give the run of V_{cap} with D for giant-giant, giant-dwarf and dwarf-dwarf collisions respectively. The results indicate that the probability of the formation of double galaxies by tidal capture is greatest for giant-giant collisions and least for dwarf-dwarf collisions. Tidal disruption of a galaxy does not occur in a single collision between two typical globular galaxies differing in masses and dimensions in accordance with the adopted mass-radius relation. A galaxy may however disrupt if it undergoes more than one collision.

Following Rood's analysis (1965) closely for the determination of the number of captures in a cluster of galaxies from the values of V_{cap} we find that, in a compact spherical cluster of galaxies of radius 1 Mpc, containing one thousand giant galaxies, having an isotropic Maxwellian distribution of velocities with velocity dispersion of

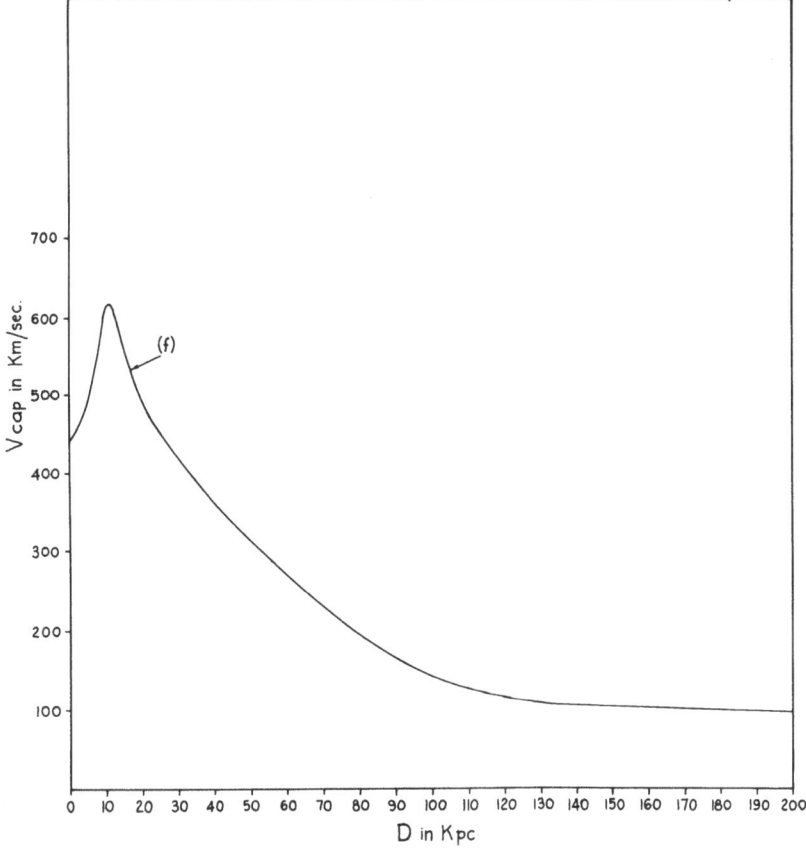

Fig. 5.   The run of V_{cap} with D for case (f).

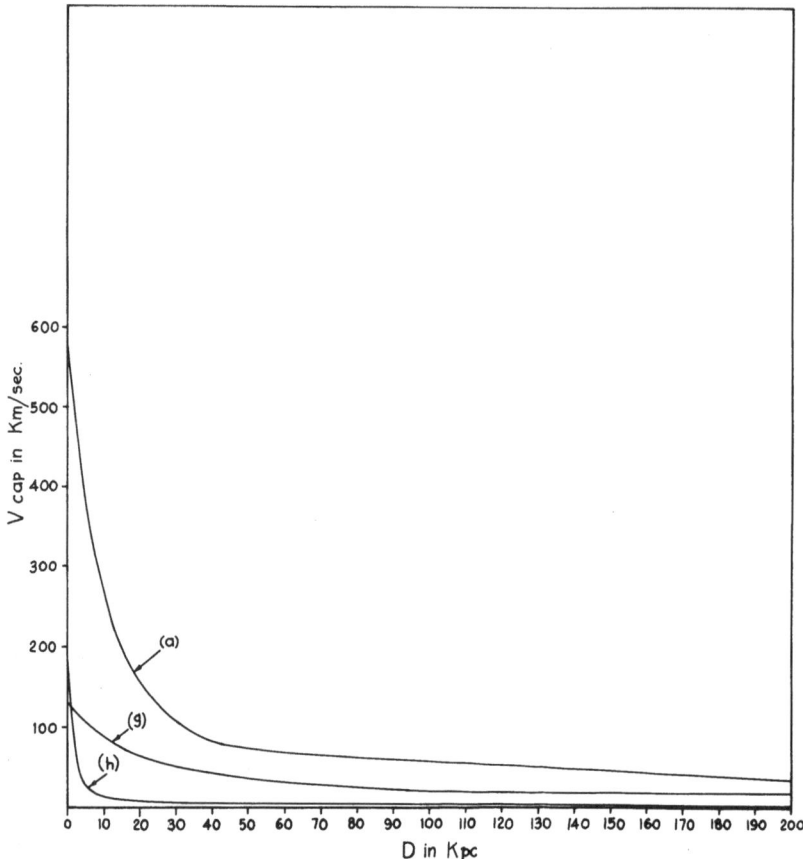

. Fig. 6.   The run of V_{cap} with D for (a) giant-giant collisions (g) giant-dwarf collisions, and (h) dwarf-dwarf collisions.

500 km s^{-1} (this is the velocity required by the Virial theorem for stability), there will be about 2000 interpenetrating collisions in 10^{10} yr and 20 of these will result in the formation of double galaxies by tidal capture. The Coma cluster is somewhat less compact than the model of the cluster considered and the velocity dispersion of the galaxies in the Coma cluster is about 1800 km s^{-1}. Hence in the case of the Coma cluster the number of tidal captures is less by two orders of magnitude.

References

Alladin, S. M.: 1965, *Astrophys. J.* **141**, 768.
Chandrasekhar, S.: 1939, *An Introduction to the Study of Stellar Structure*, University of Chicago Press, Chicago, Ch. 4.
Fish, R. A.: 1964, *Astrophys. J.* **139**, 284.
Potdar, A. and Ballabh, G. M.: 1974, *Astrophys. Space Sci.* **26**, 353.
Potdar, A.: 1974, 'Tidal Force Effects in Clusters of Galaxies', Osmania University (Ph.D. Thesis).
Rood, H. J.: 1965, 'The Dynamics of the Coma Cluster of Galaxies', University of Michigan (Ph.D. Thesis).
Sastry, K. S. and Alladin, S. M.: 1970, *Astrophys. Space Sci.* **7**, 261.

Sastry, K. S.: 1973, 'Change in the Gravitational Energy of Galaxies due to Collisions and its Consequences', Osmania University (Ph.D. Thesis).
Spitzer, L.: 1958, *Astrophys. J.* **127**, 17.
Toomre, A.: 1974, in J. R. Shakeshaft (ed.), 'The Formation and Dynamics of Galaxies', *IAU Symp.* **58**,
347.

DISCUSSION

Biermann: Don't your results let us understand very nicely why many clusters leave the two most massive galaxies form a binary? Have the two massive galaxies a high probability of combining? Isn't there many cluster galaxies where there are two very massive ones?

Alladin: The two massive galaxies do have a high probability of combining into a double system. This could be the reason why in clusters the two most massive galaxies form a binary.

Gott: The central binary pair of massive galaxies may be more probably formed via two body relaxation effects.

King: Will your captured pairs not continue to interact and spiral into each other?

Alladin: The double galaxies formed by tidal capture will revolve around each other with mean separations that will continually decrease in time. Ultimately the components of a double galaxy will form a single stellar system. I would guess that it would generally take time equal to two or three periods of revolution of the galaxies for a single system to form. Recent work by Cox and Toomre shows that the tidal friction in a slow head-on collision of galaxies is enormous.

Wielen: How is the loss of 'external' energy distributed within each galaxy? Does it somehow resemble the distribution of the gain in internal energy, which increases strongly with the distance of a star from the centre of its galaxy, or is it essentially constant over the whole galaxy?

Alladin: Due to the loss of external energy the relative motion of the centres of mass of the two galaxies will be retarded. It is conceivable that different parts of the two galaxies may be retarded at different rates. I think much computational labour would be needed to estimate this differential effect.

Innanen: Is it possible to computer simulate self-consistently the motion of two groups of say, 200 particles which have an encounter?

Aarseth: This problem is well within the reach of N-body calculations. Some simple calculations of colliding clusters with small particle numbers have already been made (Aarseth, S. J. and Hills, J. G.: 1972, *Astron. and Astrophys.* **21**, 255).

Miller: Some years ago, in the early days of the big N-body calculation that gave spiral patterns we ran a problem in which two clumps of stars were allowed to fall together. The two were allowed to interpenetrate. After the collision there were two clumps moving apart, with a third clump at rest between the two, and some bridges connecting the two. The motion was not followed beyond this stage.

Van Albada: It should be kept in mind that the problem of penetrating galaxies is a collisionless one: what matters is the star-galaxy interaction. N-body calculations with $N \sim 200$ would suffer from star-star interaction.

Heggie: A student of Professor Hockney's has recently carried out computations, in three dimensions with some tens of thousands of stars, of the interaction between two galaxies. These are self-consistent calculations.

PART II

FLATTENED SYSTEMS

CROSSFERTILIZATION BETWEEN PLASMA, STELLAR DYNAMICS AND HYDRODYNAMICS

M. R. FEIX

Groupe de Recherches Ionosphériques, Université d'Orléans, Orléans, France

We present results on four different mediums characterised by their 'density conservation' in a two dimensional space (phase space for unidimensional plasma and self gravitating systems, configuration space for two dimensional Navier Stokes fluid and guiding center rod plasma).

We review the different problems (purely collective versus individual) and introduce the distinction between Lagrangian schemes (superparticles) and Eulerian schemes.

The steady states structures for these fluids and their open (plasma, hydrodynamic) versus closed (stellar system, accelerator beam, hydrodynamic) topologies are discussed. We then turn to their stability properties and show the condition for the presence of marginal adiabatic mode. The 'double water bag' system is fully studied and interesting analogies for completely trapped systems (accelerator beam and self gravitating systems) are pointed out.

1. Introduction

From the experimenter's point of view mediums like galaxies, plasmas and regular fluids look completely different. For the theoretician the problems look much more similar and the computational physicist can't resist to see what happens if he changes the sign of the density in Poisson law! The purpose of this paper is to compare the properties of different mediums, the mathematical structures of the problems and the different methods used to solve them. It is hoped that astronomers and plasma physicist will look onto one each other problems and solutions and that some crossfertilization will result.

2. The Different Mediums

There is no point here in describing the stellar dynamics systems! Plasma Physics systems are now well known by the astronomers. Let us point out that accelerator beams provide model of one species plasma with all particles trapped and boundaries limits quite similar to the stellar self gravitating systems. Of course there must be a focusing device to balance the space charge repulsion. Plasma and self gravitating systems form what we will call phase space fluids to remind that their evolution should be described in phase space.

Now we consider two new mediums. The first is a regular, incompressible and inviscid fluid described by the Navier Stokes equation. Moreover we restrict our studies to *bidimensional fluids*. We will see why in a moment. Navier Stokes equations can be written in the x, y space

$$\frac{\partial \mathbf{u}}{\partial t}+(\mathbf{u}\cdot\mathbf{V})\,\mathbf{u}+\frac{1}{\varrho}\,\mathbf{V}p=0$$

$$\mathbf{V}\cdot\mathbf{u}=0,\tag{1}$$

where ϱ is the constant density. We introduce the vorticity $\boldsymbol{\omega}=\mathbf{V}\times\mathbf{u}$. Due to the x, y space dependence of the problem $\boldsymbol{\omega}$ has only the third (along z) component different from zero. We call ξ that component. Taking the curl of (1) it is easily shown that ξ obeys

$$\frac{\partial \xi}{\partial t}+\mathbf{V}\cdot(\xi\mathbf{u})=\frac{\partial \xi}{\partial t}+\mathbf{u}\cdot\mathbf{V}\xi=0.\tag{2}$$

Since $\mathbf{V}\cdot\mathbf{u}=0$ we can deduce the two components of the velocity from a scalar potential $\varPhi$ with

$$\mathbf{u}=\frac{\partial \varPhi}{\partial y}\,\mathbf{e}_x-\frac{\partial \varPhi}{\partial x}\,\mathbf{e}_y,\tag{3}$$

where $\mathbf{e}_x$, $\mathbf{e}_y$, $\mathbf{e}_z$ are the unit vectors along ox, oy, oz. Taking into account $\xi e_z=\mathbf{V}\times\mathbf{u}$ and Equation (3) we get

$$\varDelta\varPhi+\xi=0.\tag{4}$$

Equations (2) and (4) are now the new 'model equation' for hydrodynamics. Equation (2) indicates that ξ is a constant along the trajectory. This trajectory can be deduced from a fictious 'Electric field' $\mathbf{E}$ with $E_x=-\partial\phi/\partial x$ and $E_y=-\partial\phi/\partial y$ where $\varPhi$ satisfies the usual Poisson equation. The only difference is that the dynamics of the vortex elements is given by $U_x=-E_y$ and $U_y=E_x$ instead of the usual $\dot{\mathbf{U}}=(e/m)\,\mathbf{E}$. These properties have been extensively used by Zabusky and Deem (1971) and by Christiansen and Zabusky (1973) to compute the nonlinear evolution of bidimensional fluids and formation of 'Vortex Streets'. In these calculations the Lagrangian point of view was used consisting in representing the vortex density ξ by many points carrying each an elementary 'charge of vorticity' (eventually with a negative sign). One of the advantage of such a method is the possibility of using fast 'Poisson solver' algorithm developped for plasma and stellar dynamics studies.

Another interesting bidimensional medium is a plasma immersed in a very strong magnetic field, homogeneous along the direction of this field. The transverse motion is studied. In that case the velocity of a particle is given by the so called 'drift approximation' with

$$\mathbf{u}=\frac{\mathbf{E}\times\mathbf{B}}{B^2}\tag{5}$$

which can be written $U_x=E_y/B$, $U_y=-E_x/B$. Except for the trivial factor $(-1/B)$ these are exactly the motion equation of the vortex elements considered above. On the other hand the continuity equation can be written

$$\frac{\partial n}{\partial t} + \nabla \cdot n\mathbf{u} = 0. \tag{6}$$

From the fact that $E = -\nabla\Phi$ and taking into account (5) we deduce that $\nabla \cdot \mathbf{u} = 0$ and consequently

$$\frac{\partial n}{\partial t} + \mathbf{u} \cdot \nabla n = 0 \tag{7}$$

which is the equivalent of (2) with a conservation of the density along the trajectories in the x, y space. Of course Φ satisfies the Poisson equation

$$\Delta\Phi + en = 0 \tag{8}$$

in strict identity with the vorticity Poisson equation (4).

The fundamental property of these 4 mediums (one dimensional plasma, one dimensional stellar system, bidimensional fluid and plasma) is that in the two dimensional space describing the evolution we have conservation of the density $\{\xi(x, y) \, n(x, y)$ or $f(x, v)\}$. The other fundamental property is that the dynamics is obtained through a potential Φ obeying Poisson law. Both Eulerian or Lagrangian schemes can be used. By Langrangian scheme we mean representing the density by many 'elementary particles' each carrying a constant elementary 'charge' and 'mass'. We can also use Eulerian scheme. Although we will come back to the duality of these two schemes later on we point out a general discussion of the different models in Feix (1971).

3. Problems

3.1. COLLISIONLESS VERSUS COLLISIONAL REGIME

The first level of description for plasmas and stellar dynamics systems is obviously the Boltzmann 'collisionless' equation (called Vlasov by the plasma physicists). This equation describes the 'violent relaxation' of stellar systems. In plasma also it describes the fast evolution (especially the different instabilities) and more studies are needed in its nonlinear solution. The final state could be the one predicted on statistical argument by Lynden Bell although computer studies have shown that the mixing of the phase space elements is not always complete and that bunch of particles do not obey the Lynden Bell statistics.

The next level of description i.e. the slow relaxation to thermodynamics equilibrium involving a small Fokker Planck term is treated by astronomers and plasma physicists with different philosophies. The plasma physicists consider mostly homogeneous plasma relaxing toward maxwellian distribution. The astronomers are specially interested in the evolution of the globular clusters; starting from a steady state solution of the Vlasov equation they let it change slowly through collisions.

Let us point out that this last problem raises very difficult questions, one of them being the possibility of time scale separation. Besides the question of the exact expression for a Fokker Planck term in an inhomogeneous medium (solved only for homogeneous neutral plasma through the Lenard Balescu operator) the solution of the problem implies an equation of the form

$$\frac{\partial f}{\partial t} + \mathbb{V}f = \varepsilon \mathbb{F}f, \tag{9}$$

where $\mathbb{V}$ and $\mathbb{F}$ are respectively the Vlasov and Fokker Planck operator (ε in front of $\mathbb{F}$ indicates that this operator is in principle very small). In the neutral homogeneous plasma the term $\mathbb{V}(f)$ is initially zero and remains zero although f is slowly changing through the Fokker Planck term. This is of course not true for the globular cluster and multiple time scale technique should be used to solve (9). See Cole (1968).

Of course the direct integration of the N-body dynamical equation takes care of all these problems in principle. Nevertheless the problem of 'graininess' is still there. In all calculations the graininess effects are very much enhanced with respect to reality since the physical systems contain much more particles than the number which can be treated on the computer. If we are not interested in these 'graininess effects' we can try to decrease them. An interesting solution was introduced in plasma physics by Dawson and Hsi (1968) with the concept of 'cloud particles' i.e. of rigid structures of finite dimension larger than the interparticle distance but smaller than the wavelength of interest. On the other hand we can be interested in these graininess effects and we must know how they scale with N. This is a difficult question especially if we are interested in apparition of complex structures as the binaries stars where obviously triple correlation must play a role to allow the transition from open trajectories to close trajectories. These questions can be solved by careful comparison between computer simulation with different N and a better understanding of the hierarchy of time scale for different effects – a very difficult theoretical problem.

Plasma physicists are actually very interested in 'graininess' effects since some astrophysical plasma and some of the laser created fusion plasma have a very high density. For some of these plasma nD^3 the number of electrons in the Debye cube can be of the order of unity or smaller. Although this problem has no equivalent in the self gravitating N-body problem it is worth mentioning the analytical and computer work on this question of one dimensional ('plane stars' with an uniform mass density) systems. The great advantage is the existence of analytical theory for canonical ensemble both for plasma (Lenard, 1961) and self gravitating system (Ribicky, 1972) and the possibility of performing exact computation (Feix, 1969) in the $2N$ dimensional phase space for reasonably large time. It is the opinion of the author that such computations already initiated by Feix and Hohl (1968) will be extremely useful for a better understanding of foundations of statistical physics and its application to gravitational system (presence of isolating integral of motion, time average vs. ensemble average, difference for small N between microcanonical and canonical ensemble, etc. ...).

3.2. COLLISIONLESS EULERIAN MODELS

After this lengthy discussion of 'graininess' effects (but this is a central issue) we must mention that plasma physicists have developped models where they work directly with the Vlasov equation (representing a phase space fluid) although of course the Lagrangian superparticle model is still used to solve that equation with attempt to decrease the finite graininess effect through 'cloud particle' or filtering of high wave-number structure of the electric field.

Eulerian model are interesting if we want to study the stability properties of steady state equilibria solution of the Vlasov equation. Up to now most of the results deals with one dimensional plane geometry, a model obviously more realistic and more popular in plasma physics than in stellar dynamics. Among some of the advantages of the Eulerian method is the fact that it is easier to make comparison with analytical theory (by turning on or off different terms which are neglected in the theoretical treatment). A discussion of the Eulerian model can be found in Chapters 2 and 3 of Alder and Fernbach (1970). Among Eulerian models a special mention should be made of the Multiple Water Bag. In Alder and Fernbach (1970), p. 88 Water Bag model is described. In this paper the philosophy is that we study a phase space fluid described by two levels of the possible phase space density. These two levels phase space density cannot of course describe the complexity of all possible velocity distribution but are sufficient te describe some of the strong nonlinear effects of two counter-streaming plasma. While Alder and Fernbach (1970) deals mainly with strong turbulence effects Doremus and Feix (1972) and Finzi *et al.* (1974) respectively for a self gravitating gas and a plasma study the possible steady states and their stability properties for one dimensional plane geometry.

Another point of view is introduced when we let the number of bags become large. Then we deal with a discrete element scheme for the Vlasov equation. Navet and Bertrand (1971) have shown that for finite time we recover all the properties of a continuous velocity distribution provided we take a sufficient number of bags. More-over the time of validity increases linearly with this number. Very useful numerical properties are associated with this model, especially in the computation of properties of antennas immersed in an homogeneous plasma (a problem with no counterpart in stellar dynamics).

Indeed the Multiple Water Bag model is an extremely organised one where all the poles can be easily located and followed as a function of the different parameters. Details for grids and antennas problems can be found in Noyer *et al.* (1974).

4. Results on Stability of Inhomogeneous Plasma Self Gravitating Gases and Vortex Equilibria

4.1. STEADY STATES SOLUTION FOR PHASE SPACE FLUIDS

We sketch briefly results given in more details in Doremus and Feix (1972) and Finzi *et al.* (1974) and present the new analogy with the vortex problem.

Considering plane one dimensional geometry we know that the phase space density for a steady state is a function of the total energy E

$$f = f(E) = F(\tfrac{1}{2}mV^2 + \Phi) = F(E).$$

We will restrict our investigations to steady state of the double water bag type. Figure 1 precises the notation, it gives the velocity profile at point $x=0$ (we will suppose a symmetry around that point). We consequently have $V_1^+ = -V_1^- = V_1$

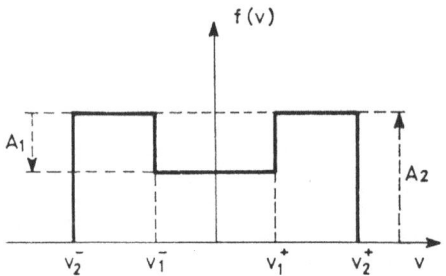

Fig. 1.   The double 'Water Bag' distribution function with negative phase density of the inner bag.

$V_2^+ = -V_2^- = V_2$ and define

$$\beta = V_1^2/V_2^2, \qquad \alpha = -A_1/A_2. \tag{10}$$

Moreover at $x=0$ the potential energy is supposed to be minimum. For a self gravitating gas α and β are the two parameters characterising the problem. We do not consider the case $A_1 > 0$ which is just a peculiar case of a decreasing $F(E)$ a case known to be always stable (Kulsrud and Mark, 1970; Feix et al., 1971). Consequently the parameter space is the square $0 < \alpha, \beta < 1$.

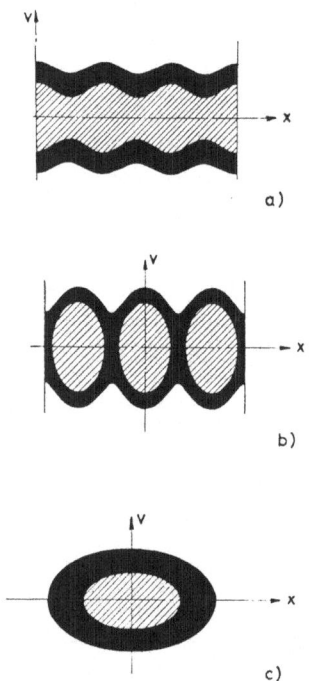

Fig. 2. Plasma.   Phase space representation of the three possible types of steady states: (a) 'type 1' solutions (b) 'type 2' solutions, (c) 'type 3' solutions.

Besides α and β a third parameter γ is needed to characterise the plasma, γ is the ratio of ion density to electron density at point $x=0$. It can be shown that γ must be bigger than 1. In this problem ions are considered as motionless and providing a fixed, homogeneous background. This is a less academic case than it looks at first sight since it corresponds to an accelerator beam with one species of repulsing particles with a focusing force linear in x.

Figure 2 gives the three possibles type of steady states. Type 1 and 2 solutions are possible for a plasma only; type 3 for a plasma and a stellar gas. The existence of periodic nonlinear steady states is a characteristic of the plasma.

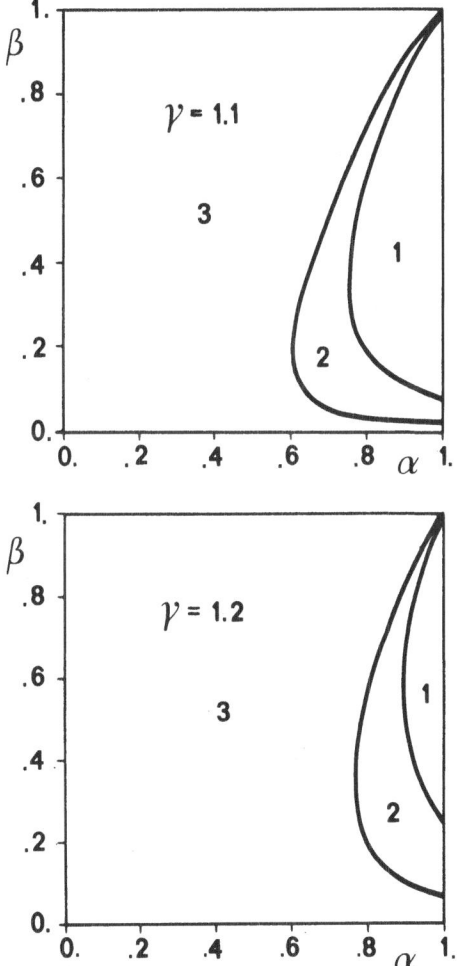

Fig. 3.   Plasma.The (α, β) parameter space for $\gamma=1.1$ and $\gamma=1.2$. Locations of type 1, 2 and 3 solutions.

Consequently any point of the square $0<\alpha, \beta<1$ is a possible steady state of type 3 for a self gravitating system. For a plasma we must consider the α, β square for a given γ. Figure 3 gives for $\gamma=1.1$ and $\gamma=1.2$ the limits of the three types solutions; when γ

increases it can be shown that region 1 and 2 tend to disappear simultaneously through the corner $\alpha = \beta = 1$ and they do not exist any more for $\gamma > 2$.

4.2. STABILITY OF THE DIFFERENT STEADY STATES FOR PLASMA AND SELF GRAVITATING SYSTEM

4.2.1. *Marginal Adiabatic Mode Theory*

Let us consider a steady state characterised by values of the parameters α, β, γ. The question is to know if it is stable. If we consider type 1 solution we immediately recognize that we are in a case where two streams instability are likely to take place. From our knowledge of the homogeneous plasma case we guess that the instability will disappear through the point $\omega = 0$. (ω being imaginary and changing sign if we vary one of the parameter around its critical value). Since, close to the critical value, both the real and imaginary part of ω are small (for two streams the real part of ω is strictly zero) we call such a mode a marginal adiabatic (i.e. slowly varying) mode. Of course other modes can cross the stability – instability border in the ω plane around a finite value of ω. Such modes called overstables must be also studied.

The theory of the marginal adiabatic mode has been given by Bertrand *et al.* (1972) and a detailed discussion is given in Doremus (1973) and Finzi (1973). If A_j and $V_j(x)$ are the characteristic of bag j the stability of the marginal adiabatic mode is given by the signs of the eigenvalues of the operator

$$\left\{ k^2(x) - \frac{d^2}{dx^2} \right\} \Psi = \lambda \Psi. \tag{11}$$

For a plasma with

$$k^2(x) = (e^2/m\varepsilon_0) \sum_{j=1}^{N} 2A_j/V_j(x) \tag{12}$$

if all λ are positive no marginal adiabatic mode is possible. If one is negative the system is unstable.

For types 1 and 2 solutions the boundaries conditions are $(d\Psi/dx)_{x = \pm \infty} = 0$.
For type 3 solution the boundaries conditions are

$$(d\Psi/dx)_{x = \pm L} = 0, \tag{13}$$

where $\pm L$ are the space boundaries of the most external (most energetic) contour.

The interesting point is that (11) was recognized as already playing an important role in the solution of the steady state. Sticking to type 3 solution we can show that the steady electric field E_0 satisfies

$$\left\{ k^2(x) - \frac{d^2}{dx^2} \right\} E_0(x) = 0 \tag{14}$$

with $k^2(x)$ given by (12) as in (11). Boundaries conditions are now (N_i ion density)

$$dE_0/dx = -eN_i/\varepsilon_0. \tag{15}$$

The difference of boundaries conditions between (12) and (15) implies that none of the eigenvalues of (11) can be negative. To demonstrate this property we apply the Rayleigh Ritz method.

In such a method we compute

$$R(\varphi) = \frac{\int \varphi A \varphi \, dx}{\int \varphi^2 \, dx}, \tag{16}$$

where φ is a trial function satisfying the same boundaries conditions as Ψ and A is the operator $k^2(x) - d^2/dx^2$. If we can find that for such trial function $\int \varphi A \varphi \, dx$ is always positive no marginal instability mode will appear. We introduce the trial function

$$\varphi = \alpha(x) E_0(x). \tag{17}$$

In (17) $\alpha(x)$ must be always defined. Indeed $E_0(x)$ for type 3 solution never cancels except for $x=0$ where we are free to take $\varphi=0$ (since Ψ is a potential). Consequently $\alpha(x)$ has no discontinuity and we can integrate by part. Introducing (17) in $\int \varphi A \varphi \, dx$ and taking (14) into account we find

$$\int \varphi A \varphi \, dx = \int_{-L}^{L} \left(\frac{d\alpha}{dx}\right)^2 E_0^2 \, dx - \alpha \frac{d\alpha}{dx} E_0^2 \Big]_{-L}^{L} \tag{18}$$

but for $x = \pm L$ we must have $d\varphi/dx = 0$. Differentiating (17) we get

$$\frac{d\alpha}{dx} E_0 + \alpha \frac{dE_0}{dx} = 0 \tag{19}$$

and

$$\int \varphi A \varphi \, dx = \int_{-L}^{L} \left(\frac{d\alpha}{dx}\right)^2 E_0^2 \, dx + \alpha^2 E_0 \frac{dE_0}{dx} \Big]_{-L}^{L}. \tag{20}$$

From the configuration at points $x = \pm L$ (where we find only ions since all the electrons are trapped inside this region) we see that $E_0 \, dE_0(dx]_{x=L}$ is positive and $E_0 \, dE_0/dx]_{x=-L}$ is negative; (20) is positive and no marginal adiabatic mode can be present in type 3 plasma equilibria.

For type 3 self gravitating system it has been shown by Feix et al. (1971) that a marginal adiabatic mode always exists but that this mode is a trivial displacement mode and consequently does not correspond to a possible instability. The complete trapping of electrons or stars in type 3 equilibria precludes the existence of marginal adiabatic instability.

4.2.2. Overstable Modes

Overstability for type 3 equilibria can of course occur. Skipping all details given in
Doremus and Feix (1972), Finzi *et al.* (1974), Doremus (1973), and Finzi (1973) we
just show in the parameter square α, β the zone of stability for type 3 equilibria. It is
quite interesting to notice on Figure 4 (self gravitating system) and Figure 5 (plasma

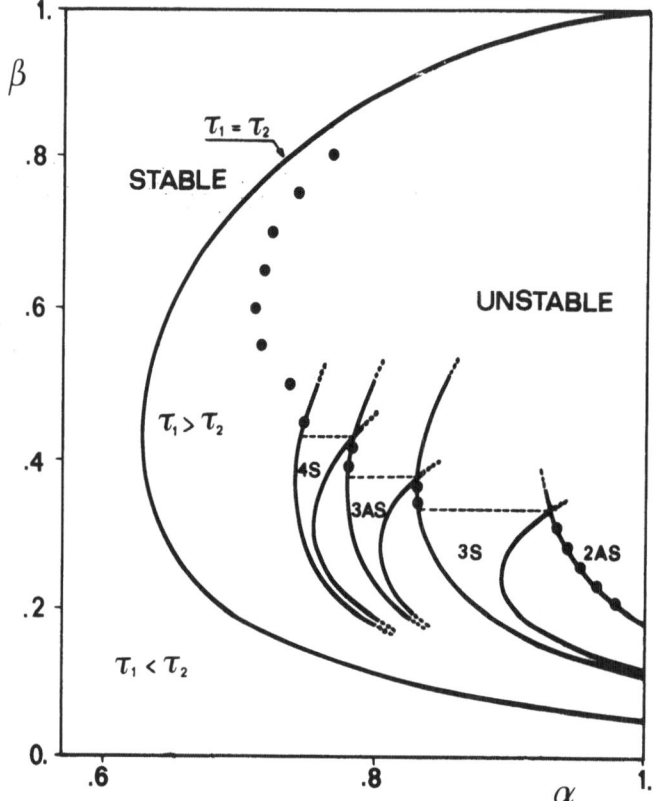

Fig. 4.   Self gravitating systems: regions of instability in the (α, β) parameter space.

or accelerator beam) the similarity of the curves delineating the different unstable
modes. Especially for the double water bag the region situated on the left of the curve
$\tau_1 = \tau_2$ (where τ_1 and τ_2 are the periods of oscillation of particles respectively on the
inner and the outer bag) are stable although such a region corresponds to $\tau_1 < \tau_2$ for
self gravitating system and $\tau_1 > \tau_2$ for plasma. In the lower part of the pictures the
appearance of narrow bands of instability is a consequence of the two bags approximation and would 'disappear' in a completely continuous treatment. How such
zones will evolve with a high bags number is an open question but preliminary results
indicate that the filaments will become thinner and thinner and will in the limit
constitute a nul measure set.

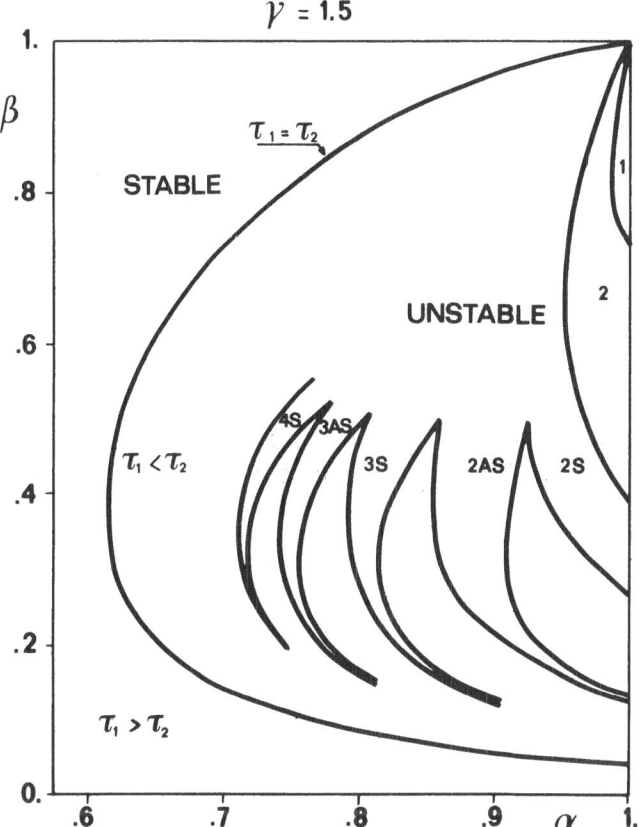

Fig. 5 Plasma.   Regions of instability in the (α β) parameter space. $\gamma = 1.5$.

4.2.3. *Type 1 and 2 Stability*

There is an interesting question since computer experiments (Finzi, 1972; Morse and Nielson, 1969) have shown that an initially homogeneous two stream unstable plasma evolves toward type 2 equilibrium and during sometimes such equilibrium were believed stable. But further computer run have exhibited a slow destruction of such equilibria through a coalescence of neighbouring holes. In this respect computer run involving both longer plasma and longer time are needed. For the two bags distribution we have shown (Feix, 1973) that all type 1 and most of type 2 equilibria are unstable. It is more than likely that all type 2 equilibria are in fact unstable but the considered trial functions of the Rayleigh Ritz methods are probably not general enough to treat all the cases.

4.3. STABILITY OF VORTEX EQUILIBRIA

We consider basically two kinds of equilibria corresponding to the following flows.

First a plane counterstreaming flow with a velocity $U_y(x, y) = 0$ and $U_x(x, y) = U(y)$, $U(y)$ being an odd function in y. This corresponds to a vorticity profile which can be

approximated by N 'bags'

$$\xi(y)=\sum_{j=1}^{N} A_j[H(y+y_j)-H(y-y_j)]. \tag{21}$$

In (21) H is the Heaviside function $H(x)=1$ if $x>0$ and $H(x)=0$ if $x<0$. A linearised analysis is carried on the motion of contours $y_j^+=y_j+\delta_j^+(x,t)$ and $y_j^-=-y_j+\delta_j^-(x,t)$; supposing a time space variation of δ_j^+ and δ_j^- in $\exp i(\omega t-kx)$. We finally end up with an eigenvalues equation of the following type (we give a 3 bags example the generalisation of which is pretty obvious). We introduce the velocity values.

$$\begin{aligned} U_1&=(A_1+A_2+A_3)\,y_1\\ U_2&=A_1y_1+(A_2+A_3)\,y_2\\ U_3&=A_1y_1+A_2y_2+A_3y_3\end{aligned} \tag{22}$$

and suppose that the velocity profile is increasing monotonically from $-U_3$ to U_3, i.e. that all A_j are positive.

$$\begin{array}{cccccc}
1-\frac{2}{A_1}(\omega+ku_1) & [1+1] & [2-1] & [2+1] & [3-1] & [3+1]\\
[1+1] & 1+\frac{2}{A_1}(\omega-ku_1) & [2+1] & [2-1] & [3+1] & [3-1]\\
[2-1] & [2+1] & 1-\frac{2}{A_2}(\omega+ku_2) & [2+2] & [3-2] & [3+2]\\
[2+1] & [2-1] & [2+2] & 1+\frac{2}{A_2}(\omega-ku_2) & [3+2] & [3-2]\\
[3-1] & [3+1] & [3-2] & [3+2] & 1-\frac{2}{A_3}(\omega+ku_3) & [3+3]\\
[3+1] & [3-1] & [3+2] & [3-2] & [3+3] & 1+\frac{2}{A_3}(\omega-
\end{array}$$

where $[i\pm j]$ are short notations to designate the functions $\exp(-k|y_i\pm y_j|)$.

It can be shown that the poles go by pair ω and $-\omega$ and that their motions is described on Figure 6.

For $k\to\infty$ phase velocities are $\pm U_1$, $\pm U_2$ and $\pm U_3$; then all roots remain located between two neighbouring velocities except for the first one which goes through the value 0 for a given $k=k_c$ and becomes imaginary for $k<k_c$ introducing an instability through a marginal 'adiabatic mode' in close analogy with the type 1 and type 2 equilibrium.

It is now interesting to see what happens when we take instead of a plane flow a rotating one corresponding to a cylindrically symmetric equilibrium with a vorticity

distribution ξ (r is the radial coordinate)

$$\xi = \sum_{j=1}^{N} A_j H(r_j - r).$$ (24)

The case $N = 1$ corresponding to the Rankine vortex.

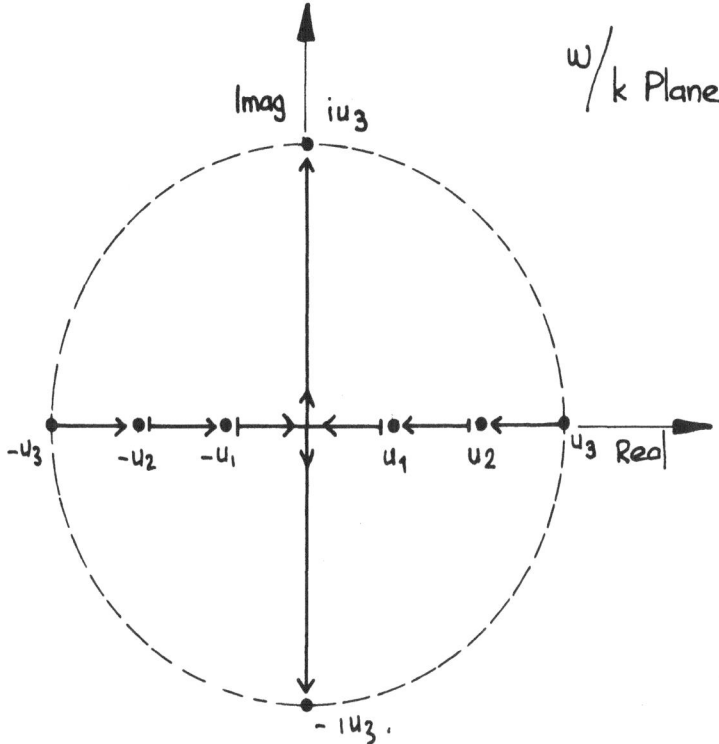

Fig. 6 Vortex Fluid.   Plane geometry. Evolution of the phase velocities in the ω/k plane. Illustration for 3 bags with all $A_j > 0$.

Again we suppose all the $A_j > 0$ and notice the corresponding velocity profile

$$U_i(r) = \sum_{j=1}^{i-1} \frac{A_j}{2} \frac{r_j^2}{r} + \sum_{j=i}^{N} \frac{A_j}{2} r$$ (25)

for $r_{i-1} < r < r_i$.

A linearized motion around the equilibrium is studied with

$$r_i(\theta, t) = r_i + \delta r_i \exp i(\omega t - m\theta)$$ (26)

m taking an integer value. The eigenfrequencies matrix becomes

$$
\begin{array}{ccc}
1 + \dfrac{2}{A_1}(\omega - m\Omega_1) & [1, 2]_m & [1, 3]_m \\[2mm]
[1, 2]_m & 1 + \dfrac{2}{A_2}(\omega - m\Omega_2) & [2, 3]_m \\[2mm]
[1, 3]_m & [2, 3]_m & 1 + \dfrac{2}{A_3}\omega - m\Omega_3),
\end{array}
$$ (27)

where $[i, j]_m$ stands for $(R_i/R_j)^{|m|}$ if $R_i < R_j$ or $(R_j/R_i)^{|m|}$ if $R_i > R_j \cdot \Omega_1, \Omega_2, ..., \Omega_N$ are the angular frequency of the fluid element on contour 1, 2, etc. ... We have

$$2\Omega_1 = A_1 + A_2 + A_3$$

$$2\Omega_2 = A_1 \left(\frac{R_1}{R_2}\right)^2 + A_2 + A_3$$

$$2\Omega_3 = A_1 \left(\frac{R_1}{R_3}\right)^2 + A_2 \left(\frac{R_2}{R_3}\right)^2 + A_3.$$

The positiveness of the A_i implies that $\Omega_1 > \Omega_2 > \Omega_3 > \cdots > \Omega_N$. Now in analogy with the plane flow case we study the behaviour of the eigenfrequencies when m goes from $+\infty$ to 1. Now all the poles can be shown to be on the real axis and if we look at the ω/m axis the angular phase velocities can be shown to vary from Ω_1, Ω_2 and Ω_3 for $m = +\infty$ to Ω_2, Ω_3 and 0 for $m = 1$. Figure 7. The marginal $m = 1$ $\omega = 0$ mode is simply

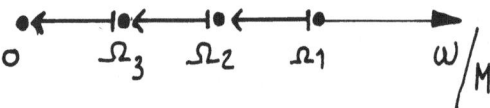

Fig. 7 Vortex Fluid.   Cylindrical geometry (Generalised Rankine Vortex). Evolution of the angular phase velocity on the ω/m axis. Illustration for 3 bags with all $A_j > 0$.

a displacement mode for the total vortex and consequently no instability will occur through the point $\omega = 0$. If all the A_j are positive the flow will be stable but if the vorticity profile exhibit a minimum, overstability can occur. In analogy with the plasma case the complete 'trapping' of the vorticity distribution has killed the marginal unstable adiabatic mode which just becomes in the limit $m = 1$ a trivial displacement mode. We notice here that the bidimensional vortex distribution present analogies both with the plasma and the stellar system. The infinite plane system with no stellar dynamic counterpart exhibiting unstable roots while the cylindrical one as the self gravitating gas exhibit a displacement mode (while the plasma case has a $\omega = \omega_p = = (n_i e^2/m\varepsilon_0)^{1/2}$ mode corresponding to a global oscillation of the electron population in the neutralising fixed ion background.

5. Conclusion

Plasma physicists and astronomers face the same N-body long range force problem. But while the plasma machines impose certain symmetry which can simplify the computations the astronomers attack the complete N-body problem taking into account both short range and long range interaction. As a consequence the plasma physicists should look more carefully into these binaries phenomena that he has a certain tendency to neglect (hydrodynamics is full of problems where the existence of a small viscosity coefficient brings very singular phenomena often presented as para-dox). May be the Vlasov equation for some geometries (spherical particularly) will also exhibit singular behaviour. Now that long confinement time machines are under construction methods to compute the long time behaviour of our plasma will be needed (including collision, recombination and ... hopefully fusion). The astronomer's Monte-Carlo method is obviously a candidate.

The plasma physicists on the other hand have developped Eulerian schemes very efficient for the stability studies (as the Multiple Water Bag and the various models involving Fourier, Hermite and Tchebycheff transforms on configuration and velocity spaces). The advantage of such models is that the analysis can be carried further on, simplifying the numerical work (the Multiple Water Bag model being a good example).

May we invite the astronomers also to get out of the real world to play with one dimensional and two dimensional models (not to mention the $1\frac{1}{2}$)? Especially if one wishes to check a tricky hypothesis, sometimes the analytical computation and the numerical simulation can be very much simplified without modifying the essence of the approximation. A good example is the computation of correlation properties in plasma where concept as the test particle is completely independent of the dimensionality (but results do depend of it).

To end this paper I will make some propaganda for a general study of models and especially of fully discretised models. The Multiple Water Bag is a first step but is not discretised in space neither, of course, in time.

We should study what are the qualitative properties (for example stability properties) which are retained in the discretisation process.

Many questions are still opened for mathematical physicists: numerical analytical method, multiple time scale system of differential equations, use of Formal algebraic manipulating language, etc. .....

After many years where new interesting results were rather easily obtained at the beginning of computational physics it may be time to get new and more powerful tools!

Acknowledgements

The author thanks for fruitful discussions on these topics Drs E. Jamin, M. Navet and M. L. Noyer. The work on the vortex stability has been done in cooperation with them.

References

Alder, B. and Fernbach, S.: 1970, *Methods in Computational Physics*, Vol. 9, Academic Press, New York, London.
Bertrand, P., Dorémus, J. P., Baumann, G., and Feix, M. R.: 1972, *Phys. Fluids* **15**, 1275.
Christiansen, J. P. and Zabusky, N. J.: 1973, *J. Fluid Mech.* **61**, 219.
Cole, J. D.: 1968, *Perturbation Methods in Applied Mathematics*, Blaisdell, New-York.
Dawson, J. M. and Hsi, C. G.: 1969, *Proceeding APS Topical Conference on Numerical Simulation of Plasma*, LA 3990, Los Alamos.
Dorémus, J. P.: 1973, 'Contribution à l'étude de la stabilité des fluides de l'espace des phases', Université de Nancy I (Thèse de Doctorat).
Dorémus, J. P. and Feix, M. R.: 1972, *Astron. Astrophys.* **20**, 259.
Feix, M. R.: 1969, in G. Kalman and M. R. Feix (ed.), *Non Linear Effects in Plasmas*, Gordon and Breach, New York, London, Paris, p. 151.

Feix, M. R.: 1971, in M. Tilston and M. Sauzade (eds.), *Progress in Radio Science*, URSI, Bruxelles, Vol. 3, p. 271.

Feix, M. R.: 1973, *Some Problems and Methods in Computational Plasma*, Culham SRC Symposium on Turbulence and Non Linear Effects in Plasma, to be published.

Feix, M. R., Dorémus, J. P., and Baumann, G.: 1971, *Astrophys. Space Sci.* **13**, 478.

Feix, M. R. and Hohl, F.: 1968, *Bull. Astron., Ser. 3*, **2**, 289.

Finzi, U.: 1972, *Plasma Phys.* **14**, 327.

Finzi, U.: 1973, 'Instabilité des Structures non linéaires dans les plasmas: ondes BGK', Université de Paris XI (Thèse de Doctorat).

Finzi, U., Dorémus, J. P., Holec, J., and Feix, M. R.: 1974, *Plasma Phys.* **16**, 189.

Kulsrud, R. M. and Mark, J. W. K.: 1970, *Astrophys. J.* **160**, 471.

Lénard, A.: 1961, *J. Math. Phys.* **2**, 682.

Morse, P. M. and Feshbach, H.: 1953, *Methods of Theoretical Physics*, McGraw Hill, New-York.

Morse, R. L. and Nielson, C. W.: 1969, *Phys. of Fluids* **12**, 2418.

Navet, M. and Bertrand, P.: 1971, *Phys. Letters* **34A**, 117.

Noyer, M. L., Navet, M., and Feix, M. R.: 1975, 'The Multiple Water-Bag Model for Forced Oscillations in a Warm Isotropic Plasma', *J. Plasma Phys.* (in press).

Ribicky, G. B.: 1972, *Astrophys. Space Sci.* **14**, 56.

Zabusky, N. J. and Deem, J. S.: 1971, *J. Fluid Mech.* **47**, 353.

DISCUSSION

Ipser: In the unstable case you talked about, involving the inverted distribution as a function of energy, are you dealing with spherical systems?

Feix: No. It is a plane problem and this is an overstable instability (which means that the instability appears through an $\omega \neq 0$ mode).

Severne: One could mention another domain of plasma physics which may prove very directly relevant to stellar dynamics. In the study of laser induced fusion, one has to analyse spherical systems in strong fields and which are strongly inhomogeneous.

A comprehensive theoretical program has been initiated at the Université Libre de Bruxelles.

Feix: O.K. Let us hope we will get money for such studies!

ELLIPSOIDAL STELLAR SYSTEMS

C. HUNTER

The Florida State University, Tallahassee, Fla., U.S.A.

Abstract. This paper gives a unified account of exact solutions for ellipsoids of uniform density. An approximate solution for a slightly spheroidal system of non-uniform density is given in the final section.

1. Introduction

Only a very limited number of exact analytical self-consistent solutions for collisionless stellar systems are known that are neither spherically symmetric nor circular disks. Such solutions are the simplest possible models for barred and elliptical galaxies. All the known non-axisymmetric solutions have gravitational potentials that are quadratic functions of the spatial coordinates. Quadratic potentials are given by ellipsoids of uniform density.

The most general ellipsoidal figures of equilibrium of an incompressible fluid of uniform density are the so-called Riemann ellipsoids. Their possible structures and their stability have been studied for the past 200 years, and have given considerable insight into the dynamics of self-gravitating systems. A comprehensive account of this work, with references, is given in Chandrasekhar (1969). Stellar dynamic analogues of these figures are described in the present work.

2. Triaxial Systems

Consider an ellipsoid of uniform density ϱ_0 with outer boundary

$$\frac{x^2}{a_1^2}+\frac{y^2}{a_2^2}+\frac{z^2}{a_3^2}=1,\tag{1}$$

that is rotating about the z-axis with angular velocity $-\Omega$. A density distribution function that gives a self-consistent solution of the collisionless Boltzmann equation for this figure relative to axes that rotate with the figure, is

$$f=-\frac{\varrho_0 a_1 a_2 \Lambda^2}{\pi^2 a_3 k_\alpha k_\beta (2\pi G\varrho_0 A_3)^{1/2}}\frac{\mathrm{d}}{\mathrm{d}I}[(1-I)^{-1/2}H(1-I)],\tag{2}$$

where

$$I=\frac{x^2}{a_1^2}+\frac{y^2}{a_2^2}+\frac{z^2}{a_3^2}+\Lambda^2\left[\frac{a_2^2(\dot{x}+\theta y/a_2^2)^2}{k_\alpha^2 k_\beta^2}+a_1^2(\dot{y}-\theta x/a_1^2)^2\right]+\frac{\dot{z}^2}{2\pi G\varrho_0 a_3^2 A_3}.\tag{3}$$

Here G is the gravitational constant, the dots denote time differentiations, and k_α, k_β, Λ^2 and θ are all constants. Their definitions, as well as details of most of the work described in this section, are given in Hunter (1974). However, the notation for the semi-axes and for the coefficients occurring in the gravitational potential such as A_3 have been changed so as to conform with that of Chandrasekhar (see Chapter 3).

Because the function H in Equation (2) is the Heaviside step function, and differentiation of $H(1-I)$ gives the delta function $-\delta(1-I)$, the distribution function (2) is unphysical since it consists of a positive delta function component and a negative step function component. Hence most orbits must be populated with stars of negative mass if the figure is to be maintained. The total mass of matter at any point in space is positive because of the domination of the delta function component of f.

The uniqueness of the solution (2) of the collisionless Boltzmann equation for the given figure has been established rigorously only for the case $\Omega=0$ of no rotation. In Section 3, it is derived again as a solution of the form $f=f(E, J)$ for the spheroidal case, and it seems likely that the solution is unique in the general case. Its unphysical nature must arise from the fact that an excessively large number of stars in high energy orbits are needed to maintain a uniform density right up to the outer boundary. The negative masses are then needed to keep the density uniform in the interior. The present dynamical picture is less general than that of the fluid Riemann ellipsoids that can rotate about an axis other than a principal one but, if more general stellar systems are possible, they will presumably have the same unphysical nature as expression (2).

Negative masses are avoided only in two interesting special cases of Equation (2). The first is that discovered by Freeman (1966a) of $\Omega \to (2\pi G\varrho_0 A_1)^{1/2}$. It corresponds to a maximum rate of rotation that the figure can endure because gravitational and centrifugal forces are then in precise balance on the longer x-axis ($a_1 > a_2$). Any small increase in rotation rate would cause particle orbits to become unstable. Mathematically $k_\alpha \to 0$ in this limit, $\theta \to 2\Omega a_2^2$, and the corresponding limiting form of f must be found. It is in fact

$$f = \frac{\varrho_0 \delta(\dot{x}+2\Omega y) \, \delta(1-J')}{\pi \mu \beta a_2 a_3 [2\pi G\varrho_0 A_3]^{1/2}}, \tag{4}$$

where

$$J' = \frac{x^2}{a_1^2} + \frac{y^2}{a_2^2} + \frac{z^2}{a_3^2} + \frac{(\dot{y} - 2\Omega a_2^2 x/a_1^2)^2}{a_2^2 \mu^2 \beta^2} + \frac{\dot{z}^2}{2\pi G\varrho_0 a_3^2 A_3}, \tag{5}$$

and

$$\mu\beta = (2\pi G\varrho_0)^{1/2} [(3 - 4a_2^2/a_1^2) A_1 + A_2]^{1/2}. \tag{6}$$

Every populated orbit now touches the outer boundary of the ellipsoid. The particles comprising the ellipsoid have no random motions in the x-direction and have a mean motion $\mathbf{u} = (-2\Omega y, 2\Omega a_2^2 x/a_1^2, 0)$ relative to the ellipsoid.

These figures of Freeman form a two-parameter family of distinct models as they are possible for all values of the ratio of the semi-axes $a_2/a_1 < 1$, and for all values of a_3/a_1. In the spheroidal limit $a_2 \to a_1$, $\mu\beta \to 0$, and the random motions in the y-direction also disappear. Then all the particles describe circular orbits in the $x-y$ plane with angular velocity 2Ω. In this limit, the distribution function f can be shown to become

$$f = \frac{\varrho_0 \delta(\dot{x} + 2\Omega y) \, \delta(\dot{y} - 2\Omega x) \, H(1 - J'')}{\pi a_3 \left[(2\pi G \varrho_0 A_3)(1 - J'') \right]^{1/2}},$$ (7)

where

$$J'' = \frac{x^2 + y^2}{a_1^2} + \frac{z^2}{a_3^2} + \frac{\dot{z}^2}{2\pi G \varrho_0 a_3^2 A_3}$$ (8)

The solution given by Bisnovatyi-Kogan and Zel'dovich (1970, §4) is precisely this special case of Freeman's solution. Although there are no random motions in the x and y-directions, there are random motions which give an anisotropic 'pressure' to support it in the z-direction. Morozov, Polyachenko and Shukman (1974) have recently discussed the stability of these special solutions. It does not seem likely that these solutions are stable, since individual orbits are only marginally stable. Also, mean motions in the figures must be relatively large, so that the figures are liable to the instabilities discussed by Ostriker and Peebles (1973).

The second and dynamically more varied class of solutions without negative masses are the thin disk limits in which $a_3 \to 0$ and the ellipsoid is collapsed into a plane perpendicular to the axis of rotation. Then the limit of the distribution function (2) is

$$f = \frac{3M\Lambda^2 H(1 - J)}{4\pi^2 k_\alpha k_\beta (1 - J)^{1/2}},$$ (9)

where M is the mass of the disk, and

$$J = \frac{x^2}{a_1^2} + \frac{y^2}{a_2^2} + \Lambda^2 \left[\frac{a_2^2 (\dot{x} + \theta y / a_2^2)^2}{k_\alpha^2 k_\beta^2} + a_1^2 (\dot{y} - \theta x / a_1^2)^2 \right].$$ (10)

These elliptical disks have a surface density distribution

$$\sigma = \frac{3M}{2\pi a_1 a_2} \left(1 - \frac{x^2}{a_1^2} - \frac{y^2}{a_2^2} \right)^{1/2},$$ (11)

which tends to zero at the outer edge, avoiding the need for negative masses.

These solutions which were also discovered by Freeman (1966b), form a two-parameter family of dynamically distinct figures given by the ranges $0 \leqslant a_2/a_1 \leqslant 1$, for the semi-axes and $0 \leqslant \Omega^2/\Lambda^2 \leqslant 1$ for the rotation rate $[\Lambda^2 = \lim_{a_3 \to 0}(2\pi G \varrho_0 A_1)]$. The quantities a_2/a_1 and Ω^2/Λ^2 are two basic dimensionless parameters for these systems. The mean flow velocity is now $\mathbf{u} = (-\theta y/a_2^2, \theta x/a_1^2)$, and θ can range between $\frac{2}{3}$ and -2. Hence it can be a circulation either in the same sense, or in the opposite sense, to that of the rotation of the figure itself. The quantity $t = -T_{\text{mean}}/W$, where T_{mean} is the kinetic energy in the form of mean motions and W is the gravitational energy, can vary between the maximum permissible limits 0 and 0.5. According to Ostriker and Peebles, there is instability to bar-like disturbances if t is greater than 0.14 or thereabouts.

The slow evolution of these disks as they lose mass can also be studied. They tend either to become more circular, or else to become less circular, but non-rotating.

Interestingly, they also tend to evolve in the direction of decreasing t, and hence of increasing stability to bar-like disturbances.

3. Stellar Maclaurin Spheroids

The solutions in Section 2 are essentially found by looking for distribution functions that depend on the three isolating integrals available for uniform ellipsoids. These are the z-component of the total energy, and two integrals that are connected with the amplitudes of motions in the $x-y$ plane. There is an extra isolating integral when $a_1 = a_2$, and the ellipsoid becomes a uniform spheroid and symmetric about its axis of rotation.

Because of the large number of isolating integrals now present, there must be a large number of possible solutions of the collisionless Boltzmann equation. As with any axisymmetric distribution of mass, we can look for a solution of the form $f(E, J^2)$ where $E = \frac{1}{2}v^2 - \psi$ is the total energy, v the velocity, ψ the gravitational potential, and J the angular momentum about the axis of symmetry. We now work in a fixed frame of reference.

A general method of finding such solutions has been given by Lynden-Bell (1962). It involves first expressing the density ϱ in terms of ψ and $R = (x^2 + y^2)^{1/2}$. Then the Laplace transform of $\partial\varrho/\partial\psi$ with respect to ψ must be taken, and the transform variable labelled s. The result, after multiplication by $(s/u)^{1/2}$ and setting $R^2 = s/u$, is the double Laplace transform of $g(B, t) = 4\pi(2t)^{-1/2}f(-B, 2t)$, when s and u are now regarded as the transform variables for B and t respectively. Hence a double inversion gives f. Following these steps, we write

$$\varrho = \varrho_0 H \left\{ 1 - \frac{I}{a_3^2 A_3} + \frac{\psi}{\pi G \varrho_0 a_3^2 A_3} + \frac{R^2}{a_1^2} \left[\frac{a_1^2 A_1}{a_3^2 A_3} - 1 \right] \right\}, \tag{12}$$

since the gravitational potential for the uniform spheroid is

$$\psi = \pi G \varrho_0 (I - A_1 R^2 - A_3 z^2). \tag{13}$$

The double Laplace transform of g is then obtained as

$$\varrho_0 \left(\frac{s}{u} \right)^{1/2} \exp\left\{ -\pi G \varrho_0 s \left[I - a_3^2 A_3 - \frac{s}{a_1^2 u}(a_1^2 A_1 - a_3^2 A_3) \right] \right\}. \tag{14}$$

The inversion of $u^{-1/2} \exp(k/u)$ is $(\pi t)^{-1/2} \cosh 2(kt)^{1/2}$ (Abramowitz and Stegun, 1964, formula 29.3.77). Hence the first inversion of Equation (14) is

$$\varrho_0 \left(\frac{s}{\pi t} \right)^{1/2} \exp\left\{ -\pi G \varrho_0 (I - a_3^2 A_3) s \right\} \cosh \frac{2s}{a_1} \left[\pi G \varrho_0 t(a_1^2 A_1 - a_3^2 A_3) \right]^{1/2}. \tag{15}$$

As always, negative exponential terms represent a shift of arguments, and the cosh term can be written as the sum of two exponentials. The inversion of $s^{1/2}$ leads to a generalized function, since the inversion of $s^{-1/2}$ is $(\pi B)^{-1/2}$, and the extra s factor corresponds to a differentiation. The result therefore is that

$$g(B, t)=\frac{\varrho_0}{2\pi t^{1/2}}\frac{\mathrm{d}}{\mathrm{d}B}\{[B-\pi G\varrho_0(I-a_3^2A_3)+$$

$$+\omega(2t)^{1/2}]^{-1/2}H[B-\pi G\varrho_0(I-a_3^2A_3)+\omega(2t)^{1/2}]+$$

$$+[B-\pi G\varrho_0(I-a_3^2A_3)-\omega(2t)^{1/2}]^{-1/2}H[B-\pi G\varrho_0(I-a_3^2A_3)-$$

$$-\omega(2t)^{1/2}]\},\qquad(16)$$

where

$$\omega=[2\pi G\varrho_0(A_1-a_3^2A_3/a_1^2)]^{1/2}.\qquad(17)$$

The distribution function is therefore

$$f(E, J^2)=|J|\,g(B=-E,\,t=\tfrac{1}{2}J^2)/4\pi$$

$$=\frac{-\varrho_0}{4\pi^2\sqrt{2}}\frac{\mathrm{d}}{\mathrm{d}E'}[(-E'+\omega J)^{-1/2}H(-E'+\omega J)+$$

$$+(-E'-\omega J)^{-1/2}H(-E'-\omega J)],\qquad(18)$$

where

$$E'=E+\pi G\varrho_0(I-a_3^2A_3)=\tfrac{1}{2}v^2+\pi G\varrho_0[A_1R^2+A_3(z^2-a_3^2)].\qquad(19)$$

This distribution is made up of two equal and oppositely rotating components, and either type of component on its own gives a possible solution. Thus,

$$f=\frac{-\varrho_0}{2\pi^2\sqrt{2}}\frac{\mathrm{d}}{\mathrm{d}E'}[(-E'+\omega J)^{-1/2}H(-E'+\omega J)],\qquad(20)$$

gives rise to a uniform density spheroid. It is a function only of

$$-E'+\omega J=-\tfrac{1}{2}v_R^2-\tfrac{1}{2}(v_\theta-\omega R)^2-\tfrac{1}{2}v_z^2+\pi G\varrho_0a_3^2A_3\left(1-\frac{R^2}{a_1^2}-\frac{z^2}{a_3^2}\right),\qquad(21)$$

where (v_R, v_θ, v_z) denote the components of velocity in cylindrical polar coordinates (R, θ, z). Hence the distribution function (20) represents a mean flow with angular velocity ω, which is precisely that for a fluid Maclaurin spheroid. The distribution function (20) therefore describes the stellar Maclaurin spheroid. The random velocities are distributed isotropically about the mean flow and provide the necessary pressure support. Unfortunately however, the solution is again unphysical because of widespread negative values of f, and is in fact just a particular $a_1=a_2$ case of solution (2).

4. A Non-Uniform Stellar Spheroid

Something other than a uniform density figure must clearly be considered for realistic models of galaxies. Lynden-Bell gave one exact axisymmetric solution as an illustration of his method and indeed, apart from the solution discussed in Sections 2 and 3, it appears to be the only one known. Lynden-Bell's solution is of infinite extent, and is a rotationally modified form of the $n=5$ polytrope. A wider variety of known exact solutions would be very desirable, but, unfortunately, they are not easily obtained. The basic difficulty is that a considerable degree of analytical simplicity is necessary if all the integrations involved in Lynden-Bell's method are to be performed, and such simplicity is not often attained. For instance, the spheroid with density,

$$\varrho=\varrho_0\left[1-\frac{R^2}{a_1^2}-\frac{z^2}{a_3^2}\right]H\left[1-\frac{R^2}{a_1^2}-\frac{z^2}{a_3^2}\right], \tag{22}$$

where ϱ_0 is now the central density is the most simple spheroid other than the uniform one. Its gravitational potential is still relatively simple and is (Chandrasekhar, 1969, p. 53),

$$\psi=\tfrac{1}{2}\pi G\varrho_0\{I-2A_1R^2-2A_3z^2+A_{11}R^4+2A_{13}R^2z^2+A_{33}z^4\}. \tag{23}$$

Elimination of z^2 between Equations (22) and (23) and differentiation of ϱ gives

$$\frac{\partial\varrho}{\partial\psi}=(\pi Ga_3^2)^{-1}(\beta\psi-\alpha)^{-1/2}H[(\beta\psi-\alpha)^{1/2}-\gamma], \tag{24}$$

where

$$\begin{aligned}\alpha&=IA_{33}-A_3^2-2R^2(A_1A_{33}-A_3A_{13})+R^4(A_{11}A_{33}-A_{13}^2),\\ \beta&=2A_{33}/\pi G\varrho_0, \qquad \gamma=A_3-a_3^2A_{33}-R^2(A_{13}-a_3^2A_{33}/a_1^2).\end{aligned} \tag{25}$$

Taking the Laplace transform of expression (24) with respect to ψ, and multiplication by $(s/u)^{1/2}$ gives as the function to be inverted

$$\frac{1}{Ga_3^2(\pi\beta u)^{1/2}}\exp\left(\frac{-\alpha s}{\beta}\right)\operatorname{erfc}\gamma\left(\frac{s}{\beta}\right)^{1/2}. \tag{26}$$

Since it is necessary to set $R^2=s/u$ in the expressions for α and γ, we obtain for expression (26) a formula that is more complicated than any entry in any of the standard tables. However, expression (26) can be inverted in the spherical case $a_1=a_3$, and we can progress if we limit our interest to the case of small eccentricity $e=(1-a_3^2/a_1^2)^{1/2}$. The following expansions are then valid

$$\frac{\alpha}{\beta}=\frac{4\pi G\varrho_0a_1^2}{9}\left(1-\frac{e^2}{7}-\frac{6e^2s}{35a_1^2u}\right)+O(e^4), \qquad A_{33}=\frac{2}{5a_1^2}\left(1+\frac{9e^2}{7}\right)+O(e^4), \tag{27}$$

$$\frac{\gamma}{\beta^{1/2}}=\left(\frac{4\pi G\varrho_0a_1^2}{45}\right)^{1/2}\left(1-\frac{e^2}{14}-\frac{3e^2s}{7a_1^2u}\right)+O(e^4). \tag{28}$$

Upon substitution into expression (26), and expanding for the complementary error function, we obtain

$$\frac{1}{Ga_3^2(\pi\beta u)^{1/2}}\exp\left[-\frac{4\pi G\varrho_0a_1^2s}{9}\left(1-\frac{e^2}{7}\right)+\frac{8\pi G\varrho_0e^2s^2}{105\,u}\right]\times$$

$$\times \left\{ \mathrm{erfc} \left[\frac{4\pi G\varrho_0 a_1^2 s}{45} \left(1 - \frac{e^2}{7} \right) \right]^{1/2} + \frac{4e^2 s^{3/2}}{7a_1 u} \left(\frac{G\varrho_0}{5} \right)^{1/2} \times \right.$$

$$\left. \times \exp \left[-\frac{4\pi G\varrho_0 a_1^2 s}{45} \left(1 - \frac{e^2}{7} \right) \right] \right\} + O(e^4). \tag{29}$$

The inversion with respect to u can now be carried out. We need the formula used earlier and also Abramowitz and Stegun (formula 29.3.79) that $u^{-3/2} \exp(k/u)$ inverts to $(\pi k)^{-1/2} \sinh 2(kt)^{1/2}$. This leaves us with functions of s that are the products of exponentials with either $s^{1/2}$ or $\mathrm{erfc}(ks)^{1/2}$. Hence the only new formula needed is Abramowitz and Stegun (formula 29.3.114) shifted that the inverse of $\mathrm{erfc}(ks)^{1/2}$ is $k^{1/2} H(B-k)/[\pi B(B-k)^{1/2}]$. We then obtain

$$g(B, t) = \frac{1}{2\pi^2 Ga_3^2} \left(\frac{\lambda}{5\beta t} \right)^{1/2} \left\{ \frac{H(B - 6\lambda/5 + \mu\sqrt{2t})}{(B - \lambda + \mu\sqrt{2t})(B - 6\lambda/5 + \mu\sqrt{2t})^{1/2}} + \right.$$

$$+ \frac{H(B - 6\lambda/5 - \mu\sqrt{2t})}{(B - \lambda - \mu\sqrt{2t})(B - 6\lambda/5 - \mu\sqrt{2t})^{1/2}} +$$

$$+ \frac{e}{a_1} \left(\frac{30t}{7\lambda} \right)^{1/2} \frac{d}{dB} \times$$

$$\left. \times \left[\frac{H(B - 6\lambda/5 + \mu\sqrt{2t})}{(B - 6\lambda/5 + \mu\sqrt{2t})^{1/2}} - \frac{H(B - 6\lambda/5 - \mu\sqrt{2t})}{(B - 6\lambda/5 - \mu\sqrt{2t})^{1/2}} \right] \right\} + O(e^2), \tag{30}$$

where the constants λ and μ are defined by

$$\lambda = \frac{4\pi G\varrho_0 a_1^2}{9} \left(1 - \frac{e^2}{7} \right), \qquad \mu = 4e \left(\frac{\pi G\varrho_0}{105} \right)^{1/2}. \tag{31}$$

The four components of (30) clearly form two pairs. Moreover, each pair combines naturally since, to the first order in any small quantity ε, the Taylor expansion

$$\frac{(1 - \frac{1}{2}\varepsilon) H[B - c + (\varepsilon - 1)\lambda/5]}{(B - c)[B - c + (\varepsilon - 1)\lambda/5]^{1/2}} = \frac{H(B - c - \lambda/5)}{(B - c)(B - c - \lambda/5)^{1/2}} +$$

$$+ \varepsilon \frac{d}{dB} \left[\frac{H(B - c - \lambda/5)}{(B - c - \lambda/5)^{1/2}} \right] + O(\varepsilon^2), \tag{32}$$

holds for any quantity c.

The solution, like that of the previous section, is the sum of two oppositely rotating components, both of which could individually give the full solution if doubled. Taking the first component, we have as a possible solution

$$f = \frac{\varrho_0 [1 - 5\mu J/4\lambda] H[-E - 6\lambda/5 + \frac{3}{2}\mu J]}{6\pi^2 \sqrt{2}(\mu J - \lambda - E)[-E - 6\lambda/5 + \frac{3}{2}\mu J]^{1/2}} + O(e^2). \tag{33}$$

Note that expression (33) is correct only to $O(e)$. Although the transformed expression (26) was expanded correctly to $O(e^4)$, the inversion process alters the ordering of the expansion in powers of e. For instance, the exponential in $e^2 s/u$ gives rise to effects that are $O(e)$, and an $O(e^2)$ error is introduced with the use of formula (32).

The distribution function (33) has an inverse square root singularity when

$$0 = -E - 6\lambda/5 + \tfrac{3}{2}\mu J$$
$$= \frac{\pi G \varrho_0 a_1^2}{15} \left(7 - \frac{3r^2}{a_1^2}\right)\left(1 - \frac{r^2}{a_1^2}\right) - \tfrac{1}{2}[v_R^2 + (v_\theta - \tfrac{3}{2}\mu R)^2 + v_z^2] + O(e^2), \qquad (34)$$

where $r = (x^2 + y^2 + z^2)^{1/2}$. The inverse square root is needed to maintain the correct density in the outer regions, and only orbits with $-E - 6\lambda/5 + \tfrac{3}{2}\mu J = 0$ ever reach the outer edge.

Integration of f over all the allowable velocities confirms that the density at any point in space is $\varrho_0(1 - r^2/a^2) + O(e^2)$, while the mean motion of particles consists of a rotation about the axis of symmetry with angular velocity

$$\frac{1}{R} \frac{\iiint v_\theta f \, d\mathbf{v}}{\iiint f \, d\mathbf{v}} = \frac{3\mu}{8}\left(5 - \frac{r^2}{a_1}\right) = \frac{3e}{2}\left(\frac{\pi G \varrho_0}{105}\right)^{1/2}\left(5 - \frac{r^2}{a_1^2}\right). \qquad (35)$$

This angular velocity, which, as in the case of a fluid spheroid is proportional to the eccentricity, is not uniform, and is somewhat larger at the center than at the edge of the spheroid.

Acknowledgement

This work has been supported in part by the National Science Foundation under grant GP-34279X.

References

Abramowitz, M. and Stegun, I. A.: 1964, *Handbook of Mathematical Functions*, National Bureau of Standards, Washington, D.C., Ch. 29.
Bisnovatyi-Kogan, G. S. and Zel'dovich, Ya. B.: 1970, *Astrofizika* **6**, 387.
Chandrasekhar, S.: 1969, *Ellipsoidal Figures of Equilibrium*, Yale University Press, New Haven, Conn.
Freeman, K. C.: 1966a, *Monthly Notices Roy. Astron. Soc.* **134**, 1.
Freeman, K. C.: 1966b, *Monthly Notices Roy. Astron. Soc.* **134**, 15.
Hunter, C.: 1974, *Monthly Notices Roy. Astron. Soc.* **166**, 633.
Lynden-Bell, D.: 1962, *Monthly Notices Roy. Astron. Soc.* **123**, 447.
Morozov, A. G., Polyachenko, V. L., and Shukhman, I. G.: 1974, *Astron. Zh.* **51**, 75.
Ostriker, J. P. and Peebles, P. J. E.: 1973, *Astrophys. J.* **186**, 467.

DISCUSSION

King: There is another interesting application of Maclaurin spheroids that is of great practical interest. If you write down stellar-dynamical equations for the center of a stellar system and expand everything around the origin, then consideration of the leading terms gives a relation between rotation and flattening that is exactly the same as the relation for Maclaurin spheroids.

Fackerell: One suspects that the distribution function in the last example will come out in terms of hypergeometric functions of several variables, probably Lauricella functions.

Hunter: The complications in inverting the Laplace transforms in the Lynden-Bell method are not due to the presence of hypergeometric functions. The functions involved are exponential and complementary error functions, but their arguments are more complicated than those given in current tables.

Lynden-Bell: One trouble with my method is the awkwardness of expressing the density as a function of axial radius R and potential Φ. One can avoid some of that trouble if one uses the crazy device of starting not with $\Phi(R, z)$ but with $z(R, \Phi)$. One may differentiate $z(R, \Phi)$ directly to find $\varrho(R, \Phi)$.

I am glad to see that you have managed to get more success in using the method which I found led to complicated functions in many cases.

Hunter: There are two stages at which your method may become difficult. One is that of the algebraic manipulations about which you asked. The other is that of taking the Laplace transforms and then inverting them. I think that this second stage is likely to be the harder one for more general problems.

A METHOD FOR SOLVING POISSON'S EQUATION FOR SYSTEMS WITH AXIAL SYMMETRY

T. S. VAN ALBADA

Kapteyn Astronomical Institute, Groningen, The Netherlands

In computer experiments on the dynamics of stellar systems special methods are often required for the computation of the forces to keep the problem manageable. For collisionless systems a method based on solving Poisson's equation with Fourier-series expansions has been used with success (Miller *et al.*, 1970; Hohl, 1973). The systems in these studies consist of a large number of particles moving on a rectangular grid.

For three-dimensional systems with axial symmetry a similar method can be used, based on the expansion of density and potential in Legendre polynomials (cf. Prendergast and Tomer, 1970):

$$\varrho(r, \theta) = \sum_{n=0}^{N} a_n(r) P_n(\cos \theta), \tag{1}$$

$$\psi(r, \theta) = \sum_{n=0}^{N} b_n(r) P_n(\cos \theta). \tag{2}$$

Poisson's equation, $\nabla^2 \psi = 4\pi G \varrho$, with appropriate boundary conditions, supplies the relations between a_n and b_n and the potential can be written in the form:

$$\psi(r, \theta) = -4\pi G \sum_{n=0}^{N} \frac{P_n(\cos \theta)}{2n+1} \left[r^n \int_r^\infty s^{1-n} a_n(s) \, ds + \frac{1}{r^{1+n}} \int_0^r s^{2+n} a_n(s) \, ds \right],$$

where

$$\tag{3}$$

$$a_n(r) = \frac{2n+1}{2} \int_{-1}^{1} \varrho(r, \theta) P_n(\cos \theta) \, d\cos \theta. \tag{4}$$

(The boundary conditions $\psi \to 0$ for $r \to \infty$ and $\psi \to$ constant for $r \to 0$ have been used). Given a grid in r and θ, the integrals over r in (3) can be calculated from recurrence relations. The number of operations required to evaluated ψ from (3) and (4) for a given density distribution is therefore proportional to the number of grid divisions in r, the number of divisions in θ and the number of Legendre polynomials.

We have used this scheme to study the evolution of a system of 4000 particles (similar to the systems studied by Gott, 1973); its numerical behavior is quite satisfactory.

References

Gott, J. R.: 1973, *Astrophys. J.* **186**, 481.

Hayli (ed.), Dynamics of Stellar Systems, 205–206. All Rights Reserved.
Copyright © 1975 by the IAU.

Hohl, F.: 1973, *Astrophys. J.* **184**, 353.
Miller, R. H., Prendergast, K. H., and Quirk, W. J.: 1970, *Astrophys. J.* **161**, 903.
Prendergast, K. H. and Tomer, E.: 1970, *Astron. J.* **75**, 674.

DISCUSSION

Miller: A technical question: recurrence-relations such as you mention are usually unstable, and require great care unless you are careful about starting and the direction of recurrence. How do you get stable results?

Van Albada: No direct tests of the behaviour of the recurrence relations have been made. The results of experiments with rotating homogeneous spheres agree well with the 'exact' solutions.

Ipser: Have you tried to use this method to construct self-consistent solutions for equilibrium configurations?

Van Albada: No.

Spitzer: Can you compare the computing time required by your method and the corresponding time required by the method of Dr Gott?

Van Albada: For a system with many particles, moving them takes longer than evaluating the forces. Solving Poisson's equation on a grid of 1000 cells takes about 1 s on a CDC Cyber 74–16.

DYNAMICAL MODELS OF ELLIPTICAL GALAXIES

C. P. WILSON

Hale Observatories, Carnegie Institution of Washington, California Institute of Technology,
Pasadena, Calif., U.S.A.

Abstract. Self-consistent dynamical models of elliptical galaxies have been constructed using a modification of the Prendergast-Tomer algorithm. They reproduce many of the observed properties of elliptical galaxies.

Because the relaxation time of an elliptical galaxy is greater than its age, such a system is at present in a stationary configuration. Consequently, we have attempted to represent elliptical galaxies by self-consistent models which are time-independent, rotationally symmetrical, and rotating.

The algorithm for constructing a model is very similar to that used by Prendergast and Tomer (1970): one adopts a distribution function $f(E, J)$ that depends on the integrals of motion (the energy and the component of angular momentum parallel to the axis of symmetry), integrates this function over velocities to get the density, and then solves Poisson's equation. Since the theory of the initial collapse process (e.g. Lynden-Bell, 1967) is not very rigorous, the choice of a distribution function is somewhat arbitrary. The following function has been used here because it includes the necessary degrees of freedom in a simple and reasonable fashion:

$$f(E, J) = (e^{-E} - 1 + E) \exp(\beta J - \tfrac{1}{2}\zeta^2 J^2), \qquad E < 0.$$
$$= 0 \qquad\qquad\qquad\qquad, \qquad E \geqslant 0.$$

The free parameters β and ζ permit construction of a range of models. Except for a modification of the energy dependence, which makes f go to zero continuously and smoothly at $E = 0$, this is simply a classical distribution function that is quadratic in the velocities (Eddington, 1915, and others). To solve Poisson's equation, the density and potential are expanded into Legendre series with coefficients which are functions of radius. The resulting simultaneous set of integral equations are solved numerically (Wilson, 1973).

Models have been projected onto the plane of the sky and compared with observations of NGC 3379. If the inclination of this galaxy is assumed to lie anywhere between 45° (with intrinsic axial ratio 0.6) and 90° (intrinsic axial ratio 0.85), a model can be found that represents rather well the observed variations of surface brightness and isophote shape with radius (Miller and Prendergast, 1962). These models also agree with the observed velocity dispersion (Burbidge *et al.*, 1961) and rotation curve (Peterson, 1974) within the large observational uncertainties.

All elliptical galaxies have similar radial intensity profiles which the models fit fairly well. However, the great observed variety of the flattening vs. radius curves cannot be reproduced; the models all start nearly round at the center, increase to a maximum flattening at some intermediate radius, and then become nearly round again at the boundary.

Once a dynamical model has been fitted to the photometric and kinematic properties of a galaxy, it can predict such interesting parameters as the mass-to-light ratio. For NGC 3379, we have $(\mathfrak{M}/L)_v = 3.3$ (solar units, $H = 55\ \mathrm{km\ s}^{-1}\ \mathrm{Mpc}^{-1}$). The limited accuracy of observations introduces an uncertainty of a factor of two into this value, whereas the unknown inclination and fitting to different models results in only a 5% uncertainty. The same data analyzed in the conventional manner (Morton and Chevalier, 1972), including several questionable assumptions, gives $(\mathfrak{M}/L)_v = 9.6$. The importance of a proper dynamical treatment of the data is obvious.

By using the stability criterion proposed by Ostriker and Peebles (1973), it is found that only the models with peak flattening less than E4 to E5 are stable.

References

Burbidge, E. M., Burbidge, G. R., and Fish, R. A.: 1961, *Astrophys. J.* **134**, 251.
Eddington, A. S.: 1915, *Monthly Notices Roy. Astron. Soc.* **76**, 37.
Lynden-Bell, D.: 1967, *Monthly Notices Roy. Astron. Soc.* **136**, 101.
Miller, R. H. and Prendergast, K. H.: 1962, *Astrophys. J.* **136**, 713.
Morton, D. C. and Chevalier, R. A.: 1972, *Astrophys. J.* **174**, 489.
Ostriker, J. P. and Peebles, P. J. E.: 1973, *Astrophys. J.* **186**, 467.
Peterson, C. J.: 1974, private communication.
Prendergast, K. H. and Tomer, E.: 1970, *Astron. J.* **75**, 674.
Wilson, C. P.: 1973, 'Dynamical Models of Elliptical Galaxies', University of California, Berkeley (Ph.D. Thesis).

DISCUSSION

Freeman: Can you think of any way to tell whether ellipticals are axisymmetric or genuinely triaxial?

Wilson: I don't know of any definite observational test. It is instructive to consider obviously non-axisymmetric systems, e.g. SBO's. One property in which they differ from ellipticals is that the position angle of the major axis of the isophotes varies with radius. Perhaps the constant position angle in ellipticals is telling us something about this problem.

INTEGRALS OF MOTION

G. CONTOPOULOS

University of Maryland, U.S.A.

and

University of Thessaloniki, Greece

Abstract. The properties of conservative dynamical systems of two or more degrees of freedom are reviewed. The transition from integrable to ergodic systems is described. Nonintegrability is due to the interaction of two, or more, resonances. Then one sees, on a surface of section, infinite types of islands of various orders, while the asymptotic curves from unstable invariant points intersect each other along homoclinic and heteroclinic points producing an apparent 'dissolution' of the invariant curves. A threshold energy is defined separating near integrable systems from near ergodic ones. The possibility of real ergodicity for large enough energies is discussed. In the case of many degrees of freedom we also distinguish between integrable, ergodic, and intermediate cases. Among the latter are systems of particles interacting with Lennard-Jones interparticle potential. A threshold energy was derived, which is proportional to the number of particles. Finally some recent results about the general three-body problem are described. One can extend the families of periodic orbits of the restricted problem to the general three-body problem. Many of these orbits are stable. An empirical study of orbits near the stable periodic orbits indicates the existence of 2 integrals of motion besides the energy.

1. Introduction

Ten years have passed since the first international Meeting in Stellar Dynamics, that took place in Thessaloniki, in 1964. During these ten years much progress has been made in several fields of Stellar Dynamics, which is reflected in the program of the present Meeting.

In the area of the Integrals of Motion we have now a fairly complete understanding of the behavior of 2-dimensional conservative dynamical systems, e.g. the orbits of stars in the meridian plane of an axisymmetric galaxy, or in the plane of symmetry of a spiral galaxy.

However we are still in the first steps towards an understanding of systems of many degrees of freedom.

Today I will review shortly this general area, mentioning also some recent developments and applications.

2. Two-Dimensional Systems

The most simple nontrivial two-dimensional system is that of two coupled oscillators. E.g., the motion of a star in the meridian plane of an axisymmetric galaxy can be described by the Hamiltonian

$$H = \tfrac{1}{2}(\dot{x}^2 + \omega_1^2 x^2 + \dot{y}^2 + \omega_2^2 y^2) + \text{higher order terms}, \tag{1}$$

where the y-axis is parallel to the axis of symmetry of the galaxy, while the negative x-axis intersects it perpendicularly.

Hayli (ed.), Dynamics of Stellar Systems, 209–225. All Rights Reserved.

Suppose now that there is another analytic integral

$$\Phi = \Phi(x, y, \dot{x}, \dot{y}) = \text{const}, \qquad (2)$$

besides H. Such a system is called 'integrable'.

If we solve Equation (1) for $\dot{y}$ and insert this value in Equation (2) we have the equation of a surface

$$\Phi[x, y, \dot{x}, \dot{y}(x, y, \dot{x})] = \text{const} \qquad (3)$$

in the three-dimensional space $(x, \dot{x}, y)$. This surface is a torus, on which lies every orbit whose initial point is on it.

Let us consider orbits with the same value of the Hamiltonian. By varying the constant (3) we find a family of tori, one inside the other. The innermost torus is reduced to a periodic orbit.

Most of the properties of these tori are found if we take their intersections by a 'surface of section', e.g. the plane $y = 0$. Each orbit is represented by an 'invariant curve' (Figure 1) on this surface of section, which contains the successive points of intersection $(x, \dot{x})$ of the orbit by the plane $y = 0$. A periodic orbit is represented by a finite number of invariant points. In particular the point C in Figure 1 represents the 'central' periodic orbit.

Thus the existence of a second integral of motion implies the existence of invariant curves on the surface of section.

If, on the other hand, we have numerical evidence that the successive points of intersection of many orbits lie on closed smooth curves, this is an indication (but not proof) of the existence of another integral of motion besides the Hamiltonian.

One convenient way to study the Hamiltonian system (1) is by using the actions

$$I_1 = \frac{1}{2\omega_1}(\dot{x}^2 + \omega_1^2 x^2), \qquad I_2 = \frac{1}{2\omega_2}(\dot{y}^2 + \omega_2^2 y^2) \qquad (4)$$

of the unperturbed problem (i.e. the quadratic part of the Hamiltonian (1)), and the corresponding angles. Then H is written

$$H = \omega_1 I_1 + \omega_2 I_2 + \sum I_1^{n_1/2} I_2^{n_2/2} \{ c_{(m_1 m_2)}^{(n_1 n_2)} \cos(m_1 \vartheta_1 + m_2 \vartheta_2) +$$
$$+ s_{(m_1 m_2)}^{(n_1 n_2)} \sin(m_1 \vartheta_1 + m_2 \vartheta_2) \}, \qquad (5)$$

where

$$n_1 + n_2 \geqslant 3, \qquad n_1 \geqslant |m_1|, \qquad n_2 \geqslant |m_2| \quad \text{and} \quad n_1 - m_1 = \text{even},$$
$$n_2 - m_2 = \text{even}.$$

The most simple integrable case is one in which the angles are missing from the Hamiltonian (5); thus

$$H = \omega_1 I_1 + \omega_2 I_2 + f_1(I_1, I_2), \qquad (6)$$

where f_1 is of degree three, or larger, in $\sqrt{I_i}$. This is what we call a 'normal form' of a Hamiltonian.

In such a case I_1 and I_2 are integrals of motion.

For a given value of H each value of I_1 defines a corresponding torus. In particular the value $I_1 = 0$ defines a periodic orbit. The angle along a torus is ϑ_2, while the angle around the torus is ϑ_1. These angles vary linearly in time with frequencies

$$\varpi_1 = \frac{\partial H}{\partial I_1}, \varpi_2 = \frac{\partial H}{\partial I_2}, \tag{7}$$

therefore the orbits are quasi-periodic.

The ratio of the frequencies (7) is called the rotation number

$$\text{Rot} = \frac{\varpi_1}{\varpi_2} = \frac{\partial H/\partial I_1}{\partial H/\partial I_2}. \tag{8}$$

This is found empirically as the average angle between the successive intersections of an orbit, as seen from C, in units of 2π. Namely (Figure 1)

$$\text{Rot} = \lim_{n \to \infty} \frac{\widehat{1C2} + \widehat{2C3} + \cdots}{n}. \tag{9}$$

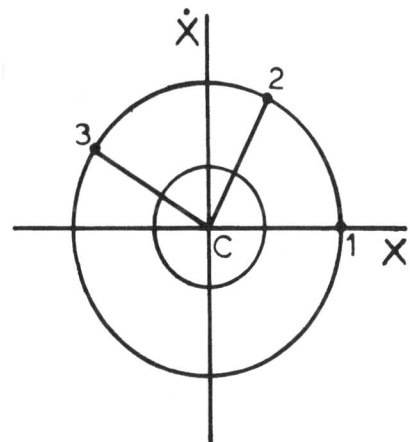

Fig. 1.   Invariant curves (schematically) around an invariant point, C, representing a periodic orbit. The successive points of intersection 1, 2, 3, ... define empirically a 'rotation number' (see text).

The rotation number is a function of H and I_1. If the value of H is fixed it can be expressed as a function of x, along the x-axis (Figure 2). All orbits on the same torus have the same rotation number.

If the rotation number is rational, say $\frac{2}{3}$, the orbit closes after three revolutions, therefore it is represented by three invariant points on the surface of section. The corresponding torus is filled with periodic orbits of the same type. Therefore all the points of the invariant curve are starting points of periodic orbits.

A more complicated case is a Hamiltonian containing one trigonometric term, besides the terms with I_1, I_2; e.g.

$$H = \omega_1 I_1 + \omega_2 I_2 + f_1(I_1, I_2) + f_{23}(I_1, I_2)\cos(3\vartheta_1 - 2\vartheta_2). \tag{10}$$

This is also an integrable case. In fact the combination

$$J_{23} = 2I_1 + 3I_2 \tag{11}$$

is an integral of motion. However the actions I_1, I_2 are no more integrals of motion. In this case we find only two periodic orbits with rotation number $\varpi_1/\varpi_2 = \frac{2}{3}$, one

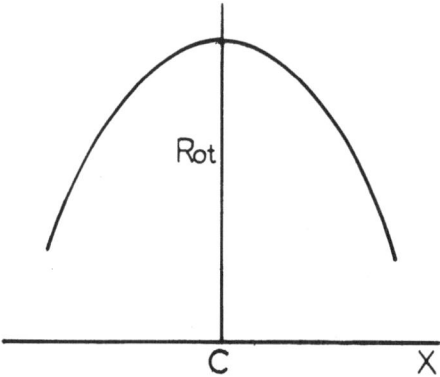

Fig. 2.   The rotation number as a function of x (x is the intersection of an invariant curve by the x-axis) (schematically).

stable and one unstable (Figure 3). The stable periodic orbit is represented by three invariant points on the surface of section, which are surrounded by sets of islands. The orbits represented by islands are called tube orbits. In phase space they lie on tori surrounding the stable resonant periodic orbits, i.e. these tori are like tubes closing after three revolutions.

The outermost islands in Figure 3 go through the three invariant points, which

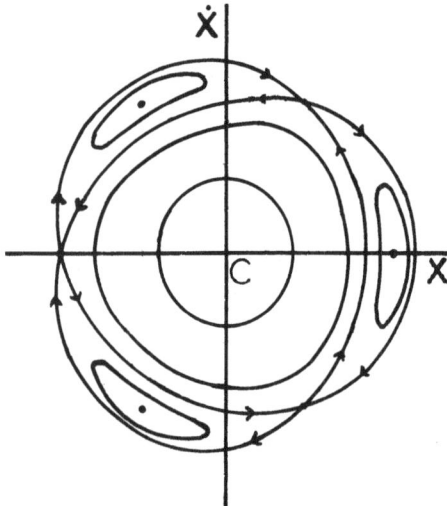

Fig. 3.   Regular invariant curves (closing around the central invariant point C) and islands. In the integrable case the outermost islands (separatrices, or asymptotic curves) join the unstable invariant points (schematically).

represent an unstable periodic orbit. These curves are called 'separatrices' as they separate the sets of islands from the regular invariant curves, which close around the central point C. They are also called 'asymptotic curves', because they represent orbits approaching asymptotically the unstable periodic orbits as $t \to \infty$, or $t \to -\infty$.

The corresponding points of intersection of every asymptotic orbit also approach asymptotically the three unstable invariant points, as indicated by the arrows in Figure 3.

The positions of the stable and unstable invariant points can be found approximately (for small energies) from Equation (8) applied to the 'unperturbed' Hamiltonian (6). We must notice here that Equation (8) may not have (real) solutions for I_1. For a given ratio of the unperturbed frequencies ω_1/ω_2 (different from $\frac{2}{3}$) we do not have real solutions for small enough H.

If we have a similar Hamiltonian

$$H = \omega_1 I_1 + \omega_2 I_2 + f_1(I_1, I_2) + f_{25}(I_1, I_2) \cos(5\vartheta_1 - 2\vartheta_2), \tag{12}$$

this is also integrable, but the integral is now

$$J_{25} = 2I_1 + 5I_2, \tag{13}$$

therefore quite different from the above integral (11). In the present case $\text{Rot} = \frac{2}{5}$, therefore we have five sets of islands and the corresponding asymptotic curves.

Now let us introduce a more general Hamiltonian with two, or more, trigonometric terms of different type, e.g.

$$H = \omega_1 I_1 + \omega_2 I_2 + f_1(I_1, I_2) + f_{23}(I_1, I_2) \cos(3\vartheta_1 - 2\vartheta_2) +$$
$$+ f_{25}(I_1, I_2) \cos(5\vartheta_1 - 2\vartheta_2). \tag{14}$$

This Hamiltonian has some common characteristics with both Hamiltonians (10) and (12).

If ω_1/ω_2 is different from $\frac{2}{3}$ and $\frac{2}{5}$, for small enough energies no resonant periodic orbits appear at all. For larger energies both types of periodic orbits appear (3-periodic and 5-periodic), one type first and the other later (for somewhat larger H) depending on the value of ω_1/ω_2. The corresponding sets of islands are then well separated by 'regular' invariant curves (Figure 4). These sets of islands are approximately described by the Hamiltonians (10) and (12) respectively.

However, for even larger energies, the islands of the Hamiltonians (10) and (12) overlap. Then some invariant curves of the Hamiltonian (14) are not closed any more, but form very complicated patterns. The successive points of intersection seem scattered at random (Figure 5). In such cases we speak about 'dissolution' of invariant curves.

This 'interaction of resonances' appears even for energies much smaller than that needed for overlapping of the original resonances. Its effects are the following:

(a) Whenever the rotation number of the unperturbed Hamiltonian (6) is rational there is no more an infinity of periodic orbits, but only two periodic orbits, one stable and one unstable. The stable orbit is represented by a number of invariant points surrounded by islands (see, e.g., the 7 islands of Figure 4).

(b) We can define a new rotation number around each stable invariant point. When this new rotation number is rational we have second order islands around the main islands (Figure 4). In the same way we have third order islands, etc.

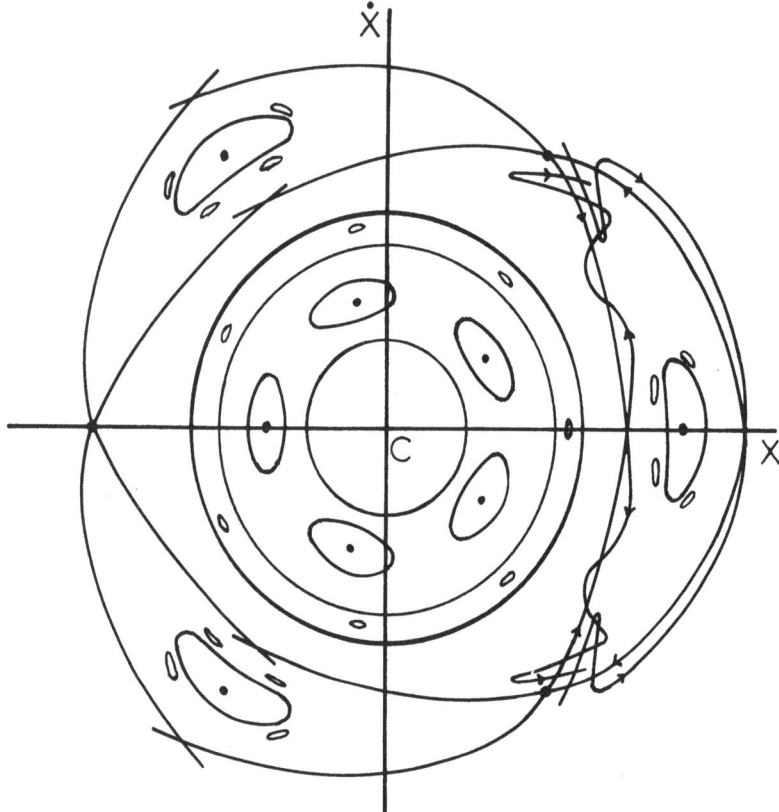

Fig. 4. Interaction of two resonances (schematically). Although the sets of 3 and 5 islands are separated by invariant curves closing around the central invariant point C, secondary islands are formed around the origin (e.g. 7 islands), and around the 3 main islands (e.g. 4 islands). Furthermore the asymptotic curves intersect each other an infinite number of times at homoclinic points.

(c) The asymptotic curves emanating from the invariant points of the same unstable periodic orbit do not join, any more, the successive invariant points, but intersect each other along an infinity of points, which are called 'homoclinic points' (Figure 4) (Poincaré, 1899). Such points generate doubly asymptotic orbits, approaching the same periodic orbit asymptotically, as $t \to \infty$, and $t \to -\infty$.

(d) At the same time the asymptotic curves from one invariant point intersect the asymptotic curves from invariant points of different multiplicities. Such points of intersection are called 'heteroclinic points' (Poincaré, 1899). The corresponding orbits are asymptotic to two different periodic orbits as $t \to \infty$ and $t \to -\infty$.

Heteroclinic points in our Galaxy were found by Martinet (1974).

An indication that the appearance of heteroclinic points is quite general is the fact that the rotation curves calculated empirically show discontinuities near the unstable periodic orbits (Figure 6) (Contopoulos, 1967). The rotation number changes abruptly in a small region near an unstable point and at the boundaries of the regions of islands. In these regions there is an infinity of resonances, which interact with each other.

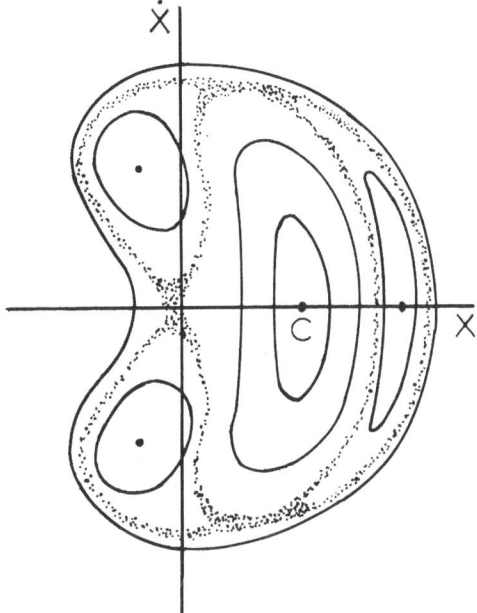

Fig. 5.   The 'dissolution' of invariant curves appears first near the unstable invariant points, where the asymptotic curves form an intricate net. For larger energies the region of dissolution increases.

A proof that an infinity of heteroclinic orbits follows the appearance of homoclinic points was given recently by Moser (1973).

The above four effects are the basic characteristics of nonintegrability. The most conspicuous, however, is the fourth characteristic, which shows clearly the interaction of resonances. It is obvious that in any region that we have interaction of resonances we cannot define closed invariant curves.

On the other hand even in nonintegrable systems there are closed invariant curves. Moser (1962) and Arnold (1963) proved rigorously the existence of invariant surfaces in phase space for small enough energies (or, what is equivalent, small enough perturbations). Their intersections by a surface of section are closed invariant curves around the stable invariant points. If the energy is small enough, the set of regular invariant curves has almost the totality of measure, although between any two invariant curves there are regions containing islands and nets of intersecting asymptotic curves.

In practical applications we find empirically that good invariant curves exist in most problems of actual interest, even for large enough energies.

Whenever the Hamiltonian is given in the form (1), or (5), the higher order terms are small for small energies. Thus the resonance effects are small, except if the ratio ω_1/ω_2 is exactly rational. One can perform canonical transformations of variables that bring the Hamiltonian in a normal form (Whittaker, 1904; Birkhoff, 1927) i.e. H is expressed

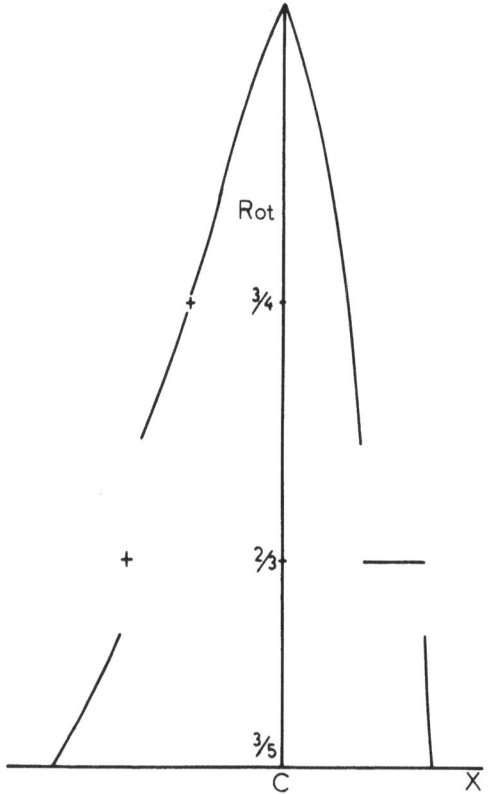

Fig. 6. The discontinuities of the rotation curve near the unstable invariant points (+) indicate that we have interaction of several neighboring resonances. The corresponding asymptotic curves intersect each other along heteroclinic points.

as a function of the new actions $\tilde{I}_1$, $\tilde{I}_2$ only. Thus we have an integral of the form

$$\tilde{I}_1 = I_1 + \text{higher order terms}, \tag{15}$$

besides the Hamiltonian, where the higher order terms are series in the original variables. A similar integral was developed in Stellar Dynamics, where it is known as a 'third' integral (Contopoulos, 1960).

The series (15) in general diverges; however, it represents asymptotically the regular invariant surfaces, whose existence was proved by Moser and Arnold.

On the other hand such a series is not applicable near resonances, because it contains an infinity of terms with small divisors of the form $(m\omega_1 - n\omega_2)$. These small divisors are responsible for the appearance of islands on the surface of section.

Near each resonance one can construct a different form of the 'third' integral. E.g., if the rotation number is near $\frac{2}{3}$, we can find an integral

$$J_{23} = 2J_1 + 3J_2 + \text{higher order terms.} \tag{16}$$

Such a form is applicable also away from all resonances. Near a different resonance a different integral is needed. A computer program, developed a few years ago (Contopoulos, 1966), calculates the algebraic form of the third integral near every resonance of interest, or away from all resonances, for any Hamiltonian of the form (1), or (5).

If the Hamiltonian contains more than one type of resonant terms (i.e. terms with more than one value of n/m) the formal integral (15) contains an infinity of small divisor terms, that do not appear in the Hamiltonian (5). This explains why in Figure 4 we see 7 islands (and, in fact, one finds infinite types of resonant islands) although the Hamiltonian contains only terms of multiplicity 3 and 5.

The resonance effects become larger as the value of the Hamiltonian increases. It was found (Contopoulos, 1967) that if the ratio ω_1/ω_2 is near the integer n/m (i.e., if the difference $|(\omega_1/\omega_2) - (n/m)|$ is smaller than a quantity of $O(H)$ we find m islands. The area covered by these islands is of the order of $H^{(n+m-4)/4}$. At the same time Equation (8) gives approximately the positions of the various islands. Therefore the theory of the 'third' integral allows the calculation of the value of H for which two nearby resonances interact, and produce a 'dissolution' of invariant curves.

This interaction, and the corresponding dissolution, become much stronger for larger H. This is more marked if the interacting resonances correspond to relatively large m and n. Therefore there is a kind of threshold energy, H_d, above which the dissolution is very conspicuous. In specific problems we can plot the proportion of the

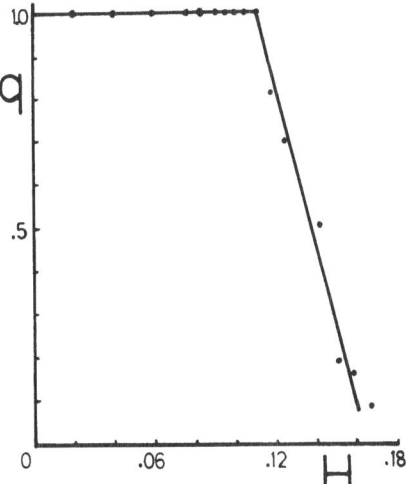

Fig. 7.   The proportion, q, of the space covered by good invariant curves vs. the energy, H, in a case studied by Hénon and Heiles (1964). The transition from near integrability to near ergodicity is rather abrupt.

surface of section covered with good invariant curves versus the energy. Thus we find a curve like the one of Figure 7. We see that the transition from cases with very small dissolution to cases where the dissolution is practically complete is rather abrupt.

Thus we can state the following. For energies H much smaller than the threshold energy, H_d, the system behaves like an integrable one. In such a case we have practically only regular invariant curves (closing around the central invariant point) and, possibly, one main set of islands, corresponding to one particular resonant term in the Hamiltonian, if ω_1/ω_2 is near a rational number n/m.

The islands corresponding to other resonances are very small and in many problems they may be disregarded. In fact ω_1/ω_2 can be approximated by an infinity of rationals n/m, but for most of them n and m are large, therefore the corresponding islands are extremely small.

On the other hand for energies much larger than H_d there are practically no closed invariant curves at all and the system is almost ergodic. This means that most orbits approach arbitrarily closely the greatest majority of points on the surface of constant energy. The points of intersection are scattered in a random way on the surface of section.

The remaining problem is whether real ergodicity is ever reached, or whether there remain always small regions covered with closed invariant curves.

In particular we can check if all periodic orbits are unstable, because otherwise the system cannot be ergodic.

The study of highly perturbed dynamical systems (for large H) indicates the following (Contopoulos, 1970). If we calculate the stability index (the trace of the monodromy matrix) of various families of periodic orbits as a function of H we find curves like those of Figure 8. Every particular family becomes eventually unstable,

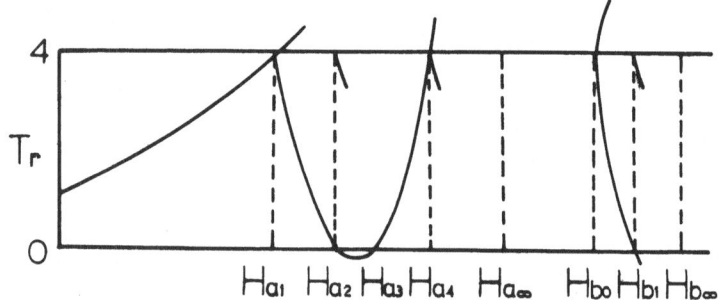

Fig. 8.   Stability diagram (trace vs. energy) of some families of periodic orbits (schematically). If the trace is between 0 and 4 the orbits are stable, otherwise unstable. The orbits of the same group of families (a, or b) can be transformed to each other continuously by varying the energy H. However two different groups are quite independent.

but a new stable family starts whenever the characteristic curve crosses the boundary of the stable region at an angle (not perpendicularly), e.g. at the points H_{a1}, H_{a2}, ...
However the range of values of energy for which these new families remain stable is smaller and smaller. Thus it seems that the total range of stability of the group of

families generated from the original one (the one that appears also for $H=0$) is finite. Beyond the limit $H_{a\infty}$ all the families of this type become unstable.

However there are new families independent of the above, which appear only beyond a certain minimum value of $H=H_{b0}$. The corresponding characteristic curve intersects the boundary of the stability region in Figure 8 perpendicularly, generating two families of periodic orbits, one stable and one unstable. The stable family becomes again unstable at H_{b1}, generating a new stable family, etc.

There is again a limit, $H_{b\infty}$, beyond which all the families of this type are unstable. The situation reminds the spectrum of a star, with groups of lines, like the Lyman and Balmer series.

The question remains open whether there is an upper limit of the values $H_{a\infty}, H_{b\infty}, \ldots$ beyond which all periodic orbits are unstable. In such a case systems with large enough energy would be probably genuinely ergodic. However, for large enough H usually most orbits are escaping.

If, on the other hand, there exist always stable families, for all H, there are always small islands of stability and there is never real ergodicity. In such a case the regions of stability would be extremely small for large enough H, therefore we may speak of 'practical ergodicity' in the sense that every orbit can approach any point on the energy surface to a distance $\leqslant \delta$, where δ is a small, but not arbitrarily small number.

There are further problems of interest concerning systems that are ergodic or 'practically ergodic'. One question is how long does it take for an initial distribution to get randomized. It is known that orbits starting along an asymptotic curve from an unstable periodic orbit deviate exponentially, like $e^{\lambda t}$. However the parameter λ differs from one orbit to the other. In many cases we define an average value $\bar{\lambda}$ empirically, by calculating several orbits. However a similar system with different energy may have a smaller or large $\bar{\lambda}$, therefore it may reach randomness slower or faster than the first, although both systems are ergodic.

These examples indicate that we have to be particularly careful in applying the methods of statistical mechanics in particular systems. Because (a) the system may be integrable or approximately integrable, in which case the assumption of equal a priori probability has to be applied to a space of fewer dimensions than the energy surface, and (b) even if the system is ergodic, or very nearly ergodic, the time needed for a particular quantity to get randomized may be long, or it may depend on the initial distribution.

Such problems are of even greater interest in cases of many degrees of freedom.

Further reviews of the problems connected with the transition from integrable to ergodic systems are given by Walker and Ford (1969), Chirikov (1971), Galgani and Scotti (1972), Ford (1973), and Contopoulos (1973).

3. Many Degrees of Freedom

In the case of many degrees of freedom we can also divide the systems into integrable, ergodic, and intermediate.

3.1. INTEGRABLE SYSTEMS

A trivial example is a set of n uncoupled oscillators. A nontrivial case is the Toda lattice. This is a set of particles, x_i, on a string, attracted by its neighbours according to the law

$$f_i = \exp\left[-(x_i - x_{i-1})\right].$$

Therefore we have

$$\ddot{x}_i = f_i - f_{i+1} = \exp\left[-(x_i - x_{i-1})\right] - \exp\left[-(x_{i+1} - x_i)\right].$$

This example is extremely interesting because it seems of a quite general type. One might think that such a force law is so different from the linear law that characterizes the uncoupled oscillators, that one should expect ergodicity. However, Hénon (1974a) proved rigorously that this system is integrable.

3.2. ERGODIC SYSTEMS

Sinai (1970) proved rigorously that the hard sphere gas is ergodic. In this case the orbits are straight lines until a collision occurs, when we have perfect reflection. This result of Sinai is particularly important and it provides a rigorous foundation of statistical mechanics, at least for the hard sphere gas. It looks probable that it is also valid in cases of purely repulsive forces. However it is doubtful whether it applies in cases where we have attractive forces, or if the forces are attractive for large distances and repulsive for small distances.

3.3. INTERMEDIATE

A well known intermediate case is the celebrated Fermi-Pasta-Ulam problem (1955). This is the case of N particles on a string attracted with non linear forces

$$f_i = -k^2(x_i - x_{i-1}) - \alpha(x_i - x_{i-1})^2.$$

This problem can be reduced, by a linear transformation of variables, to a set of N coupled oscillators. Fermi, Pasta and Ulam expected that, if one mode is excited initially, the energy would be shared by all modes because of the non-linearity. However, it was found, by numerical experiments, that only a few modes shared part of the energy. The exchange of energy between the various modes took place in an almost periodic way. Therefore the system was definitely not ergodic.

A great literature has developed following this unexpected result. Among other developments I should mention the representation of a continuous string (i.e. the case when the number N tends to infinity) by a Kortewert-de Vries equation by Kruskal and Zabusky, which led to the idea of solitons (Zabusky, 1967). One important result in this area was the discovery of an infinity of conserved quantities for this problem (Miura, Gardner and Kruskal 1968). Recently Zakharov and Faddeev (1972) proved that the Kortewert-de Vries equation can be considered as an integrable problem of infinite degrees of freedom and this explains the conservation laws of Kruskal and his associates.

Many numerical experiments were made in the case of N finite with different laws of attraction, and different energies. In general, it was found that there is a threshold of energy, above which the behavior of the system is ergodic.

In particular many numerical experiments were made with a Lennard-Jones inter-particle potential

$$V_{i, i-1} = 4\varepsilon\left[\left(\frac{1}{x}\right)^{12} - \left(\frac{1}{x}\right)^{6}\right],$$

where $x = x_i - x_{i-1}$, and ε is the depth of the potential well, i.e. the escape energy from the minimum of potential.

The force is attractive for large distances, while it is repulsive for small enough distances. This potential is assumed to govern the interaction of molecules in a fluid or in a solid, therefore it is of particular interest in physics. Stoddard and Ford (1973) studied this problem numerically in cases of rather large energies and found practically complete stochasticity. However, Galgani and Scotti (1972) found that if the energy is small enough there is no stochasticity at all, but the problem is governed by N integrals of motion.

Their study consists of two parts. First they calculated formal integrals of motion for systems of N-degrees of freedom and checked numerically how well they are conserved if they are truncated after the terms of a given degree. For this purpose they extended to many degrees of freedom the method of Contopoulos (1966) to calculate the 'third integral' in a two-dimensional system, by performing algebraic manipulations with the help of a computer. Their first results derived for systems of 4 degrees of freedom are very encouraging. In fact for small enough energies the N integrals are conserved better and better as higher and higher order terms are added in their expressions. On the other hand for large energies the numerical values of the truncated formulas vary considerably, and do not improve by adding higher order terms. This indicates that the system is then ergodic rather than integrable.

The second method is purely numerical. They find the deviations of two systems which are very close to each other initially. This deviation is linear in time if the energy is small enough, while it is exponential if we go beyond a certain threshold. The transition is rather abrupt, and allows to define a threshold energy, H_d.

The most important result of Galgani and Scotti is the dependence of the threshold energy on the number of degrees of freedom, N. The function H_d/N runs as shown in Figure 9.

For $N=2$ it seems that H_d is infinite. The system behaves as integrable, even for large energies. For $N=4$ we have $H_d/N \simeq 4$ but as N increases H_d/N tends to 1 in the particular units used. It is remarkable that the value H_d/N remains near 1 even if N becomes as larger as 500.

If this result is valid also when $N \rightarrow \infty$ (and, in my opinion, the fact that $H_d/N \simeq 1$ from $N=8$ to $N=500$ is a very good indication for that) then we can derive many interesting conclusions. In particular one can define a zero-point energy, which is usually considered as a pure quantum-mechanical effect. This allows one to derive a

classical analogue of Planck's law, as it was first done by Einstein and Stern (1913).
Galgani has drawn some far reaching, but very interesting conclusions concerning a
classical foundation of quantum mechanics from these studies.

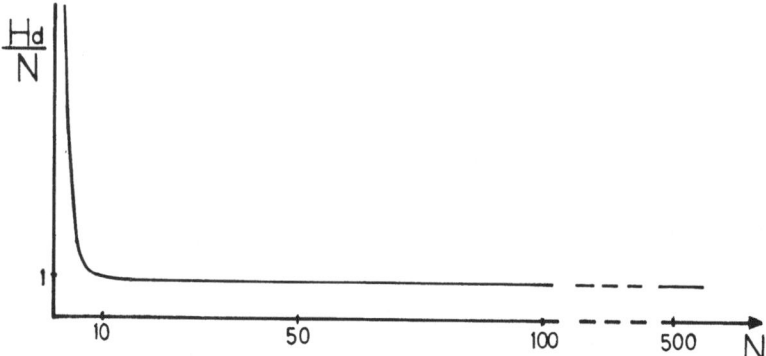

Fig. 9.   The ratio H_d/N (transition energy divided by the number of particles) as a function N for a
Lennard-Jones interparticle potential (after Galgani and Scotti).

Another problem of more than two degrees of freedom, where new important results
have been found quite recently, is the general three body problem. I will report here
on some recent work by Hadjidemetriou in Thessaloniki. Hadjidemetriou uses a
rotating frame of reference, whose origin is the center of mass of the bodies m_1 and m_2,
while the x-axis contains always these two bodies; thus the planar three-body problem
is reduced to a system of three degrees of freedom, with coordinates x_1, x_3, y_3. Then it
is proved that all families of periodic orbits of the restricted three-body problem can be
extended to the general three-body problem.

The theoretical proof was implemented by numerical calculations for increasing
values of the mass of the third body. Eventually families of periodic orbits were found
for three equal masses. A study of the stability of these orbits indicates that there may
be some rather open triple systems, which are stable. Similar results were found by
Hénon (1974b).

Further, Hadjidemetriou studied the behavior of non periodic orbits in the vicinity
of stable periodic orbits. At every intersection of an orbit by the plane $y_3 = 0$ the values
of $x_1, \dot{x}_1, x_3, \dot{x}_3, \dot{y}_3$ were calculated. If there are two more integrals of motion besides
the energy, one can eliminate two of these quantities and find the equation of an
invariant surface in a reduced three-dimensional space, say $(x_1, \dot{x}_1, x_3)$. This corresponds to an invariant curve on a surface of section in the 2-dimensional case.

The successive points on the invariant surface can be joined by a continuous curve.
The projection of such a curve on the plane $(x_1, \dot{x}_1)$ is given in Figure 10. One can then
define three mean periods. The first, P_1, is the period between two successive intersections with the plane $y_3 = 0$. The second, $P_2 \sim 50 P_1$, is the period of one loop in
Figure 10, while the third, P_3, of the order of some $100 P_2$, is the period needed for the
loops to complete a libration, or rotation. It is remarkable that the 3 periods above are
of quite different orders of magnitude.

The loops of Figure 10 were drawn empirically by joining the projections of the successive points of intersection of an orbit by the surface $y_3 = 0$. The fact that these loops are smooth curves indicates the existence of two integrals of motion besides the energy in the general three body problem.

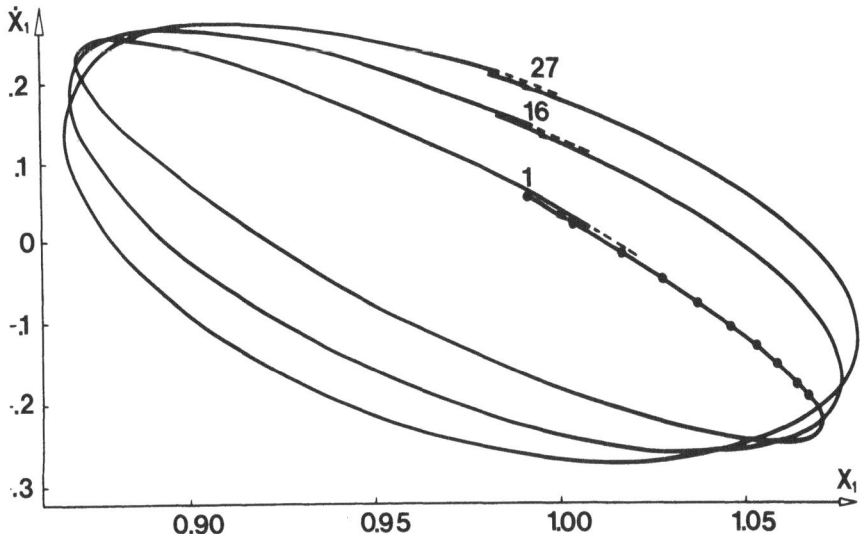

Fig. 10. Projection on the plane $(x_1, \dot{x}_1)$ of the curve joining the successive intersections of an orbit by the surface $y_3 = 0$. Only the 1st, 16th and 27th loops are marked, and the 10 first points on the 1st loop (after Hadjidemetriou).

In one particular case it was found that the curve on the plane $(x_1, \dot{x}_1)$ is closed, i.e. successive loops coincide. This phenomenon is not well understood. It means either that the libration period, P_3, is extremely large, or that there are four integrals of motion in this special problem, which is rather improbable.

The above phenomena appear whenever the third body is far from the two other bodies. If the distance of the third body is small the successive points of intersection are no more on smooth loops. However in many cases the orbits do not escape to infinity, although the energy integral does not provide such a restriction. On the other hand orbits far away from periodic orbits eventually escape.

The fact that no escape appears for a large set of orbits and for long times indicates that no Arnold diffusion (Arnold and Avez, 1967) is operative in these systems, or that its time scale is very large.

An extension of this study to systems of more than three bodies has already started in Thessaloniki. This research opens important new possibilities for the general N-body problem. By studying the stable families of periodic orbits one will be able to separate the regions of stability in each case. In such regions one has to take into account the effects of $3N$ integrals of motion, instead of the 10 classical ones. This will allow a better statistical treatment of N-body systems.

I cannot make many predictions about the future developments in these very interesting areas of many degrees of freedom. But I can be sure of one thing. That they will keep us busy for the whole next decade.

References

Arnold, V. I.: 1963, *Uspekhi Mat. Nauk* **18**, 91; Transl. Air Force Systems Command, Wright Patterson Air Force Base, Ohio.

Arnold, V. I. and Avez, A.: 1967, *Problèmes Ergodiques de la Mécanique Classique*, Gauthier-Villars, Paris.

Birkhoff, G. D.: 1927, *Dynamical Systems*, American Math. Soc., Providence. R.I.

Chirikov, B. V.: 1971, *Nucl. Phys. Inst. Siberian Sect. USSR Acad. Sci.*, Rep. No. 267; CERN Transl. 71–40.

Contopoulos, G.: 1960, *Z. Astrophys.* **49**, 273.

Contopoulos, G.: 1966, *Astrophys. J. Suppl.* **13**, 503.

Contopoulos, G.: 1967, *Bull. Astron.* Ser. 3, **2**, 223.

Contopoulos, G.: 1970, *Astron. J.* **75**, 108.

Contopoulos, G.: 1973, in L. Martinet and M. Mayor (eds.), *Dynamical Structure and Evolution of Stellar Systems*, Swiss Society of Astronomy and Astrophysics Third Advanced Course, Geneva Observatory, p. 52.

Einstein, A. and Stern, O.: 1913, *Ann. Phys.* **40**, 551.

Fermi, E., Pasta, J., and Ulam, S.: 1955, Los Alamos Sci. Lab. Rep. LA-1940.

Ford, J.: 1973, *Adv. Chem. Phys.* **24**, 155.

Galgani, L. and Scotti, A.: 1972, *Rev. Nuovo Cimento* **2**, 189.

Hénon, M.: 1974a, *Phys. Rev. Letters* (in press).

Hénon, M.: 1974b, *Celes. Mech.* **10**, 375.

Hénon, M. and Heiles, C.: 1964, *Astron. J.* **69**, 73.

Martinet, L.: 1974, *Astron. Astrophys.* **32**, 329.

Miura, R. M., Gardner, C. S., and Kruskal, M. D.: 1968, *J. Math. Phys.* **9**, 1204.

Moser, J.: 1962, *Nachr. Akad. Wiss. Göttingen*, Math. Phys. Kl., 1.

Moser, J.: 1973, *Stable and Random Motions in Dynamical Systems*, Princeton Univ. Press.

Poincaré, H.: 1899, *Les Méthodes Nouvelles de la Mécanique Céleste*, Gauthier-Villars, Paris.

Sinai, Ya.: 1970, *Russian Math. Surveys* **25**, 137.

Stoddard, S. D. and Ford, J.: 1973, *Phys. Rev. A* **8**, 1504.

Walker, G. H. and Ford, J.: 1969, *Phys. Rev.* **188**, 416.

Whittaker, E. T.: 1904, *Analytical Dynamics of Particles and Rigid Bodies*, Cambridge Univ. Press.

Zabusky, N. J.: 1967, in *Nonlinear Partial Differential Equations*, Academic Press, New York.

Zakharov, V. E. and Faddeev, L. D.: 1972, *Funct. Anal. and Appl.* **5**, 280.

DISCUSSION

Miller: In these solutions of the three-body problem, if you somehow bound the relevant part of the phase space, what fraction of that part is filled with these 'non-ergodic' orbits?

Contopoulos: It is still too early to give numerical results. My impression is that the set of bounded orbits is small. However as this set is not infinitesimal one cannot be sure what will be the final evolution of a particular triple system by taking initial conditions at random. This is why Dr Hadjidemetriou started a systematic exploration of the regions of stability.

Lecar: Is there an analog to the triangular points in the problem with three equal masses?

Contopoulos: Yes. Motions near the triangular equilibrium points in the planar general three body problem have been studied already (see e.g., Siegel, C. L. and Moser, J. K.: 1971, *Lectures on Celestial Mechanics*, Springer, Berlin, p. 113).

Lynden-Bell: With such widely different periods involved as 1, 50 and 500 are not the apparent integrals directly related to the adiabatic invariants?

Contopoulos: Yes. I believe that the new integrals are, in fact, to be considered as adiabatic invariants.

Froeschle: I ask further explanations about the appearance of 'wild behaviour' in the model used by Galgani and Scotti.

Scotti: The threshold turns out to be one only because of the particular units employed.

Galgani: I think the question of Dr Froeschlé will be answered by the following clarification. The figure discussed here refers to a class of computations with a fixed kind of initial conditions, for example equipartition of energy among the normal modes and phases zero. Then the unique remaining parameter is the value of the energy, which just appears in Figure 7 as abscissa (what is reported in ordinate has already been said).

I would like to add now some considerations. Obviously a graph as the one referred to above should be given for all classes of initial conditions, which is a formidable problem, actually that of exploring a surface of $2N-1$ dimensions if N is the number of particles. In this connection I can only say that many classes of initial conditions have been considered and in all of them the threshold was well defined and did not apparently vary from case to case. This, I believe, gives meaning and support to the statement that there is something as a threshold energy which is proportional to N. Naturally we expect that the situation be much more complicated and even we attach great importance, from the point of view of principle, to the circumstance, that the statement given above should not be taken literally. Indeed the presence of nonstochastic regions on every energy surface is of fundamental importance in determining the statistics, i.e. the dynamically correct invariant measure, on the stochastic region. But this, I hope, is work for the near future, while the rough statement given above was directed to disprove the often expressed opinion that the Kolmogorov-Arnold-Moser theorem guaranteeing the existence of invariant tori, or, as we say, of ordered motions, should be of no interest for physical systems of many degrees of freedom.

ON THE SENSITIVITY OF THE
BEHAVIOUR OF STELLAR ORBITS TO THE PARAMETERS
OF SIMPLE GALACTIC MASS MODELS

LOUIS MARTINET

Observatoire de Genève, Sauverny, Switzerland

Abstract. Some conjectures are given about the sensitivity of the behaviour of the stellar orbits (appearance of resonant orbits, possible interaction of resonances etc.) to the set of parameters characterizing a simple realistic 'spheroid + nucleus' mass model of our Galaxy. Connections with a systematic exploration of orbits in Schmidt mass models are mentioned. Finally we present arguments in favour of the existence of a third integral for the whole solar neighbourhood old stellar populations, independently of the chosen model.

1. Introduction

In recent years, we have extensively studied the orbital properties of old stellar populations in our Galaxy represented by an axisymmetric stationary potential $\Phi(\varpi, z)$ (Martinet and Hayli, 1972 (I); Mayer and Martinet, 1973 (II); Martinet and Mayer, 1975 (III)). In fact these objects can give us information on the early stages of the dynamic galactic evolution. But the point is: What are the changes in the orbits when the proposed force function changes so that the rotation curve or the acceleration perpendicular to the galactic plane, as well as the constants used in the model, take all permissible values within the limits set by the errors of observation? For our systematic exploration of galactic orbits, we used both versions of a galactic model given by Schmidt in 1956 and 1965 (noted I and II henceforth). We shall not discuss here what should be the 'best' mass model of our Galaxy. Until recently, the Schmidt models were those which permitted the computation of numerous orbits without excessive expense of time. The main types of features we have found in our exploration also appear in Innanen's models which seem to answer better the criteria of the 'best model' (Innanen *et al.*, 1972). It is not easy to answer the above question by comparing the results from both versions of the Schmidt potential, because for the first we use an interpolation formula and the 2nd includes a nucleus, a very flattened spheroid and an outer shell. Here we shall use some results which we have obtained in the three papers mentioned before as a guide and as a checking for a first approach of a more general study which consists of seeing the influence of different accessible parameters which characterize a galactic mass model on the existence of important resonances and on the possible appearance of 'wild' behaviour of orbits. Having a care for using the 'best' model, we ought to include in a complete theory a superposition of non homogeneous spheroids but the calculations which follow seem sufficient in order to understand the main effects, at least for a first approach.

Hayli (ed.), Dynamics of Stellar Systems, 227–235. All Rights Reserved.

2. Oscillations in the Meridian Plane

Let us consider the classical expansion for the density law corresponding to the potential $\Phi(\varpi, z)$ for a flattened spheroid,

$$\varrho(\alpha) = \frac{p_{-2}}{\alpha^2} + \frac{p_{-1}}{\alpha} + p_0 + p_1\alpha + p_2\alpha^2 + \cdots \quad \text{with} \quad \alpha = \sqrt{\varpi^2 + (1-e^2)^{-1}z^2}$$

(1)

to which corresponds, for the rotation velocity $\Theta_c(\varpi)$, the expansion

$$\frac{\Theta_c^2}{\varpi^2} = \frac{v_{-2}}{\varpi^2} + \frac{v_{-1}}{\varpi} + v_0 + v_1\varpi + v_2\varpi + \cdots$$

(2)

by using the relation

$$\Theta_c^2 = 4\pi G\sqrt{1-e^2} \int_0^{\varpi} \frac{\varrho(\alpha)\, \alpha^2\, d\alpha}{\sqrt{\varpi^2 - \alpha^2 e^2}}$$

with e = eccentricity of the spheroid.

It is easy to compute the ratio $(A_c/B_c)^{1/2}$ of the infinitesimal oscillation frequencies in a meridian orbital plane, where the motion takes place, controlled by the 'reduced' potential

$$U(\varpi, z) = \frac{J^2}{\varpi^2} + \Phi(\varpi, z)$$

J being the angular momentum.

Expanding U around $\varpi = \varpi_c$, $z = 0$, where ϖ_c is defined by

$$\frac{J^2}{\varpi_c^3} = \left(\frac{\partial \Phi}{\partial \varpi}\right)_{\varpi = \varpi_c}$$

we have

$$U(\varpi, z) = U_0 + \tfrac{1}{2}\left(A_c(\varpi - \varpi_c)^2 + B_c z^2\right) - \varepsilon(\varpi - \varpi_c)\, z^2 - \eta(\varpi - \varpi_c)^3 + \cdots, \quad (3)$$

where

$$A_c = \frac{3J^2}{\varpi_c^3} + \left(\frac{\partial \Phi}{\partial \varpi}\right)_{\substack{\varpi = \varpi_c \\ z = 0}} = -\left(\frac{3K_\varpi}{\varpi} + \frac{\partial K_\varpi}{\partial \varpi}\right)_{\substack{\varpi = \varpi_c \\ z = 0}}$$

(4)

and

$$B_c = -\left(\frac{\partial^2 \Phi}{\partial z^2}\right)_{\substack{\varpi = \varpi_c \\ z = 0}} = \left(\frac{K_\varpi}{\varpi} + \frac{\partial K_\varpi}{\partial \varpi}\right)_{\substack{\varpi = \varpi_c \\ z = 0}} + 4\pi G\varrho(\varpi_c).$$

(5)

With the previous relations and with

$$J_n = \frac{e^n}{\sqrt{1-e^2}} \Big/ \int_0^{\varpi} \sin^{n-1}\vartheta\, d\vartheta$$

$$\frac{A_c}{B_c}=\frac{\dfrac{2v_{-2}}{\varpi_c^2}+\dfrac{3v_{-1}}{\varpi_c}+4v_0+5v_1\varpi_c+6v_2\varpi_c^2+\dfrac{C}{\varpi_c^3}}{\dfrac{J_1v_{-2}}{\varpi_c^2}+\dfrac{(J_2-1)\,v_1}{\varpi_c}+(J_3-2)\,v_0+(J_4-3)\,v_1\varpi_c+(J_5-4)\,v_2\varpi_c^2+\dfrac{C}{\varpi_c^3}},$$

$$(6)$$

where the term C/ϖ_c^3 has been introduced in order to take into account the possible existence of a massive nucleus. Two features must be emphasized: (a) Without the presence of a nucleus or in regions where the influence of such a nucleus is not important, the ratio $(A_c/B_c)^{1/2}$ is approximately constant on a large range of ϖ_c, as long as e is not too far from 1 (thus this property is connected with the mode of construction of the models as well as with the eccentricity of the spheroids); (b) The presence of a massive nucleus introduces a significant variation of $(A_c/B_c)^{1/2}$ essentially for small ϖ_c These features appear in Figure 1 which shows the influence of e and of the mass of the nucleus on the ratio $(A_c/B_c)^{1/2}$ for a model in which only v_{-1} and v_1 are different from zero. As shown below, $(A_c/B_c)^{1/2}$ is also modified if we change the value of the mass density at the Sun, $\varrho_\odot$ (which is used by Schmidt to fix the value of e). It is easy to repeat the discussion for a density law corresponding to another choice of the coefficients $v_i\neq0$ (i.e. v_0 and v_{-1}). The main features mentioned below remain unchanged.

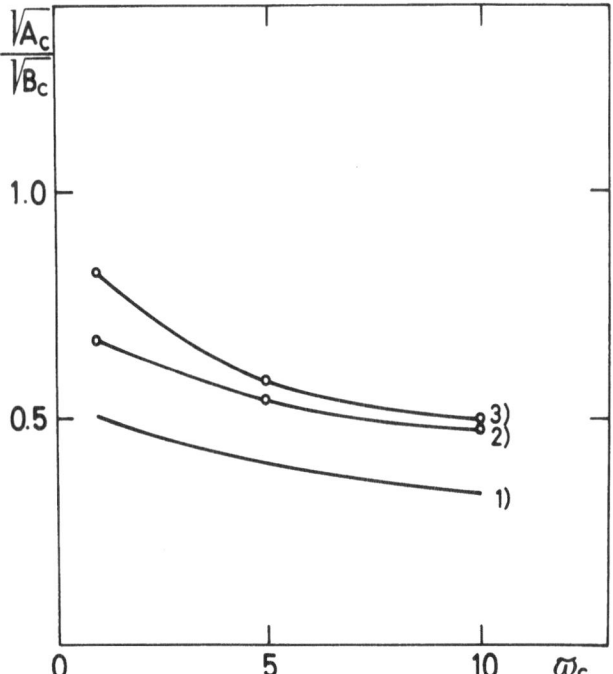

Fig. 1.   Ratio of oscillation frequencies in the meridian plane $(A_c/B_c)^{1/2}$ vs $\varpi=\varpi_c$ for a 'nucleus+spheroid' model characterized by a rotation curve $\Theta_c=\sqrt{v_{-1}\varpi+v_1\varpi^3+(C/\varpi)}$ for different values of the parameters: (a) $v_{-1}=10000$, $v_1=-40$, $C=30000$ and $e=$ eccentricity of the spheroid $=0.9988$; (b) $v_{-1}=10000$, $v_1=-40$, $C=30000$ and $e=0.995$; (c) $v_{-1}=10000$; $v_1=-40$, $C=120000$, $e=0.995$.

Figure 2 gives $(A_c/B_c)^{1/2}$ vs ϖ_c for the models existing in the literature (Einasto, 1970; Innanen, 1971). The quasi-invariance of $(A_c/B_c)^{1/2}$ on a large range of ϖ_c is explained well by the discussion above, which also allows to interpret the differences between the curves for $\varpi_c < 10$ kpc as connected with differences in the eccentricity of the spheroid and in the mass of the nucleus. (From this point of view, Einasto's and Schmidt's models with a nucleus and a spheroid, the surface boundary of which passes through the Sun, are directly comparable in this range of ϖ_c: Einasto's spheroid is slightly less flattened and the nucleus mass is more important).

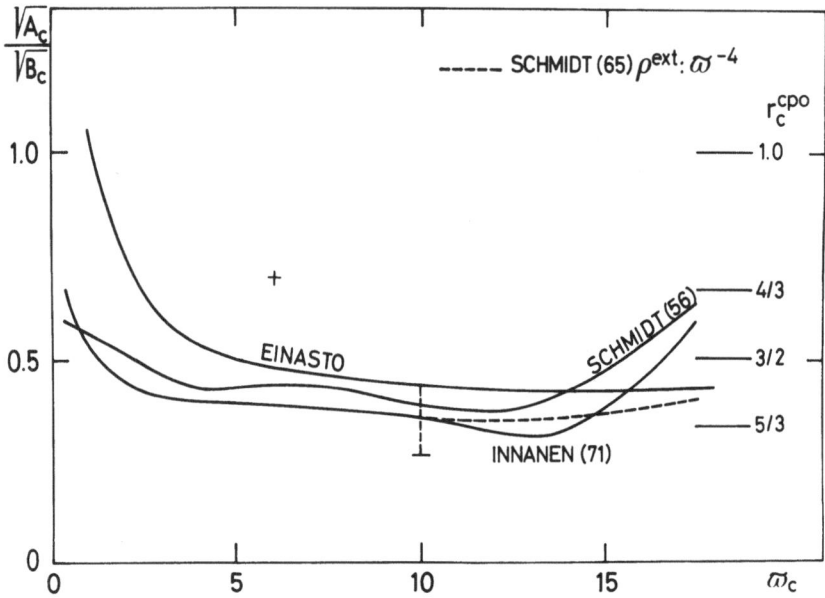

Fig. 2. $(A_c/B_c)^{1/2}$ vs ϖ_c for different usual models. For the different symbols, see the text.

The vertical bar at $\varpi_c = 10$ kpc shows the possible values of $(A_c/B_c)^{1/2}$ which we have, taking into account the 'observational' uncertainty to be considered for $\varrho_\odot$ and with the standard values of Oort Constants $(A=15, B=-10)$. The upper limit corresponds to $\varrho_\odot = 0.09 \mathcal{M}_\odot$ pc^{-3} and the lower limit to $0.27 \mathcal{M}_\odot$ pc^{-3}. As the invariance of the ratio $(A_c/B_c)^{1/2}$ on a more or less large range of ϖ_c is a general property of the models based on spheroids, we are able to guess what are the 'plateau' values of $(A_c/B_c)^{1/2}$ which we obtain, taking account of the errors on $\varrho_\odot$. We added a point ($\times$) representing the value of $(A_c/B_c)^{1/2}$ which we obtain using the Oort constants recommended by Clube with B_c computed from $\varrho_\odot = 0.15 \mathcal{M}_\odot$ pc^{-3}.

We infer from the diagram that no *exact* important resonance (corresponding to $(A_c/B_c)^{1/2} = \frac{2}{3}, \frac{1}{1}$ or $\frac{4}{3}$) is obtained for ϖ_c not too small, unless rather extreme (but not excluded!) local galactic structure parameters are chosen.

3. Resonant Orbits

In non-linear problems such as the present one, the properties of the orbits do not depend critically on the exact value of $(A_c/B_c)^{1/2}$. If $(A_c/B_c)^{1/2}$ is 'near' a rational number n/m, we expect to find resonant periodic orbits closing after n oscillations along the ϖ-axis and m oscillations along the z-axis. If the energy E increases, $(A_c/B_c)^{1/2}$ is 'approached' by more and more rational numbers in the sense that $|(A_c/B_c)^{1/2} - (n/m)|$ is of the order of $E - E_c$, where E_c corresponds to the circular energy. Hence, new types of resonant periodic orbits must appear. Our numerical explorations confirm this expectation. In both Schmidt potentials, orbits with $n/m = \frac{1}{1}$ or $\frac{4}{3}$ (not very near $(A_c/B_c)^{1/2} \approx 0.4$) end up in appearing provided the energy is high enough. Here we must emphasize that these rational values are the most important because they correspond to the periodic orbits which are surrounded by very important tubes in the configuration space. In a surface of section, more precisely in the phase plane $(\varpi, \dot\varpi)$, these tubes give the most important islands, which may occupy a large part of the accessible region (see i.e. Mayer and Martinet, 1973, Figure 9). It would be extremely useful to establish the dependence of the range Δ of values of n/m which produces resonant orbits with respect to the parameters characterizing the model. Unfortunately a theory giving Δ is available only for small values of ε in the artificial 'galactic potential' $V = \frac{1}{2}Ax^2 + \frac{1}{2}By^2 - \varepsilon xy^2 - \eta x^3$ (Contopoulos, 1973), where moreover η is neglected. Thus the theoretical results are not applicable to our realistic galactic models, for which we often have rather large ε's and where η is not always negligeable with regard to ε. However theory may indicate some qualitative tendency of behaviour for some functions of interest in our problem. For example, from the treatment of the orbits by the method of surface of section, we can use the notion of 'rotation number' (Contopoulos, 1970), that means the asymptotic value of the angle between two successive points of intersection $(\varpi, \dot\varpi)$ of the trajectory with the plane $z = 0$ in the phase plane, as seen from the 'central' invariant point $\varpi = \varpi_c$, $\dot\varpi = 0$, which corresponds to the symmetric periodic orbit crossing the ϖ-axis always at the same point $(\varpi = \varpi_c, z = 0)$. For this 'central' periodic orbit, we define the rotation number r_{cp0} as the limit of the rotation number of nearly non-periodic orbits. Further we also use the so-called characteristic diagram where symmetric periodic orbits are represented by a point (E, ϖ), where ϖ is the intersection point of the given orbit in the galactic plane $(z = 0)$, with $\dot\varpi = 0$ (see Paper II). Each family of such orbits is identified by its rotation number $r = 2 - (n/m)$ if the orbit closes after n radial oscillations and m perpendicular oscillations. For E increasing from $E = E_c$ as we move along the characteristic of the central family towards the intersection with characteristics of other families of order m, we tend for r_{cp0} to rational values $n'/m = 2 - (n/m)$. The variation of r_{cp0} with E gives the rate to which different resonant periodic orbits appear. Thus the slope dr_{cp0}/dE contains important information. Theory allows us to deduce an analytical form of it submitted to the condition that ε is small but much larger than η. In fact from the formula given by Contopoulos (1968) for r_{cp0}, we immediately deduce that

$$\frac{\mathrm{d}r_{cp0}}{\mathrm{d}E} = \frac{\sqrt{A_c}}{\sqrt{B_c}} \frac{3A_c - 4B_c}{2A_c B_c^2 (A_c - 4B_c)} \varepsilon^2 \tag{7}$$

which is of the order of ε^2/B_c^3 if $(A_c/B_c)^{1/2}$ is a constant. It can be shown (Contopoulos, 1970) that the range of values of n/m for which resonant orbits appear is proportional to this quantity. This remark suggests to examine if the existence of a linear relation between Δ, defined as $r_{cp0}^c - r_{cp0}^{min}$ (where $r_{cp0}^c = r_{cp0}(E=E_c)$ and $r_{cp0}^{min} = $ minimum accessible value of r_{cp0}) and $r' = dr_{cp0}/dE$ is a general property of more realistic galactic models, at least in the range of ϖ_c where $(A_c/B_c)^{1/2}$ is approximately a constant.

The results of our numerical exploration for both Schmidt potentials show that this property is actually satisfied although the conditions of application of the theory are not filled in general. We emphasize that in both models the slope of the linear relation (r', Δ) is nearly the same. Starting from the idea that r' could give information for determining Δ in any model given we can examine what is the tendency of the sensitivity of this function to a changing of the model parameters, at least qualitatively.

Thus we express r' as a function of C, $v_i (i = -1$ and $1)$, J_n for the 'spheroid + nucleus' model chosen previously, using (2), (4), (5) and (7). Detailed results are published in Paper III.

By considering values of v_i which lead to reasonable shapes of the rotation curve $\Theta_c(\varpi)$, we may conclude (everything else being equal) that: (a) an increase of the mass of the nucleus leads to smaller values of r' for small ϖ_c; (b) if we take for instance a larger value of v_1 (with respect to Schmidt (1965) model), we have gradually larger values of $\Theta_c(\varpi)$ from $\varpi_c = 3$ kpc, which lead to a less sharp slope r' for ϖ_c increasing; (c) if we take for $\varrho_\odot$ the value $0.09 \, \mathcal{M}_\odot \, \text{pc}^{-3}$ instead of $0.15 \, \mathcal{M}_\odot \, \text{pc}^{-3}$, $r'(\varpi_c)$ becomes systematically slightly sharper. In order to be able to predict what values of Δ we can expect in a model for any given value of parameters, it would be necessary to have several numerical experiments which overlap the conditions of interest, to construct families of relations (r', Δ) parameterized either by v_{-1} or by v_1 for instance.

A large value of r' could be a condition *among others* for the appearance of regions of instability in the phase plane $(\varpi, \dot{\varpi})$ ('wild' orbital behaviour). Another factor, important in this connection, may be guessed by considering the shape of the limit curve in the so-called characteristic diagram (E, ϖ). The equation of this curve is

$$E = \tfrac{1}{2}A_c(\varpi - \varpi_c)^2 - \eta(\varpi - \varpi_c)^3 + \cdots.$$

Neglecting the terms of higher order, we observe that the ratio of both terms of the 2nd member is in the order of $\eta/A_c^{3/2}$. If $\eta/A_c^{3/2}$ is small, the curve becomes more concave and this fact can also contribute to congest the characteristics and lead to interactions of resonances and wild orbital behaviour. Of course this effect is enhanced if r' is larger. Further in the present context this conjecture is available for values of the energy not too far from $E = E_c$. However, here again, we can deduce a qualitative tendency of the sensitivity of $\eta/A_c^{3/2}$ to the parameters of our 'spheroid + nucleus' model. As for $(A_c/B_c)^{1/2}$ or r', we write $\eta/A_c^{3/2}$ for different $\varpi = \varpi_c$ as a function of the parameters C, v_i, J_n. Detailed results are published in Paper III. For the same set of values of these parameters as chosen before, we conclude:

(a) An increase of the mass of the nucleus leads to smaller values of $\eta/A_c^{3/2}$ for small ϖ_c and could promote the appearance of interactions of resonances.

(b) Gradually larger values of $\Theta_c(\varpi)$ from 3 kpc (with respect to Schmidt (1965) values) leads to smaller values of $\eta/A_c^{3/2}$ for $\varpi_c > 6$ kpc and could promote the appearance of interactions of resonances in these regions.

(c) $\eta/A_c^{3/2}$ is independent of the eccentricity of the spheroid and consequently of the value chosen for $\varrho_\odot$.

4. Numerical Exploration in Both Schmidt Potentials and Conclusions

From our systematic exploration of orbits in both Schmidt models for a grid of values of J and E ($1.5 < J < 250$ [units 10 km s^{-1} kpc], $E_c < E < -100$ [units 100 km^2 s^{-2}]) we emphasize the following points:

(1) For small J's we found always the main periodic orbits of order m, corresponding to the rotation number $r = 2 - (n/m) = \frac{4}{3}, \frac{1}{1}$ or $\frac{2}{3}$. As mentioned before, these values are related to the main islands in the phase plane $(\varpi, \dot{\varpi})$, which correspond to important tube orbits in the configuration space (ϖ, z). For large J (> 75) we found the main periodic orbits corresponding to $r = \frac{3}{2}, \frac{4}{3}, \frac{1}{1}$. In Paper III we have presented complete statistics of families of periodic and tube orbits in the models we have considered.

(2) We found some interactions of resonances revealed by the dissolution of invariant curves in the phase plane, which correspond to a rather wild (or semi-ergodic) orbital behaviour in the configuration space, particularly for small J's. Table I summarizes the ranges of r where such interactions are observed in each model. In fact, we ought to write '$\cdots < r < \max\{r\}$', where $\max\{r\}$ is the maximum value of r, which always appears of the order of r_{cp0}^c, obtained from Figure 2.

TABLE I

J	Schmidt I	Schmidt II
1.5	$\frac{1}{2} < r < r_{cp0}^c$	$\frac{1}{2} < r < r_{cp0}^c$
45	$\frac{2}{3} < r < r_{cp0}^c$	$1 < r < r_{cp0}^c$
105	$\sim 0.9 < r < r_{cp0}^c$	Possible interactions of high order ($r \sim r_{cp0}^c$)
180	$1 < r < r_{cp0}^c$	No interaction detected

These phenomena occur at different values of E (see for instance Paper I). We note that the range of interactions is generally more important in the first model. It is difficult to try to explain such differences in terms of our discussion in the previous section because of the structural differences between the models. On the contrary, we have verified the prediction concerning the influence of a more massive nucleus on quantities such as r' or $\eta/A_c^{3/2}$: If we double the mass of the nucleus in the 2nd Schmidt model, we observe that for cases where J is rather small (i.e. 20) the values of r' and $\eta/A_c^{3/2}$ graphically and numerically determined at ϖ_c given, become weaker, especially for $\eta/A_c^{3/2}$. These modifications are accompanied by a clearer tendency to the dissolution of invariant curves for values of E relatively not too far from E_c.

(3) The previous results, particularly those from Table I, reveal that anyone wanting to apply such considerations to some real galactic populations is faced by an apparently complicated situation. For instance, if we consider initial conditions corresponding to old disc or intermediate populations in the solar neighbourhood, characterized among others by $\bar{J} > \sim 150$, no 'wildness' probably appears in the orbital behaviour, whatever the mass model may be, even if the ϖ-velocity, $\dot{\varpi}$, is of the order of 150 km s^{-1}. But for local halo objects ($\bar{J} < 150$), if we consider at J given (say $J = 45$) the values of the energy for which the dissolution begins, it is possible to find that the upper limit of $\dot{\varpi}$ which corresponds to good invariant curves in the phase plane (and consequently to the existence of a 'third' integral) is 150 km s^{-1} in Schmidt I but 250 km s^{-1} in Schmidt II! We must add that halo objects form a small fraction of the nearby star population and moreover few of them satisfy the conditions of motion which lead to a semi-ergodic behaviour. We could also emphasize that the observational fact suggesting the existence of a third isolating integral, that means the observed inequality of the meridional velocity dispersions, is also evident for the more metal-deficient local halo populations as RR Lyrae with $\Delta S > 5$ or extreme subdwarfs, which have the highest velocities with respect to the Sun. Thus from a practical point of view we can consider that a semi-ergodic behaviour in the meridian plane is quite exceptional for observed objects in the solar neighbourhood. The situation is certainly more intricate for intermediate or halo populations in more central regions of the Galaxy where they are well represented. Unfortunately a confrontation of our results with real motions is impossible in such cases because kinematical data completely fail.

Finally we can think of some extensions of the first approach presented here concerning the sensitivity of the orbital behaviour to the model parameters:

(1) With the help of several numerical experiments a complete theory turning on a system of superposed heterogeneous spheroids, considered by some people as the best way of representing the Galaxy, could allow to specify the effects here conjectured.

(2) An 'up-to-date' massive halo could be included.

(3) Later, the same type of question (resonances, dissolution, etc.) could be asked with regard to a system with three degrees of freedom (a non axisymmetric galaxy with a potential $\Phi(\varpi, \vartheta, z)$. Such a study could concern old disc and perhaps intermediate galactic populations.

References

Contopoulos, G. and Hadjidemetriou, J.: 1968, *Astron. J.* **73**, 86.

Contopoulos, G.: 1970, *Astron. J.* **75**, 96.

Contopoulos, G.: 1973, in L. Martinet and M. Mayor (eds.), *Dynamical Structure and Evolution of Stellar Systems*, Third Advanced Course of the Swiss Society of Astronomy, Geneva Observatory, p. 52.

Einasto, J.: 1970, *Tartu Astron. Obs. Teated*, No. 26.

Innanen, K.: 1971, private communication.

Innanen, K. and Ryman, A. G.: 1972, *Astrophys. Space Sci.* **17**, 447.

Martinet, L. and Hayli, A.: 1971, *Astron. Astrophys.* **14**, 103 (Paper I).

Martinet, L. and Mayer, F.: 1975, in preparation.

Mayer, F. and Martinet, L.: 1973, *Astron. Astrophys.* **27**, 199 (Paper II).

Schmidt, M.: 1956, *Bull. Astron. Inst. Neth.* **13**, 15.

Schmidt, M.: 1965, *Stars and Stellar Systems* **5**, 513.

DISCUSSION

Contopoulos: What are the most important resonances, in the sense that the tube orbits occupy a large part of phase space?

Martinet: The main values of the rotation number $r = 2 - (r/m)$ which correspond to the most important islands in the phase plane are $\frac{1}{1}$ and $\frac{2}{3}$. It may happen that these islands occupy a very large part of the accessible region in this phase plane particularly for relatively small values of the angular momentum J ($\leqslant 1500$ km s^{-1} kpc).

Contopoulos: How do you think that these tube orbits might be observed?

Martinet: Tube orbits could be connected to the existence of stellar groups or drifts of stars as such as we observe in the solar neighbourhood for example. The most important tubes we described above ought rather to be related to larger structures in more extended regions than the very local solar neighbourhood and concern stellar samples with non-negligeable z-motions in certain cases.

Freeman: People have suggested that the inner parts of galactic disks could be hot, to stabilize the disks against barlike modes. Is this consistent with what you know about the third integral in realistic galactic potentials, e.g. take a star whose orbit has mean $\varpi \simeq 3$ kpc, mean $\Pi \simeq 200$ km s^{-1}, mean $Z \simeq 50$ or 100 km s^{-1}. Does this star have a third integral which keeps it in the disk (i.e. in $|Z| \leqslant 800$ pc), or does it in fact orbit into a spheroidal volume, about the galactic centre?

Martinet: From our numerical exploration in both Schmidt potentials we can say that, for initial conditions in the range which you propound, we have some evidence that a third integral effectively keeps the motion in the disk, even if the z-velocity is ~ 100 km s^{-1}. The half thickness of the boxes occupied by the orbits is of the order of 500 to 800 pc, at most 1 kpc. These results seem to be independent of the potential chosen.

RESONANT STELLAR ORBITS
IN SPIRAL GALAXIES

P. O. VANDERVOORT

University of Chicago, Chicago, Ill., U.S.A.

Abstract. This paper reviews a series of investigations of the orbits of stars in the regions of the Lindblad resonances of a spiral galaxy. The analysis is formulated in an epicyclic approximation. Analytic solutions of the epicyclic equations of motion are obtained by the method of harmonic balance of Bogoliubov and Mitropolsky. These solutions represent the resonance phenomena exhibited by the orbits in generally excellent agreement with numerical solutions.

1. Introduction

In recent years, it has become apparent that the clarification of important aspects of the dynamics of spiral structure in galaxies would require a systematic study of stellar orbits in the regions of the so-called Lindblad resonances. In a series of investigations (Vandervoort, 1973, 1975; Vandervoort and Monet, 1975), it has been possible to formulate an analytic theory of these orbits in an epicyclic approximation, to test this theory with the aid of numerical solutions of the epicyclic equations of motion, and to make use of these analytic and numerical studies in order to survey the resonance phenomena exhibited by the orbits. Apart from its applications to problems of spiral structure, the work is of interest, because it deals with a relatively simple dynamical system which exemplifies important features of the study of resonant orbits in stellar dynamics.

2. The Epicyclic Theory of the Orbits

Specifically, we consider the orbit of a star in the plane of a galaxy in which the prevailing gravitational potential is a superposition of a dominant axisymmetric component and a small non-axisymmetric perturbation of the form

$$\mathfrak{B}(\varpi, \theta) = \mathfrak{B}^{(a)}(\varpi) + \mathrm{Re}\left[\mathfrak{B}^{(1)}(\varpi) \exp(im\theta)\right]$$
$$= \mathfrak{B}^{(a)}(\varpi) + \mathfrak{B}^{(1)}(\varpi, \theta), \quad \text{(say)},$$

(1)

where m is an integer. Here ϖ and θ are the radial and azimuthal coordinates of the star, respectively, in a frame of reference rotating uniformly with angular velocity Ω. The canonical momenta conjugate to these coordinates are the ϖ-component of the velocity Π and the angular momentum h, respectively. We introduce epicyclic variables $\varpi_1, \theta_1, \Pi_1$, and h_1, by referring the motion of the star to a circular orbit in the potential $\mathfrak{B}^{(a)}(\varpi)$. Let the circular orbit have a radius ϖ_0 and an initial azimuthal coordinate θ_0, and let $\Omega_0 - \Omega$ denote its angular velocity (in the rotating frame) and h_0

its angular momentum. The epicyclic variables are defined by the relations

$$\varpi = \varpi_0 + \varpi_1, \qquad \theta = \theta_0 + (\Omega_0 - \Omega)\, t + \theta_1,$$
$$\Pi = \Pi_1, \quad \text{and} \quad h = h_0 + h_1, \tag{2}$$

where t denotes the time.

In the epicyclic approximation, the equations of motion happen to be the canonical equations associated with the Hamiltonian

$$H_1 = \tfrac{1}{2}\Pi_1^2 + \tfrac{1}{2}\omega^2 \varpi_1^2 - \frac{2\Omega_0}{\varpi_0}\, \varpi_1 h_1 + \frac{h_1^2}{2\varpi_0^2} + \mathfrak{B}^{(1)}(\varpi, \theta), \tag{3}$$

where ω is the epicyclic frequency, the arguments of $\mathfrak{B}^{(1)}$ are interpreted in accordance with Equations (2) and Π_1 and h_1 are the canonical momenta conjugate to the coordinates ϖ_1 and θ_1, respectively. The quantity

$$H_E = H_1 + (\Omega_0 - \Omega)\, h_1 \tag{4}$$

is an exact integral of these equations of motion, and this integral expresses (in the epicyclic approximation) the constancy of Jacobi's integral. The Hamiltonian character of the epicyclic equations and the existence of an exact integral of those equations which is both approximate and analogous to Jacobi's integral ensure that the epicyclic theory will provide a good model of the orbits.

For the sake of definiteness, we shall concentrate on the orbits in the region of the inner Lindblad resonance where the quantity

$$v = \frac{m(\Omega_0 - \Omega)}{\omega} \tag{5}$$

has the value $+1$. A similar treatment applies to the region of the outer resonance where $v = -1$.

The epicyclic equations are solved by the method of harmonic balance (Bogoliubov and Mitropolsky, 1961). This is a perturbation theory in the amplitude of $\mathfrak{B}^{(1)}(\varpi, \theta)$. The essence of the method is the manner in which it allows for large perturbations of the amplitudes and phases of the epicyclic motion by the resonance. The solution is written in the form

$$\varpi_1 = \varpi_{10}(t) \sin[v\omega t + \varphi(t)] + \frac{2\Omega_0}{\omega^2 \varpi_0}\, h_{10}(t) + \cdots, \tag{6}$$

$$\theta_1 = \frac{2\Omega_0}{\omega \varpi_0}\, \varpi_{10}(t) \cos[v\omega t + \varphi(t)] + \theta_{10}(t) + \cdots, \tag{7}$$

and

$$h_1 = h_{10}(t) + \cdots, \tag{8}$$

where we are showing only the dominant terms, and we are omitting the solution for Π_1 as inessential to the present discussion. The functions $\varpi_{10}(t)$, $\varphi(t)$, $h_{10}(t)$, and $\theta_{10}(t)$, satisfy certain differential equations which are determined by a requirement that the time-dependence of these functions gives rise to terms in the equations of motion which just balance the resonant parts of the perturbing forces. This procedure eliminates terms with small denominators and secular terms which appear in solutions obtained with the aid of more conventional perturbation theories.

With the solutions which are obtained for $\varpi_{10}(t)$, $\varphi(t)$, $h_{10}(t)$, and $\theta_{10}(t)$, Equations (2), (6), (7), and (8), reproduce, at least qualitatively, all of the resonance phenomena which Contopoulos (1970) has found in numerical solutions for the orbits. (1) There are solutions in which ϖ_{10}, h_{10}, and the linear combination $m\theta_{10}+\varphi$, are all constants. In this case the circular and epicyclic motions are commensurate, and the orbits are periodic. Both stable and unstable periodic orbits are represented. (2) There are solutions in which ϖ_{10}, h_{10}, and $m\theta_{10}+\varphi$, are all periodic functions of time. These solutions represent tube orbits which oscillate around stable periodic orbits. The circular and epicyclic motions are commensurate only in the sense of a long-time average. (3) Finally, there are solutions in which ϖ_{10} and h_{10} are periodic functions of time whereas $m\theta_{10}+\varphi$ is a superposition of periodic and linear functions of time. In this case, the circular and epicyclic motions are not commensurate even in the sense of a long-time average. These solutions represent non-resonant orbits.

The solution of the epicyclic equations along these lines leads to the construction of a formal, isolating integral of the motion in addition to the integral of the Jacobi type given in Equation (4). Near the inner Lindblad resonance, the new integral is of the form

$$I = h_{10} + \tfrac{1}{2}m\omega\varpi_{10}^2 + \cdots \tag{9}$$

through its dominant terms.

3. The Comparison of Analytic and Numerical Solutions of the Epicyclic Equations of Motion

We have compared the analytic solutions of the epicyclic equations with numerical solutions in case

$$\mathfrak{B}^{(1)}(\varpi) = A \exp(ik\varpi) \tag{10}$$

in the perturbation in Equation (1), where the amplitude A and the wavenumber k are complex constants. This choice of the form of the perturbation reduces the Hamiltonian given in Equation (3) to a well-defined model problem suitable for systematic numerical tests of the analytic theory. Moreover, it is of a form which is appropriate, at least locally, for a tightly-wound spiral pattern. For the latter reason, the calculations performed in the comparison of analytic and numerical solutions also provide a useful survey of the resonance phenomena.

It is actually the surfaces of section derived from corresponding families of analytic and numerical solutions which have been compared. In the case at hand, a surface of section is essentially the (ϖ, Π)-plane for a fixed value of Jacobi's integral (Contopoulos, 1970). A given orbit is represented in the surface of section by points defined by the pairs of values (ϖ, Π) which occur when the azimuthal coordinate θ returns to a specified value θ_i, say. If the orbit is one for which the equations of motion admit a second isolating integral, in addition to Jacobi's integral, then these points lie on a simple closed contour called an invariant curve.

In Figure 1, we compare the analytic and numerical versions of a particular surface of section. This example is representative of the situations which might occur in the Galaxy. The important parameters of the problem have been assigned the values $\Omega_0 = 0.65$, $m = 2$, $v = 0.93$, and $k = 20 - 2i$, where we have adopted ϖ_0 as the unit of length and ω^{-1} as the unit of time. The value of A is chosen so that the amplitude of the spiral component of the field is 5% of the strength of the axisymmetric component. The choice of the values of ϖ_0 and Jacobi's integral is such that $H_E = 0$ (see Equation (4)).

The main features of this surface of section are three stable periodic orbits, their associated families of tube orbits, and two unstable periodic orbits. The analytic and numerical versions agree very well in the representation of these features. Solid contours represent well-defined invariant curves in both versions. All orbits have well-defined invariant curves in the analytic theory in virtue of the existence of the second isolating integral. While most orbits are found numerically to have well-defined invariant curves, a few do not. In the numerical version of the surface of section, orbits without invariant curves are represented schematically by dashed contours. The first three dashed contours, as we count outward from the central periodic orbit, represent orbits which form islands in the surface of section. One such system of island, corresponding to the innermost dashed contour in Figure 1, is shown in Figure 2. The remaining two dashed curves in Figure 1 represent invariant curves which have dissolved as a consequence of their proximity to the unstable periodic orbit which lies between them. The two dissolved invariant curves are shown in Figures 3 and 4.

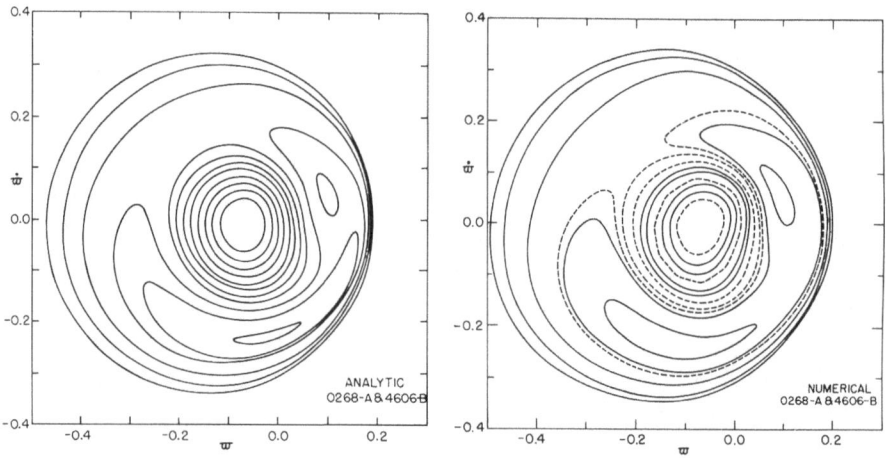

Fig. 1.   The surface of section described in the text as derived from analytic and numerical solutions of the epicyclic equations. Note that in this and the following figures the ϖ-axis has been labeled with the values of $\varpi - \varpi_0$ $(= \varpi - 1$ in the adopted system of units).

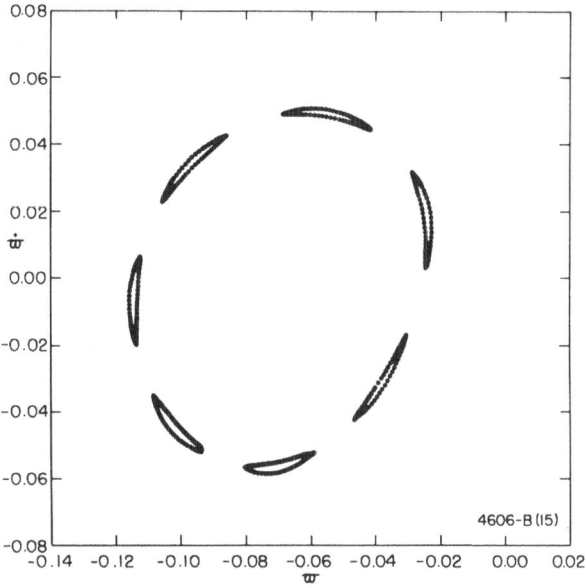

Fig. 2.   An example of a system of islands.

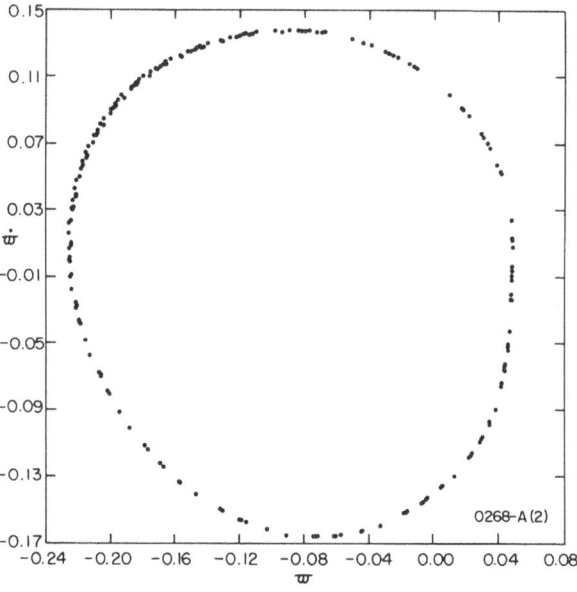

Fig. 3.   An example of the dissolution of an invariant curve.

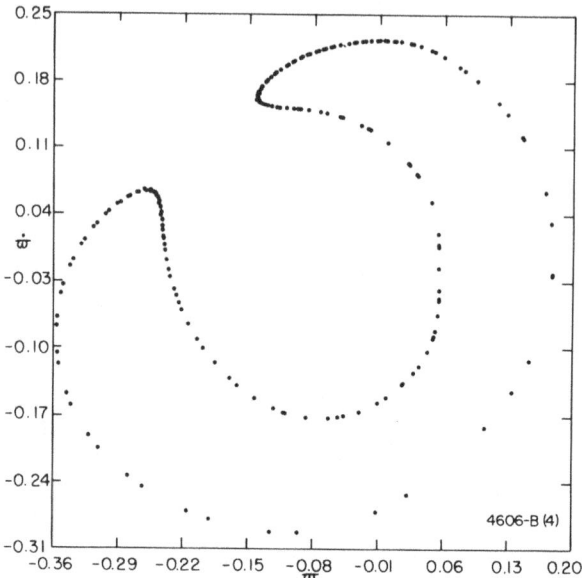

Fig. 4.   An example of the dissolution of an invariant curve.

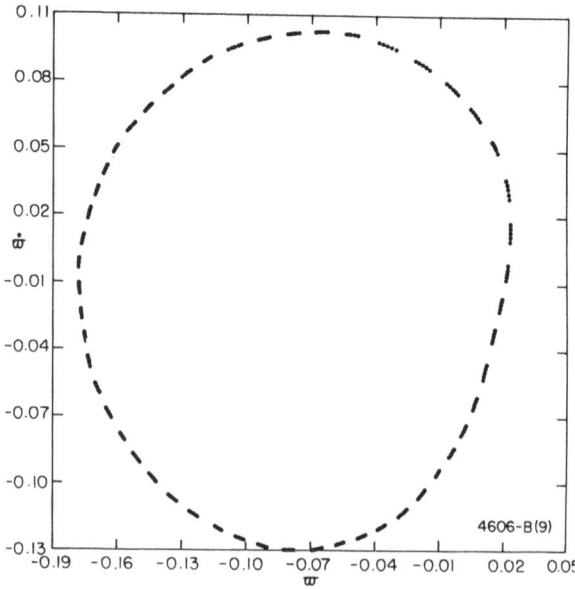

Fig. 5.   An example of a well-defined invariant curve.

Finally, we show in Figure 5 the points in the surface of section representing an orbit which is judged to have a well-defined invariant curve.

The results described above are typical of what have been obtained under a wide range of conditions. However, two points should be noted. First, it is more commonly the case that a surface of section will contain only two stable periodic orbits, their associated families of tube orbits, and only one unstable periodic orbit. Secondly, when we reduce the amplitude of the perturbation by a factor 10, as may be appropriate if the spiral structure is damped in the resonance region, the numerical results show no detectable tendency for the dissolution of invariant curves or for the formation of islands.

4. Concluding Remarks

(1) The solutions of the epicyclic equations obtained by the method of harmonic balance account for the main resonance phenomena in a simple and natural manner. The agreement of analytic and numerical solutions is very good, especially when one considers that the analytic solutions used in these comparisons were truncated at the lowest order of approximation.

(2) The construction of two isolating integrals of the motion is an important aspect of the theory. It is in terms of such integrals that one would hope to incorporate the resonance phenomena into the dynamics of spiral structure.

(3) The tendency for the dissolution of invariant curves and for the formation of islands is more fully developed than might have been expected for the perturbations considered. However, these phenomena do not appear to be so fully developed that they would vitiate applications of the analytic theory to the dynamics of spiral structure.

(4) The qualitative agreement of the present results, all obtained at the level of the epicyclic approximation, with the properties of numerical solutions of the exact equations of motion (Contopoulos, 1970) confirm the validity of the epicyclic orbits as models of the exact orbits. However, more quantitative comparisons are of practical interest, and they will be made.

Acknowledgements

The numerical parts of these investigations have been done in collaboration with Mr David G. Monet, who deserves the highest praise for the effort, initiative, and enthusiasm, which he has brought to the work. Our research has been supported in part by the National Science Foundation through Grant GP-17639 to the University of Chicago.

References

Bogoliubov, N. N. and Mitropolsky, Y. A.: 1961, *Asymptotic Methods in the Theory of Non-Linear Oscillations*, Gordon and Breach, New York, §14.
Contopoulos, G.: 1970, *Astrophys. J.* **160**, 113.
Vandervoort, P. O.: 1973, *Astrophys. J.* **180**, 739.

P. O. VANDERVOORT

Vandervoort, P. O.: 1975, *Astrophys. J.* (to be submitted).
Vandervoort, P. O. and Monet, D. G.: 1975, *Astrophys. J.* (to be submitted).

DISCUSSION

Contopoulos: I was glad to check that Dr Vandervoort's results about the new integral near the inner Lindblad resonance agree with mine (*Astrophys. J.* **160**, 113, 1970). Perhaps the most simple way to describe these resonance phenomena is by using action-angle variables (introduced in spiral structure theory by Kalnajs and Lynden-Bell). In these variables the Hamiltonian takes the form

$$H = \omega_1 I_1 + \omega_2 I_2 + f_{21}(I_1, I_2) \cos(\theta_1 - 2\theta_2) + \text{higher order terms.}$$

Near the resonance we can use canonical changes of variables to eliminate all trigonometric terms except those containing the combination $(\theta_1 - 2\theta_2)$. Thus we find a resonant integrable case of the type discussed earlier this morning.

It is easy to see that the combination $J_2 = I_2 + 2I_1$ is an integral of motion. This is the appropriate adiabatic invariant near resonance, while I_1 and I_2 separately are not.

I will discuss this problem further at the Paris Meeting next week.

Vandervoort: The problem is one in two degrees of freedom. Therefore, the epicyclic motion is described in terms of two amplitudes and two phases. However, the construction of the second integral involves only a linear combination of the phases, and this is precisely your linear combination of angle variables.

Lynden-Bell: I would like to point out the physical reason for the constant of the motion at resonances. If one looks in the axes rotating with the spiral wave the resonant orbits exactly close and the near resonant orbits move slowly. The circulation around the orbits is rapid compared with the slow movement of the near resonant orbit so the action corresponding to that circulation is adiabatically invariant. This adiabatic invariant is the constant of the motion. Born in his book *The Mechanics of the Atom* shows a beautiful method of treating all these problems in Angle and Action variables and combining them into slow and fast changing variables near resonances. This is the best way of treating this problem.

Contopoulos: I would only like to add that the appropriate adiabatic invariant is a different combination of I_1 and I_2 at each resonance.

Colin: Have you made some identical work for corotation resonance?

Vandervoort: No, I have not. I suspect that the method might encounter difficulties near the corotation point.

IS THERE A THIRD INTEGRAL OF MOTION?

P. PIŞMIŞ

Instituto de Astronomía, Universidad Nacional Autónoma de Mexico, Mexico

Abstract. It is argued that in a galaxy like ours a third integral of motion, a third independent argument in the distribution function, should exist if the potential function has to satisfy a third condition imposed on it, namely symmetry with respect to a plane. Orbit computations of single stars in a symmetric potential of the kind (Martinet and Hayli, 1971) indicate that a third integral seems to exist for Population I stars while it ceases to exist for Population II objects. This situation is explained by the author as follows. We state that a third integral should exist for all populations alike if the Boltzmann equation is interpreted within the statistical context for which it is valid. When applied to a complex system like the galaxy the third integral of the Boltzmann equation, which holds for an elementary volume in phase space, will also hold for a single particle if the latter is representative of the behavior of the element of volume as in Population I (coherent motion) whereas it will not necessarily hold for a single star of Population II; in the latter population the elementary volume, containing the same number of stars, does not represent the behavior of the element of volume during the motion of this in phase space.

Reference

Martinet, L. and Hayli, A.: 1971, *Astron. Astrophys.* **14**, 103.

THE COLLAPSE AND FORMATION OF GALAXIES

R. B. LARSON

Yale University Observatory, New Haven, Conn., U.S.A.

1. Introduction: The Problems

The ultimate aim of studies of the dynamics of stellar systems is to gain an understanding of why they have the structures that are observed, how they might have formed, and how they might evolve with time. In the case of galaxies, unlike smaller astronomical systems such as stars and star clusters, the problem of understanding their present structure is inseparable from the problem of understanding their formation since, as has long been clear, the two-body relaxation time in galaxies is much longer than the age of the universe, so that properties such as the radial density distribution and the flattening of elliptical galaxies cannot have changed significantly since they were formed. Other features of galaxies, such as the presence of spiral patterns or nuclear activity, may be of a more transitory nature and not closely related to the formation process, but such phenomena probably involve mainly the gas component of galaxies and may be regarded as perturbations on the underlying stellar system, whose basic structure must still be understood in terms of the formation process.

In considering those properties of galaxies which should be explained by a theory of the formation process, it seems most profitable to concentrate first on those characteristics which show the greatest degree of regularity or similarity from case to case. Thus it is natural to consider first the elliptical (E) galaxies, whose structures appear to be very smooth and regular and whose radial surface brightness distributions $I(r)$ are very similar to each other (Liller, 1960, 1966; van Houten, 1961; King, 1975). Thus a model of galaxy formation should be able to explain first the radial brightness profiles of E galaxies, and should provide a basis for understanding why $I(r)$ is so similar for different E galaxies. A second readily measurable property of the E galaxies for which considerable data exist is the isophotal ellipticity $\varepsilon(r)$ as a function of radius. Unlike the surface brightness profile $I(r)$, however, the ellipticity profile $\varepsilon(r)$ shows considerable variability from case to case; thus a theory of galaxy formation should also be able to explain typical $\varepsilon(r)$ profiles, while still allowing the possibility of variations in $\varepsilon(r)$ from case to case.

Many galaxies contain a prominent or (visually) dominant disc component, and in such cases it is necessary to explain the existence and the relative prominence of the disc and spheroidal (halo and/or nuclear bulge) components in a galaxy; again, considerable variations exist in the relative prominence of disc and spheroidal components (Freeman, 1970), and this should ultimately be understandable on the basis of a theory of galaxy formation. On the other hand, the characteristic exponential structure claimed for many galactic discs may not be so significant for theories of

Hayli (ed.), Dynamics of Stellar Systems, 247–269. All Rights Reserved.

galaxy formation, since the structure of galactic discs can be substantially altered by the occurrence of bar and/or spiral instabilities (see the contributions of Bardeen and Hohl to this symposium.)

In recent years it has become clear that a very general characteristic of both elliptical and spiral galaxies is the presence of radial composition gradients; the abundances of the heavy elements are always greatest at the center, and decrease markedly with radius. This effect is observed in the radial gradients in the strength of the CN band and other metallic absorption features in the spectra of E galaxies and the bulge components of spirals (e.g. McClure, 1969; Spinrad *et al.*, 1971, 1972); in the compositions of H II regions in the discs of spiral galaxies (e.g. Shields, 1974); in the radial gradient in the CN abundance of stars in the disc of our Galaxy (e.g. Janes, 1975); and in the radial variation of the composition of globular clusters in our Galaxy (Arp, 1965) and M31 (Hartwick and Sargent, 1974). In addition, the presence of composition differences between the various stellar populations in our Galaxy has long been recognized, and generally interpreted in terms of the collapse and simultaneous metal enrichment of the protocloud from which our Galaxy formed. A similar collapse or infall process might also provide a natural explanation for the composition gradients observed in other galaxies, in which case models for the formation of galaxies should be able to reproduce the observed abundance gradients. Finally, models for galaxy formation should be able to explain the observed correlation between the metal abundances and the masses of elliptical galaxies; various investigations (e.g., Sandage, 1972; Faber, 1973) have shown that the metal abundances of E galaxies increase systematically with increasing mass.

It may also be hoped that, in addition to explaining the various characteristics of galaxies mentioned above, a complete theoretical picture of the galaxy formation process will provide a better understanding of the origin of several other important phenomena which seem to be related in some way to the formation or early evolution of galaxies. For example, there is now considerable evidence that the quasar phenomenon is related in many cases to an early stage in the evolution of galactic nuclei. Also, the still incompletely understood problem of the origin of spiral structure may eventually be clarified by a better understanding of the formation and evolution of galaxies. At any rate, another criterion for deciding between possible contending models of galaxy formation is the extent to which the models may allow such additional phenomena to be understood.

2. Possible Models of Galaxy Formation

2.1. Gravitational collapse vs nuclear ejection

At present, since almost nothing is known in detail about how galaxies form, it is necessary to start by considering some possible models based on simple but plausible hypotheses. Hopefully, by comparing the predictions of the various models with the observed properties of galaxies, we can then get some idea about which types of models best account for the observations. Up to now, all of the models for which detailed calculations have been made have been based on the hypothesis that galaxies form by the gravitational collapse or condensation of diffuse matter in regions of the universe which are dense enough to separate out from the expanding background and recollapse. This general picture of galaxy formation is implied by conventional big bang cosmologies in which it is assumed that the universe began with a nearly uniform density but with small superimposed density fluctuations which grew in amplitude through gravitational instability and eventually collapsed into galaxies (e.g., Peebles, 1972). The gravitational instability picture is attractive because it offers the possibility of explaining within a common framework not only the formation of galaxies but also the origin of the observed clustering of galaxies on various scales; as Peebles (1974a) has shown, the spatial distribution of galaxies is consistent with a gravitational instability origin for all scales of clustering, including the galaxies themselves as the smallest 'clusters'. The general success of the gravitational instability picture in accounting for the large scale distribution of matter in the Universe is persuasive evidence in favor of the gravitational collapse picture of galaxy formation.

A second important reason for considering collapse models for galaxy formation is the evidence from correlations between the metal abundances, kinematics, and spatial distributions of stars in our Galaxy that our Galaxy formed by a collapse process (Eggen *et al.*, 1962). Subsequent work has further substantiated the general result that the oldest and most metal poor stars and star clusters have the most extended spatial distributions, whereas younger and/or more metal rich objects are increasingly concentrated toward the galactic center or the galactic plane. As has already been noted, similar correlations between metal abundance and spatial position exist in other galaxies, and model collapse calculations (Larson, 1974a) show that these observations are quite naturally explained by a collapse model of galaxy formation.

While the collapse picture in some form is widely accepted, an alternative hypothesis has occasionally been advocated which proposes that galaxies or parts of galaxies have been formed by the creation of matter and its ejection from nuclear singularities or 'white holes'. However, apart from the fact that many of the apparently strange phenomena (such as quasars) which this concept has been invoked to explain now seem better accounted for on more conventional grounds, this hypothesis suffers several severe drawbacks as a possible basis for understanding galaxy formation. The first is that, since it lies outside the realm of known physics, one cannot use any known physical principles to construct a model of galaxy formation based on the nuclear ejection hypothesis. Secondly, even if a theory could be found to explain the ejection of such large amounts of mass from galactic nuclei, the ejection velocity would still have to be very finely adjusted to make the ejected matter fill a galaxy-size volume without escaping to infinity. Finally, an apparently fatal difficulty is that the ejected matter cannot possibly carry the amount of angular momentum possessed by typical galaxies, yet still remain bound in a galaxy. This is a general difficulty with all theories which attempt to account for galactic phenomena as a result of nuclear ejection of matter; the ejected matter cannot carry any significant angular momentum, so that to the extent that the phenomena in question involve angular momentum, it must have some other origin. By contrast, gravitational collapse or infall theories encounter no such *a priori* difficulty in accounting for substantial amounts of angular momentum in galaxies.

2.2. SIMPLE COLLAPSE VS CLUSTERING OR ACCRETION

If we accept, for the reasons discussed above, that galaxies form by some sort of gravitational collapse or condensation process, we must then consider in more detail just how the pre-galactic material becomes collected together under gravity to form a galaxy. The simplest possible picture is that a galaxy begins as a discrete, roughly spherical, and nearly uniform cloud of gas which, after it has stopped expanding, is well isolated from its surroundings and collapses more or less all together to form a galaxy. Because of its simplicity, this picture has served as the basis for most of the model calculations which have been undertaken to describe the collapse of protogalaxies. However, it is probably a considerably over-idealized view of the real situation, in which a protogalaxy is probably not such a simple and well-defined object. It is possible, for example, that a protogalaxy may originate through the gravitational clustering of smaller objects, or that it may interact continually with its surroundings and grow from small beginnings by accretion of more matter. For example, it has been suggested (e.g., Peebles and Dicke, 1968) that the first condensations to form in the expanding universe actually have the mass of globular clusters, and that protogalaxies later develop as clusters of these globular-cluster sized fundamental units. A possible mathematical description of such a gravitational clustering process has been given by Saslaw (1972).

Whatever may be the most correct description of the galaxy formation process, it is likely to involve at least some elements of continuing accretion or infall of matter in a forming galaxy. This is because even in the most idealized case of an isolated uniform spherical cloud, collapse calculations (both for interstellar clouds and for protogalaxies) always show a nonhomologous collapse characterized by the development of a dense core surrounded by an extended envelope of residual gas which continues to fall into the central condensation until some effect such as an outflowing wind intervenes to halt the infall. More realistically, we would in general expect a forming galaxy to be surrounded by additional uncondensed material with which it continues to interact, eventually accreting at least part of it; this is suggested, for example, by the numerical simulations by Peebles (1971, 1972) of galaxy or cluster formation by gravitational instability, which show that the forming protogalaxies or protoclusters do not immediately become separated from their surroundings but continue to interact gravitationally with surrounding matter even after they have begun to collapse. Also, Oort (1970) has shown via simple models that one should in general expect ambient matter to continue to fall into a galaxy for a long time after it has formed. Tidal interactions between neighboring young galaxies may also continue to exert important effects on their structure and dynamics, and galaxies may even collide inelastically and merge to form larger systems (Toomre and Toomre, 1972; Alladin *et al.*, 1975).

In view of the possibilities mentioned above, it can hardly be expected that simple idealized collapse models will provide a fully complete or correct description of the galaxy formation process. Perhaps it is reasonable to hope that the simple models which can now be treated numerically will adequately describe the formation of the simplest systems, namely the elliptical galaxies; systems with prominent disc components, spiral structure, nuclear activity, etc., may have formed in a more complex manner involving, for example, continuing interactions with or accretion of surrounding matter.

2.3. STELLAR DYNAMICS VS GAS DYNAMICS

We now consider the possibilities for constructing 'simple collapse models' intended first to represent the formation of E galaxies. The conventional general view concerning the formation and evolution of galaxies is that a galaxy begins as a diffuse cloud of gas which somehow condenses and over a period of time becomes transformed into stars, leaving in the end a system composed mainly of stars. This type of picture certainly seems required to explain the evolution of the disc of our Galaxy, where it is clear that star formation has gone on continuously over the history of the system. In the case of the elliptical galaxies, however, all indications are that there has been little or no star formation since early in their history (Larson and Tinsley, 1974), so that star formation in the E galaxies must have taken place relatively rapidly. Since nothing is known directly about how soon the star formation in E galaxies is completed, one can consider as a simple limiting case the possibility that star formation is completed at a very early stage, even before significant collapse has taken place. In this case one has only to consider the dynamics of a system of stars, and the complexities of attempting to treat the gas dynamics are avoided. Since the collapse of a system of stars is the simplest problem to study and has received a fair amount of attention, we shall consider in the next section the results that have been obtained for purely stellar collapse models. Later sections will consider the probable necessity of including a gas component in a more complete model for the formation of E galaxies, and also the results that have been obtained with models which treat the dynamics of both gas and stars.

3. Stellar Collapse Models

It has been remarked by King (1966) and others that, despite the fact that two-body relaxation processes cannot be important in elliptical galaxies, these galaxies nevertheless 'look relaxed'; that is, their radial surface brightness distributions resemble those predicted on the basis of two-body relaxation for models of star clusters. A possible relaxation effect which might account for the apparently 'relaxed' structure of E galaxies has been described by Lynden-Bell (1967), who pointed out that the rapid changes in gravitational potential associated with the initial collapse and approach to equilibrium of a stellar system should cause large changes in the energies of the individual stars, thus producing a 'violent relaxation' effect. Assuming that this violent relaxation effect is sufficient to completely redistribute the stars in energy and bring about an approach to the most probable distribution, Lynden-Bell argued that the resulting system would be basically an isothermal sphere with a Maxwellian velocity distribution modified by a cutoff at the escape velocity. The particular form of the velocity cutoff derived by Lynden-Bell was equivalent to the 'lowered Maxwellian' approximation previously used by King (1966) as the basis for his star cluster models, and since these models were able to represent also the structure of E galaxies, it was argued that the structure of E galaxies could be explained as a result of violent relaxation. Recently, Wilson (1975) has generalized the King models to the case of axisymmetric rotating systems, and has shown that these models are capable of representing quite well a number of the basic properties of E galaxies.

However, the extent to which the violent relaxation concept and Lynden-Bell's statistical mechanics are applicable to elliptical galaxies is still somewhat unclear, in view of the fact that various numerical experiments designed to test the predictions of violent relaxation theory for one-dimensional systems (e.g., Lecar and Cohen, 1972) have shown only partial agreement with the theory: in all cases, significant numbers of particles are thrown into high energy 'halos' where relaxation is ineffective and the experimental energy distributions disagree with predictions. This result suggests that violent relaxation and Lynden-Bell's statistics may not apply to the halos of elliptical galaxies.

Recently, a more direct test of the applicability of violent relaxation theory to explaining the structure of E galaxies has been provided by the detailed numerical calculations by Gott (1973) of the collapse of an axisymmetric rotating system of stars, represented numerically by a system of 2000 discrete mass rings. Gott found that these stellar collapse calculations yielded models qualitatively resembling E galaxies in their structure, and also qualitatively consistent with the predictions of violent relaxation theory. In particular, these models have ellipticities and ellipticity profiles $\varepsilon(r)$ that are roughly consistent with those typically observed for E galaxies. However, the density distributions in the outer envelopes of E galaxies are not correctly reproduced by these models; at large radii, the density distributions in these stellar collapse models are approximately of the form $\varrho \propto r^{-4}$, whereas the observed* density distributions in E galaxies are considerably less steep, varying from about $\varrho \propto r^{-2.7}$ at intermediate radii to $\varrho \propto r^{-3}$ (Hubble's law) at large radii. Thus it appears that violent relaxation alone is not able to explain the observed density distribution in the envelopes of E galaxies. Furthermore, the stellar collapse models do not reproduce the highly condensed nuclei observed in the giant E galaxies (although this may in part be because Gott's calculations had to be cut off at small radii because of numerical difficulties.)

More recently, Gott (1975) has extended the stellar collapse models to the more

---

* Assuming, as usual, a constant mass-to-light ratio.

realistic case where a protogalaxy is treated not as an isolated object but as a perturbation in an expanding cosmological medium. Ambient material which is gravitationally bound to the central density perturbation will then have its expansion retarded and eventually reversed, and will continue to fall into the forming galaxy for some time after the central perturbation has collapsed. The importance of continuing infall of matter as a natural consequence of the galaxy formation process has also been discussed by Oort (1970) on the basis of a slightly different picture whereby galaxies are assumed to develop from primordial velocity perturbations rather than density perturbations, and by Larson (1972a, 1974a) on the basis of hydrodynamical collapse calculations (see below). If the infalling matter is in the form of stars, as assumed by Gott, then a much more extended stellar envelope can be built up than in an isolated system where infall effects are ignored. For his 'infall' models, Gott found a density distribution of the form $\varrho \propto r^{-2.8}$, in good agreement with that observed in the envelopes of E galaxies. Thus the infall models offer a more promising possible explanation of the structure of E galaxies; we note, however, that the density distribution in these models is no longer determined primarily by violent relaxation effects, but depends essentially on the infall effect and thus might be expected to depend to some extent on the ambient conditions present when a galaxy forms.

It is interesting that some of the gas-dynamical models of Larson (1974a) which were calculated with an expanding boundary yielded rather similar results, the density distribution at large radii being given approximately by $\varrho \propto r^{-3}$ independently of the detailed model assumptions. These expanding boundary models are basically similar to the Gott infall models in that the central region collapses first, while material farther from the center continues to expand for a longer period of time before turning around and falling back into the forming galaxy. Since the infalling matter is essentially in free fall, the dynamics is not much affected by whether it consists of stars only, as in the Gott models, or of a mixture of gas and stars, as in the Larson models; this may account for the fact that the resulting envelope density distributions are fairly similar in all of the 'infall' models. The fact that these results seem to resemble closely the density distributions in the envelopes of E galaxies suggests that the similar structure of E galaxies could be a result of cosmological infall effects, as proposed by Gott, but as will become evident, this explanation is not unique.

The rotating infall model computed by Gott (1975) has an ellipticity profile $\varepsilon(r)$ which agrees approximately with that observed for the E4 galaxy NGC 4697, which is interesting because the same model also gives a good fit to the outer part of the observed surface brightness profile $I(r)$ of NGC 4697. These calculations do not offer any direct explanation of the observed differences in ellipticity profiles between different E galaxies, but Gott suggests that this could be due to structural irregularities in the initial protogalaxies and to the resulting irregularities in the angular momentum distribution induced by tidal torques. Unfortunately, attempts to compare theoretical rotation curves with observations have so far been largely frustrated by the paucity of reliable data, but there appears to be an unexplained discrepancy between the rotation curve predicted by Gott's model of NGC 4697 and the observed rotation curve of this galaxy (Bertola and Capaccioli, 1975), which shows smaller velocities than expected theoretically for any dynamical model with the same ellipticity.

In summary, the available stellar collapse models appear to enjoy some success in explaining the flattening of E galaxies and, when infall effects are incorporated, the light distribution in the envelopes of E galaxies. However, the stellar collapse models do not reproduce the highly condensed nuclei observed in giant E galaxies, nor do they offer any explanation of the observed composition gradients. For this it still appears necessary to assume that the formation of at least the central regions of E galaxies involves gaseous condensation and metal enrichment processes.

Before considering models containing a gas component, we note some additional stellar dynamical processes that could contribute to the development of dense cores in E galaxies. First, if an E galaxy begins as a cluster of smaller objects or mass concentrations (e.g., proto-globular clusters) or if it rapidly fragments into such objects, then two-body relaxation effects between these objects could cause some evolution in the structure of the system, just as in a cluster of stars. However, since it seems unlikely that a major part of the mass could remain concentrated in small objects for the ~ 10–20 crossing times required for two-body relaxation effects to produce a dense core, it also seems unlikely that this could be a major effect determining the structure of E galaxies, but it could play a contributing role. Secondly, Tremaine *et al.* (1975) have shown that the dynamical friction effect experienced by massive globular clusters moving in a smooth background of halo stars could cause them to spiral inward and pile up at the center to form a small dense nucleus. These authors suggest that this process might explain the apparently distinct small 'nucleus' at the center of M31, which has a radius of ~ 7 pc and appears to have a slightly smaller metal abundance than its immediate surroundings (Spinrad *et al.*, 1972).

4. Models Containing Gas

Even if the primordial gas in a protogalaxy is quickly exhausted and plays no role in the collapse process, it is still necessary to consider the dynamics of the recycled gas lost from evolving stars. If this recycled gas is not all lost from the galaxy, at least some of it will fall toward the center and form new stars there. Spitzer (1971) has outlined schematically how a sequence of stages of stellar mass loss, infall of gas toward the center, and formation of a new generation of stars in a region of smaller size and higher density might eventually lead to the formation of a highly condensed, metal rich nucleus. If the system has substantial angular momentum the recycled gas will presumably condense instead into a disc, and as Ostriker and Thuan (1975) have shown, galactic evolution models based on this assumption can reproduce successfully the observed distribution of stellar metal abundances in the solar neighborhood. It seems very likely, however, that primordial gas must also have been present along with the recycled gas from halo stars, since the primordial gas in a protogalaxy is almost certainly not totally converted into stars before the system has begun to collapse, as assumed in the stellar collapse models. Indeed, the disc of our Galaxy must have formed at least partly from unprocessed primordial gas if the observed local deuterium abundance is of cosmological origin, as currently seems likely. Thus we are led to consider the collapse of systems of both gas and stars, in which gaseous dissipation effects play an important role in producing the final highly centrally condensed or flattened structure of the system.

In general, it is expected from experience with calculations for collapsing interstellar clouds that gaseous cooling or dissipation effects will lead quite naturally to a highly centrally condensed density distribution. However, the dynamics of the gas in a collapsing protogalaxy is difficult to treat in any precise way because the gas is probably quite 'turbulent', i.e. characterized by large density inhomogeneities and a more or less chaotic state of motion. The presence of turbulent motions in the protocloud from which out Galaxy formed seems required to account for the observed large velocity dispersions of the halo stars, many of which even have retrograde orbits. There are many ways in which such motions could have originated, including residual primeval turbulence, early supernova explosions or other violent events, collisions of pregalactic gas clouds, and the action of gravitational forces on an irregular mass distribution. Since the stars that form in a collapsing protogalaxy inherit the motions of the gas from which they form, it is important to keep track of the velocity dispersion of the turbulent gas and how it changes with time as a result of the collapse and dissipation processes. This means that it is necessary to specify the rate at which the turbulent motions are dissipated by inelastic collisions between the different randomly moving fluid elements. In addition, if we assume that star formation continues for a significant period of time during the collapse, it is necessary to specify the rate at which gas is transformed into stars. Needless to say, not enough is known about the physical processes involved to allow the gaseous dissipation and star formation rates to be calculated in any *a priori* way, and therefore it is necessary to resort to simple parameterized models which hopefully represent the essential features in a qualitatively correct fashion, while still leaving a few undetermined parameters which can be adjusted to achieve the best agreement with the observations. Clearly, one cannot hope in this way to produce models which are quantitatively accurate in all respects, but one can at least hope to gain some insight into the essential features of galaxy formation by collapse. The situation is perhaps analogous to the earliest days of the study of stellar structure, when many of the basic physical processes were not well understood but it was nevertheless possible to make important progress by using simple polytropic models.

One possible approach to estimating the turbulent dissipation rate, adopted by Larson (1969) and Brosche (1970), is to assume that the protogalactic gas is entirely concentrated in discrete spherical clouds occupying about one-tenth of the volume and possessing individual random motions with an isotropic velocity distribution. The rate of dissipation of the kinetic energy of cloud motions is then straightforwardly calculated in terms of the velocity dispersion and mean free path of the clouds. In order to estimate the size of the clouds, Larson (1969) assumed that the cloud mass is equal to the Jeans mass at a temperature of 10^4 K, in which case the cloud masses are initially of the order of 10^9 $M_\odot$; Brosche (1970), on the other hand, estimated that a protogalaxy typically consists of ~ 10 clouds, assuming that the typical angular momentum of a galaxy is the sum of the purely random angular momenta of its constituent clouds. In either case, the dissipation time is predicted to be comparable to or not much longer than the free-fall time, so that the collapse of a protogalaxy takes place within a few free-fall times, i.e. a few times 10^9 yr. Both investigations also arbitrarily adopted the common assumption that the star formation rate is proportional to a power of the gas density; Brosche assumed that the star formation rate varies as the square of the gas density, whereas Larson allowed both the exponent and the coefficient of proportionality in the power law to vary and studied the effects of these parameters on the resulting stellar density distribution, obtaining the best agreement with observations with an exponent of about 1.8.

Unfortunately, it is difficult to justify physically all of the assumptions described above, particularly the assumption that the star formation rate varies as a power of the gas density. Possibly some justification for this type of assumption could result if supernova explosions provide a sufficient amount of energy to strongly heat the gas in a protogalaxy and thus control the star formation rate through a negative feedback effect. As has been shown by Larson (1974b), there is a critical star formation rate proportional to the 1.76 power of the gas density such that if the actual star formation rate exceeds this critical value all of the supernova energy goes into heating the gas, whereas for lower star formation rates most of the supernova energy is radiated away and the heating effect is considerably reduced. At present the various numerical quantities involved are too uncertain to permit a definite conclusion concerning the possible effect of supernovae in controlling the star formation rate in collapsing protogalaxies, but it is intriguing that the *form* of the star formation rate predicted on this basis agrees with what is needed to produce realistic collapse models.

A simpler approach to specifying the gaseous dissipation and star formation rates has been adopted in some of the more recent model calculations of Larson (1974a). Since it is difficult to justify the assumption that neighboring gas elements or clouds in a protogalaxy have independent random motions, and since a protogalaxy may in reality contain systematic streaming motions on various scales up to that of the galaxy itself, it is possible that the dissipation time scale is more closely related to the time scale for collisions between such large (galaxy-size) gas streams, i.e. to the dynamical or free-fall time for the system. A simple example of a situation where this would be the case is the free infall of gas from the outer part of a protogalaxy into the dense inner region, where it collides with other gas and is stopped. Calculations have therefore been made with the assumption that the dissipation time is proportional to the local free-fall time, the constant of proportionality being left as a free parameter.

The star formation rate may also depend on the dynamical time scale: for example, if star formation is caused primarily by the compression of gas in the shock fronts produced when large scale gas streams collide, then the time scale for transformation of gas into stars is related to the time scale for processing of the gas through these shock fronts, which in turn is related to the dynamical time scale for the system. Accordingly it has been assumed by Larson (1974a) that the star formation time scale is proportional to the local free fall time, with an undetermined constant of proportionality of order unity. The assumption that star formation is related to the large scale dynamics of a galaxy receives some support from the observation that star formation generally occurs in spiral arms, which evidently reflect large scale patterns in the dynamics of the gas in a galaxy, being currently interpreted as shock fronts accompanying spiral density waves. A similar type of assumption for the star formation rate was found by Oort (1974) to be successful in explaining the radial variation of the gas density and star formation rate in spiral galaxies. Undoubtedly the dependence of the star formation rate on physical parameters is much more complicated that a simple dependence on the dynamical time scale, since many small scale processes, including the effects of massive stars and supernovae, must almost certainly play a role. However, in view of our still very limited understanding of all the relevant processes, it is probably not worthwhile at present to introduce a much more elaborate representation of the star formation rate than the simple possibilities mentioned above.

5. Results for Spherical Gas-Dynamical Models

Here we shall briefly review the properties of the spherical collapse models calculated by Larson (1969, 1974a) on the basis of the assumptions described above. In each case the calculations begin with a uniform gaseous protogalaxy, and the system begins to collapse nearly in free fall, continually transforming gas into stars as it does so. As expected, the gas component of the protogalaxy condenses more and more strongly toward the center as the collapse proceeds; thus successive generations of stars are on the average formed in regions of successively smaller size and higher density, and the galaxy is built up from the outside in. Since the infalling gas becomes progressively more enriched in heavy elements as it flows toward the center, a composition gradient is naturally and inevitably established in models of this type.

As was found by Larson (1974a), it is possible with plausible choices of the parameters to obtain models which reproduce quite closely the surface brightness distribution of the El galaxy NGC 3379 measured photoelectrically by Miller and Prendergast (1962), and also the very similar mean $I(r)$ profile for 14 elliptical galaxies as determined by King (1975). It is noteworthy that, unlike the stellar collapse models, these gas-dynamical models reproduce not only the surface brightness distribution in the envelope but also the dense cores of E galaxies. However, while the possibility of obtaining detailed agreement with the surface brightness observations shows that the gas-dynamical models are at least consistent with these observations, it is still not possible to discriminate between different models of this type; for example, when the parameters are appropriately adjusted, the differences between predictions based on the 'power law' star formation rate and those based on the 'dynamical time scale' star formation rate are smaller than the differences between the observed brightness profiles of different E galaxies. When we recall that the stellar infall models of Gott are also able to reproduce the observed envelope brightness profiles, it is evident that we cannot on this basis alone decide in detail between the various mechanisms that may be responsible for establishing the structure of E galaxies, particularly in the outer regions.

A further point of comparison with the observations is in the predicted radial variation of metal abundance $Z(r)$. Unfortunately, the available observations refer only to the innermost 1 or 2 kpc of elliptical galaxies, and cannot be compared very quantitatively with the predictions because of the lack of accurate calibrations. The models of Larson (1974a) all predict essentially the same behavior for $Z(r)$ in the innermost 2 kpc of elliptical galaxies, and this predicted $Z(r)$ is qualitatively in agreement with the observations; this provides strong evidence for the formation of at least the inner regions of E galaxies by a gaseous infall process. Different gas-dynamical models make different predictions for $Z(r)$ in the outer parts of E galaxies, but here there are no direct observations, and the indirect measure of $Z(r)$ provided by the UBV colors is unfortunately uncertain, due to disagreements between different observers. One of the most important ways of discriminating between models for the formation of elliptical galaxies will be to obtain measurements of the metal abundances in the outer envelopes of E galaxies.

The predicted stellar velocity dispersions are even more difficult to compare with observations, owing to the complete lack of measurements except at the centers of a number of E galaxies. If it becomes possible to measure velocity dispersions in the envelopes of E galaxies, this will provide another important way of discriminating between models.

Although this result cannot yet be related to observations of elliptical galaxies, the gas-dynamical models provide a 'bonus' in that they yield a prediction of the frequency distribution $N(Z)$ of stellar metal abundances at each radius. An important result is that in the innermost few hundred parsecs of the models most of the stars at any given radius have approximately the same metal abundance, and relatively metal poor stars are quite rare. This result comes about because the continuing infall of gas in this region allows a nearly time-independent $Z(r)$ distribution to be established, and most of the stars are formed after this approximate steady state has been attained. The predicted frequency distribution of stellar Z values near the center of these models resembles that observed among stars in the solar neighborhood, and this suggests that the disc of our Galaxy was formed by a continuing gas infall process, like the central regions of these models. The results described below for rotating models also appear to point toward an infall picture for the formation of discs in galaxies.

The effect of supernova explosions on the early evolution of these models has been estimated by Larson (1974b). It is found that supernova explosions can deposit a substantial amount of energy in the gas component of a forming galaxy, causing some of this gas to be driven out of the galaxy in a 'galactic wind'. The importance of supernova-driven gas loss increases with decreasing galactic mass, and in a forming galaxy whose mass is less than $\approx 10^9$–10^{10} $M_\odot$, much or all of the residual metal-enriched gas which would otherwise have condensed at the center to form a dense metal-rich nucleus may instead be expelled from the system. Model calculations incorporating this effect predict, in qualitative agreement with the observations, that the less massive galaxies should show less prominent nuclei and less prominent composition gradients; furthermore, a variation of overall metal abundance with galactic mass is predicted which agrees qualitatively with the observed $Z(M)$ relation for elliptical galaxies. Thus it appears that the effect of supernova explosions on the gas dynamics can explain, at least qualitatively, the observed mass dependence of the structure and composition of E galaxies. This result provides additional evidence for the importance of gas-dynamical processes in the formation of E galaxies.

We note that the effect of supernovae in expelling most of the gas from small protogalaxies and perhaps disrupting them completely may be of even more fundamental importance in influencing the mass spectrum of galaxies. Peebles (1974a) has noted that the spatial distribution of galaxies shows no preferred scale of clustering, and suggests that the galaxies themselves may simply represent the smallest surviving scale of clustering, appearing as single objects rather than clusters of smaller objects because gas-dynamical effects have destroyed the smaller scales of clustering. One such 'gas-dynamical effect' might be the effect of supernova-driven gas loss in destroying most of the smaller systems. Suppose, for example, that a galaxy of typical mass $\sim 10^{11}$ $M_\odot$ begins as a cluster of smaller systems (eg., proto-globular clusters) of mass $\lesssim 10^9$ $M_\odot$; then supernova explosions will expel most of the initial gas from these smaller systems, and many of them may be completely disrupted. However, because of its higher escape velocity, the larger system of mass $\sim 10^{11}$ $\mathcal{M}_\odot$ may retain much of the debris from the smaller systems and thus survive as a single galaxy. This argument would predict that most of the mass in galaxies should be in systems with masses greater than $\approx 10^9$–10^{10} $M_\odot$, which appears to be consistent with the observations.

6. Rotating Gas-Dynamical Models

In view of the general success of the spherical gas-dynamical models in explaining those properties of E galaxies that are not related to their rotation, it is of considerable interest to see whether with the addition of rotation such models can also explain such properties as the ellipticity profiles $\varepsilon(r)$ of E galaxies and the presence of disc components in spiral and S0 galaxies. Calculations with such rotating models, intended mainly to provide more realistic models for flattened and rotating E galaxies, are currently in progress, and the most significant (if still somewhat tentative) results so far obtained with these models will be briefly reviewed here; detailed results will be published in a later paper.

A model calculation must begin with a specification of the initial conditions for a collapsing protogalaxy; in particular, we are concerned here with the initial state of rotation in a system with significant angular momentum. We shall not attempt to discuss the much-debated question of the origin of the angular momentum of galaxies, but merely mention some of the possibilities which have been suggested. Peebles (1974b) has pointed out that it is difficult to understand how the angular momentum of galaxies could originate directly from primeval vorticity in the early universe, and has advocated the point of view that protogalaxies acquired their angular momentum through the action of tidal torques (e.g., Peebles, 1971). Other authors (e.g., Icke, 1973; Binney, 1974) have attempted to understand the rotation of galaxies as having resulted from motions generated in a collapsing protocluster of galaxies. Silk and Lea (1973) have pointed out that inelastic collisions and coalescence of randomly moving primordial gas clouds or protogalaxies could plausibly produce systems with the angular momenta characteristic of galaxies. Finally, there is the possibility that large scale turbulent motions could be generated by processes such as supernova-driven winds or quasar activity occurring at about the time of galaxy formation. Evidently, there is no lack of possible mechanisms for generating rotational motions, and it is even possible that different galaxies may have acquired their angular momenta in different ways. However, all of the proposed mechanisms seem likely to leave a protogalaxy with a more or less irregular structure and state of motion, and no simple prescription can be given for the initial rotational motion of a collapsing protogalaxy.

If it is indeed the case that protogalaxies are chaotic in structure and motion, then the observed regularities in the structure of galaxies must have been produced by processes occurring during the collapse. Thus one may hope that, if the model calculations realistically describe the physical processes occurring during the collapse, one can still obtain reasonably realistic and representative models by starting with idealized initial conditions such as a uniform, spherical, and uniformly rotating protocloud. At any rate, until the details of galaxy formation are better understood, there seems to be little point at present in exploring any but the simplest models, since more complex models would contain more parameters than we would know what to do with. In this spirit, all of the model calculations so far undertaken have assumed an initially uniform and uniformly rotating protogalaxy.

Calculations of the collapse of rotating protogalaxies have been made using a two-fluid generalization (to represent both the gas and stellar components) of the axisymmetric computer code used by Larson (1972b) to calculate the collapse of a rotating cloud. The first calculations to be undertaken consisted essentially of repeating the spherical collapse calculations with the addition of rotation but without otherwise modifying the simple hydrodynamic treatment previously used for the gas and stars in a protogalaxy; this means in effect that the basic flow is assumed to be laminar and inviscid, so that each fluid element conserves its angular momentum during the collapse. Models were calculated using both of the approaches described previously for estimating the gaseous dissipation and star formation rates. However, all of these calculations produced qualitatively the same unsatisfactory result. The first stars to form are distributed in a spheroidal halo, as might be expected, but as the residual gas continues to condense inward it develops a more and more flattened distribution; this is reflected in the ellipticity $\varepsilon(r)$ of the resulting stellar system, which increases strongly and monotonically toward decreasing radii. In fact, the central part of the resulting system consists of a remarkably flat, uniform, and uniformly rotating disc, with a radius of the order of some kpc but without any central nuclear condensation. This disc would certainly be unstable to non-axisymmetric modes (see the contributions of Bardeen and Hohl to this symposium) and might eventually evolve into a centrally condensed structure resembling those observed in spiral or barred spiral galaxies, but we would still be left with the problem of understanding the formation of *elliptical* galaxies with realistic ellipticity profiles that do not rise sharply to very high values at small radii.

The trouble with the inviscid hydrodynamical models described above is almost certainly that the neglect of viscosity effects and angular momentum transport is not justified. Indeed, the basic conceptual picture of a highly turbulent protogalaxy implies a large associated 'turbulent viscosity' (e.g., Saslaw, 1971); for example, if a protogalaxy is imagined to consist of discrete randomly moving gas clouds, the system of gas clouds will possess a kinetic viscosity just like a classical molecular gas, except that here we have very large 'molecules' (i.e., the gas clouds) and very long mean free paths, so that the viscosity effect is correspondingly larger. In general, whatever the detailed nature of the gas motions and the dissipative processes, we might expect that associated with the gaseous dissipation will be a viscosity effect which will tend to transport angular momentum, and that the time scale for significant transport of angular momentum will be of the same order as the dissipation time scale.

Given a cloud model for the turbulent gas in a protogalaxy, it is straightforward to calculate the kinetic viscosity coefficient in terms of the mean free path of the gas clouds, and to substitute this into the appropriate generalized form of the classical Navier-Stokes equations describing the flow of a viscous fluid. If a discrete cloud model is not used but the dissipation time scale is instead related directly to the free fall time, as in the models of Larson (1974a), the viscosity coefficient can instead be expressed in terms of the mean free time of the colliding gas elements, which is approximately the same as the dissipation time. Both of these approaches have been used in the rotating collapse models described below, and they yield qualitatively similar results, albeit with significant quantitative differences. In either case, however, the use of the Navier-Stokes equations, in which the viscous stress term is proportional to the gradient of the angular velocity, provides only a crude first approximation to reality, since in the present situation the mean free paths of the fluid elements are not small compared with the size of the system, as assumed in classical kinetic theory. As an example of a system with very large mean free paths, consider an equilibrium stellar disc where, even in the presence of differential rotation, there is no viscous stress and no transfer of angular momentum, so that the Navier-Stokes equations are clearly not applicable. A more correct treatment of the viscosity effect in a turbulent protogalaxy would require solving explicitly for the relevant moments of the velocity distribution, and this would substantially increase the already great computing effort required. An attempt was made to circumvent this extra calculation by approximating the relevant moments of the velocity distribution, and this produced results qualitatively similar in most respects to those obtained with the Navier-Stokes equations, but possibly describing more realistically the formation of a highly flattened subsystem in which the viscosity is expected to be relatively small; these results will be described briefly below in Section 7. In all cases, the effect of the viscosity is to transport angular momentum *outward*.

When viscosity effects are incorporated in any of the ways described above, the collapse calculations yield much more realistic models for elliptical galaxies. The outward transport of angular momentum allows the gas component to develop a highly centrally condensed distribution without becoming as highly flattened near the center as in the calculations without viscosity. The radial variation of the resulting stellar density distribution is similar to that previously obtained for the spherical models, and the ellipticity $\varepsilon(r)$ now shows much more moderate variations with radius, as is observed in elliptical galaxies. The predicted rotation curves always rise rapidly to a maximum a few kpc from the center and then fall off gradually with increasing radius, closely resembling in form the rotation curves measured for early-type spiral galaxies by Roberts and Rots (1973).

It is worthwhile to consider in more detail the ellipticity profiles $\varepsilon(r)$ predicted by the models, since at present they provide the main point of comparison between these rotating models and the observations. In general, we expect the ellipticity profile $\varepsilon(r)$ to depend both on the initial angular momentum of the protogalaxy and on the effect of viscosity in redistributing angular momentum during the collapse. From the results of calculations made with different values of the parameters and different ways of treating the viscosity effect, it appears that the ellipticity of the outer parts of the resulting system depends mainly on the initial angular momentum of the protogalaxy, whereas the ellipticity of the inner regions depends primarily on the way in which viscosity redistributes angular momentum during the collapse. When the viscosity coefficient is calculated from the cloud model of Larson (1969) and the generalized Navier-Stokes equations are used, the resulting ellipticity profile is nearly constant at large radii and decreases near the center, a behavior which is not unusual for E galaxies. If the viscosity coefficient is reduced by a factor of two the resulting $\varepsilon(r)$ profile is moderately peaked at intermediate radii, a characteristic which again is fairly common among E galaxies. Models which are not based on the Navier-Stokes equations but on a particular simple approximation to the moments of the velocity distribution produce $\varepsilon(r)$ profiles which instead of decreasing at small radii tend to increase near the center.

In view of the dependence of $\varepsilon(r)$ on details of the gas dynamics which are not well understood, detailed agreement between predicted and observed $\varepsilon(r)$ profiles should probably not be expected and has not been sought, but it seems significant that a variety of more or less realistic profiles can be obtained with plausible values for the angular momentum and viscosity parameters. This result provides support for the general idea that turbulent gas dynamics plays an important role in the formation of galaxies, and offers a possibility of explaining the considerable variations among the observed $\varepsilon(r)$ profiles of elliptical galaxies. We have already noted that it is not to be expected that all protogalaxies have the same initial angular momentum distribution; nor is it to be expected that the internal dynamics would be the same in all cases, since different amounts of 'turbulence' could lead to differences in the effectiveness of viscous transport of angular momentum during the collapse. Thus the variations in observed $\varepsilon(r)$ profiles could reflect differences in the initial conditions as well as in the dynamics of the collapse. When more data become available, it will be interesting to see whether the variations in $\varepsilon(r)$ correspond to differences in the kinematic properties of E galaxies; if so, the E galaxies may not be quite so simple and homogeneous a class of objects as has usually been supposed.

At the least, it seems established by the results described in this section that it is not valid to assume, as has often been done, that each fluid element conserves its angular momentum during the collapse of a protogalaxy; the structure of E galaxies can only be explained if some redistribution of angular momentum takes place during the collapse. Consequently, the final distribution of angular momentum in a galaxy depends not only on the initial conditions but also on the dynamics of the collapse and formation process, and inferences that galaxies form from protoclouds with the same distribution of angular momentum are probably not justified.

7. Formation of Galactic Discs

As yet, there have been no model calculations which realistically represent the formation of a spiral galaxy with a prominent disc component, so our final remarks concerning the formation of galactic discs must necessarily be somewhat conjectural. Nevertheless, it appears that the rotating gas-dynamical models described above provide some useful information or constraints on the possible ways in which discs might form.

In all of these rotating models, the residual gas remaining during the later stages of the galaxy formation process gradually settles into a more and more flattened distribution, finally forming a thin, centrifugally supported disc in the equatorial plane. (Near the center of the disc, the velocity dispersion of the gas remains large compared with the rotational velocity, so that the density distribution there remains nearly spheroidal and not highly flattened.) Thus during the later stages of the galaxy formation process the stars are formed in a more and more flattened distribution, finally ending with the formation of a thin disc-like 'Population I' component in the equatorial plane. This flat component contains only a small fraction ($\lesssim 10\%$) of the total mass, but in the models with higher angular momentum its presence begins to have a noticeable effect on the shape of the stellar iso-density contours, causing them to become slightly pointed in the equatorial plane. This effect is exactly what is observed in the isophotes of the more flattened E galaxies: for systems flatter than about E5, the isophotes become noticeably pointed at the ends of the major axis (e.g., Liller, 1966; King, 1975). The model results are qualitatively consistent with these observations, so that both models and observations suggest that a disc component begins to appear in the E galaxies with higher angular momentum, i.e. those flatter than about E5. Indeed, Liller (1966) suggests that all such systems should be classified as S0 rather than E galaxies.

It is interesting to remark in passing that the presence of a flattened subsystem in the models is always much more evident in the composition distribution and the kinematics than in the density distribution. Just as is the case for the Population I component of our Galaxy, the incipient disc component found in the models has a higher metal abundance, larger rotational velocity, and smaller velocity dispersion than the halo component, and all of these quantities show quite marked gradients in the direction perpendicular to the equatorial plane. The contours of equal metal abundance, for example, are quite remarkably flattened toward the equatorial plane, much more so than the stellar iso-density contours. This result suggests that the presence of disc components in flattened E and S0 galaxies may be more conspicuous in their composition and kinematics than in the light distribution, and suggests ob-servations to look for these effects. Observational evidence for a composition gradient perpendicular to the plane of the edge-on S0 galaxy NGC 4762 has been presented by Freeman at this symposium.

From the results described above, we can understand how a disc component might form in a rotating galaxy from residual gas left over after the formation of a spheroidal halo. However, in none of the models so far calculated does the disc component exceed a minor fraction ($\lesssim 10\%$) of the total mass. Recycled gas from halo stars, not yet explicitly incorporated in the models, would contribute to the disc mass but would not add more than another $\simeq 10\%$ of the total mass. If only this small amount of residual gas is involved in the formation of a disc, it is not even certain that in reality a disc would be formed at all, since this residual gas can easily be swept out of the galaxy by a supernova-driven galactic wind (Larson, 1974b), or possibly even by an intergalactic wind. After some experimentation with the parameters, we have tentatively concluded that simple collapse models of the type described above cannot explain the formation of galaxies with prominent disc components, the reason for this being simply that if the models are to be capable of explaining *elliptical* galaxies, then star formation must be reasonably efficient during the initial collapse and most of the gas must be transformed into stars before there has been time for it to settle into a very flat disc.

Perhaps it is reasonable that the simplest model, i.e. one that begins by assuming a uniform, spherical, and uniformly rotating protocloud should be capable of ex-plaining only the simplest galaxies, i.e. the elliptical galaxies; this is probably all that we are entitled to expect. However, since these models do produce incipient disc components, they suggest how a less idealized model might in general yield a galaxy with a more prominent disc component. Basically, what is required to form a more massive disc is a larger amount of leftover primordial gas which does not parti-cipate in the formation of a spheroidal halo but condenses later into a disc. If the halo component of a galaxy is formed by the free fall collapse of a roughly spherical pre-galactic density perturbation, as in the above models, then we require that there be additional ambient primordial gas around the central perturbation which does not participate in the initial free fall collapse but continues to fall in later for a more extended period of time. Such a situation is entirely plausible, and is in fact predicted by calculations which take into account the effects of cosmological expansion, as was already pointed out in connection with the infall models discussed in Section 3.

Gott (1975) assumed that the ambient infalling material would all be efficiently converted into stars, in which case it would contribute to building up the halo of the galaxy but would not form a disc. However, it is possible that once a halo of significant mass has already formed, star formation would be inhibited in the gas which continues to fall in later, especially if the density of the infalling gas is lower than the average density of the system; then the tidal disruptive forces due to the rest of the system will dominate over local self-gravitational forces in the gas, and local collapse of the gas to form stars will presumably be inhibited (unless large density enhancements are already present in the gas.) The gas will then continue to swirl around and dissipate its energy until it eventually settles into a disc, and star formation can finally take place once the local gas density in the disc has become higher than the mean density of the system (or, more precisely, once the Goldreich and Lynden-Bell (1965) criterion for local gravitational instability in a gas disc is satisfied.) Depending on the amount of ambient gas which continues to fall into a forming galaxy in this way and condense into a disc, it would seem possible to eventually build up a disc of almost any mass relative to the already formed halo component.

Thus, while no realistic models have yet been computed and many of the details remain to be clarified, there appears to be no difficulty in principle in understanding how galaxies might form with varying proportions of disc mass to halo mass; the simple collapse of a roughly uniform, spherical protocloud yields a nearly pure halo system, i.e. an E galaxy, whereas a more complicated formation process involving the accretion of varying amounts of leftover primordial material will in general result in a spiral or S0 galaxy. The models described above, together with the results described by Brahic at this symposium, indicate that the halo component of a galaxy forms relatively rapidly (within $\simeq 10^9$ yr), whereas the formation of a disc is a more gradual process; several rotation periods are required for the gas to settle down into a flat disc, and the disc may continue to be built up over an extended period of time by continuing infall of matter. Indeed, if there is a significant amount of uncondensed intergalactic matter remaining at the present time, the discs of spiral galaxies may still be slowly gaining mass by accretion (Oort, 1970; Larson, 1972a). We note that the present picture for the formation of galactic discs is consistent with the theoretical indications (Ostriker and Peebles, 1973; Bardeen, 1975) that the discs of spiral galaxies may be embedded in halos of at least comparable mass, since the present picture requires a halo of significant mass to be present initially; we cannot yet, however, draw any quantitative conclusions about halo masses on the basis of theories of galaxy formation.

Finally, we note that if galactic discs are really built up over a period of time by the infall of leftover primordial gas, as suggested here, then this picture offers little basis for explaining the characteristic approximately exponential structure of galactic discs as a consequence of the initial conditions, since we expect no particular regularities in the initial spatial distribution or angular momentum distribution of the matter that eventually makes up the disc. Thus, to the extent that galactic discs show any regularities in their structure, such regularities must result from the dynamical evolution of the discs themselves. Such dynamical evolution is certainly expected to occur as a result of the bar and/or spiral mode instabilities which occur in a disc when its velocity dispersion is small (Bardeen, 1975; Hohl, 1975). Calculations of the dynamics of discs show that an unstable disc characteristically develops a central bar with trailing spiral arms, and that the associated gravitational torques transport angular momentum outward and cause the disc to develop a more centrally condensed structure. According to Hohl (1971), the structure finally attained is approximately an exponential disc. Even the observed presence of spiral structure in galaxies implies the presence of gravitational torques which will tend to transport angular momentum outward and slowly change the structure of the disc. It is probably in terms of such processes that the structure of galactic discs is ultimately to be understood, rather than in terms of the initial conditions or the details of the collapse process.

8. Epilogue

It should be evident from the foregoing review that theorists are still groping in considerable darkness in attempting to understand the formation of galaxies. Major uncertainties remain in our understanding of the basic physical processes involved, particularly with regard to gas-dynamical processes and star formation. Until we have a better understanding of the efficiency of star formation in collapsing protogalaxies, for example, it will be difficult to determine the relative importance of gas-dynamical and stellar-dynamical processes, and it will be difficult to decide between the different possible models which can be constructed to give a good fit to the surface brightness distribution of galaxies like NGC 3379. It will be of great importance to obtain more observational data on the properties of galaxies, particularly the composition distribution and the kinematics in regions other than the nuclei of elliptical galaxies, since such data should eventually provide a powerful means of discriminating between models.

Another avenue of progress will be to look for observable phenomena which are similar to, or perhaps even identical to, those which take place during the formation of galaxies. Thus the processes of gas dynamics and star formation currently taking place in the disc of our galaxy should have a good deal in common with processes which took place at a much earlier stage of evolution, and it is clear that a better understanding of current star formation processes will be most useful, if not prerequisite, for understanding star formation in forming galaxies.

The most intriguing and perhaps most promising possibility is that of actually observing galaxy formation processes taking place in relatively nearby systems. The theoretical models strongly suggest the possibility that in some cases residual protogalactic gas may continue to condense into a galaxy long after the initial collapse process is completed and thus help sustain continuing star formation, either steadily or in occasional bursts. Some well known galaxies which may have experienced recent gas infall and bursts of star formation are M82, which contains many relatively young stars and a large amount of chaotically distributed gas and dust; NGC 5253, an elliptical galaxy which contains several knots of gas and young stars, and which has recently attained notoriety as a prolific producer of supernovae (an elliptical galaxy in the process of formation?); and NGC 5128, a giant spherical galaxy surrounded by a conspicuous dark band of dust, gas, and young stars (a disc system in the process of formation?) Detailed studies of such systems should eventually increase substantially our understanding of gaseous condensation and star formation processes in forming galaxies.

References

Alladin, S. M., Potdar, A., and Sastry, K. S.: 1975, this volume, p. 167.
Arp, H. C.: 1965, *Stars and Stellar Systems* **5**, 401.
Bardeen, J. M.: 1975, this volume, p. 297.
Bertola, F. and Capaccioli, M.: 1975, this volume, p. 373.
Binney, J.: 1974, *Monthly Notices Roy. Astron. Soc.* **168**, 73.
Brosche, P.: 1970, *Astron. Astrophys.* **6**, 240.
Eggen, O. J., Lynden-Bell, D., and Sandage, A. R.: 1962, *Astrophys. J.* **136**, 748.
Faber, S. M.: 1973, *Astrophys. J.* **179**, 731.
Freeman, K. C.: 1970, *Astrophys. J.* **160**, 811.
Goldreich, P. and Lynden-Bell, D.: 1965, *Monthly Notices Roy. Astron. Soc.* **130**, 125.
Gott, J. R.: 1973, *Astrophys. J.* **186**, 481.
Gott, J. R.: 1975, *Astrophys. J.* (in press).
Hartwick, F. D. A. and Sargent, W. L. W.: 1974, *Astrophys. J.* **190**, 283.
Hohl, F.: 1971, *Astrophys. J.* **168**, 343.
Hohl, F.: 1975, this volume, p. 349.
Icke, V.: 1973, *Astron. Astrophys.* **27**, 1.
Janes, K. A.: 1975, in press.
King, I. R.: 1966, *Astron. J.* **71**, 64.
King, I. R.: 1975, in preparation.
Larson, R. B.: 1969, *Monthly Notices Roy. Astron. Soc.* **145**, 405.
Larson, R. B.: 1972a, *Nature* **236**, 21.
Larson, R. B.: 1972b, *Monthly Notices Roy. Astron. Soc.* **156**, 437.
Larson, R. B.: 1974a, *Monthly Notices Roy. Astron. Soc.* **166**, 585.
Larson, R. B.: 1974b, *Monthly Notices Roy. Astron. Soc.* **169**, 229.
Larson, R. B. and Tinsley, B. M.: 1974, *Astrophys. J.* **192**, 293.
Lecar, M. and Cohen, L.: 1972, in M. Lecar (ed.), *Gravitational N-Body Problem*, D. Reidel Publ. Co., Dordrecht-Holland, p. 262.
Liller, M. H.: 1960, *Astrophys. J.* **132**, 306.
Liller, M. H.: 1966, *Astrophys. J.* **146**, 28.
Lynden-Bell, D.: 1967, *Monthly Notices Roy. Astron. Soc.* **136**, 101.
McClure, R. D.: 1969, *Astron. J.* **74**, 50.
Miller, R. H. and Prendergast, K. H.: 1962, *Astrophys. J.* **136**, 713.
Oort, J. H.: 1970, *Astron. Astrophys.* **7**, 381.
Oort, J. H.: 1974, in J. R. Shakeshaft (ed.), 'The Formation and Dynamics of Galaxies', *IAU Symp.* **58**, 375.
Ostriker, J. P., and Peebles, P. J. E.: 1973, *Astrophys. J.* **186**, 467.
Ostriker, J. P., and Thuan, T. X.: 1975, *Astrophys. J.* (in press).
Peebles, P. J. E.: 1971, *Astron. Astrophys.* **11**, 377.
Peebles, P. J. E.: 1972, *Comments Astrophys. Space Phys.* **4**, 53.

Peebles, P. J. E.: 1974a, *Astrophys. J. Letters* **189**, L51.

Peebles, P. J. E.: 1974b, in J. R. Shakeshaft (ed.), 'The Formation and Dynamics of Galaxies', *IAU Symp.* **58**, 55.

Peebles, P. J. E. and Dicke, R. H.: 1968, *Astrophys. J.* **154**, 891.

Roberts, M. S. and Rots, A. H.: 1973, *Astron. Astrophys.* **26**, 483.

Sandage, A. R.: 1972, *Astrophys. J.* **176**, 21.

Saslaw, W. C.: 1971, *Monthly Notices Roy. Astron. Soc.* **152**, 341.

Saslaw, W. C.: 1972, *Astrophys. J.* **177**, 17.

Shields, G. A.: 1974, *Astrophys. J.* **193**, 335.

Silk, J. I. and Lea, S. M.: 1973, *Astrophys. J.* **180**, 669.

Spinrad, H., Gunn, J. E., Taylor, B. J., McClure, R. D., and Young, J. W.: 1971, *Astrophys. J.* **164**, 11.

Spinrad, H., Smith, H. E., and Taylor, B. J.: 1972, *Astrophys. J.* **175**, 649.

Spitzer, L.: 1971, in D. J. K. O'Connell (ed.), *Nuclei of Galaxies*, American Elsevier, New-York, p. 443.

Toomre, A. and Toomre, J.: 1972, *Astrophys. J.* **178**, 623.

Tremaine, S. D., Ostriker, J. P., and Spitzer, L.: 1975, *Astrophys. J.* **196**, 407.

Van Houten, C. J.: 1971, *Bull. Astron. Inst. Neth.* **16**, 1.

Wilson, C. P.: 1975, *Astron. J.* **80**, 175.

DISCUSSION

Lynden-Bell: I am interested in the actual value of your kinematic viscosity and its ratio to the typical angular momentum per unit mass of a star. Is this ratio 10^{-3} or what?

Larson: I'm not sure exactly how the kinematic viscosity that I assumed compares with the angular momentum momentum per unit mass of the stars, but in any case it is a large effect: in terms of the mean free path of the colliding fluid elements, the mean free path is of the same order as the radius, and in terms of the time scale for significant redistribution of angular momentum, this time scale is comparable with the collapse time.

Hunter: I agree that some mechanism for the transfer of angular momentum is very important for this problem. What justification can you give for supposing the relevant mechanism to be a turbulent viscosity, rather than gravitational torques that arise when axisymmetry is broken? Other work tells us of the importance of bar instabilities, and these could transfer significant angular momentum. Also, do you have a turbulent viscosity for the stars as well as for the gas?

Larson: Perhaps I should clarify first that I use the terms 'turbulence' and 'turbulent viscosity' merely as convenient labels, without implying any close connection with turbulence as understood (?) in other contexts. The physical picture that I have in mind is that a protogalaxy consists of some collection of gas elements, in general having a chaotic distribution and state of motion. Then, as was pointed out some time ago by McCrea in connection with his floccule theory of star formation, the gas elements which happen because of their random motions to have little angular momentum will fall toward the center, where they will remain if their motions are rapidly dissipated. On the other hand, the elements with large angular momentum will remain in the outer part of the system, and the net result will be a redistribution of angular momentum in which angular momentum is transported outward. It seems to me that any system with random motions will show this effect, and this is what I have attempted to simulate in my fluid dynamical calculations. This separation of the low- and high-angular momentum material will be important during the initial collapse, but probably not after the gas has settled into a flattened layer in approximate centrifugal equilibrium. Then non-axisymmetric instabilities to bar and/or spiral modes would likely set in and might well lead to further important outward transport of angular momentum.

It might even be that gravitational torques could be important during the initial collapse, since differential rotation during the collapse could wind the density inhomogeneities into spirals, associated with which would be gravitational torques. However, I don't know how this effect could be estimated or taken account of in my models.

The models that I have calculated so far have no viscosity for the stars, although such an effect would probably exist also for the stars. My guess is that the viscosity effect for the stars would be less important than for the gas.

Brosche: What are the assumptions on star formation in your models?

Larson: I tried two different approaches to treating the star formation rate: one was to relate the star formation rate to a power of the gas density, in which case a star formation rate proportional to about the

1.8 power of the gas density is found to give models in good agreement with observations; and the second approach, possibly better in that it involves fewer free parameters, was to assume that the time scale for transformation of gas into stars is proportional to the dynamical time scale for the system. The latter approach may be roughly justified on physical grounds if star formation is initiated by large scale dynamical processes such as the compression of the gas in shock fronts caused by collisions between gas clouds or streams. This point of view would be consistent, for example, with current attempts to explain star formation in spiral shock fronts in spiral galaxies.

Freeman: Turbulence plays a fundamental part in your picture and your results. Do you have any physical external reason for invoking it?

Larson: The strongest reason for assuming that the gas had large chaotic motions is perhaps an 'internal' one and is the fact that halo stars in our Galaxy are observed to have a very large velocity dispersion (including many retrograde orbits), and this is most reasonably explained in terms of large random motions in the gas from which the stars formed. The 'external' reasons are harder to make precise, but the various mechanisms which have been proposed for imparting large scale motions (eg. rotation) to galaxies all seem likely to produce a more or less chaotic state of motion in a protogalaxy. This is true in varying degrees of tidal torques, primordial turbulence, motions generated during the collapse of protoclusters, and random motions generated (eg. by supernova explosions) during the formation of galaxies. If any (or all) of these processes operate, it would seem rather difficult to produce a protogalaxy whose internal motion is extremely regular, such as pure uniform rotation with no random motions superimposed.

King: The $Z(r)$ observations that you mentioned were made with an old generation of equipment. A new generation now exists that will be able to extend the curve into the region that you want.

Miller: Could you provide a guide to the observers as to what observations are most effective in discriminating among various models? It appears that the models differ most strongly in the faint outer regions, and that details near the center are less sensitive.

Larson: It is unfortunately true that some of the important observational tests to discriminate between models involve the properties of the faint outer envelopes of galaxies. For example, if it turns out that the metal abundance is constant in the outer envelopes of elliptical galaxies, then a model in which star formation is rapid compared to the collapse time (such as Gott's models) might be viable, whereas if the metal abundance Z continues to decrease with r even in the halo, as seems to be indicated by recent observations of NGC 3379 by Burkhead and Kalinowski, it seems clear that a model involving the collapse and simultaneous metal enrichment of a gas cloud is required. Also, different models make somewhat different predictions for the kinematics of the outer regions of galaxies, while they may predict much the same thing in the central regions. In general, it is my impression that observations of the composition and kinematics of galaxies and especially of their variation with radius may provide more information that is useful for discriminating between models than observations of the light distribution.

ON THE FORMATION OF ELLIPTICAL GALAXIES

J. R. GOTT III

California Institute of Technology, Dept. of Astronomy, Pasadena, Calif., U.S.A.

Abstract. Simple collapse and violent relaxation models produce sharply cut off envelopes with $\varrho(r) \propto r^{-4}$, whereas actual elliptical galaxies have more extended envelopes approximately given by $\varrho(r) \propto r^{-3}$ as implied by Hubble's Law. Numerical models are presented, showing that when cosmological infall effects are included, galaxies are produced with more extended envelopes having a $\varrho(r) \propto r^{-2.8}$ dependence in excellent agreement with Hubble's Law. We have also computed a realistic rotating model of an E5 galaxy which includes infall effects and where the angular momentum of the galaxy is gained through tidal interactions with neighboring protogalaxies. The envelope displays a Hubble dependence but at great distances shows a cutoff due to the tidal effects. The model is compared to the E5 galaxy NGC 4697 and provides an outstandingly good fit to the observations. Except for simple scaling, this excellent fit is produced without the use of any free fitting parameters.

1. Introduction

I would like to present a different point of view, namely that elliptical galaxies are produced by a dissipationless collapse and relaxation process.* According to current cosmological theories galaxies begin as small density perturbations present at the epoch of hydrogen recombination at a redshift of $z \sim 1000$. Because of the excess density in the perturbation the density in the protogalaxy exceeds the critical density $[\varrho > \varrho_{ci} = 3H_i^2/8\pi G]$ and is therefore gravitationally bound. The protogalaxy initially is expanding with the uniform Hubble expansion $\mathbf{V} = H_i \mathbf{r}$, but because it is gravitationally bound, it will expand to reach a maximum radius, stop momentarily and then begin to collapse. If star formation in the early protogalaxy [perhaps occurring in early globular clusters as proposed by Peebles and Dicke (1968)] is sufficiently rapid so that star formation is essentially complete by the time it reaches the point of maximum collapse, then one has a dissipationless collapse and relaxation process.

The dissipationless collapse and relaxation process for a stellar system offers several attractive features in explaining elliptical galaxies. First, the Lynden-Bell (1967) violent relaxation process is expected to operate and lead naturally to the relaxed elliptical isophotes observed in elliptical galaxies. In such numerical models (Gott, 1973) the equilibrium structure is reached after only a couple of free fall times and the isophotes of the equilibrium galaxy are indeed nicely elliptical. Secondly, the dissipationless collapse picture explains in a natural way the observed flattenings of elliptical galaxies.

Consider a nonrotating sphere of stars with no initial kinetic energy and a radius of R_0, corresponding to the point of maximum expansion. Let E_b be the total binding energy which is conserved in the dissipationless collapse. Then initially $-E_b = T + W$ with $W = -(3GM_{gal}^2/5R_0) = -E_b$ and $T = 0$. Let it collapse and undergo violent

---

* The results presented here will be discussed in greater detail in a forthcoming paper, Gott (1975).

Hayli (ed.), Dynamics of Stellar Systems, 271–285. All Rights Reserved.

relaxation. When virial equilibrium is reached, $2T+W=0$ and $-E_b=T+W$ so $W=-2E_b$, or twice its original value, and the equilibrium system will be spherical with a characteristic radius of $\sim\frac{1}{2}R_0$. Now consider a rotating galaxy of radius R_0 and $\Omega=\Omega_0$ so that it is initially holding itself up against gravity in the equatorial plane. This is surely an upper limit on the amount of rotation we would expect cosmologically, and is an amount which can be produced from tidal interactions with neighboring protogalaxies as we shall see below. Let this galaxy collapse. It continues holding itself up in the equatorial plane, leaving $r\cong R_0$ in that direction. Rotation does not inhibit collapse perpendicular to the plane, so to first approximation the collapse proceeds perpendicularly as in the spherical case to give $r=\frac{1}{2}R_0$. These approximations indicate an equilibrium galaxy with a 2 to 1 axial ratio (i.e., an E5 galaxy). Thus, with a dissipationless collapse picture we expect elliptical galaxies to show intrinsic flattenings approximately in the range E0–E5. Significantly, elliptical galaxies are observed to have intrinsic flattenings in the range E0–E5 with just a few E6 galaxies and no E7 or flatter galaxies. The fact that elliptical galaxies have intrinsic flattenings in just the range predicted by the dissipationless picture is strong evidence for this mechanism. When gas is present, as in spiral galaxies, it is quite dissipative forming disk systems with characteristic flattenings of 10 or even 20 to 1.

The simple models of the collapse and relaxation of a spherical cloud of stars thus give equilibrium models having many of the properties of elliptical galaxies. The main difficulty with this picture is that the simple models considered in Gott (1973) produce elliptical galaxies with envelopes in which the density falls off according to $\varrho(r)\propto r^{-4}$. Observed elliptical galaxies have envelopes that are much more extended than this. The envelopes of elliptical galaxies can be modeled well by Hubble's Law (1930), which gives a projected surface density of the form $I(r)=I_0/[1+(r/a)]^2$ where a is a constant. This requires $I(r)\propto r^{-2}$ in the envelope and thus $\varrho(r)\propto r^{-3}$. Other N-body calculations that study the collapse and relaxation of a spherical cloud of stars show the same problem. Models by Peebles (1970), Hénon (1964), and Aarseth (1973) all produce sharply cut off envelopes with the r^{-4} dependence. In the present paper we will show that the simple models ignore cosmological infall effects and that when these effects are included we can explain the main features of elliptical galaxy envelopes.

2. Cosmological Infall

The dynamics of cosmological infall have been discussed in detail in relation to infall into clusters of galaxies by Gott and Gunn (1971) and Gunn and Gott (1972). We shall present here just the primary results needed for elliptical galaxies. In accordance with the standard big bang cosmology we envision the protogalaxy beginning as an isothermal density perturbation present at the epoch of recombination at redshift $z_i\sim 1000$. Initially there is a uniform Hubble flow ($V=H_i r$). Because of the excess density, the material in the protogalaxy is gravitationally bound. Its expansion is slowed, so the perturbation reaches a maximum radius and begins to collapse, completing the collapse at time T_c. Material outside the perturbation proper can also be gravitationally bound and suffers infall into the galaxy at $t > T_c$.

We assume that at the epoch of recombination there exists a spherical region of radius R_i which has uniform density slightly higher than the surrounding region. Such a perturbation structure is in the spirit of a white noise spectrum of density fluctuations (i.e., Poisson distribution of points) as studied by Press and Schechter (1974) and Peebles (1974). In such a white noise situation the density in the region just outside a known positive density fluctuation is, on the average, just equal to the average density. Press and Schechter (1974), using this model and standard dynamics, were able to derive a theoretical luminosity function for galaxies in good agreement with the observations. Thus, we adopt the condition that the density outside the perturbation proper should be of average density. Let ϱ_{ci} be the critical density at the epoch $z_i = 1000$

$$\varrho_{ci} = \frac{3H_i^2}{8\pi G}. \tag{1}$$

For densities less than the critical density the expansion is unbounded; for densities in excess of ϱ_{ci} the material will expand to a maximum radius and collapse. Let ϱ_{ei} be the external density which we set equal to the mean density in the universe at epoch z_i. From Gunn and Gott (1972) one can show that

$$\varrho_{ei} = \varrho_{ci} \cdot \frac{2q_0(1+z_i)}{(1-2q_0) + 2q_0(1+z_i)}. \tag{2}$$

We consider pressureless cosmologies with no cosmological constant, so the deceleration parameter q_0 is given by

$$3H_0^2 q_0 = 4\pi G \varrho_0, \tag{3}$$

where the subscript '0' refers to the present. If q_0 is greater than $\frac{1}{2}$ the universe is closed; if $0 < q_0 < \frac{1}{2}$, the universe is open with unbounded expansion. Let ϱ_i be the density in the perturbation proper $r < R_i$ and define $\varrho_+ = \varrho_i - \varrho_{ci}$. To an accuracy of several percent for small ϱ_+ relative to ϱ_{ci} we find

$$T_c \simeq \frac{\pi}{H_i} \left(\frac{\varrho_{ci}}{\varrho_+} \right)^{3/2}, \tag{4}$$

where T_c is just the age of the universe when the perturbation completes its collapse. The perturbation proper is of uniform density initially and remains of uniform density throughout its expansion and collapse. The initial perturbation has a mass $M_0 \simeq 4\pi \varrho_{ci} R_i^3 / 3$. One can show (Gunn and Gott, 1972) that for $t > T_c$ the mass of the galaxy including infall is given by

$$\frac{M(t)}{M_0} = 1 + \frac{(t/T_c)^{2/3} - 1}{1 + (1 - 2q_0)\left(\dfrac{H_0 t}{2\pi q_0} \right)^{2/3}}. \tag{5}$$

With $q_0 = \frac{1}{2}$, $M(t) = M_0 \cdot (t/T_c)^{2/3}$ and the infall material can be considerable. This infall material is less tightly bound and should produce a more extended envelope. As we shall see below this effect is precisely sufficient to produce elliptical galaxy models in agreement with the observations.

From Equation (5) it is clear that the infall depends on the value of q_0. However, for $t \ll H_0^{-1}$ it is clear that in all cosmologies the infall approximates closely that of the $q_0 = 0.5$ case. This is because densities in all cosmological models approach asymptotically the critical density at early epochs as indicated by Equation (2). In particular the model we will compare with NGC 4697 has a collapse time for the central perturbation of $T_c \sim 3.3 \times 10^7$ yr, which combined with $H_0 = 55$ km s^{-1} Mpc^{-1} gives $H_0 T_c = 0.0019$ so even if q_0 is as low as $q_0 = 0.02$ (see Gott et al., 1974) Equation (5) shows that for times shortly after $t = T_c$ the infall is within about 6% of that predicted by the $q_0 = \frac{1}{2}$ model. Thus, the collapse times of elliptical galaxies are typically short enough so that the infall and resulting envelope structure are approximated well by the $q_0 = \frac{1}{2}$ model. This is significant not only because it makes the results essentially independent of the cosmological model but because it produces ellipticals with essentially the same envelope structure relatively independently of their surroundings. If the open cosmological picture of Gott et al. (1974) is correct then ellipticals in the field see a local cosmology with $q_0 = 0.025$, while ellipticals in bound clusters experience an effective $q_0 > \frac{1}{2}$. The above results indicate that these differences will not result in great differences in infall structure.

3. Spherical Systems

Two spherical models are computed using 1000 stars with a variant of Hénon's spherical shell technique. Model I is the simple collapse of a uniform density sphere of stars with no infall. This model is thus representative of the class of simple models like those of Hénon (1964), Peebles (1970), and Gott (1973). The calculation is started at the point of maximum expansion.

Model II includes the effects of cosmological infall. As discussed in Section 2 it is sufficient to consider a single case with $q_0 = \frac{1}{2}$. For the initial perturbation $r < R_i$ we adopt $\varrho = \varrho_{ci} + \varrho_+$ such that T_c computed by Equation (6) is $T_c = \pi (R_0^3/2GM_0)^{1/2}$ matching the T_c of model I. For $r > R_i$ we adopt $\varrho = \varrho_{ci}$. The initial conditions are set at the epoch of recombination at $z = 1000$; one can easily show that the results are independent of this particular epoch as long as ϱ_+ is chosen so as to yield the correct T_c and $H_i \ll T_c$. We assign 100 stars of mass 0.01 M_0 each to the initial perturbation and 900 stars of mass 0.01 M_0 to the adjacent surrounding volume. The trajectory of each star is computed analytically via the formulas in Gunn and Gott (1972) up until $t = \frac{1}{2} T_c$, when the calculation is continued numerically using the shell technique mentioned above.

We note that these elliptical galaxies have enough stars so as to be essentially collisionless systems (i.e., the two-body relaxation time is significantly longer than the times over which the system is studied). We have made several checks to assure that our systems are indeed essentially collisionless (see Gott, 1975), and that the only significant relaxation is the Lynden-Bell violent relaxation.

The model I results are quite similar to those found in Gott (1973). By time $t = \frac{3}{2} T_c$ the system had reached its equilibrium structure. The model I results displayed in Figures 1 and 2 are for the equilibrium structure and refer to time averages over the period $t = 2T_c$ to $3T_c$.

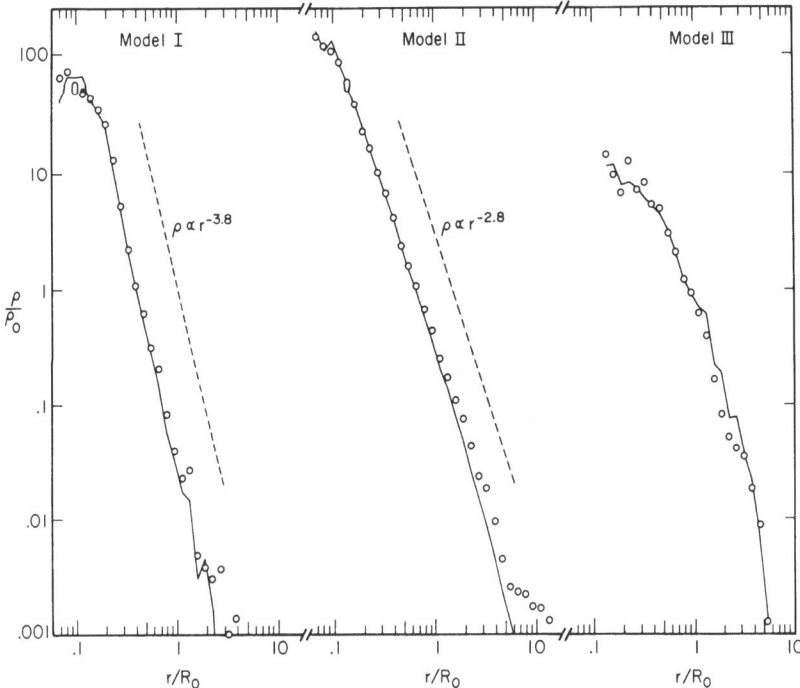

Fig. 1. Stellar density in the galactic plane as a function of radius for the equilibrium models. The open circles represent the actual runs of density with radius while the solid lines represent that expected from satisfaction of the stationary collisionless Boltzmann equation. The density $\varrho_0 = 3M_0/4\pi R_0^3$ is a standard density which is the same for each of the models. Model I is a standard spherical collapse model without infall, model II includes cosmological infall and thus has a much more extended envelope. The dotted lines give the best power law fit to each of the envelopes. Model II with $\varrho \propto r^{-2.8}$ agrees closely with $\varrho \propto r^{-3}$ predicted from satisfaction of Hubble's Law $I = I_0/[1 + (r/a)]^2$. Model III is a rotating model whose rotation has been derived through tidal interactions with neighboring protogalaxies. It includes both cosmological infall effects and tidal interaction effects.

In Figure 1 the run of density versus radius is presented. The solid line represents the expected run of density expected from satisfaction of the stationary collisionless Boltzmann equation.

$$-\frac{\partial \Phi}{\partial r} = -\frac{\langle V_\phi^2 \rangle}{r} + \langle V_r^2 \rangle \left[\frac{\partial \ln \varrho}{\partial r} + \frac{\partial \ln \langle V_r^2 \rangle}{\partial r} + \frac{1}{r} \left(1 - \frac{\langle V_\theta^2 \rangle}{\langle V_r^2 \rangle} \right) + \right.$$
$$\left. + \frac{1}{r} \left(1 - \frac{\langle (V_\phi - \langle V_\phi \rangle)^2 \rangle}{\langle V_r^2 \rangle} \right) \right]. \tag{6}$$

There is excellent agreement between the observed density and the solid line showing that the system is essentially collisionless and that it has reached an equilibrium structure. In the envelope, the best fit is given by $\varrho \propto r^{-3.8}$ as indicated by the dashed line in Figure 1.

In model II the infall, given by $M(t) = M_0 \cdot (t/T_c)^{2/3}$, goes as a power law with the infall rate depending only on the mass accumulated up to that time. Thus for $t \gg T_c$ there are no intrinsic scales to the problem. The solution is what Press and Schechter (1974) would refer to as selfsimilar. With no intrinsic scale present, it is easy to show that the expected envelope structure should be a power law in the density. A numerical experiment is then needed to determine the slope of this power law. An important factor is the relaxation of each shell of stars as it suffers infall. The results presented in Figure 1 show that the equilibrium envelope of model II has indeed a remarkably straight power law dependence with $\varrho \propto r^{-2.8}$ in virtually exact agreement with the Hubble's Law envelope dependence $\varrho \propto r^{-3}$. Model II satisfies the stationary collisionless Boltzmann equation, showing that it has reached a stationary equilibrium configuration. For $r > 5R_0$ there is a sharp departure from satisfaction of the stationary collisionless Boltzmann equation. This is due to the fact that the outer shells of stars have not yet suffered infall and are still traveling outward. The outermost shell of stars at radius $r = 14R_0$ has a Hubble expansion that has been only slightly slowed. The run of density with radius presented in Figure 1 is taken

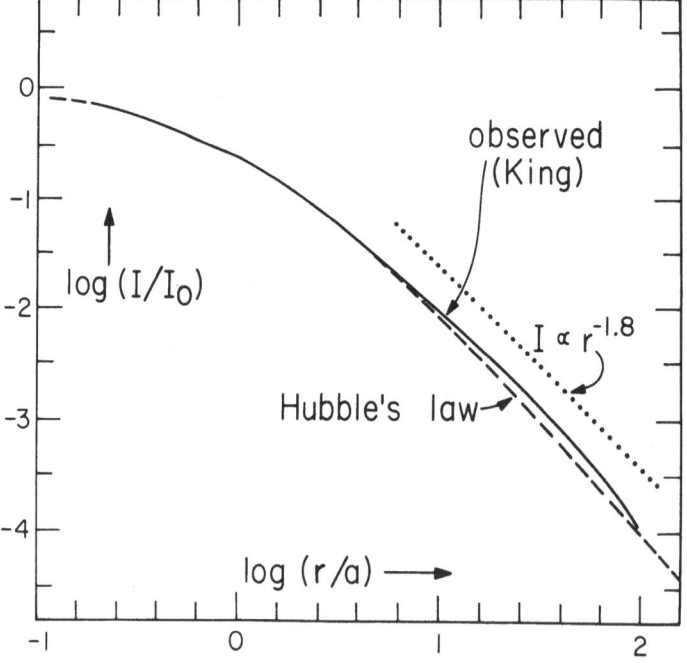

Fig. 2. Surface brightness as a function of projected radius. The solid line represents an average of 13 observed galaxies for which King has done accurate photometry. For comparison the dashed curve gives Hubble's Law $I = I_0/[1 + (r/a)]^2$. The dotted curve $I \propto r^{-1.8}$ gives the slope expected for an envelope where $\varrho \propto r^{-2.8}$ as in model II. Thus model II, which includes infall, gives an envelope structure in good agreement with the observations.

at time $t \sim 6\,T_c$ (i.e., a time average from 5.5 to 6.5 T_c). As time goes on, the background cosmological density continues to go down, and more and more of the envelope separates out continuing to show the same power law dependence.

Let us compare the results for model II with those for observed elliptical galaxies. In Figure 2 is plotted a mean observed curve of light intensity (flux arcsec^{-2}) vs radius for 13 individual galaxies from photometry obtained by King (private communication). The curves for the individual galaxies have been scaled both in central intensity and radius. For comparison, the mean observed curve has been fit (Wilson, 1973) with the best fit of Hubble's Law. The radial distances are measured in terms of the constant a in Hubble's Law. The fit of Hubble's Law to the observed data is certainly outstanding, particularly in the central regions and the fit in the envelope is also quite close. It is noteworthy also that the curves of $I(r)$ for the 13 individual galaxies all lie quite close to the mean curve, differing from it less in all cases than the maximum difference between the mean curve and Hubble's Law. This is one of the extraordinary properties of elliptical galaxies, that they exhibit envelopes with nearly identical radial structures. Model II, which includes infall, predicts $\varrho(r) \propto r^{-2.8}$ and $I(r) \propto r^{-1.8}$. For comparison we have drawn a dotted line showing the slope $I \propto r^{-1.8}$. It can be seen that this slope is in excellent agreement with the data over the decade and a half in radius of interest. Thus the infall picture provides a ready explanation for the observed density falloff in the envelopes of elliptical galaxies.

4. Rotating Systems

4.1. INTRODUCTION

Peebles (1971) proposes that galaxies acquire their angular momentum through tidal interactions. The original protogalaxy is undoubtedly irregular in shape and thus the tidal action of the neighboring protogalaxies exerts a torque on it. Peebles (1971) has shown by numerical experiments that the angular momentum transfer is accomplished primarily around the time the protogalaxy reaches its point of maximum expansion. He also shows that the maximum amount of angular momentum acquired by the protogalaxy is (to a factor of order unity) $J = (2/5)\,M_0 R_0^2 \Omega_0$ where $\Omega_0 = = (GM_0/R_0^3)^{1/2}$ is the angular velocity required to just prevent collapse in the equatorial plane. By application of simple virial theorem arguments earlier it was shown that the maximally rotating galaxy $(\Omega = \Omega_0)$ is left with a flattening of at most 2 to 1, i.e., an E5 galaxy. Thus, it was argued that one should expect to see elliptical galaxies in the range E0 to approximately E5, which agrees quite well with the observations.

4.2. METHOD AND INITIAL CONDITIONS

We wish to study the collapse and relaxation of a protogalaxy which has been set into rotation by tidal interactions and where infall effects are included. We first define a spherical coordinate system (r, θ, ϕ) centered on the protogalaxy and having an axis $(\theta = 0)$ equal to the proposed rotation axis of the protogalaxy. We start with 2000 stars, 200 stars for the initial perturbation and 1800 stars in the outlying infall region (so $m_s = 0.005\ M_0$). We start by giving the stars radial positions and velocities at $t = \frac{1}{2}T_c$ just as done for model II. We give the galaxy an irregular shape by confining the stars to $0 < \phi < \pi/2$ and $\pi < \phi < \frac{3}{2}\pi$. Thus, at the point of maximum expansion the initial perturbation has density $\varrho = 2\varrho_0$ in two opposite quadrants and density $\varrho = 0$ in the other two quadrants. Likewise the infall material is confined to these quadrants.

The tidal effects of neighboring protogalaxies are effective for only a short time near $t = \frac{1}{2}T_c$ (assuming that all the protogalaxies have similar collapse times). At earlier times the other protogalaxies are just small density fluctuations $\delta\varrho/\varrho \ll 1$ and Peebles' (1971) calculations show the angular momentum transfer to be small. By $t = \frac{1}{2}T_c$ the protogalaxies have separated out and are behaving essentially as point masses but as time goes on their tidal effects decrease since the distances between the protogalaxies are growing as $d \propto t^{2/3}$ and the tidal force goes as d^{-3}. Thus, as Peebles' calculations show, the angular momentum transfer occurs primarily around $t = \frac{1}{2}T_c$. Our initial conditions are taken to start at $t = \frac{1}{2}T_c$ so we adopt the following simple model for the tidal interactions of the neighboring protogalaxies. We define the perturbing potential

$$\Phi_{\text{tidal}} = -2(GM/d^3)(x^2), \tag{7}$$

where $x = r \sin\theta \cos\phi$. One may think of this perturbing potential arising from the action of two protogalaxies of mass M and distance d on either side of the galaxy in question (i.e., at $\theta = \pi/2$, $\phi = 0$ and $\theta = \pi/2$, $\phi = \pi$). We have used for simplicity here just the lowest order spherical harmonic. The perturbing potential should act for a time $\Delta t \sim T_c$ approximately centered on $t = \frac{1}{2}T_c$. Thus each star should receive a kick in velocity due to the tidal forces given by

$$\Delta \mathbf{V}(r, \theta, \phi) \simeq \Delta t \cdot \nabla \Phi_{\text{tidal}}. \tag{8}$$

We thus compute the velocity kick given each star due to the tidal forces and add that extra velocity to the stars original velocity as defined from the initial conditions of model II. Since Δt is assumed to be short we will not alter the positions of the stars. Since the tidal forces die out rapidly after $t \sim \frac{1}{2}T_c$ we will assume their entire effect is represented by the velocity kicks and set the disturbing potential equal to zero after $t = \frac{1}{2}T_c$. Because of the irregular shape of the protogalaxy with the stars confined to opposite quadrants, it is easy to show that a net angular momentum is imparted to the protogalaxy. The value of the tidal potential is set, see Equation (7) at a value $(M/d^3) = 0.25(M_0/R_0^3)$. This value is sufficient to produce an E5 galaxy, as we shall see. The value above was chosen on the basis that it would produce a reasonable amount of rotation. In Gott (1975) we compare this value with what one would calculate a priori given a random distribution of neighboring protogalaxies similar to the one we are considering.

The result of this calculation gives a root mean square tidal torque corresponding to $(M/d^3) = 0.15\ (M_0/R_0^3)$. Considering that Peebles (1969, 1971) estimates that such calculations are accurate to no more than a factor of 4, the above value is in excellent agreement with the value $(M/d^3) = 0.25 \ (M_0/R_0^3)$ adopted for model III on the basis of producing the required amount of rotation. The present picture is that all proto-galaxies experience an rms tidal field due to neighboring protogalaxies; when the ir-regular shape of the protogalaxy is aligned properly with the tidal field (as is model III) the maximum net torque is applied and a maximally rotating galaxy is produced. Since model III produces an E5 galaxy it is clear that production of elliptical galaxies in the range E0 to E5 is consistent with the tidal interaction hypothesis.

With the initial conditions as given, the collapse will be non-axisymmetric and individual stars, seeing a non-axisymmetric potential, will either gain or lose angular momentum. Unfortunately such a three dimensional calculation is beyond the range of present computers. Instead we simulate the mixing of angular momenta produced by the violent relaxation by giving the stars extra random velocities in the model II initial conditions and then fix the system to be axisymmetric. The non-axisymmetric relaxation phase, mimicked by the extra random velocities, should lead to an axi-symmetric final state so our approximation of an axisymmetric potential is reason-able.

The collapse and violent relaxation phase can now be followed by the axisym-metric potential method of Gott (1973).

4.3. RESULTS

The system was allowed to collapse and relax and the central region reached an equilibrium configuration by $t = 2T_c$ (where all times are measured from the beginning of the universe). The envelope infall continued after this. The outer portions of the envelope have escape energy and are lost from the galaxy, i.e., stolen away by the neighboring protogalaxies. The infall is thus cut off and the envelope settles down by about $t = 10 \ T_c$. We take a time average over $10.5 \ T_c < t < 14.5 \ T_c$ for the following data.

From Figure 1 it can be seen that the equilibrium galaxy (model III) closely satis-fies the stationary collisionless Boltzmann equation. The envelope is extended like that of model II due to the infall, the envelope retaining the basic $\varrho \propto r^{-2.8}$ depen-dence but the tidal effects introducing a cutoff at large radii.

We wish to compare the results of model III with the best studied E5 galaxy NGC 4697. The fact that there are no observed elliptical galaxies appreciably flatter than E5 means that the galaxy must be seen essentially edge on and projection effects need not be considered. Figure 3 shows the model III data compared with the actual run of $I(r)$ obtained by King (private communication) for NGC 4697. The fit is certainly remarkably good. It is important to remember that, except for simple scaling, *no* free fitting parameters have been used to achieve this fit. (We have made only one rotating model. Started with reasonable initial conditions, it is compared with the best studied galaxy of its type. Thus, there has been no adjustment of the model parameters to improve the fit.) Figure 4 shows a comparison of the runs of ellipticity with radius. The scale has been set already by the intensity vs radius fit of Figure 3, so even the scale in Figure 4 is not adjustable. The data for model III is more scattered here due

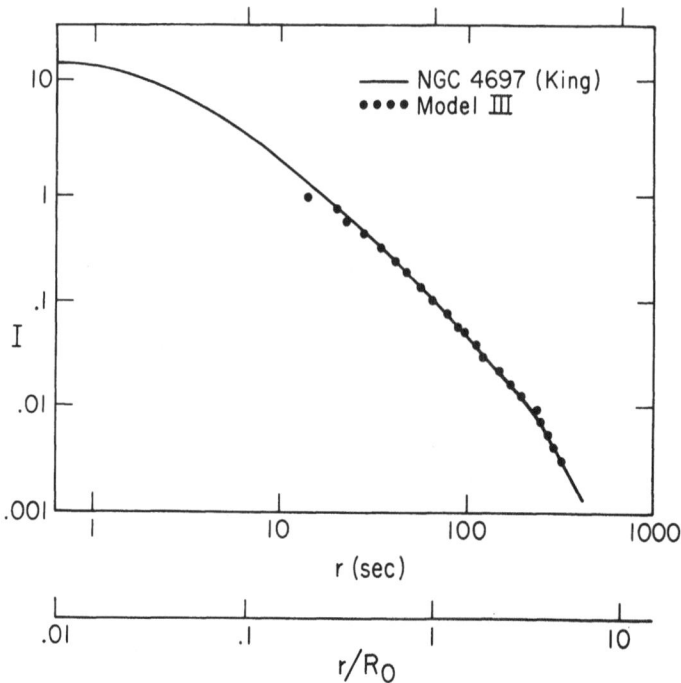

Fig. 3.   A comparison of the results of model III and NGC 4697. The solid line plots surface brightness vs projected radius for NGC 4697 as obtained by King (private communication). The dots show similar surface brightness vs projected radius data for model III. It is noteworthy that except for simple scaling this excellent fit is obtained without the use of any free fitting parameters.

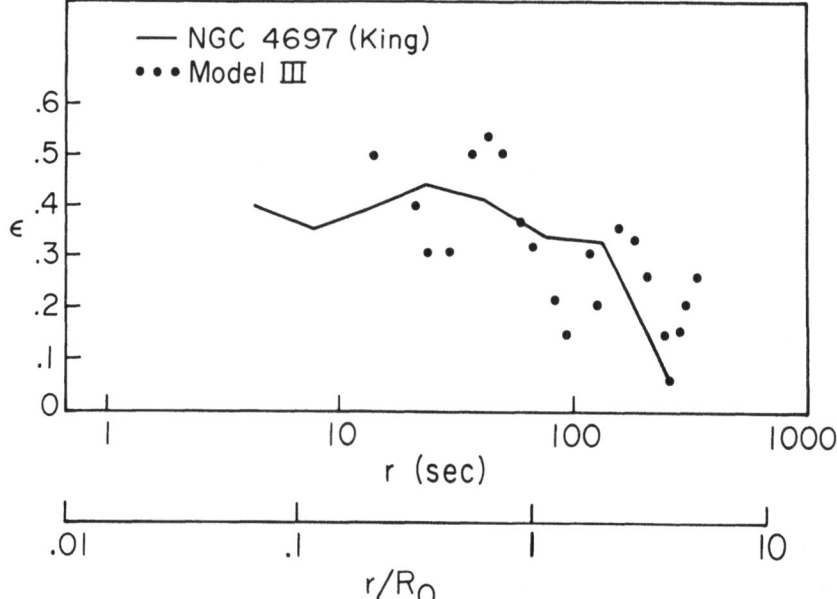

Fig. 4.   A comparison of ellipticity $\varepsilon=(a-b)/a$ vs radius in model III and NGC 4697. The two radial scales are preset by the fit of Figure 3. Although the model III data contain considerable statistical scatter, it is clear that in the mean it closely parallels the data from NGC 4697.

to the unavoidably small number of stars in the envelope. However, given this random scatter, again the fit is quite good.

4.4. DISCUSSION

Wilson (1973) by bringing together considerable photometric data by King has shown that (1) elliptical galaxies regardless of intrinsic flattening have curves of intensity versus radius $[I(r)]$ that are remarkably similar (i.e., Figure 2); (2) curves of ellipticity versus radius $[\varepsilon(r)]$ show no particular pattern; in some cases the ellipticity increases, then decreases, in some it starts high and decreases, and in some it increases steadily outward. The present theory accounts easily for these two remarkable facts. From the previous sections it can be seen that the radial structure of the galaxy is determined virtually entirely by the infall effects and the tidal interactions which are similar for all protogalaxies. The ellipticity of the galaxy on the other hand is determined by the degree of irregularity of the original protogalaxy. If the original protogalaxy is spherically symmetric, then a spherical equilibrium galaxy is produced; if the original protogalaxy is irregular in shape (like model III where stars are confined to opposite quadrants) then the tidal action of neighboring protogalaxies can produce a rotating galaxy. The radial structure which is determined effectively by the binding energies of the different stars is unaffected by the irregularities in shape (ϕ, θ distribution) and depends through infall and tidal action primarily on the radial distribution. So Wilson's result number (1) is to be expected. It is reasonable that the irregularity in shape may be a function of radius. If the perturbation proper is spherical but the infall material is confined to quadrants, we will produce a galaxy with nearly circular isophotes near the center and more elliptical isophotes in the envelope. On the other hand, if the initial perturbation is irregular and the infall is spherically symmetric, one would have a galaxy with ellipticity decreasing outwards. Combine these effects with projection effects and it is clear that a variety of runs of eccentricity versus radius can be produced in agreement with Wilson's result (2). Despite the fact that a variety of runs of ellipticity are possible the runs of ellipticity for model III and NGC 4697 are still in good agreement. Two factors make this result plausible. First, an E5 galaxy represents the *maximum* possible flattening, so it is reasonable that the irregularities in shape might be of unit strength both in the perturbation proper and the envelope as well (as in model III). Secondly, since it is E5, we know we are seeing the galaxy virtually edge on (as assumed for model III) so there are no projection effects.

5. Conclusions

In the present paper we present detailed models for the formation of elliptical galaxies via a dissipationless collapse and violent relaxation process with cosmological infall effects included.

It is relevant to contrast this picture with Larson's (1974) picture that elliptical galaxies might be formed through some slow dissipative process. Such models envisage the elliptical galaxy as composed of gas slowly dissipating its energy and turning itself into stars. Because of the dissipation, stars formed later and later form a denser and denser core. Thus, the envelope of the galaxy is built from the outside in. Larson (1974) has produced an excellent fit to the E1.5 galaxy NGC 3379, using a gaseous dissipation timescale $\tau_{\text{diss}} = 1.19\ t_f$, and a star formation timescale $\tau_{\text{star}} = 1.11\ t_f$ (t_f = free fall time), however, this model of NGC 3379 which achieves the best fit includes no infall.

There are several important factors that make one favor dissipationless infall models over dissipation models with no infall. (1) The observed flattening of ellipticals is predicted by the dissipationless theory, as we have discussed. It would seem that any slow dissipation process capable of producing an E5 galaxy could also produce an E8 galaxy. (2) We have argued strongly that infall must be present because the protogalaxy must be matched to a cosmological background solution. Infall is important both for galaxies in clusters and field galaxies. The present work shows that without dissipation infall produces an envelope in agreement with Hubble's Law, this would indicate that when infall is included, as it always should be, dissipation is unimportant. (3) Dissipation models must have $\tau_{\text{diss}} \sim t_f$ to work. For $\tau_{\text{diss}} \sim 0.1\ t_f$ the models would be far too centrally condensed. The fact that spiral galaxies which contain appreciable amounts of gas are highly flattened, argues that gas, when it is present, is indeed highly dissipative. A self-supporting cloud of hot gas of galactic mass ($10^{11}\ \mathfrak{M}_\odot$) and galactic dimensions (10 kpc) has a bremsstrahlung cooling time an order of magnitude smaller than the free fall time. The gas density in such a gaseous protogalaxy is similar to the present gas density in the solar neighborhood. If the gas is composed of cold clouds with large translational velocities then the cloud-cloud collision time will be much shorter than a dynamical time if clouds in the protogalaxy have properties similar to those observed in the interstellar medium today. For example, clouds currently travel only 10^7 yr between collisions as compared with the dynamical time of 10^8 yr (see Spitzer, 1968). These considerations suggest that $\tau_{\text{diss}} \lesssim 0.1\ t_f$ is more realistic for gaseous dissipation than $\tau_{\text{diss}} \sim t_f$.

Let us now discuss the effects of supernovae in sweeping gas out of the galaxy (cf. Mathews and Baker, 1971). It is this effect that prevents elliptical galaxies formed via the dissipationless process from forming disks due to mass loss from evolved stars. Mathews and Baker (1971) show that in the central regions the gas density can be high enough to provide sufficient cooling to overbalance the supernovae and allow trapping and dissipation. Such a mechanism may indeed allow reprocessing of material in the very innermost regions and give rise to the cyanogen gradients in the nuclei of elliptical galaxies, observed by Spinrad et al. (1972). Alternatively, Tremaine et al. (1974) have proposed a mechanism for forming dense galactic nuclei from old globular clusters that have spiraled in toward the center through dynamical friction. The existence of many compact galactic nuclei (even in M31 and M32), which seem to be completely independent dynamical entities, is a warning that they may require special mechanisms for formation. Thus, the existence of cyanogen gradients in the very innermost regions of elliptical galaxies should not be used to infer that reprocessing and dissipation are important in the formation of the galaxy as a whole.

To repeat, dissipationless collapse and relaxation models have a number of strong arguments supporting them. (1) The intrinsic flattenings of elliptical galaxies (E0 to E5 with just a few E6's) are in excellent agreement with that predicted from dissipationless collapse with simple virial theorem arguments. (2) Relaxed isophotes and velocity distributions can be produced naturally with the Lynden-Bell violent relaxation process. (3) With infall included, as it must be, elliptical galaxies are found to have envelopes in good agreement with Hubble's Law. (4) A detailed model including infall and tidal interactions with neighboring protogalaxies produces an excellent fit to the observed rotating galaxy NGC 4697. (5) With the Peebles' (1971) tidal interaction picture, and the present theory of dissipationless collapse, it is possible to understand why elliptical galaxies have such similar radial structures and yet exhibit a variety of runs of ellipticity versus radius. The radial structure is determined primarily by infall and tidal effects which are similar for all galaxies, but the run of ellipticity versus radius depends heavily on the shape of the original protogalaxy which varies from galaxy to galaxy. (6) Finally, the dissipationless collapse model for ellipticals provides a natural scenario for the production of elliptical and spiral galaxies as distinct classes. When star formation is essentially complete before the point of maximum collapse, a dissipationless collapse and relaxation occurs leaving an elliptical galaxy with flattening E0 to $\sim$E5. Such a galaxy will show stars at all later times with little or no gas because gas from mass loss of stars will be swept out continually by supernovae. On the other hand, if significant gas is left over at the point of maximum collapse, it, being highly dissipative, will quickly form a flat disk. Thus, spiral galaxies should have both spheroidal and disk components, representing the material processed into stars before the collapse and that which was not. Once a galaxy has an appreciable disk, cooling will be sufficient to prevent supernovae from disrupting it and it will accumulate additional mass from stellar mass loss in the spheroidal component. This view is in complete agreement with the work of Ostriker and Peebles (1973) on the dynamics of spiral galaxies and with the work of Ostriker and Thuan (1974) on metal abundances in spiral galaxies.

All in all, the dissipationless collapse picture explains many of the observed features of elliptical galaxies and provides a natural explanation for the formation of elliptical and spiral galaxies as distinct types.

Acknowledgements

It is a pleasure to thank Drs Chris Wilson, Richard Larson, James E. Gunn, Ivan King, Gus Oemler, and Mr Paul Schechter for helpful conversations and comments. I especially thank Dr King for allowing use of his data prior to publication. This research was supported in part by the National Science Foundation [GP-36687X].

References

Aarseth, S. J.: 1973, Cited in Oemler, A. 1973.
Gott, J. R., III and Gunn, J. E.: 1971, *Astrophys. J. Letters* **169**, L13.

Gott, J. R., III: 1973, *Astrophys. J.* **186**, 481.
Gott, J. R., III, Gunn, J. E., Schramm, D. N., and Tinsley, B. M.: 1974, *Astrophys. J.* (in press).
Gott, J. R. III: 1975, submitted for publication.
Gunn, J. E. and Gott, J. R. III: 1972, *Astrophys. J.* **176**, 1.
Hénon, M.: 1964, *Ann. Astrophys.* **27**, 83.
Hubble, E.: 1930, *Astrophys. J.* **71**, 231.
King, I. R.: 1974, private communication.
Larson, R. B.: 1974, *Monthly Notices Roy. Astron. Soc.* **166**, 585.
Lynden-Bell, D.: 1967, *Monthly Notices Roy. Astron. Soc.* **136**, 101.
Mathews, W. G. and Baker, J. C.: 1971, *Astrophys. J.* **170**, 241.
Oemler, A.: 1973, Ph.D. Thesis, California Inst. of Technology.
Ostriker, J. P. and Peebles, P. J. E.: 1973, *Astrophys. J.* **186**, 467.
Ostriker, J. P. and Thuan, T. X.: 1974, in press.
Peebles, P. J. E. and Dicke, R. H.: 1968, *Astrophys. J.* **154**, 891.
Peebles, P. J. E.: 1970, *Astron. J.* **75**, 13.
Peebles, P. J. E.: 1971, *Astron. Astrophys.* **11**, 377.
Peebles, P. J. E.: 1974, *Astrophys. J. Letters* **189**, L51.
Press, W. H. and Schechter, P.: 1974, *Astrophys. J.* **187**, 425.
Spinrad, H., Smith, H. E., and Taylor, D. J.: 1972, *Astrophys. J.* **175**, 649.
Spitzer, L.: 1968, *Diffuse Matter in Space*, Wiley, New York.
Tremaine, S., Ostriker, J. P., and Spitzer, L.: 1974, in press.
Wilson, C. P.: 1973, Ph.D. Thesis, Univ. of California, Berkeley.

DISCUSSION

King: It seems to me that we, as practitioners of dynamics, are making good progress in understanding the reasons for the existence and nature of the different types of galaxies. Both the preceding speakers have distinguished elliptical galaxies from disks, however, according to whether some considerable part of the original material remains in gaseous form. I think that this a crucial question in the nature of galaxies, and that we should go home and tell our colleagues in other fields that the key to understanding the nature of galaxies is first to understand the process of star formation.

Larson: Do I understand correctly that everything that you have said applies only to the outer most half or two-thirds of the density profile of NGC 4697, and that your models say nothing about the formation of the core regions of elliptical galaxies? If so, your models do not explain *everything* with no free parameters.

Gott: My model is designed to follow the envelope well, but because of cost constraints does not give information about very small scales in the center. In my model the velocity distribution has become nearly isothermal in the inner regions so that an isothermal model for the center would be indicated. Such models like those of King give a very good fit to the center parts of Hubble's law. Thus, one would expect the central region of the model to also give a good fit to the observations if one followed into the center. When isothermal conditions are established by the violent relaxation one has always fit the central parts of elliptical galaxies well. The problem in the past has always been the shape of the envelope. Also when discussing the formation of dense galactic nuclei it must be remembered that there may be additional special mechanisms involved.

Lecar: How is your model tied to the cosmology? For example, do you require that you form galaxies before clusters of galaxies?

Gott: Galaxies are larger than the Jeans mass at recombination so they and larger structures such as clusters can form as density perturbations. In the present picture galaxies which we observe to have the shortest crossing times form first, then having clusters of galaxies forming from statistical aggregates of galaxies.

Bertola: Using your model can you fit also the De Vaucouleurs' law for the luminosity profile of ellipticals?

Gott: A comparison of Hubble's law and De Vaucouleurs' law shows that they are quite close. De Vaucouleurs' law has a slight curvature to it that one can also see in the model results. In the model one knows the total mass of the galaxy and this agrees to within 10% of the extrapolated mass based on De Vaucouleurs' law. So that would indicate that it was a good fit.

Innanen: Do these models lead in a natural explanation to a higher velocity dispersion in the radial direction compared to the axial dispersion?

Gott: For the rotating model, because of the angular momentum picked up by stars in the envelope due to the tidal interaction, the random velocity distributions were roughly although not accurately isotropic.

Pişmiş: Obviously there still remains a free parameter – at least one – in order to decide the fate of a galaxy, that is its type. I have often wondered whether the total mass of a galaxy may not be an important parameter in the evolutionary path followed by the galaxy. We know little about fragmentation mechanisms. But it is conceivable that in some way the distribution law of the fragments may be related to the total mass of the system, and that the distribution of fragment sizes affects the process of evolution.

NUMERICAL STUDY OF A
GRAVITATING SYSTEM OF COLLIDING PARTICLES;
FORMATION AND DYNAMICS OF DISCS

A. BRAHIC

Observatoire de Paris, Meudon, Université Paris VII, Paris, France

Abstract. The study of gravitating systems of colliding particles has many potential astrophysical applications, for instance the dynamics of Saturn's ring, the formation of the solar system, the flattening of protogalaxies and the evolution of galactic nuclei. We consider numerically a three-dimensional system of particles moving in the gravitational field of a central mass point and interacting through inelastic collisions. After a very fast flattening, the system forms a disc of finite thickness: this disc spreads slowly, and collisions still occur. A central condensation is formed and there is an outward flux of angular momentum. The energy which is continually lost in the inelastic collisions is obtained at the expense of the bodies which fall inwards.

There can be little doubt that collisions between 'macroscopic bodies' are of frequent occurrence in the Universe. All kinds of quite different objects undergo such collisions – these may range from interstellar clouds to small solid bodies in the solar system; it is therefore important to understand the past and present contribution of collisions to the overall evolution of the system in which they take place.

It has been known for a long time that inelastic collisions tend to flatten any system (Poincaré, 1911).

The configuration of a flat disc around a central body or bulge is found in many different contexts, for example spiral galaxies, the solar system, Saturn's ring, planetary satellites, etc. It may well be that the disc is often produced by the effect of contraction on a rotating mass of gas: nevertheless, it is interesting to see just what could be the effect of other mechanisms – in particular of inelastic collisions.

Therefore, in view of the many possible astrophysical applications, it would seem important to investigate gravitating systems of colliding particles. A full dynamical study has never been completed. A numerical study seems more suitable for this than an analytical formulation based on the Boltzmann equation, because such a formulation involves too many approximations of uncertain effect, even though a number of interesting conclusions can be derived from it for some specific cases.

Calculations of this sort have so far only been carried out by Ulam (1968), who was interested in the nuclei of galaxies, and by Trulsen (1972a, b) who studied the dynamics of 'jet streams'. Corresponding numerical experiments have stimulated significant progress in molecular dynamics (Alder and Wainwright, 1959, 1960; Rahman, 1964; Verlet, 1967, 1968).

We hope thereby to throw some light on a number of still unanswered questions:

What is the evolution of such a system in the limit of very large times?

What is the time scale of evolution under the effect of collisions?

Does the system reach an equilibrium state? or not?

What is the energy balance?

Hayli (ed.), Dynamics of Stellar Systems, 287–295. All Rights Reserved.

Are inelastic collisions sufficient to explain the formation and dynamics of some discs in the Universe? or do they constitute merely a secondary process?

With the help of Michel Hénon, I am studying systematically these kinds of systems by numerical simulation. This work is not only a way of finding the winning strategy in a game of three-dimensional billiards, but finds an immediate application to the as yet poorly understood dynamics of Saturn's ring (Jeffreys, 1947; Cook *et al.*, 1972; Brahic, 1974b, 1975). Note that this is a case where on the one hand, the input of the theory is relatively free of uncertainties, and on the other hand, detailed observations will be soon available, so that observation and theory can be compared fruitfully. Of course, many other astrophysical potential applications suggest themselves, for example:

The dynamics of colliding cloudlets, such as have been invoked to explain the formation of the solar system disc, and the disc of planetary satellites (McCrea, 1960; Woolfson, 1964; Urey, 1966).

Another example could be the dynamics of protostars in a cluster: it has been suggested that collisions of protostars in a gas cloud which has just fragmented could affect their subsequent evolution (Arny and Weissman, 1973).

One might also study collisions in galactic nuclei (Spitzer and Saslaw, 1966; Spitzer and Stone, 1967; Sanders, 1970).

To explain the generation of X-ray sources in close binary systems, Prendergast and Burbidge (1968), Schwartzman (1971), Pringle and Rees (1972), and others introduce a model in which matter form a differentially rotating disc around a compact object – either a neutron star or a black hole. A somewhat similar situation, but on a much larger scale, is discussed by Lynden-Bell (1969) and by Lynden-Bell and Rees (1971), with application to quasars and active galactic nuclei in which gas clouds collide to form a nuclear disc.

Finally, we note also that Brosche (1970) assumed that a protogalaxy consists of several randomly moving clouds. The galaxy loses energy through collisions and therefore shrinks. Brosche made an approximate qualitative and analytical model of this process and found that the Hubble sequence could be interpreted as an angular momentum sequence at constant mass. He also noted that this kind of model could be refined by an N-body calculation.

I am currently studying a sequence of numerical models. The system under study is evidently an N-body system, but, for the time being, my first models in no way utilise N-body techniques. I intend to include some kind of N-body integration – using Aarseth's method (1973), for example – at a future date.

The pure dynamics of a gravitating system of colliding particles is by no means obvious. In order to understand first the basic mechanics of the process, I consider the simplest model. Attraction between particles has been neglected; and so particles orbits are keplerian around a central mass point. Positions and velocities at any given time are obtained from Kepler's equation. In a collision, the grazing component of velocity is conserved and the perpendicular component is multiplied by a coefficient k which lies between 0 and -1; -1 corresponding to the elastic case. The

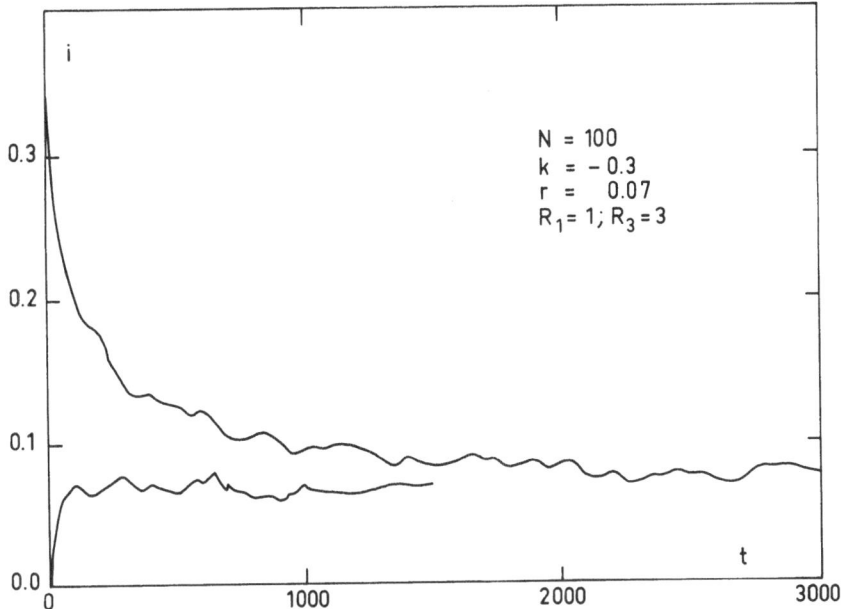

Fig. 1. Variations as a function of time of the mean inclination. The number of particles is equal to $N = 100$ and their radius to $r = 0.07$. Initial trajectories are all ellipses, which lie between two spheres of radius $R_1 = 1$ and $R_2 = 3$ respectively and centered on the central mass point, and with inclinations lying between 0 and 0.5 rad (for the upper curve) and between 0 and 10^{-3} rad. (for the lower curve). In the case of the upper curve, after 5257 collisions ($t = 3000$), 1 body out of 100 has escaped and 23 out of 100 have fallen on the central body.

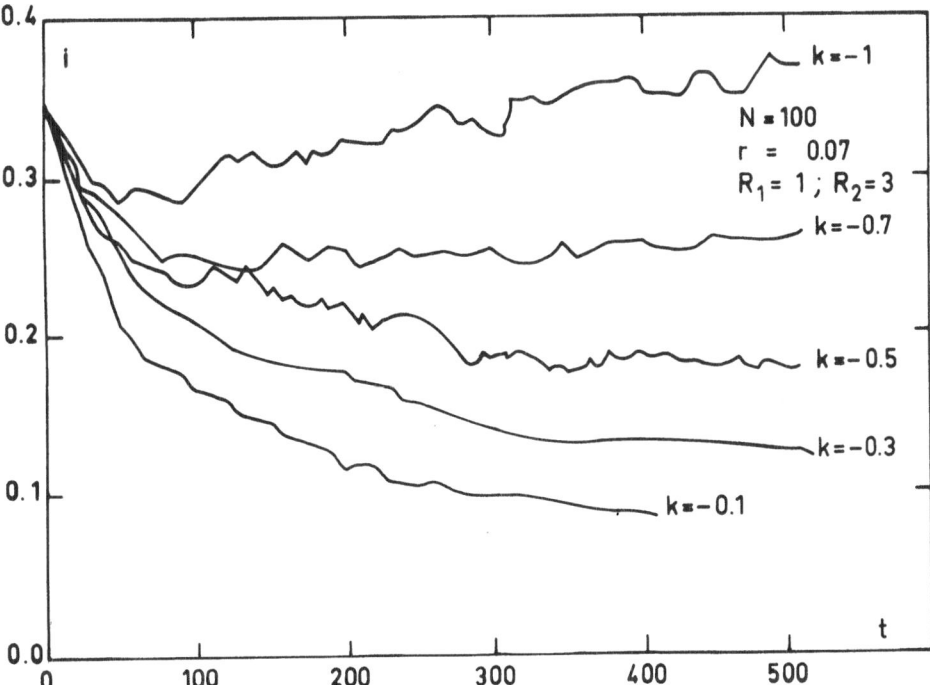

Fig. 2. Variations as a function of time of the mean inclination for different values of the rebound coefficient k. The initial inclinations of the orbits are all distributed between 0 and 0.5 rad.

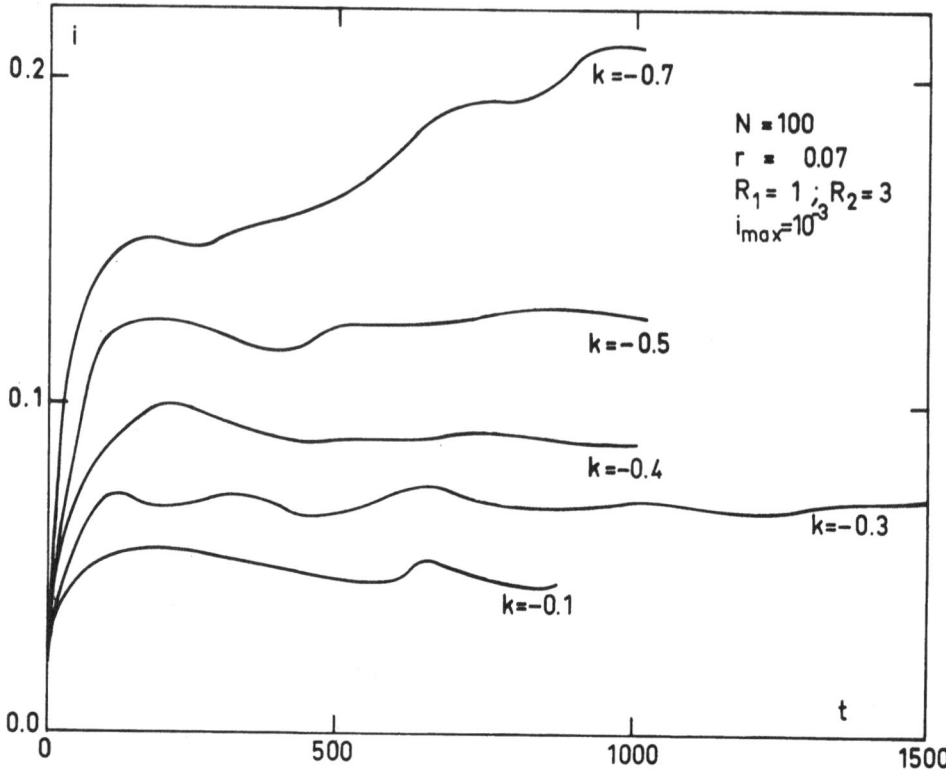

Fig. 3. Variations as a function of time of the mean inclination for different values of the rebound coefficient k. The initial inclinations of the orbits are all distributed between 0 and 10^{-3} rad.

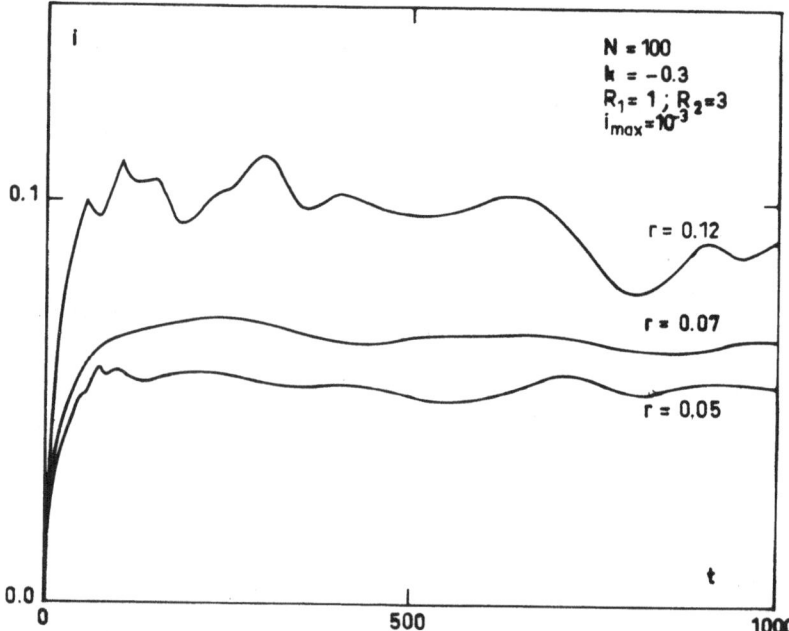

Fig. 4. Variations as a function of time of the mean inclination for different values of the size r of the particles. The initial inclinations of the orbits are all distributed between 0 and 10^{-3} rad.

initial conditions were set up by selecting at random the six elements of the keplerian orbit in such a way that trajectories were all ellipses lying between two spheres centered on the central mass point and with inclinations lying between 0 and some maximal value. We have assumed that particles on hyperbolic trajectories escape at once and, for technical reasons, that particles near to the centre are captured by the centre of mass. The kinetic rotational energy of the bodies has been neglected. The principal difficulty of the rather intricate calculations is to know whether two particles will in fact collide or not. By appropriate scaling, a few hundred particles suffice to simulate more realistic systems. Indeed, the time scale of evolution is inversely proportional to the number of particles and to their geometrical cross-section.

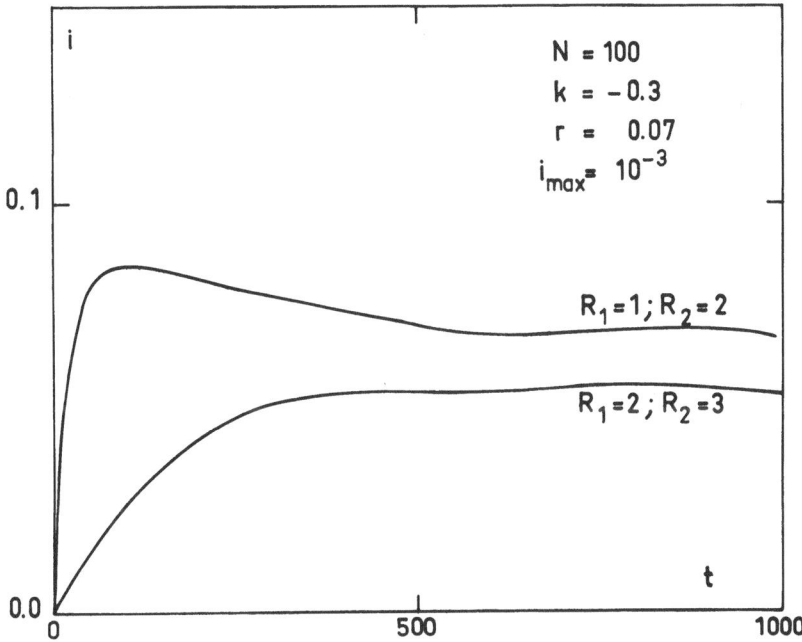

Fig. 5. Variations as a function of time of the mean inclination for different values of the scale size of the disc. The initial inclinations of the orbits are all distributed between 0 and 10^{-3} rad and initial trajectories are all ellipses, which lie between two spheres of radius R_1 and R_2 respectively and centered on the central mass point.

The preliminary results were presented at the IAU symposia last year (Brahic, 1974a, b) and a number of more up to date results were presented this year (Brahic, 1975) within the context of the dynamics of Saturn's ring. I shall give only some examples of the results, repeating a few already presented, and summarise the most important points.

Figure 1 shows the mean inclination as a function of time in a typical case. As one would expect, the system is considerably flattened quickly – in less than about ten collisions per particle – but, contrary to what is often stated, collisions do not reduce the thickness of the disc to one layer of particles; after a fast flattening, the system reaches a quasi-equilibrium state (see the upper curve of Figure 1) in which the thickness is finite. This is presumably a consequence of the keplerian motion, which introduces differential rotation. Even if particles are in circular orbits, there are still collisions. Part of this residual velocity is transferred in vertical motion. It is exactly like viscosity which produces heat.

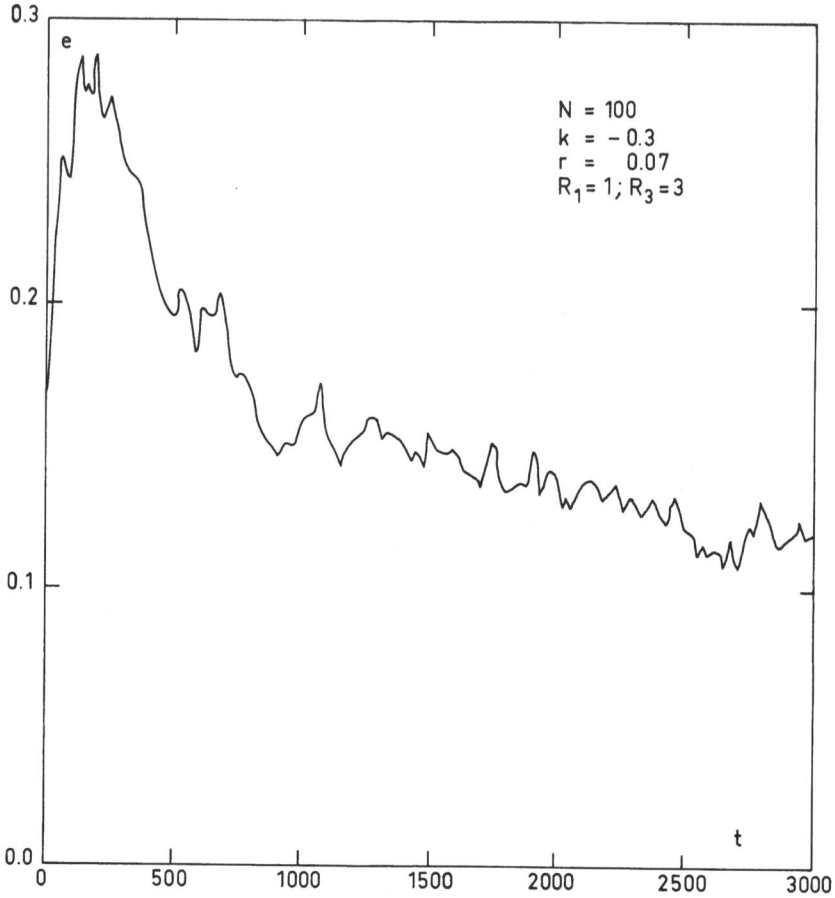

Fig. 6.   Variations as a function of time of the mean eccentricity. The initial values of N, r, R_1, R_2, etc. are the same as for the upper curve of Figure 1.

If we start with a system almost completely flat (see the lower curve of Figure 1), the system is growing and reaches quickly the same limit; indeed, in a system where particles have finite collision cross-section, collisions necessarily introduce movements out of any plane as well as radial ones. We can start with a very small value of the mean inclination i and study the limit as a function of the parameters of the problem.

The results are very sensitive to the rebound coefficient k (see Figures 2 and 3). If k lies between -0.5 and -1, there is no important flattening of the system.

Figures 4 and 5 show how the ratio r/R – where r is the particle size, and R is a scale size of the disc – affects the results.

Figure 6 shows the mean eccentricity as a function of time for a typical case. The initial rise is due to the fact that thermal equilibrium is established between radial and vertical velocities. After this the orbits tends to become increasingly circular but the mean eccentricity does not reach 0. This behaviour is similar to that of the mean inclination.

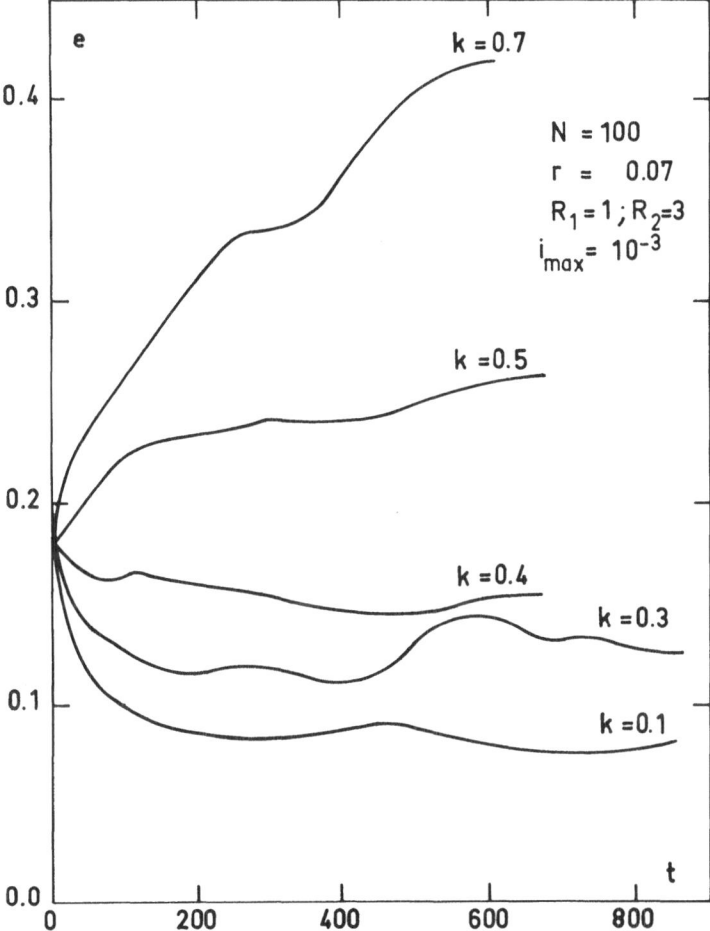

Fig. 7.   Variations as a function of time of the mean eccentricity for different values of the rebound coefficient k.

Figures 7 and 8 show the role of the rebound coefficient k and of the mean size of the particles respectively.

We have also started a study of other models; in particular if we consider particles of different masses and dimensions, the first results indicate that the very massive bodies were confined to a thinner disc than the light ones. Subsequent models will explore different kinds of collision regimes, fragmentation and coalescence and also the use of different potential field.

For the time being, the most interesting result is that an initial phase during which

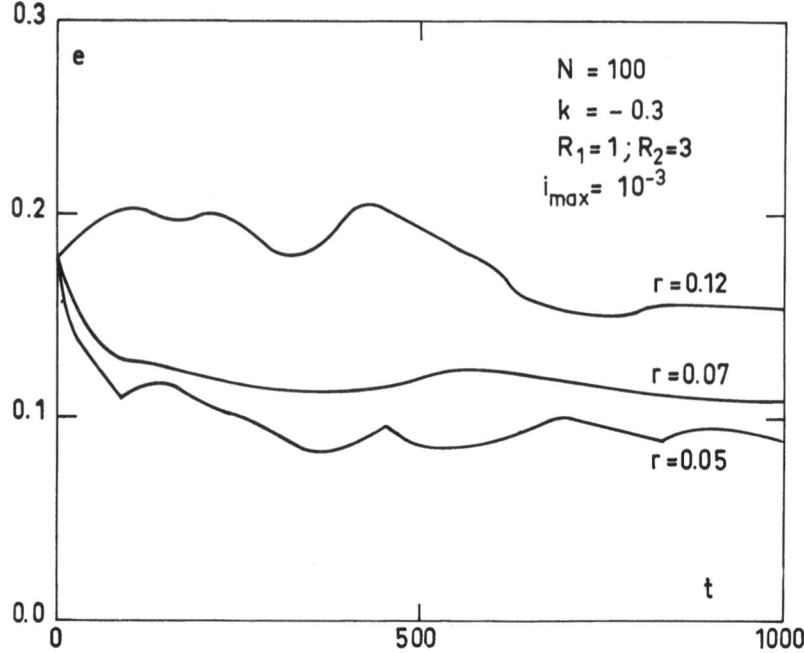

Fig. 8.   Variations as a function of time of the mean eccentricity for
different values of the size r of the particles.

the system flattens rapidly is followed by a phase in which collisions still occur: the thickness of the disc thus formed is finite and the disc spreads very slowly: many particles move inwards forming a central condensation while some particles move outwards. There is an outward flux of angular momentum and the energy which is continually lost in the inelastic collisions is obtained at the expense of the bodies which fall inwards.

References

Aarseth, S. J.: 1973, *Vistas in Astronomy* **15**, 13.
Alder, B. J. and Wainwright, T. E.: 1959, *J. Chem. Phys.* **31**, 459.
Alder, B. J. and Wainwright, T. E.: 1960, *J. Chem. Phys.* **33**, 1439.
Arny, T. and Weissmann, P.: 1973, *Astron. J.* **78**, 309.
Brahic, A.: 1974a, in J. R. Shakeshaft (ed.), 'The Formation and Dynamics of Galaxies', *IAU Symp.* **58**, 173.
Brahic, A.: 1974b, in Y. Kozai (ed.), 'The Stability of the Solar System and of Small Systems', *IAU Symp.* **62**, 83.
Brahic, A.: 1975, in J. A. Burns (ed.), 'Planetary Satellites', *IAU Colloq.* No. 28; *Icarus* (in press).
Brosche, P.: 1970, *Astron. Astrophys.* **6**, 240.
Cook, A. F., Franklin, F. A., and Palluconi, F. D.: 1973, *Icarus* **18**, 317.
Jeffreys, H.: 1947, *Monthly Notices Roy. Astron. Soc.* **107**, 263.
Lynden-Bell, D.: 1969, *Nature* **223**, 690.
Lynden-Bell, D. and Rees, M. J.: 1971, *Monthly Notices Roy. Astron. Soc.* **152**, 461.
McCrea, W. H.: 1960, *Proc. Roy. Soc.* **A256**, 245.
Poincaré, H.: 1911, *Bull. Astron.* **28**, 251.
Prendergast, K. H. and Burbidge, G. R.: 1968, *Astrophys. J. Letters* **151**, L83.
Pringle, J. E. and Rees, M. J.: 1972, *Astron. Astrophys.* **21**, 1.
Rahman, A.: 1964, *Phys. Rev.* **136**, 405.
Sanders, R. H.: 1970, *Astrophys. J.* **162**, 791.

Schwartzmann, V. F.: 1971, *Soviet Astron.* **15**, 377.
Spitzer, L. and Saslaw, W. C.: 1966, *Astrophys. J.* **143**, 400.
Spitzer, L. and Stone, M. E.: 1967, *Astrophys. J.* **147**, 519.
Trulsen, J.: 1972a, *Astrophys. Space Sci.* **17**, 241.
Trulsen, J.: 1972b, *Astrophys. Space Sci.* **18**, 3.
Ulam, S. M.: 1968, *Bull. Astron. 3ème* serie **3**, 265.
Urey, H. C.: 1966, *Monthly Notices Roy. Astron. Soc.* **131**, 199.
Verlet, L.: 1967, *Phys. Rev.* **159**, 98.
Verlet, L.: 1968, *Phys. Rev.* **165**, 201.
Woolfson, M. M.: 1964, *Proc. Roy. Soc.* **A282**, 485.

DISCUSSION

Lecar: In the asteroid belt, there has been collisions and both the eccentricities and inclinations are large and approximately equal. In your model, are the final Z-dispersion velocities equal to the final r-dispersion velocities?

Brahic: Yes, the final i and e dispersions are of the same order of magnitude.

Lynden-Bell: What is the final value of the e-dispersion as a function of the parameters of the problem?

Brahic: It depends on the rebound coefficient k and on r/R where r is the particle size and R is a scale size of the disc. For $k = -0.3$, $r = 0.07$ and $R = 2$, $e \simeq 0.1$ which is the same order of magnitude as i. There is some kind of equipartition between the mean inclinations and the mean eccentricities.

Contopoulos: In some of your slides (e.g. when $k \leqslant -0.7$) there was a slow increase of inclination and eccentricity. How long does this increase continue?

Brahic: For $-1 < k < -0.5$, I cannot really distinguish flattening from a quasi-equilibrium state: instead, what I see are rather large non-periodic fluctuations. I observe no important flattening for almost elastic collisions.

GLOBAL INSTABILITIES OF DISKS*

J. M. BARDEEN

Yale University, New Haven, Conn., U.S.A.

Abstract. Current understanding of the stability of gas and stellar disks suggests very strongly that local stability to axisymmetric modes is not sufficient for global stability. A global instability to a bar mode will develop unless the rotational kinetic energy is sufficiently small compared with the random kinetic energy for the system as a whole. A disk as cool as the galactic disk near the Sun can survive only if most of the mass of the Galaxy is in a 'hot' component, such as a central bulge and/or an extended halo. We review the theoretical evidence for this conclusion coming from analytic results for simple gas and stellar disks, from numerical simulations of stellar disks, and from numerical calculations of the stability of gas disks. Some new results on the precise form of dynamic bar instabilities of gas disks with and without halos are reported.

1. Introduction

An understanding of the structure and dynamics of spiral galaxies depends in an important way on questions of global stability. Conventional fits to rotation curves, based on the observed light distribution, assume all the mass outside the 'central bulge' type of spheroidal component lies in a 'cool' disk with only enough velocity dispersion to satisfy the Toomre criterion for local stability against axisymmetric ring modes. However, attempts to simulate the stellar dynamics of these cool disks have consistently found strong instabilities to bar modes. The non-linear effect of the bar is to increase the velocity dispersion until it is comparable with the circular velocity over much of the disk. The final state is typically characterized by a ratio t of kinetic energy of rotation to gravitational potential energy of only 0.14 or so, compared with the value 0.5 for a zero velocity dispersion disk (Ostriker and Peebles, 1973). Ostriker and Peebles argue that a cool disk is stable against bar modes only if a substantial part of the mass of the galaxy is in the form of a more or less spherical halo, so that for the system as a whole t has an upper limit in the range 0.1–0.2.

No direct observational evidence for such halos has been found (Freeman, 1975). They must be composed of objects with very large mass to light ratios, such as extreme M dwarfs or black holes. Ostriker *et al.* (1974) have assembled some indirect evidence for very massive halos extending to many times the radius of the visible galaxy. Of course, only the mass within the outer edge of the disk is relevant to our discussion of stability. One should also keep in mind that the observational evidence for disks being cool is rather meagre; in the vicinity of the Sun the galactic disk is at best only marginally stable according to Toomre's criterion (see Schmidt, 1975), but in other galaxies one must appeal to the apparent small ratio of thickness to radius or argue that spiral density waves are only possible in rather cool disks.

My discussion will focus on the theoretical arguments and results of numerical simulations as they affect questions of global stability. The most directly relevant

* Supported in part by National Science Foundation Grant GP-36317

calculations are the numerical simulations of stellar disks carried out by Miller *et al.*
(1970), Hohl (1971, 1975), and Ostriker and Peebles (1973). I will only mention some
of the general implications of Hohl's calculations, since he will describe them in
detail in his review paper. The analytic results of Kalnajs (1972) on stability are of
great help in checking and interpreting the simulations, but so far they have not
been extended to realistic galaxy models.

Bar instabilities seem to be a general property of rapidly rotating, self-gravitating
systems, with many qualitative and some quantitative similarities between stellar
disks and gaseous disks. Furthermore, the only obvious way for a stellar disk to form
is from a gas disk at least as thin as the final stellar disk. Therefore, this review will
include some discussion of gas disks; in particular, I will present the results of some
of my calculations of the instability of two-dimensional gas disks to bar and two-
armed spiral modes. The linearized equations for dynamical perturbations are
integrated in time until, if the disk is unstable, one mode dominates and gives a steady,
uniform pattern speed and growth rate.

I will begin by reviewing the known analytic results on the stability of gas and
stellar disks, both for global and local instability and then discuss the numerical
calculations.

2. Analytic Results

Only for very special configurations can one solve analytically for the gravitational
potential, both for the axisymmetric equilibrium and for general perturbations. The
classic example is the sequence of Maclaurin spheroids, uniform density and uniformly rotating spheroids of incompressible perfect fluid, along with the associated
sequences of ellipsoids which have various combinations of internal motion and
uniform rotation (see Chandrasekhar, 1969). For a given density and mass the
eccentricity of a Maclaurin spheroid goes form zero to one and the ratio t of kinetic
energy of rotation to gravitational potential energy goes from zero to one-half as the
angular momentum increases from zero to infinity.

In the gas case one distinguishes between secular instability and dynamic instability. In the absence of any dissipative mechanism such as viscosity both the total
angular momentum and the circulation ($\int \mathbf{v} \cdot \mathbf{dl}$ around a closed curve comoving with
the fluid) are conserved in any dynamic process. Furthermore, an axisymmetric perturbation conserves the angular momentum of each ring of matter.

A perturbation in which one of the dynamic constraints is violated is a secular
perturbation. For instance, with only viscosity present the circulation need not be
conserved and angular momentum can be transferred between neighboring rings of
matter even when the perturbation is axisymmetric, but the total angular momentum
still remains constant. On the other hand, gravitational radiation reaction by itself
preserves the circulation but allows the total angular momentum to change. The
time scale for growth of a secular perturbation is governed by the amount of dissipation present and often is large compared with the characteristic dynamic time scale
of the system. The concept of secular instability is meaningful only if the original

equilibrium configuration is not affected by the dissipation. One expects secular instability to set in before dynamic instability, since a neighboring lower energy configuration may not be reachable by a perturbation consistent with all the dynamical constraints.

The first instability along the Maclaurin sequence is a secular instability to a bar mode at $t \geqslant 0.1376$. As excited by viscosity the marginally unstable mode is an ellipsoidal deformation which corotates with the matter (a Jacobi ellipsoid). As excited by gravitational radiation reaction the ellipsoidal shape remains fixed in space, a Dedekind ellipsoid (Chandrasekhar, 1970). The non-linear evolution of the secular instability for t a little greater than 0.1376 has been solved by Press and Teukolsky (1973) for viscosity and by Miller (1973) for gravitational radiation.

Only for $t > 0.2738$ are the Maclaurin spheroids dynamically unstable. The marginally unstable mode is a bar which rotates at one-half the angular velocity of the matter. At $t = 0.3589$ the spheroid first becomes secularly unstable to an axisymmetric redistribution of angular momentum and matter, either in the direction of a 'central bulge' surrounded by a thin disk or in the direction of a ring-like distribution of matter. Finally, the first onset of an axisymmetric dynamic instability is at $t = 0.4574$ (see Bardeen, 1971).

A class of stellar dynamic analogues to Maclaurin spheroids has been studied by Kalnajs (1972), who was able to separate the normal modes and solve for their characteristic frequencies. The models are infinitesimally thin disks with the same surface density as a function of radius as the Maclaurin spheroids and with a range of possible velocity distributions in the plane of the disk which are obtained by superimposing a particular type of distribution function with only one free parameter, the mean angular velocity of rotation Ω. The stability to the two-armed bar mode analogous to the mode that dominates the non-axisymmetric instability of the Maclaurin spheroids, Kalnajs shows, depends only on the weighted average of Ω in a composite model. The bar mode is unstable if $\langle \Omega \rangle$ exceeds $(125/486)^{1/2} = 0.507$, in units such that the angular velocity Ω_0 of the zero velocity dispersion disk with the same mass and radius is one. Since $\langle \Omega \rangle$ is the mean angular velocity of the matter in the composite model, the parameter t at marginal instability has the unique value $t = 0.1286$. Thus $t \leqslant 0.1286$ is a necessary condition for overall stability of this class of models. Unfortunately, the velocity distribution has some peculiar properties, and even aside from the question of the importance of shear, one should be cautious in extrapolating this result to more realistic stellar disks.

Kalnajs also considers how a halo which supplies a rigid component to the potential modifies the stability of the disk. For instance his model B_1 without a halo has $t = 0.1735$ and is unstable to three modes; the most rapidly growing instability is to the (2, 2) bar mode. With a halo corresponding to a uniform density sphere equal in radius to the disk, the model which has the same distribution of stellar orbits as B_1 is marginally stable when the halo mass is 0.199 of the disk mass or 0.166 of the total mass. For the system as a whole $t = 0.145$. His model B_3, which without a halo has $t = 0.271$, is stabilized by a halo mass equal to $\frac{2}{3}$ the mass of the disk or 40% of the total mass. The effect of the halo is to multiply the response by a factor equal to the fraction of the equilibrium gravitational force due to the disk.

There is no distinction between secular and dynamic instability in a collisionless stellar system. The conservation of density in phase space is a constraint on the evolution of the system, but one which can allow a considerable amount of relaxation and conversion of bulk motion into random motion to occur (see Lynden-Bell, 1967). There are no macroscopic constraints analogous to conservation of vorticity of a perfect fluid. It is perhaps not surprising, then, that the bar instability for the stellar disk occurs at about the same value of t as the secular instability of the Maclaurin spheroid. The pattern angular velocity of the marginally unstable mode is 0.4564, roughly equal to the mean angular velocity of the stars, so the marginally unstable mode of the stellar disk resembles the Jacobi mode of the Maclaurin spheroid.

Typical galactic disks are highly centrally condensed and are far from uniform rotation. The only generally applicable analytic stability criteria are the 'local' criteria which apply to axisymmetric ring modes or tightly wound spiral modes in which the radial wavelength of the perturbation is small compared with the scale of radial inhomogeneities. In this limit one treats a local region of the disk as an infinite plane sheet.

The Toomre criterion (Toomre, 1964) was derived for infinitesimally thin stellar disks with a Gaussian radial velocity dispersion c_r. It predicts instability when c_r falls below

$$(c_r)_{\text{crit}} = \frac{3.36\,G\sigma}{\kappa},$$

where κ is the local epicyclic frequency and σ is the local surface density. The wavelength of the marginally unstable mode is

$$\lambda_{\text{crit}} = 3.5(c_r)^2_{\text{crit}}/G\sigma$$
$$= 40\,G\sigma/\kappa^2.$$

For a completely self-gravitating disk λ_{crit} is comparable with the radius of the disk unless the surface density in the region being considered is small compared with the average surface density. Therefore, the conditions necessary for the local criterion to be used with confidence will not be satisfied unless the disk contains a relatively small part of the total mass of the system. The bulk of the mass could either be in a 'central bulge' or 'spheroidal' component or in an extended halo component. This point has also been made by Vandervoort (1970) in connection with a study of non-linear density waves.

A local stability criterion for gas disks was derived by Goldreich and Lynden-Bell (1965). If a measure of the mean density in the disk,

$$\bar{\varrho} = \int \varrho^2 \, dz / \int \varrho \, dz,$$

exceeds a numerical coefficient times κ^2/G the disk is unstable. The numerical coefficient ranges from 0.23 for an isothermal equation of state to 0.56 for an incompressible fluid equation of state. The critical wavelength at marginal instability is a few times the mean thickness of the disk $W = \sigma/\bar{\varrho}$. The ratio λ_c/W ranges from 4.5 for an isothermal equation of state to 10.4 for an incompressible fluid equation of state.

The exact result for marginal instability of a Maclaurin spheroid to axisymmetric modes is $\varrho = 0.61 \; \kappa^2/G$ when W is 0.0465 times the diameter of the disk. However, it seems likely that the local criterion will work less well when σ and κ^2 vary strongly with radius.

Even when such a local criterion does apply it gives at most a necessary condition for stability. It says nothing about the stability of modes which are not tightly wound, particularly the bar modes which are the dominant instabilities of the uniformly rotating disks, both gas and stellar.

Nevertheless, it is the local type of analysis which has formed the basis for the density wave theory of spiral structure developed by Lin and Shu (1964, 1966). They and coworkers have obtained dispersion relations and included effects of resonances. Results of Toomre (1969) on the group velocity of density waves have raised serious question as to whether they can be considered persistant normal modes of the stellar system. Further discussion of density waves as such seems outside the scope of this review.

3. Numerical Simulations of Stellar Disks

Direct N-body calculations are impractical for N larger than 500. It is questionable whether 500 stars are adequate to represent a galaxy, since the two-body relaxation time is not too much larger than the dynamical time scale. Certainly 500 stars are not enough to represent any detailed structure in the galaxy. These considerations have prompted the development of approximate schemes which allow calculations with much larger numbers of stars.

For instance, Miller and Prendergast (1968) devised an approximation based on a discrete phase space in which stars jump between integer values of position and velocity. A fast Fourier transform technique was used to solve for the gravitational potential. Miller, Prendergast, and Quirk (1970) applied this method to infinitesimally thin disks made up both of 'stars' and 'gas', with about 10^5 total particles. The gas component was kept cool by reducing the relative velocities of the gas particles or 'clouds' at each point in configuration space at every time step. Without 'gas' present the stars formed a 'hot' pressure-supported system. With gas some moderately persistent spiral patterns developed which were much less prominent in the stars than in the gas. In some cases bars developed as transient phases in the evolution of the system. No attempt was made to look in detail at the conditions necessary to stabilize the disk, but typically there was a substantial background of hot stars once things settled down to a steady state. Ostriker and Peebles (1973) quote a range of 0.130–0.135 for the final value of t in the MPQ simulations.

More elaborate and more accurate simulations of infinitesimally thin two-dimensional purely stellar disks have been carried out by Hohl over the last several years (Hohl, 1971, 1975). The configuration space is broken up into cells and the mass density averaged over each cell before the potential is calculated. However, unlike MPQ the equations of motion for the individual stars are solved exactly for an infinitesimally thin disk with velocity dispersion only in the plane. The simulations have been tested against the Kalnajs analytic models. Most of the models calculated have contained 10^5 stars, which is large enough for two-body encounters to have a negligible effect (Hohl, 1973). Hohl found that an initially uniformly rotating disk with just enough velocity dispersion to satisfy the Toomre stability criterion was violently unstable to a bar. Outside the corotation radius, where the pattern speed of the bar equaled the circular velocity of the stars, the stars were pushed outward; inside the corotation radius the stars drifted inward. In the final steady state the surface density was a steep exponential function of radius inside the corotation radius; outside the corotation radius the disk was also exponential, but with a considerably larger scale length. The ratio Q of the radial velocity dispersion to the Toomre critical velocity dispersion in the final state varied from about 2 near the center to 5 or 6 in the outer part of the disk. Ostriker and Peebles (1973) quote a final value of $t \simeq 0.14$. Any transient spiral patterns quickly decayed as the velocity dispersion increased, but a bar did persist indefinitely.

In an attempt to generate a cooler stable disk Hohl symmetrized this final steady state to remove the bar and then tried artificially reducing the velocity dispersion of a certain fraction of the stars peroidically. However, this caused a new bar instability. With steady cooling an open spiral pattern did persist in the outer part of the disk surrounding a central bar. Heating of the disk by Landau damping of the spiral pattern compensated for the cooling and kept the disk rather hot.

The most striking feature of these calculations (Hohl, 1971) was the relaxation to an exponential type of disk structure, somewhat like that observed in many spiral galaxies (Freeman, 1970). This suggests that the exponential surface density reflects an epoch of dynamic relaxation in the disk, rather than the initial angular momentum distribution in the gas cloud out of which the disk formed. A phase of bar instability may also greatly increase the central condensation of the disk and assist in the formation of a condensed nucleus of the galaxy.

More recently Hohl has run experiments with rigid background potentials representing 'halo' populations superimposed on the self-gravity of the disk (Hohl, 1975). A halo potential corresponding to the mass distribution in the Schmidt (1965) model of the Galaxy seemed to suppress the bar instability if the halo mass was comparable to that of the disk, but substantial heating of the disk still took place if the initial Q was one, suggesting that less prominent instabilities were still present. The steady-state Q was in the range 2–3. A less centrally condensed halo of comparable mass allowed a bar to develop, but the heating was less with a final Q of 1.5–2. At present these results are only suggestive, but combined with my results on gas disks reported in Section 4, one can conclude that a halo less centrally condensed than the disk seems to be more effective in stabilizing the disk than a halo corresponding to a massive 'central bulge' component.

Hopefully in the near future it will be possible to make more realistic simulations of this type which allow for a finite thickness of the disk and the presence of a genuine gas component. However, it seems unlikely that the results on global stability will be altered significantly.

An attempt to model a galactic disk with a finite thickness has been carried out by Ostriker and Peebles (1973). Their N-body calculation has N in the range 150–500 and usually $N = 300$. The forces are calculated by summing over particle pairs; force is cut off at an interparticle distance about 0.05 times R, the initial radius of the disk. The initial conditions typically have all the stars in the equatorial plane evenly divided between intervals of $0.1R$ in radius and distributed randomly in each interval. The stars are given an initial radial velocity dispersion scaled to satisfy the Toomre criterion by a margin of 20% everywhere except near the centre. The initial z-velocity dispersion was set equal to the initial axial-velocity dispersion obtained from the epicyclic approximation. Finally, the total velocity, including the circular velocity, was rescaled to satisfy the equilibrium virial theorem.

Such a system developed a strong bar instability. The velocity dispersion increased rapidly in the plane, but after an initial adjustment the vertical velocity dispersion and the root mean square value of z for the disk as a whole increased more slowly. The parameter t, initially about 0.35, seemed to approach an asymptotic constant value of about 0.14.

Some runs included a rigid halo potential corresponding to a spherical mass distribution with about the same central concentration as the disk,

$$M_H(r) = (1.1)^2 \frac{r^3 M_H}{R(r + 0.1R)^2} \qquad 0 \leqslant r \leqslant R$$
$$= M_H \qquad r > R.$$

A halo mass M_H greater than or equal to the mass of the disk was sufficient to remove the violent bar instability, but the velocity dispersion still did increase slowly. For the system as a whole the parameter t seemed to level off at about 0.17 for $M_H/M_D = 0.5$ and 1.0. Once the halo mass dominates that of the disk t becomes very insensitive to the velocity dispersion of the disk; a better measure of the velocity dispersion necessary for stability would be the Toomre parameter Q. It does seem that with $M_H/M_D \gtrsim 2$ or 3 the initial disk is basically stable.

Ostriker and Peebles claim their results are insensitive to the value of N in the range 150–500 and therefore that two-body relaxation effects are not important. While this probably is true for the violent bar instability of the cool disk without a halo, the precise point at which the halo stabilizes the initial disk is unclear because a slow increase of velocity dispersion could be due either to a mild instability or to two-body relaxation. The cut-off of the halo at the initial radius of the disk is rather artificial, particularly since the radius of the disk increases substantially during the dynamical evolution. The value of t is very sensitive to the amount of mass in the outer part of the halo, but the stability is not likely to be sensitive to the halo mass outside the great bulk of the mass of the disk. For this reason, Ostriker and Peebles' claim that $t \simeq 0.14$ is a universal number for marginal bar instability, even with a halo, should not be taken too seriously.

Taken together the numerical stimulations and the Kalnajs analytic results constitute a very strong, even overwhelming case that without a massive halo a stable stellar disk must have a much higher velocity dispersion in the plane than that required for local axisymmetric stability. The numerical simulations generally have not used self-consistent stationary solutions to the Liouville equation as initial conditions, so they are not true stability calculations. However, the stationary final states presumable do correspond to stationary solutions of the Liouville equation and attempts to gradually cool these differentially rotating disks did not succeed in reducing the total velocity dispersion by a large amount. Therefore, it seems unlikely that single-component stellar disks can be globally stable with t much greater than 0.14 or a mass-weighted average of Q^{-1} greater than 0.3 or so. However, it is still very important to extend the exact stability analysis to at least some differentially rotating disks and to construct good equilibrium stellar dynamical models of disks with various degrees of central condensation to serve as initial conditions for numerical simulations. The one possibility that has not been ruled out by the numerical simulations to date is that a very hot 'central bulge' in which Q is much larger than one could allow a surrounding disk containing the bulk of the mass (as observed from rotation curves) to be stable with $Q \simeq 1$. My experiments with gas disks make this alternative seem unlikely.

A less realistic version of a stellar disk, but one which is susceptible to a fairly complete stability analysis, uses a modified gravitation interaction potential,

$$\phi_{ij} = -\frac{m_i m_j}{(r_{ij}^2 + a^2)^{1/2}}, \tag{1}$$

to soften the gravitational interaction. The stars move in circular orbits in the equatorial plane of the unperturbed disk. Miller (1971) first noticed that the modified gravity, which with small values of a is often used in N-body calculations to prevent large accelerations in close encounters, can by itself stabilize the axisymmetric ring modes of a zero-velocity-dispersion disk. The WKB dispersion relation relating frequency ω and radial wave number k is

$$\omega^2 = \kappa^2 - 2\pi G \sigma k e^{-ak}. \tag{2}$$

This is qualitatively similar to the dispersion relation of Lin and Shu (1966) in that $\omega^2 \to \kappa^2$ in the high wave number limit as well as at zero wave number. The dispersion relation (2) predicts local axisymmetric stability if

$$a > a_{\text{crit}} = \frac{2\pi G \sigma}{e \kappa^2}. \tag{3}$$

Miller (1974) and Erickson (1974) have found this WKB estimate agrees rather well with more exact numerical stability calculations for the axisymmetric modes of some simple disk models. Erickson also looks in detail at the whole spectrum of axisymmetric modes. However, the modified gravity fails to stabilize against some non-axisymmetric modes even when a is comparable with effective radius of the disk (Erickson, 1974); global stability does seem to require a real velocity dispersion.

4. Global Instabilities of Gas Disks

Gas disks are simpler to work with than stellar disks in that fewer dynamic variables describe the response of the matter. As mentioned in the introduction, they also have a direct relevance to the existence of disk galaxies. A stellar system cannot relax into a disk. Therefore, the disk stars must have formed from gas that was already in a thin disk. The original gas disk must not have been too unstable or it would not have lasted in a quiescent state long enough to form stars with small velocity dispersion.

A completely realistic gas disk would still require an elaborate computer simulation to follow its dynamical evolution, since it would involve two or three-dimensional hydrodynamics depending whether the calculation is restricted to small perturbations of an equilibrium configuration or follows the full non-linear development of instabilities.

A relatively simple type of approximate stability analysis is based on the tensor virial equations developed by Chandrasekhar and Lebovitz (1962) and adapted for use on rapidly rotating stellar models by Tassoul and Ostriker (1968). The fluid displacements are constrained to be linear functions of the Cartesian spatial coordinates,

$$\xi_i = A_{ij}(t) \, x_j. \tag{4}$$

Assuming a harmonic time dependence, one can derive a set of algebraic equations for the coefficients A_{ij} which has solutions for certain characteristic frequencies ω_k. The modes which have an angular dependence corresponding to a $l=2$, $m=\pm 2$ spherical harmonic are the 'bar' modes and in fact are the exact bar modes of a Maclaurin spheroid.

The tensor virial equations predict dynamical instability when the characteristic frequencies of the bar modes become complex. Since in general one does not expect the exact eigenfunction for a differentially rotating, centrally condensed star to have the form of equation (4), one does not expect the tensor virial estimate of, say, the value of t at the point of marginal dynamical instability to be exact. There is no minimum principle for non-axisymmetric modes (Lynden-Bell and Ostriker, 1967), so the sign of the error is not known.

As far as secular stability goes, Ostriker and Tassoul (1969) argue that any stationary, non-axisymmetric, differentially rotating configuration with triplanar symmetry must satisfy virial relations which in the limit of axisymmetry make one of the bar mode frequencies of the axisymmetric star calculated with the tensor virial equations equal to zero. In other words, there must be a zero frequency tensor virial mode at any point of bifurcation to a sequence of triplanar configurations along a sequence of axisymmetric rotating stars. One can show from the variational principle for differentially rotating stars of Lynden-Bell and Ostriker (1967) that secular instability will set in when a certain energy integral first becomes negative for some trial displacement and furthermore (see Friedman and Schutz, 1975) that a zero-frequency Dedekind-type mode does exist when some non-trivial trial displacement corresponding to an axial eigenvalue $m = 2$ minimizes the energy integral at a value of zero. There is no guarantee that a Jacobi-type mode with a non-zero pattern angular velocity exists at this point. Taken together, these arguments imply that the tensor virial method locates the point of marginal secular instability exactly.

The tensor virial stability analysis has been applied by Ostriker and various collaborators to a variety of rapidly rotating stellar models calculated by the self-consistent field method of Ostriker and Mark (1968). The most systematic study has been carried out for the differentially rotating polytrope models of Ostriker and Bodenheimer (1973). At the point of marginal secular instability they find $t \simeq 0.138$ for all their models, while dynamical instability seems to set in at $t \simeq 0.26$. The models cover a fairly wide range of central condensation and angular momentum distribution.

None of the Ostriker-Bodenheimer models are as thin as typical galactic disks and the general validity of the tensor virial estimate of $t \simeq 0.26$ for dynamical instability is open to question. Also, the effect of halos on the instability of gas disks deserves attention. Therefore, I have developed a simplified treatment of the dynamics of gas disks perturbed away from axisymmetry in which the vertical structure of the disk is suppressed, either by assuming the disk is infinitesimally thin and the 'pressure' only acts in the plane of the disk or by solving for the vertical structure of the disk analytically and inserting this into the horizontal equilibrium and dynamics.

In the case of the infinitesimally thin disk we further simplify things by assuming that the horizontal stress per unit length p is a universal function of the surface density σ both in the equilibrium disk and in the dynamics of a given fluid element, $p = p(\sigma)$. Then the dynamical equations governing motions in the plane of the disk can be written.

$$\frac{d\mathbf{v}}{dt} = \nabla(U - \psi(\sigma)), \tag{5}$$

where

$$\psi(\sigma) = \int_0^\sigma \sigma^{-1} (dp/d\sigma) \, d\sigma \tag{6}$$

and U is the (positive) gravitational potential in the plane of the disk obtained by solving the Laplace equation outside the disk with boundary condition

$$\left. \frac{\partial U}{\partial z} \right|_{z=0^+} = -2\pi G\sigma. \tag{7}$$

Equation (5), linearized about the equilibrium disk, has been used by several authors as a simple context in which to discuss local density waves (e.g. Hunter, 1972, 1973; Feldman and Lin, 1973).

To first order in the thickness of an ordinary isotropic-pressure gas disk in which there is a polytropic relation between pressure and volume gas density the equation governing the horizontal dynamics has the same form as Equation (5), if in Equation (5) $U = U_0$, the potential in the plane of an infinitesimally thin disk with the same surface density as the finite thickness disk. To zeroth order in the thickness the vertical acceleration is small compared with the vertical gravitational potential gradient,

$$\ddot{z} = \mathcal{O}\left[(GM/R^3)\, z \right] = \mathcal{O}(z/R)\, G\sigma.$$

Assume vertical equilibrium, then, and also assume that the thickness is small compared with the scale of horizontal variations in the surface density. The vertical equilibrium gives

$$U(r, \phi, z) = \int_0^p \frac{\mathrm{d}p}{\varrho(p)} + U_1(r, \phi), \tag{8}$$

where U_1 is the value of U at the surface of the disk, $z = z_1(r, \phi)$. In the equation governing horizontal dynamics, the net horizontal gravitational and pressure force per unit mass is $\nabla U_1(r, \sigma)$, independent of z. To first order in z the potential at $z = z_1$ is the same as the potential at $z = z_1$ outside the zero thickness disk,

$$U_1(r, \phi) = U_0(r, \phi) + \frac{\partial U}{\partial z} z_1 =$$

$$= U_0(r, \phi) - 2\pi G\sigma z_1. \tag{9}$$

For a polytrope of the form

$$\varrho = K\, p^{\frac{n}{n+1}}, \tag{10}$$

Harrison and Lake (1972) find that z_1 for an infinite plane sheet is

$$z_1 = \frac{\sigma}{2K} \left(\frac{\pi G}{2} \sigma^2 \right)^{-n/(n+1)} \zeta_1, \tag{11}$$

where ζ_1 is a numerical coefficient depending on n. The equation for motions in the plane then has the form of Equation (5) with $\psi \propto \sigma^{2/(n+1)}$. The pressure forces and the reduction in U due to the finite thickness contribute to ψ in comparable amounts.

The correction to the horizontal dynamics second order in z requires taking into account vertical accelerations as well as the second order corrections to $U_1(r, \phi)$ which depend on the radial variation of U. Equation (5) can represent a thin isotropic pressure gas disk only as long as the scale of radial variation of both the equilibrium and perturbed surface density is large compared with the thickness. A local stability analysis based on Equation (5) will give only roughly the same result as the

Goldreich-Lynden-Bell local axisymmetric stability criterion, since the wavelength at marginal stability is only a few times the thickness.

My numerical analysis of global instabilities has been applied to disks whose horizontal dynamics is described by Equation (5). Only some of the results will be discussed here; the details will be published elsewhere.

From now on all equations and quantities will be written in units such that the outer radius of the disk $R=1$, the mass of the disk $M_D=1$, and the gravitational constant $G=1$.

Only first-order perturbations of equilibrium models will be considered. The models discussed in most detail will have a $\psi(\sigma)$ of the form

$$\psi = \beta(\pi\sigma)^\alpha. \tag{12}$$

The potential U in the plane of the disk is found by separating variables in oblate spheroidal coordinates (Hunter, 1963, 1965), so it is convenient to make the square of the angular velocity of the zero-pressure equilibrium model, Ω_0^2, a polynomial in η^2, where

$$\eta = (1-r^2)^{1/2}. \tag{13}$$

The corresponding potential U_D in the plane of the disk is found by integrating

$$\Omega_0^2 = \frac{1}{\eta}\frac{dU_D}{d\eta}; \tag{14}$$

the surface density derived from U_D is a polynomial containing only odd powers of η. The Ω_0^2 polynomial is adjusted to make $\sigma \propto \eta^3$ as $\eta \to 0$ at the rim. Then the angular velocity of the 'warm' disk Ω, given by

$$\Omega^2 = \frac{1}{\eta}\frac{d}{d\eta}(U-\psi(\sigma)), \tag{15}$$

is a regular function of η^2 at the rim if α in Equation (12) equals $\frac{2}{3}$ or $\frac{4}{3}$, corresponding to a finite thickness disk with polytropic index $n=2$ or $\frac{1}{2}$, respectively. The potential U in (15) may include a contribution U_H from a 'halo',

$$U = U_D + U_H. \tag{16}$$

If the halo is spherical, the halo mass inside radius r is

$$M_H(r) = r^3\Omega_H^2 = r^3\left[\frac{1}{\eta}\frac{dU_H}{d\eta}\right]. \tag{17}$$

A remarkable property of the choice $\alpha=\frac{2}{3}$ in Equation (12) is that for reasonable disk models in which σ is roughly exponential over a substantial fraction of the radius, the fractional correction to Ω_0^2 produced by the finite pressure is remarkably uniform as the surface density changes by a couple of orders of magnitude. This allows a choice of β which gives a substantial average reduction in Ω^2 from Ω_0^2 without Ω^2 becoming negative near the center or near the rim. Only for $\alpha = \frac{2}{3}$ was I able to increase the pressure enough to stabilize the disk without a halo.

The structure of one standard disk model, Model C, is shown in Figure 1. The central surface density is 16.61 times the average surface density. The disk is roughly an exponential disk with a scale height of 0.16 in radius. The velocities of rotation for the zero-pressure disk and for the disk with $\alpha = \frac{2}{3}$ and $\beta = 0.4601$ are also plotted.

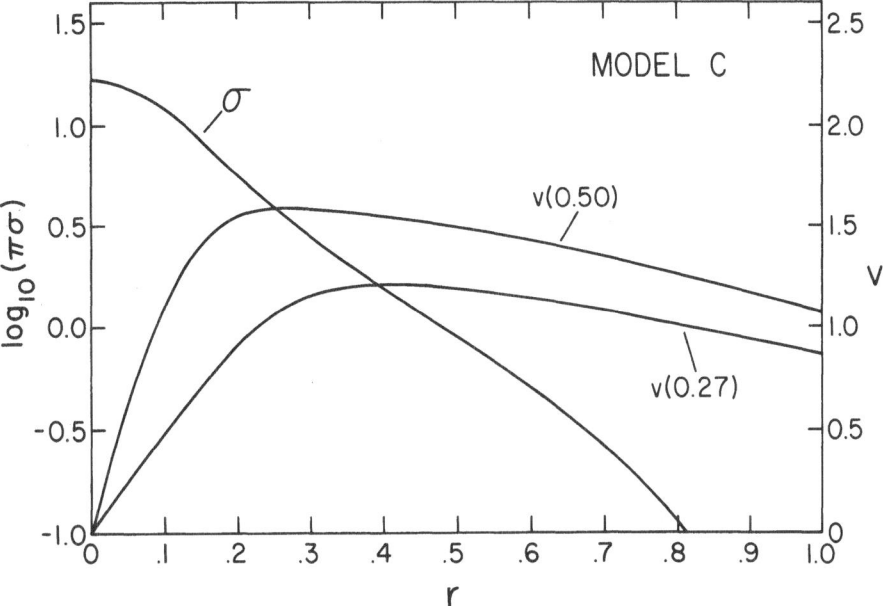

Fig. 1. The surface density σ and rotational velocity of Disk C are plotted. The rotational velocity $v(0.50)$ is for the zero pressure disk and the rotational velocity $v(0.27)$ is for a disk almost dynamical stable to the bar mode.

The ratio of kinetic energy of rotation to gravitational potential energy, t, labels the velocity curves; it is the value estimated for an isotropic pressure, finite thickness disk. The infinitesimally thin disk with $\beta = 0.4601$ has $t = 0.2341$.

The equations governing dynamic perturbations are obtained by expanding Equation (5) to first order in the deviation from equilibrium. It is instructive to consider harmonic time dependence and, of course, a particular axial harmonic; so the perturbed quantities all contain a factor $e^{im\phi - i\omega t}$. Then the perturbations in the velocity are algebraically related to the perturbations in the force per unit mass,

$$[\kappa^2 - (\omega - m\Omega)^2]\,\delta v_r = -i(\omega - m\Omega)\frac{d}{dr}(\delta U_{eff}) + \frac{2im\Omega}{r}(\delta U_{eff}) \tag{18}$$

and

$$[\kappa^2 - (\omega - m\Omega)^2]\,\delta v_\phi = -\frac{\kappa^2}{2\Omega}\frac{d}{dr}(\delta U_{eff}) + \frac{m(\omega - m\Omega)}{r}(\delta U_{eff}), \tag{19}$$

where

$$\delta U_{eff} = \delta U_0 - \frac{d\psi}{d\sigma}\,\delta\sigma. \tag{20}$$

These equations are supplemented by the conservation of mass equation

$$i(\omega - m\Omega)\,\delta\sigma = -\frac{1}{r}\frac{d}{dr}(\sigma r\,\delta v_r) + \frac{im\sigma}{r}\,\delta v_\phi. \tag{21}$$

In Equations (18) and (19) the 'epicyclic frequency' κ is defined by

$$\kappa^2 = 2\Omega\left(2\Omega + r\frac{d\Omega}{dr}\right); \tag{22}$$

it is not the epicyclic frequency for a circular test particle orbit.

A special cancellation on the right-hand side of one of the above equations is necessary to avoid a singularity in the eigenfunction at a radius where the pattern angular velocity of the perturbation $\Omega_p \equiv \omega/m$ equals the local angular velocity of the gas Ω (corotation resonance) or at a radius where

$$\omega/m = \Omega_p = \Omega + \kappa/m \quad \text{(outer Lindblad resonance)}$$

or

$$\omega/m = \Omega - \kappa/m \qquad \text{(inner Lindblad resonance)}.$$

These resonances act like extra boundary conditions on the eigenfunction. One boundary condition is already used up in requiring regularity at the center (the rim is a free boundary), so more than one resonance makes it difficult for any *regular* real-frequency modes to exist.

The non-local relation between the surface density perturbation $\delta\sigma$ and the potential perturbation δU_0 prevents the direct integration of Equations (18)–(20). However, if the radial wavelength of the perturbation is small compared with the radial scaleheight of surface density and small compared with r/m (a tightly wound spiral), there is an approximate local relation

$$\delta U_0 \simeq 2\pi\,\delta\sigma/k, \tag{23}$$

where k is the radial wavenumber. In this limit Equations (18)–(21) combine to give a dispersion relation (see Hunter, 1972)

$$(\omega - m\Omega)^2 = \kappa^2 - 2\pi\sigma k + \sigma\frac{d\psi}{d\sigma}k^2. \tag{24}$$

The minimum of $(\omega - m\Omega)^2$ is at a wavenumber

$$k_c = [d\psi/d(\pi\sigma)]^{-1} \tag{25}$$

or a wavelength

$$\lambda_c = 2\,d\psi/d\sigma. \tag{26}$$

The minimum value is less than zero, implying local instability, if

$$Q^2 \equiv \frac{\kappa^2}{\pi\sigma}\frac{d\psi}{d(\pi\sigma)} < 1. \tag{27}$$

The parameter Q is the analogue of Toomre's local stability parameter for stellar disks.

The actual method used to study the global stability of these disks was to solve numerically the initial value problem. For a particular angular eigenvalue m ($m=2$ for all cases discussed here) the radial dependence of the perturbations are represented by sums over associated Legendre functions $P_l^m(\eta)$, in such a way that the boundary conditions at the center and at the rim are automatically satisfied. The dynamic equations then reduce to an infinite set of ordinary differential equations for the coefficients C_l of the Legendre functions,

$$\frac{dC_l}{dt} = \sum_{l'=m}^{\infty} M_{ll'} C_{l'}. \tag{28}$$

The coefficients of the matrix $M_{ll'}$ depend only on the equilibrium structure of the disk. In practice the Legendre expansion was truncated at about 20–30 terms (counting only even values of $l-m$), enough to represent a moderately complicated eigenfunction accurately.

Starting from arbitrary initial conditions one mode, the most rapidly growing unstable mode, will eventually dominate if the disk is in fact unstable. By first choosing a small value of β, so the growth rate is large, then using the eigenfunction obtained as the initial condition for a model with a somewhat larger value of β, and so on, one can find the value of β required for marginal instability and the marginally unstable eigenfunction without an inordinate expenditure of computer time.

Table I shows the approach to stability for the Disk C whose structure is shown in Figure 1 and for a Disk B which has the same radial variation of surface density

TABLE I

Unstable modes of two disks without halos

Model	β	t	$\langle\Omega\rangle$	Ω_p	Growth rate	r_{corot}	r_{Lind}
C	0.3336	0.34	2.804	2.24	0.82	0.56	0.75
	0.3708	0.32	2.722	1.81	0.55	0.65	0.83
	0.3891	0.31	2.680	1.58	0.44	0.71	0.88
	0.4072	0.30	2.638	1.42	0.44	0.76	0.92
	0.4250	0.29	2.595	1.32	0.40	0.79	0.96
	0.4427	0.28	2.552	1.24	0.31	0.81	1.00
	0.4514	0.275	2.530	1.21	0.217	0.82	–
	0.4601	0.270	2.508	1.17	0.08	0.84	–
	0.4636	0.268	2.500	1.13?	0	0.86?	–
B	0.4352	0.30	–	1.272	0.463	0.80	0.96
	0.4543	0.29	–	1.20	0.32	0.83	1.00
	0.4619	0.286	2.338	1.169	0.229	0.84	–
	0.4657	0.284	2.329	1.154	0.158	0.85	–
	0.4694	0.282	2.320	1.134	0.076	0.85	–
	0.4732	0.280	2.312	1.06?	0	0.89?	–

in the outer half of the disk but a central surface density only 9.26 times the average surface density. Both disks have an exponent $\alpha = \frac{2}{3}$, corresponding to a polytropic index $n = 2$ in the finite thickness interpretation. The parameter t is calculated assuming a finite thickness. The quantity $\langle \Omega \rangle$ is the angular momentum divided by the moment of inertia, the tensor virial prediction for the frequency of the marginally stable bar mode. The pattern angular velocity Ω_p is one-half the real part of the frequency (since $m = 2$) and the growth rate is the imaginary part of the frequency. The radii of the corotation resonance and the (outer) Lindblad resonance are in the last two columns.

Both disks become stable at roughly the point predicted by the tensor virial method. The form of the dominant unstable mode is a moderate trailing spiral which straightens into a bar in the limit of marginal stability. The displacements at marginal stability are close to linear functions of the Cartesian coordinates out to $r \simeq 0.6$, the part of disk containing the bulk of the mass, but deviate by a large amount in the vicinity of the corotation radius. The Eulerian perturbations $\delta\sigma$, δv_r, δv_ϕ vary smoothly through the corotation radius near marginal stability. For Disk C there is an indication that the existence of a Lindblad resonance tends to keep the disk unstable. This effect might not be expected to carry over to stellar disks, since there one expects strong Landau damping near a Lindblad resonance.

The influence of a halo on the stability and on the form of the dominant unstable mode has been studied most thoroughly for Disk C. One interesting point is the degree of central condensation of the halo which is most effective in stabilizing the disk. Models C1–C5 combine Disk C with halos all of same Mass $M_H = 1.433$. The distribution of mass in each of the halos and in the disk is shown in Figure 2. The properties of the dominant unstable modes are listed in Table II for certain values of β. In particular, note that the growth rate compared at the same value of β, $\beta = 0.2605$, is smallest for a halo considerably less centrally condensed than the disk, Model C2.

Some additional properties of Model C2 with $\beta = 0.2605$ are shown in Figure 3. The local stability parameter Q is substantially greater than one everywhere. It is smallest in the center of the disk, partially due a substantial reduction of Ω^2 below $\Omega_0^2 + \Omega_H^2$ there. The wavelength at the minimum of the local dispersion relation, λ_c, increases strongly outward, and even at the center of the disk it is somewhat larger than the characteristic scale of radial variation of the surface density.

The angular patterns of the dominant unstable modes of Models C2 and C3 at $\beta = 0.2605$ are depicted in Figure 4. The trailing spiral rather clearly will persist at marginal stability, since the growth rates are so small, as will the presence of a Lindblad resonance. The amplitude of the *fractional* perturbation in the surface density has a large peak at $r \simeq 0.21$, well within the central bar. Model C2 has a secondary maximum in $\delta\sigma/\sigma$ near the corotation radius, down by a factor of 1.7. This probably becomes a pole at marginal stability, associated with the sharp break in the pattern there in Figure 4. The mode is strongly dominated by the central bar, and is not even qualitatively similar to a local density wave.

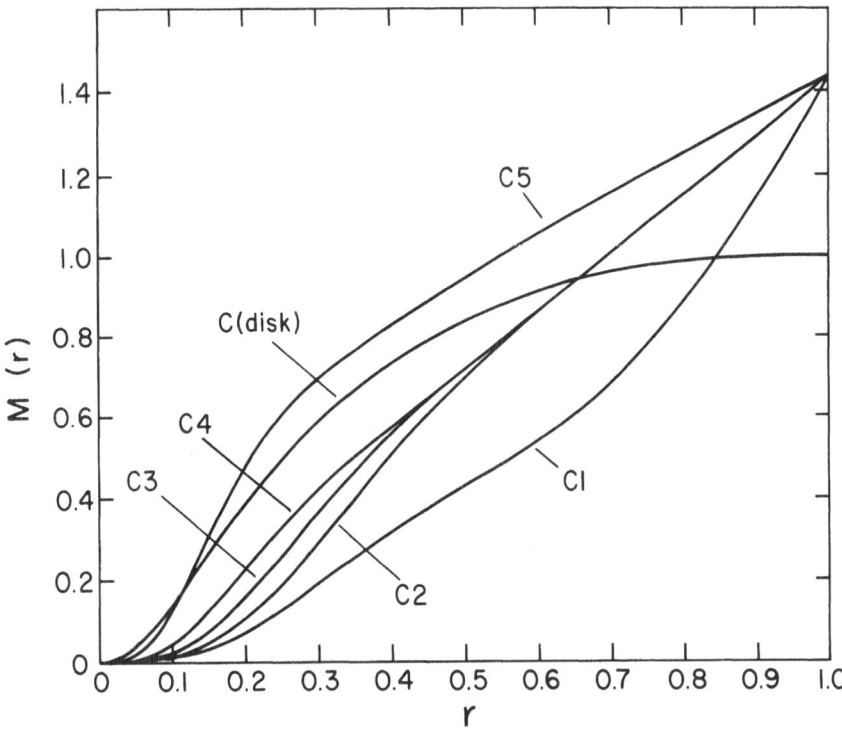

Fig. 2.   The mass distribution of the halo as a function of radius is shown for Models C1–C5 along with the mass distribution of Disk C.

TABLE II

Unstable modes of $\alpha = \frac{2}{3}$ Disk C with halo

Model	β	t	$\langle \Omega \rangle$	Ω_p	Growth rate	r_{corot}	r_{Lind}
C1	0.2605	0.2076	3.569	3.03	0.42	0.53	0.86
C2	0.2605	0.2121	3.839	2.75	0.09	0.63	0.92
C3	0.2605	0.2179	3.921	2.68	0.16	0.64	0.93
C4	0.2605	0.2252	4.003	2.862	0.266	0.61	0.88
C5	0.111	0.2658	4.640	5.7	0.8	0.36	0.55
	0.1760	–	4.557	4.3	0.5	0.46	0.67
	0.2605	0.2437	4.446	3.365	0.34	0.56	0.79

The nature of the dominant instability changes substantially if the distribution of pressure in the disk is changed to make Q and λ_c larger in the center of the disk than in the middle part of the disk. This is accomplished by changing the exponent α in Equation (12) to $\frac{4}{3}$ from $\frac{2}{3}$. Now at least some halo is necessary to approach stability while Ω^2 is still positive at the center. Models C11 and C12 still have the zero-pressure

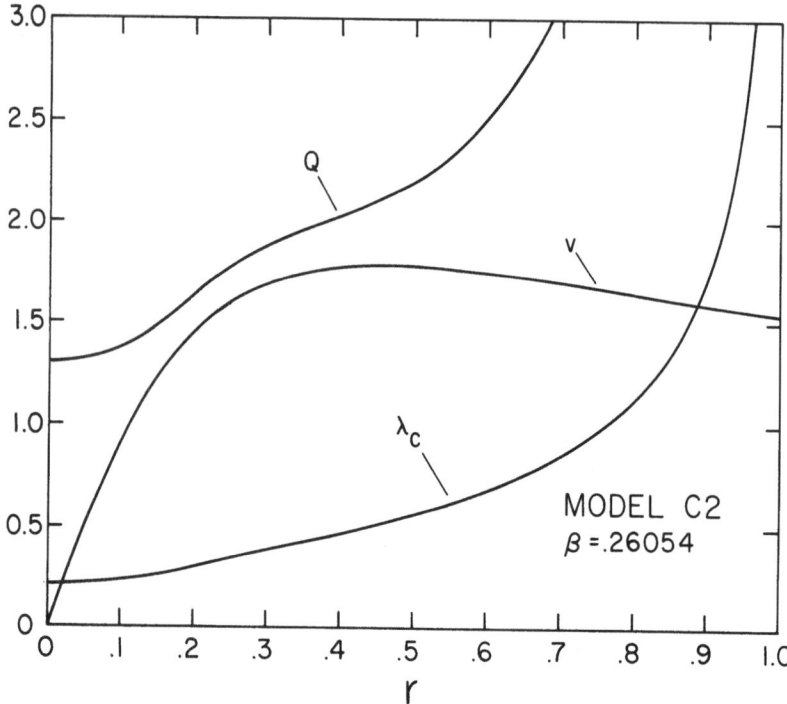

Fig. 3. The local stability parameter Q, the rotational velocity v, and the local characteristic wavelength λ_c for Model C2 at $\beta = 0.2605$.

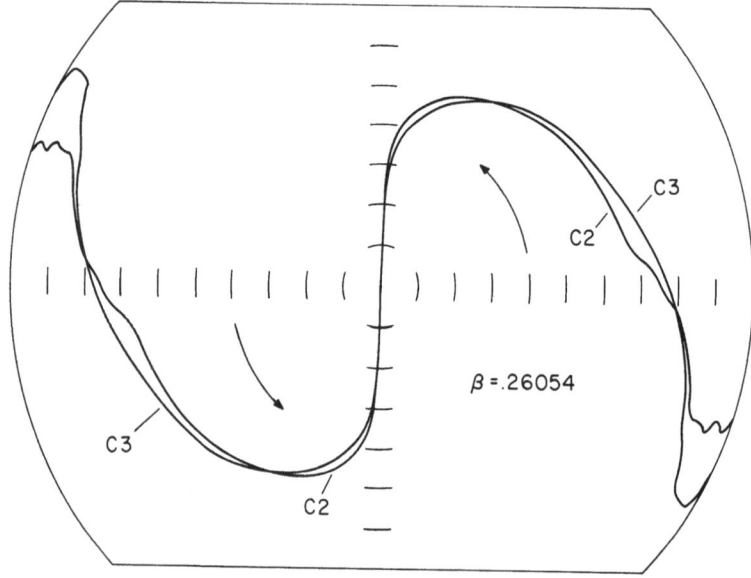

Fig. 4. The pattern of the maximum in surface density as a function of radius for Models C2 and C3 close to stability.

structure of Cisk C, but $\alpha=\frac{4}{3}$. The halo mass of C11 is the same as Models C1–C5 and is distributed in radius like Model C2 in the outer part of the disk. A strongly centrally condensed component of the halo is adjusted to keep Ω^2 reasonably large at the center. Some properties of Model C11 are shown in Figure 5. Model C12 has a halo mass $M_H=2.149$, but except for a smaller centrally condensed component to the halo is otherwise identical to Model C11.

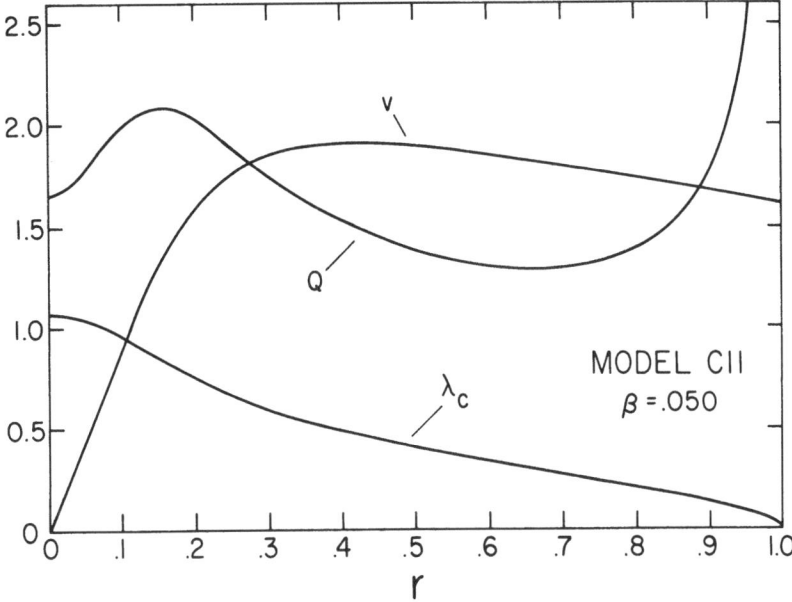

Fig. 5.   The local stability parameter Q, the rotational velocity v, and the local characteristic wavelength λ_c for Model C11 at $\beta=0.50$.

Table III lists some of the properties of the dominant unstable modes of these models. Again, t is estimated from the finite thickness interpretation of the disk and includes the halo, which is truncated at the outer edge of the disk. The minimum value of Q for each value of β is given in the third column. In all the models the radius of the minimum Q is $r=0.66$. My numerical methods were stretched to or a little beyond their limit in the case of Model C12. A spurious numerical instability

TABLE III

Unstable modes of $\alpha=\frac{4}{3}$ Disk C with halo

Model	β	t	Q_{min}	Ω_p	Growth rate	r_{corot}	r_{Lind}
C11	0.030	0.2408	0.994	3.8	0.7	0.50	0.74
	0.040	0.2327	1.150	3.2	0.5	0.58	0.84
	0.050	0.2225	1.289	2.74	0.328	0.66	0.94
C12	0.023	0.1839	0.984	3.3	$\lesssim 0.3$	0.60	0.90

of very short wavelength was present, but did not obscure the overall properties of
the pattern, which maintained itself with little change in the region where the amplitude was largest for a couple of rotation periods.

The amplitude of the dominant mode, as measured by the fractional perturbation
in the surface density, has a broad maximum in the vicinity of the corotation radius
in these models and is very small in the central bar and outside the Lindblad resonance. The absolute value of $\delta\sigma$ is largest near the inner edge of the spiral pattern.
Therefore, these modes are genuine spiral modes. The pattern for Model C11 is
shown in Figure 6 and that for Model C12 in Figure 7. The mode at $\beta=0.023$ for

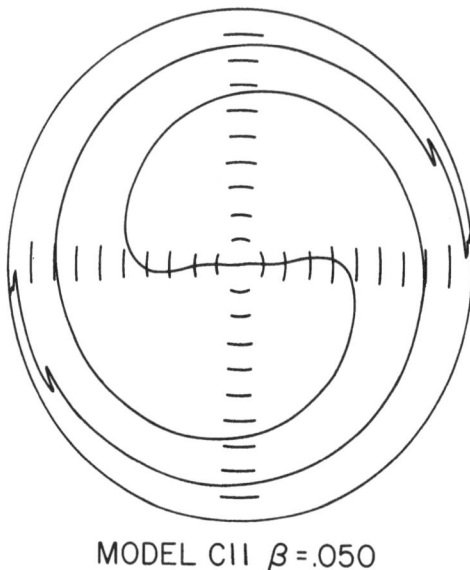

MODEL CII $\beta=.050$

Fig. 6.   The pattern of the surface density perturbation for Model C11 at $\beta=0.050$.

Model C12 corresponds roughly to a local density wave. The spacing of the spiral
pattern is roughly comparable with λ_c (0.68 times the λ_c plotted in Figure 4), and the
fairly smooth behavior near the corotation radius is consistent with the density wave
prediction, since $Q_{min}\sim 1$ there. The inner radius of the spiral pattern is roughly where
λ_c becomes comparable with the radius.

The relatively large value of Q near the center does make it easier to get a spiral
pattern, but a substantial halo is still required to damp the global instability of the
disk to the point that a local density wave analysis has some validity. The spiral
pattern in Models C11 and C12 is outside most of the mass in the disk, so in effect
these models rely on a 'central bulge' as well as the halo to stabilize the part of the
disk with Q near one.

Models with Q and λ_c roughly independent of radius over most of the disk have
also been studied, with results intermediate between the extremes quoted here. The
fractional surface density perturbation has comparable amplitudes in a central bar

and an outer spiral pattern; the latter at comparable halo masses is more open than the spiral pattern of Models C11 and C12 near marginal stability.

The spiral patterns indicate that the antispiral theorem of Lynden-Bell and Ostriker (1967), which does apply to these disks, is not a severe restriction in practice. With a substantial halo it takes only a small imaginary part in the frequency to produce a

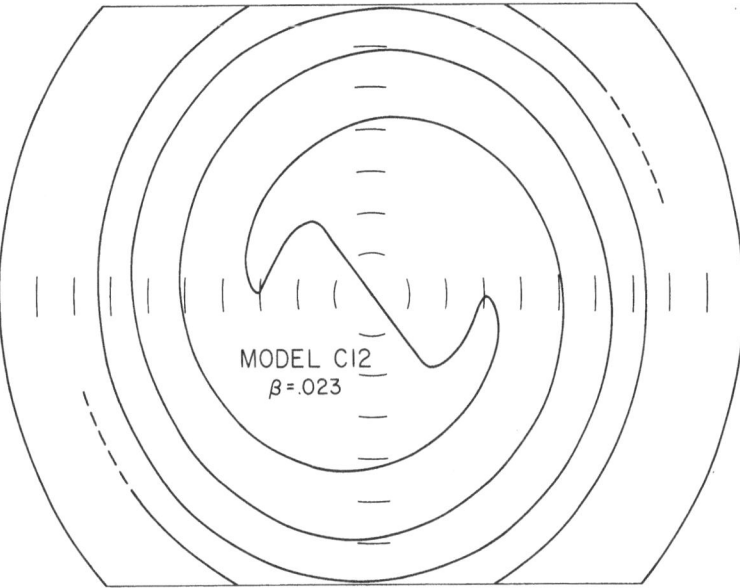

Fig. 7.   The pattern of the surface density perturbation for Model C12 at $\beta = 0.023$. The pattern is not well defined inside $r = 0.3$ and outside $r = 0.8$.

fairly tightly wound spiral without a noticeable singularity in the eigenfunction. In a stellar disk Landau damping associated with the spiral pattern can be expected to limit the growth of the instability to small amplitudes.

5. Summary and Conclusion

All lines of theoretical evidence lead to the same conclusion. Any disk which remotely resembles the disk of a spiral galaxy as represented by the neighborhood of the Sun or as contemplated in density wave theories of spiral structure will be globally unstable unless the disk contains only a rather small fraction of the total mass within its outer radius.

In the absence of a halo there seems to be a rather close correspondance between the dynamic instability of a stellar disk and the secular instability of a gas disk. Both are marginally stable when the ratio of kinetic energy to gravitational potential energy t is about 0.14. Even the small difference between the critical value of $t = 0.1376$ for the Maclaurin spheroids and $t = 0.1286$ for the Kalnajs (1972) uniformly rotating disk is probably largely due to the difference in thickness. A more precise gas analogue to the Kalnajs disk is an infinitesimally thin disk with isotropic stress only in the plane; a choice of $\psi = \beta(\pi\sigma)^2$ and a Maclaurin surface density distribution produces a zero frequency mode, corresponding to marginal secular instability, at $t = 0.1250$.

More realistic differentially rotating stellar dynamic configurations have not yet been tested directly for stability against small perturbations, but the numerical simulations do provide quite strong evidence that shear and central condensation do not change t at marginal stability very much. However, the numerical simulations should be carried out with more realistic initial conditions, particularly initial conditions which correspond to self-consistent stellar dynamic equilibrium models with a large velocity dispersion and a large Q near the center relative to that in the outer part of the disk.

A stellar disk presumably began as a gas disk. The gas disk, considering a probable large effective viscosity from turbulence and perhaps magnetic fields at least during its formation, must have been secularly stable. Therefore, it is certainly relevant that Ostriker and Bodenheimer (1973) find t at marginal secular instability is about 0.138. Also, the dynamic instability calculations for thin or infinitesimally thin gas disks reported in Section 4 of this paper do provide necessary conditions for stability which already exclude disks like those commonly used to model spiral galaxies without halos.

Just how much halo is necessary to stabilize a disk with a local stability parameter Q close to one is not yet well established. The numerical simulations have so far been carried out for only a few special cases. The indications are that a disk which does not contain a massive hot 'spheroidal component' in its center requires a halo of perhaps 3 times the mass of the disk or more. With a fairly massive halo it seems likely that the distinction between secular and dynamic instability for a gas disk becomes less important, and that the dynamic instability calculations reported in Section 4 have a more direct relevance to galactic disks. The central condensation of the halo should be somewhat less than that of the disk for an optimum overall stabilizing effect, though a more centrally condensed halo does reduce the amplitude of the central bar component of the overall unstable mode relative to the spiral component.

All in all, I do not think the halo should be considered an extension of a central spheroidal component. The observational evidence is that the halo does not contain anything remotely like a normal stellar component. It seems to me that it is much easier to conceive of a halo containing black holes than a halo containing extreme M-dwarfs, since the black holes would fit in better with current ideas on star formation in metal-deficient interstellar gas (Larson and Starrfield, 1971) and on the chemical evolution of galaxies (Truran and Cameron, 1971). In fact, there are indications some elliptical galaxies may have extended halos with large M/L ratios, perhaps largely composed of black holes (Wolfe and Burbidge, 1970).

The calculations in Section 4 and some of the numerical simulations suggest that at least the more open spiral patterns in galactic disks may be understandable as slightly unstable global modes of the disk. They also indicate that conventional density wave theory may not be applicable. In particular, the local relation between surface density perturbation and potential perturbation breaks down badly when the spiral pattern is fairly open in a centrally condensed disk. The potential perturbation in the outer part of the disk is dominated by the inner part of the spiral pattern, and does not follow the surface density perturbation. Also, if a massive halo is present the density wave fits to observed spiral patterns based on conventional galaxy models, which ignore any extended halo in obtaining the surface density of the disk from a rotation curve (scc Lin *et al.*, 1969; Tully, 1974), are invalid.

The detailed study of global instabilities of galactic disks and their possible relation to spiral structure is just beginning. Direct observational confirmation of the existence of halos will be virtually impossible if the halos are made up of black holes, so observational tests of the broad implications of the global stability analysis may well have to rely on indirect evidence, such as M/L variations obtained from rotation curves (see Roberts, 1975).

Acknowledgements

I would like to thank F. Hohl, A. Kalnajs, J. P. Ostriker and A. Toomre for discussions which have greatly increased my understanding of the questions discussed in this paper. I would also like to recognize valuable contributions of Bernard F. Schutz, Jr. to early stages of the dynamic stability analysis of thin gas disks.

References

Bardeen, J. M.: 1971, *Astrophys. J.* **167**, 425.
Chandrasekhar, S.: 1969, *Ellipsoidal Figures of Equilibrium,* Yale University Press, New Haven.
Chandrasekhar, S.: 1970, *Astrophys. J.* **181**, 497.
Chandrasekhar, S. and Lebovitz, N. R.: 1962, *Astrophys. J.* **135**, 248.
Erickson, S. A.: 1974, 'Vibrations and Instabilities of a Disk Galaxy with Modified Gravity', Massachusetts Institute of Technology (Ph.D. Thesis).
Feldman, S. I. and Lin, C. C.: 1973, *Studies in Applied Math.* **7**, 1.
Freeman, K. C.: 1970, *Astrophys. J.* **160**, 811.
Freeman, K. C.: 1975, this volume, p. 367.
Friedman, J. L. and Schutz, B. F.: 1975, 'On the Stability of Relativistic Systems', *Astrophys. J.* (to be published).
Goldreich, P. and Lynden-Bell, D.: 1965, *Monthly Notices Roy. Astron. Soc.* **130**, 7.
Hohl, F.: 1971, *Astrophys. J.* **168**, 343.
Hohl, F.: 1973, *Astrophys. J.* **184**, 353.
Hohl, F.: 1975, this volume, p. 349.
Hunter, C.: 1963, *Monthly Notices Roy. Astron. Sov.* **126**, 299.
Hunter, C.: 1965, *Monthly Notices Roy. Astron. Soc.* **129**, 311.
Hunter, C.: 1972, *Annual Reviews of Fluid Mechanics,* Annual Reviews, Palo Alto.
Hunter, C.: 1973, *Astrophys. J.* **181**, 685.
Kalnajs, A. J.: 1972, *Astrophys. J.* **175**, 63.
Larson, R. and Starrfield, S.: 1971, *Astron. Astrophys.* **13**, 190.
Lin, C. C. and Shu, F. H.: 1964, *Astrophys. J.* **140**, 646.
Lin, C. C. and Shu, F. H.: 1966, *Proc. Nat. Acad. Sci.* **55**, 229.
Lin, C. C., Yuan, C. and Shu, F. H.: 1969, *Astrophys. J.* **155**, 721.
Lynden-Bell, D.: 1967, *Monthly Notices Roy. Astron. Soc.* **136**, 101.
Lynden-Bell, D. and Ostriker, J. P.: 1967, *Monthly Notices Roy. Astron. Soc.* **136**, 293.
Miller, B. D.: 1973, *Astrophys. J.* **181**, 497.

Miller, R. H.: 1971, *Astrophys. Space Sci.* **14**, 73.

Miller, R. H.: 1974, *Astrophys. J.* **190**, 539.

Miller, R. H. and Prendergast, K. H.: 1968, *Astrophys. J.* **151**, 699.

Miller, R. H., Prendergast, K. H., and Quirk, W. J.: 1970, *Astrophys. J.* **161**, 903.

Ostriker, J. P. and Bodenheimer, P.: 1973, *Astrophys. J.* **180**, 171.

Ostriker, J. P. and Mark, J. W-K.: *Astrophys. J.* **151**, 1075.

Ostriker, J. P. and Peebles, P. J. E.: 1973, *Astrophys. J.* **186**, 467.

Ostriker, J. P., Peebles, P. J. E., and Yahil, A.: 1974, *Astrophys. J. Letters* **193**, 1.

Ostriker, J. P. and Tassoul, J. L.: 1969, *Astrophys. J.* **155**, 987.

Press, W. H. and Teukolsky, S. A.: 1973, *Astrophys. J.* **181**, 513.

Roberts, M.: 1975, this volume, p. 331.

Schmidt, M.: 1965, *Stars and Stellar Systems* **5**, 513.

Schmidt, M.: 1975, this volume, p. 325.

Tassoul, J. L. and Ostriker, J. P.: 1968, *Astrophys. J.* **154**, 613.

Toomre, A.: 1964, *Astrophys. J.* **139**, 1217.

Toomre, A.: 1969, *Astrophys. J.* **158**, 899.

Truran, J. W. and Cameron, A. G. W.: 1971, *Astrophys. Space Sci.* **14**, 179.

Tully, R. B.: 1974, *Astrophys. J. Suppl.* **27**, 449.

Vandervoort, P.: 1970, *Astrophys. J.* **162**, 453.

Wolfe, A. and Burbidge, G. R.: 1970, *Astrophys. J.* **161**, 419.

DISCUSSION

Hohl: Why is a central core/halo not effective in stabilizing the bar instability?

Bardeen: At least for these gas disks, I find that a halo as centrally condensed or more centrally condensed than the disk is less effective (for a given total mass) than a less centrally condensed halo in stabilizing the disk as a whole. The central condensation does help damp the central bar, but the pattern angular velocity is increased, which means that the outer Lindblad radius moves in and has a stronger destabilizing effect. This behavior is somewhat in conflict with what is expected from the local criterion, since the central condensation increases Q substantially at the center.

Miller: Have you allowed modes other than $m = 2$ yet?

Bardeen: My program is set up to calculate the stability of modes with any value of m, but so far all my calculation have been for $m = 2$.

Miller: An interesting problem would result if the halo could respond to the disk, rather than being rigid. Do you see a way to allow for this?

Bardeen: A crude way would be to superimpose a hot disk and a cool disk, which interact only gravitationally.

Pişmiş: Is there any restriction to the extent of the halo required for stability?

Bardeen: All the stability requires is a halo mass somewhat larger than that of the disk within the radius of the disk. The halo mass at larger radii whether it exists or not, has no effect on the dynamics of the disk.

THEORETICAL LIMITS ON THE PROPERTIES OF
LOW-VELOCITY M-DWARFS

P. BIERMANN

Universitäts-Sternwarte Göttingen, F.R.G.

Abstract. Using a version of the Toomre-criterion (Toomre, 1964) for a mixture of stellar populations, limits are derived for the kinematical properties of the newly discovered population of low-velocity M-dwarfs (Weistrop, 1972; Murray and Sanduleak, 1972); these stars are claimed to be a large mass-fraction ($\gtrsim 0.5$) of all stars in the solar neighbourhood and to have only 10 km s^{-1} velocity dispersion. They could make up for the locally missing mass (Oort, 1965; Schmidt, 1974). It is found that a configuration of stellar populations with the M-dwarfs included (having such properties) is not stable in the sense of the Toomre-criterion. The limiting stability properties are given.

1. Introduction

Using the method of moments of the Boltzmann-equation the Toomre-criterion has been rederived. Since this method leads to an infinite set of equations, cutoff conditions have to be imposed. The cutoff conditions assumed here are: (a) The temperature-asymmetry-tensor of the stellar motions (the tensor of third order) is identically zero. (b) All higher order tensor-moments are ignored. This blunt approach reproduces within 10% the limiting velocity dispersion of stellar motions for stability against axisymmetric perturbations in a flat disk; since the numbers that go into this number, the surface mass density and the epicyclic frequency, are not known to that accuracy, the difference is not significant.

The moment-method has been used by King (1965), Larson (1969), Hunter (1970), and many others. Here the criterion is generalized to many stellar populations and applied to the problem of the low-velocity M-dwarfs, in Sections 2 and 3, respectively. The results are discussed in Section 4. Superpositions of stellar systems in flat galaxies have previously been considered by Vandervoort (1970).

We assume here that the low-velocity M-dwarfs are a general population in the disk of the galaxy, and that their properties in the solar neighbourhood reflect those at large.

2. The Criterion

We define: k is the wave-number of the perturbation, α_j is the radial component of the temperature-tensor of stellar motions (the $r-r$-component) of the j stellar populations, κ the epicyclic frequency, ϱ_{j_0} the unperturbed surface density of the j stellar population, ϱ_{t_0} the total unperturbed stellar surface density, and G the gravitational constant. Then the critical wave-number is

$$k_{\text{crit}} = \frac{\kappa^2}{2\pi G \varrho_{t_0}}.$$

Hayli (ed.), Dynamics of Stellar Systems, 321–324. All Rights Reserved.

Defining

$$x = \frac{k_{crit}}{k} \quad \text{and} \quad y_j = \frac{\alpha_j k_{crit}^2}{\kappa^2}$$

one finds (for one stellar population only) an expression relating the velocity dispersion to the wavelength separating stable and unstable regions:

$$\left(3\frac{y}{x^2} - \frac{1}{x} + 1\right)\left(\frac{y}{x^2} + 4\right) - 9\frac{y}{x^2} = 0.$$

A related expression has been discussed by Hunter (1970). The maximum of the curve
$y(x)$ gives the minimum that y can have to ensure stability for all $x \in [0, 1]$:

$$y_{max} = 0.2272 \quad \text{at} \quad x = 0.60.$$

The corresponding number of Toomre (1964) is 0.2857. Generalizing to J stellar
populations the relevant equation is:

$$1 = \sum_{j=1}^{J} \frac{\varrho_{jo}}{\varrho_{to}} \frac{x}{3y_j + x^2 - \dfrac{9y_j x^2}{y_j + 4x^2}}.$$

3. Results

Wielen (1974) finds that the average velocity dispersion of all observed stars is 48
km s^{-1}. We adopt this number from Wielen and take from the model of the solar
neighbourhood of Biermann and Tinsley (1974) the numbers for the mass fractions
of the different stellar populations: 0.17 for normal stars, 0.26 for invisible remnants,
and 0.57 for the low-velocity M-dwarfs. The normal stars and the invisible remnants
are thus assumed to have 48 km s^{-1} velocity dispersion and the low-velocity M-
dwarfs 10 km s^{-1}. Three questions can now be asked from the criterion:

(a) Is the configuration stable as given? For the answer we write down a necessary
condition (index j is for the M-dwarfs here)

$$\frac{\varrho_{jo}}{\varrho_{to}} \frac{x}{3y_j + x^2 - \dfrac{9y_j x^2}{y_j + 4x^2}} \leqslant 1 \quad \text{for all} \quad x \in [0, 1].$$

This expression corresponds to the assumptions that all populations except j have
infinite velocity dispersion. (In that extreme case it becomes a sufficient condition.)
We find then that $\varrho_{jo}/\varrho_{to} \leqslant 0.33$ and thus that the configuration as given is *not* stable.

(b) If we use all mass fractions and the velocity dispersions for the normal stars
and the invisible remnants as adopted we can find the *lowest* allowable velocity dispersion for the low-velocity M-dwarfs: The answer is 21 km s^{-1}.

(c) If we use the velocity dispersions of all stellar populations as adopted, we can ask for the *maximum* mass fraction the low-velocity M-dwarfs can have: We find 0.30.

4. Discussion

Applying the generalized Toomre-criterion we find

(a) the mixture of stellar populations with kinematical properties as claimed to be observed is not stable,

(b) the lowest allowable velocity dispersion for the low-velocity M-dwarfs (at 0.57 mass fraction) is 21 km s^{-1},

(c) the largest allowable mass fraction (at 10 km s^{-1} velocity dispersion) is 0.30. Evidently, intermediate cases can be derived.

The observations are still discussed and extended (Schmidt, 1974; Thé and Staller, 1974). These M-dwarfs could be a local phenomenon and then some problems would disappear (Biermann, 1974).

The relevance of the Toomre-criterion is clearly limited (Miller, 1974); however, if it indicates instability, more detailed considerations of the same physics will not render a configuration stable, whereas the reverse cannot be said.

The form of the moment-method used here cannot be applied for non-axisymmetric perturbations. But it is a useful example of the application of the moment-method which may be helpful in further generalizations.

Putting gas into the equations just as another population would restrict the limits for the M-dwarfs even further. It remains an interesting task to really integrate over all stellar populations that can be distinguished kinematically.

Acknowledgements

I wish to thank Dr I. R. King, Dr M. Schmidt, Dr A. Toomre, and Dr R. Wielen for interesting discussions. This work was done while the author was supported by the Deutsche Forschungsgemeinschaft (Bi 191/1) at Columbia University and UC Berkeley. The visit at Berkeley was made possible through a generous invitation from Dr J. Silk. I wish to thank the departments at New York and Berkeley for their hospitality.

References

Biermann, P. and Tinsley, B. M.: 1974, *Astron. Astrophys.* **30**, 1.
Biermann, P.: 1974, *Astron. Astrophys.* **30**, 31.
Hunter, C.: 1970, *Studies in Appl. Math.* **49**, 59.
King, I. R.: 1965, *Astron. J.* **70**, 296.
Larson, R. B.: 1969, *Monthly Notices Roy. Astron. Soc.* **145**, 405.
Miller, R. H.: 1974, *Astrophys. J.* **190**, 539.
Murray, C. A. and Sanduleak, N.: 1972, *Monthly Notices Roy. Astron. Soc.* **157**, 273.
Oort, J. H.: 1965, *Stars and Stellar Systems* **5**, 455.
Schmidt, M.: 1974, this volume, p. 325.
Thé, P. S. and Staller, R. F. A.: 1974, *Astron. Astrophys.* (in press).
Toomre, A.: 1964, *Astrophys. J.* **139**, 1217.

Vandervoort, P. O.: 1970, *Astrophys. J.* **162**, 453.

Weistrop, D.: 1972, *Astron. J.* **77**, 849.

Wielen, R.: 1974, in G. Contopoulos (ed.), *Highlights of Astronomy*, Vol. 3, D. Reidel Publ. Co., Dordrecht – Holland, p. 395.

DISCUSSION

Miller: Do you mean that some fraction of high velocity stars such as we have in the solar neighborhood cannot stabilize in the sense of the Toomre criterion?

Biermann: Even for infinite velocity dispersion for all stars except the M-stars, a thin disk with 57% of the mass in M dwarfs and 10 km s^{-1} velocity dispersion is not stable.

PROBLEMS ASSOCIATED WITH THE M-DWARF POPULATION

M. SCHMIDT

Hale Observatories,
California Institute of Technology, Carnegie Institution of Washington,
Pasadena, Calif., U.S.A.

Abstract. The existence of a rich population of late M-type dwarfs with small peculiar velocities is discussed on the basis of the work of Weistrop, Murray and Sanduleak, Pesch and others. These stars probably supply the 'missing' mass in the solar neighborhood. Their small peculiar velocities create problems related to the stability of the galactic disk.

Observations seem to indicate that there exists in the solar neighborhood a dense layer of M-type dwarfs with small peculiar velocities. These stars would solve the longstanding problem of the 'missing mass'. However, their small peculiar velocities create problems so severe that theoreticians would prefer these stars not to exist. I wish to review the observations pertaining to the space density and the kinematics of the M-type dwarfs, and to discuss briefly the theoretical problems.

The first evidence for a large population of M-type dwarfs came from an objective prism survey by Sanduleak (1964). He found 1200 M-type stars over an area of 120 sq deg and used colors and magnitudes to derive a space density of M-type dwarfs that was three times that corresponding to the stellar luminosity function of Luyten (1938). These results did not attract wide attention since they were published in summary form as part of the Annual Report of the Warner and Swasey Observatory.

Several years later Donna Weistrop (1972a, b) undertook a program of star counts as a function of color in a field near the north galactic pole. She derived a luminosity function that shows an increasing excess over the luminosity function of Van Rhijn (1936), to a factor of about 10 at $M_v = 13$. Table I reproduces a table given by Veeder (1974) based on Weistrop's luminosity function. The total mass density of main-sequence stars is 0.16 $M_\odot$ pc^{-3} if we adopt the estimated values of the luminosity

TABLE I

Local mass densities of dwarfs

M_v	n (pc^{-3})	$M/M_\odot$	ϱ ($M_\odot$ pc^{-3})
<9			0.018
10	0.01	0.45	0.005
11	0.03	0.35	0.010
12	0.06	0.27	0.016
13	0.13	0.20	0.026
14	0.25[a]	0.16	0.040[a]
15	0.40[a]	0.12	0.048[a]

[a] Estimated

Hayli (ed.), Dynamics of Stellar Systems, 325–329. All Rights Reserved.

function beyond $M_v = 13$ as given in Table I. Since interstellar gas contributes about 0.03 $M_\odot$ pc^{-3} and white dwarfs around 0.02 $M_\odot$ pc^{-3} we can apparently account for the dynamical value of the local mass density (0.21 $M_\odot$ pc^{-3}) derived by Oort (1965) on the assumption that the 'missing mass' is distributed like the interstellar gas. Weistrop found, in fact, that the density distribution of the fainter stars resembles that of interstellar gas. The corresponding scale height of only about 110 pc suggests that these stars have peculiar velocities of only around 10 km s^{-1}.

I would like to illustrate the observational evidence for such a large number of late M-type dwarfs by deriving their expected numbers on the basis of Van Rhijn's luminosity function. We use an exponential density distribution $D(z) \sim \exp(-z/z_s)$ with a scale height z_s of 300 pc, which represents approximately Weistrop's $D(z)$ for G and K-type dwarfs. Mean absolute visual magnitudes M_t as a function of color $(B-V)$ based on available trigonometric parallaxes (see Weistrop, 1972b, for references) are given in Table II. These absolute magnitudes are based on apparent mag-

TABLE II

Observed and predicted star counts

$B-V$	1.40–1.50			1.50–1.60			1.60–1.70		
M_t	9.5			11.0			12.5		
M_0	10.0			11.5			13.0		
Spect. type	dM0–3			dM4			dM5		
	Obs.	Pred.	$\dfrac{\text{Obs.}}{\text{Pred.}}$	Obs.	Pred.	$\dfrac{\text{Obs.}}{\text{Pred.}}$	Obs.	Pred.	$\dfrac{\text{Obs.}}{\text{Pred.}}$
$M_v = 13.5$	4	3 ⎫	1.3	9	0.7 ⎫	6	2	0.1 ⎫	22
14.5	15	12 ⎭		13	2.7 ⎭		9	0.4 ⎭	
15.5	37	37	1.0	20	10	2	27	1.6	17
16.5	70	83	0.8	68	32	2	54	6	9
17.5	133	254	0.5	108	97	1	79	20	4

nitude-selected samples. The mean absolute magnitude in a volume of space M_0 is fainter by $\sigma^2 (A(m))^{-1} (dA(m)/dm)$ magnitudes (Malmquist, 1927). For $\sigma = \pm 0.6$ mag. the Malmquist correction amounts to 0.5 mag. We adopt a gaussian distribution of absolute magnitude around M_0 with a dispersion of 0.75 mag., to take into account the variation of the mean absolute magnitude within the interval of $B-V$. We used a classical $(m, \log r)$ table technique (Schouten, 1914) to derive the predicted star counts.

Table II gives the predicted star counts as well as those observed by Weistrop over an area of 13.5 sq deg in three $B-V$ color ranges. The results can be summarized in two conclusions: (1) At bright apparent magnitudes, $m_v \simeq 14$, the observations show an excess that increases steeply with color; while for $B-V = 1.45$ there is little discrepancy, we find the ratio observed/predicted to be a factor of 6 for $B-V = 1.55$, and a factor of 20 for $B-V = 1.65$, and (2) the ratio observed/predicted decreases for increasing apparent magnitude at a given color. Conclusion (1) suggests that the actual luminosity function shows an increasing excess over Van Rhijn's function for

$B - V > 1.5$ or $M_v > 11$, while (2) indicates that the scale height for these stars is smaller than the assumed value of 300 pc. These conclusions, based on a direct confrontation of the expected and observed numbers in Table II, are qualitatively similar to those obtained by Weistrop.

The main uncertainty in the densities is related to the absolute magnitudes. If we adopt for the three color ranges of Table II absolute magnitudes M_0 of 10, 11, and 12 instead of 10, 11.5, and 13, respectively, then the predicted star counts would change by factors of $\frac{2}{3}, \frac{4}{3}$, and $\frac{8}{3}$, respectively. Apparently such a substantial revision of the absolute magnitude calibration would do little to explain observed excesses in the counts by factors of 6 and 20, for $B - V = 1.55$ and 1.65, respectively.

An entirely different approach was taken by Murray and Sanduleak (1972) who investigated proper motions of 21 Sanduleak survey stars near the galactic north pole. They found: (1) none of the stars could be giants since all had distinctly non-zero motions, (2) the dispersion of velocities parallel to the galactic plane corresponds to about ± 10 km s^{-1}, and (3) the mean parallax together with the statistics of the Sanduleak stars results in a number density of 0.23 stars per cubic parsec down to $M_v = 13$, in good agreement with Weistrop's luminosity function.

The small velocity dispersion found by Murray and Sanduleak is compatible with the small vertical scale height derived by Weistrop. The derivation of the space density from the proper-motion work has been done in a somewhat approximate fashion and I believe that the results based on the star counts deserve rather more weight. Nevertheless, the agreement between the two investigations is remarkable.

Pesch (1972) undertook a spectroscopic study of M stars from the Sanduleak objective prism survey. He considered only stars that showed no proper motion larger than $0.''05$ per year. Only two of the 17 stars fainter than $V = 12.5$ were found to be giants. The published $B - V$ colors allow an estimate of absolute magnitude and distance for each of the dwarfs. I find that the 12 dwarfs within 100 pc have transverse velocities less than around 15 km s^{-1}. These no-proper-motion stars constitute about half of the Sanduleak stars according to Pesch. All these results are compatible with those of Murray and Sanduleak.

The transition from the large to the small scale height appears to occur at a $B - V$ color of around 1.50. The corresponding spectral type of M2 or M3 is somewhat uncertain since colors and spectral types do not correlate well for early M-type dwarfs. Few stars later than M3 are contained in the McCormick objective-prism survey (Vyssotsky, 1956, and earlier papers) and this may explain the apparent discrepancy noted by Gliese (1974) between the velocity dispersions of the Pesch stars and the McCormick stars. Similarly, the McCarthy-Treanor (1964) objective-prism survey of a region near the Pleiades contains no stars later than M3. Hence, the relatively large velocity dispersions between 17 and 32 km s^{-1} derived by B. F. Jones (1972) for M stars in this survey may be understood.

D. H. P. Jones (1973) has carried out a photoelectric narrow-band classification of M-type stars found in an objective-prism survey near the south galactic pole by McCarthy et al. (1964). He finds a space density only one-fifth of that found by

Murray and Sanduleak, and derives a dispersion of peculiar velocities of ± 36 km s^{-1}. The disagreement with the results of Weistrop and of Murray and Sanduleak is remarkable. B. F. Jones (1972) had noted that the McCarthy Pleiades survey is only one-third complete. If this applies to the South Pole survey too, then a substantial correction is in order. As noted by D. H. P. Jones (1973) the McCarthy stars are concentrated to early M subtypes whereas the Sanduleak stars have a more even distribution of subtype. Hence, the McCarthy survey may be particularly incomplete for later types, which are the ones that show a high density and small peculiar velocities.

I have not considered in this review investigations based on catalogs of stars with large proper motion since these carry little information about the slow-moving late M-type dwarfs.

The total evidence points to the existence of a rich population of late M-type dwarfs with small velocity dispersion. The small velocity dispersion leads to two problems, however. The first problem concerns the age of these objects. It has been generally assumed that the large velocity dispersion of the K-type dwarfs is caused by gravitational encounters with large gas clouds (Spitzer and Schwarzschild, 1953) or perhaps density waves connected with spiral structure. If the late M-type dwarfs are as old as the K-type dwarfs they should have suffered the same dynamical history. The discrepancy in the velocity dispersion suggests that the late M-type dwarfs are much younger, perhaps around one billion years. However, these stars carry two-thirds of the local mass density and it seems inconceivable that so large a fraction of the interstellar gas condensed into stars as recent as one billion years ago. Hence, we have to assume that the Spitzer-Schwarzschild mechanism does not operate and then we have to find a new explanation for the observed increase in the velocity dispersion along the main sequence from types A to K.

The second problem arising from the existence of the slow-moving M-type dwarfs concerns the stability of the disk against axi-symmetric perturbations. For a surface density of around 60 solar masses per pc^2 the stability criterion established by Toomre (1964) requires a velocity dispersion of at least 28 km s^{-1}. This is much higher than the observed value around 10 km s^{-1} and as shown by P. Biermann in the preceding communication there is no possibility to satisfy this criterion with a mixture of populations.

In summary, the solution of one problem (that of the missing mass) has created two new problems. Further studies, both observational and theoretical, are called for. Perhaps in this confrontation between observation and theory, new insights in the nature and evolution of our Galaxy will be generated eventually.

References

Gliese, W.: 1974, *Astron. Astrophys.* **34**, 147.
Jones, B. F.: 1972, *Monthly Notices Roy. Astron. Soc.* **159**, 3P.
Jones, D. F. P.: 1973, *Monthly Notices Roy. Astron. Soc.* **167**, 19P.
Luyten, W. J.: 1938, *Publ. Astron. Obs. Univ. Minnesota* **2**, No. 7.

Malmquist, K. G.: 1927, *Lund Obs. Medd., Ser. II*, No. 46.

McCarthy, M. F., Bertiau, F. C., and Treanor, P. J.: 1964, *Ricerche Astron. Vaticana* **6**, 571.

McCarthy, M. F. and Treanor, P. J.: 1964, *Ricerche Astron. Vaticana* **6**, 535.

Murray, C. A. and Sanduleak, N.: 1972, *Monthly Notices Roy. Astron. Soc.* **157**, 273.

Oort, J. H.: 1965, *Stars and Stellar Systems* **5**, 455.

Pesch, P.: 1972, *Astrophys. J.* **177**, 519.

Sanduleak, N.: 1964, *Astron. J.* **69**, 720.

Schouten, W. J. A.: 1918, dissertation, Groningen University.

Spitzer, L. and Schwarzschild, M.: 1953, *Astrophys. J.* **118**, 106.

Toomre, A.: 1964, *Astrophys. J.* **139**, 1217.

Van Rhijn, P. J.: 1936, *Publ. Kapteyn Astron. Lab. Groningen*, No. 47.

Veeder, G. J.: 1974, *Astrophys. J. Letters*, **191**, L57.

Vyssotsky, A. N.: 1956, *Astron. J.* **61**, 201.

Weistrop, D.: 1972a, *Astron. J.* **77**, 366.

Weistrop, D.: 1972b, *Astron. J.* **77**, 849.

DISCUSSION

Lecar: How does this data look expressed as number per unit mass (e.g. as compared with Salpeter function $n(m) \simeq m^{-2.3}$)?

Schmidt: The overall character of the luminosity function is like that of a Salpeter power law but there are systematic deviations as a function of mass, I believe.

Lecar: What masses are associated with $M_0 = 10$, 11.5, 13?

Schmidt: Of the order of half a solar mass.

Spitzer: Should one at least consider the possibility that these excess stars are not so far away as giants would be but somewhat further away than assumed? This assumption would place the stars somewhat above the main sequence and would give them a more nearly normal dispersion of velocities.

Schmidt: If these stars were well above the main sequence, then it would be surprising that they have not shown up in the parallaxe work, on which the relation between M_v and $B - V$ is based.

King: Recently two Berkeley students, David Koo and Richard Kron, have determined radial velocities for 5 of the 21 Murray-Sanduleak stars. Their velocity dispersion is 10 km s^{-1}. This tends to confirm the low velocity dispersion that has been suggested. In conjunction with the proper motions of Murray and Sanduleak, this also tends to confirm the conventional absolute magnitudes that they assumed. These stars have no Hα emission, which suggests that they are not young. Their spectra have dwarf characteristics.

Other programs are under way at Berkeley, extending Weistrop's study to fainter magnitudes, and measuring proper motions for the same stars. In using lists of stars found by spectral surveys, one should ask what spectral region was used. Blue surveys, like the McCormick survey, discriminate against late-M dwarfs, because of the steep spectral gradient at Ca I 4227, on which the classification depends. Sanduleak's survey was done in the red and was thus better able to pick up late M's.

THE ROTATION CURVES OF GALAXIES

M. S. ROBERTS

National Radio Astronomy Observatory*, Green Bank, W. Va., U.S.A.

Abstract. Currently available data on rotation curves are reviewed. For curves derived from optical measurements the distribution of the ratios: the last measured point on a rotation curve to the optical radius of the galaxy has a median value of $\sim\frac{1}{2}$ if Reference Catalogue radii are used and $\sim\frac{1}{3}$ if Holmberg radii are used. It is the absence of easily measurable H II regions that so severely limits the extent of these rotation curves. Accordingly, little can be said of the dependence of V_c on R for large R, where R is comparable to a Holmberg radius. The assumption that a rotation curve approaches a Keplerian curve after passing its peak rotational velocity implies a strongly concentrated and limited extent of the mass distribution within a galaxy. This assumption is not supported by 21-cm observations of the velocity field within a galaxy. Because of the greater extent of H I compared to measurable optical (blue) surface brightness, rotation curves may be defined to much larger radii from 21-cm observations. The median value of the above ratio for 14 galaxies is 1.3. At least 7 of these galaxies show an essentially constant rotational velocity at large R, while 5 galaxies have a slowly decreasing $V_c(R)$. For both types of curves, a significant surface mass density at large R is required, and a large ($\gtrsim 100$) mass-to-luminosity ratio is indicated. Such values are consistent with a late dwarf M star population (the most common type of star in the solar neighborhood) in the outer regions of a galaxy.

The first detections of rotation within a galaxy (Slipher, 1914; Wolf, 1914) were made a decade before the definite establishment of the extragalactic nature of many of the nebulae (Hubble, 1924). In the intervening 60 years rotation curves for fewer than 100 galaxies have been measured, a rate of less than two per year. Before the advent of image tubes (and the vast majority of the available data are from this earlier era), the process was difficult and time consuming. An early effort to gather information on the rotation of the inner region of a number of galaxies was made by Mayall (Mayall, 1960; Mayall and Lindblad, 1970). In obtaining spectra of galaxies for redshift measurements he purposefully avoided trailing the image of the galaxy on the spectrograph slit and was able to measure line inclinations for over 50 systems.

But it is the detailed variation of velocity with position over as large an angular extent as possible that is desired. Until the late 1950's such information was available for only a few galaxies. At this time a major observing effort was initiated by the Burbidges. Collaborators and independent observers added to this effort, and much of the information on rotation curves that we now have dates from this era. Another technique for studying the kinematics of galaxies was developing during this same period: observation of the 21-cm spectral line with radio telescopes. This approach has been remarkably improved over this last 15 years through the development of low-noise receivers, the construction of large-filled aperture telescopes, and the application of aperture synthesis techniques.

Examples of all types of galaxies show rotation. For spirals this has long been recognized. Irregular-type galaxies (Ir I) have a chaotic appearance but both optical and 21-cm observations show a well-ordered radial velocity field and well-defined rota-

---

* Operated by Associated Universities, Inc., under contract with the National Science Foundation.

Hayli (ed.), Dynamics of Stellar Systems, 331–340. All Rights Reserved.

tion curves (e.g., NGC 1569, NGC 6822, Ho II, SMC, WLM). The irregulars of type II, such as M82 (Mayall, 1960), also show rotation. There are several examples of rotation within elliptical galaxies, although few such systems have been studied. NGC 4621 (E5) and NGC 4697 (E6) both show inclined spectral lines indicating rotational velocities of 13 and 29 km s^{-1} (arcsec)$^{-1}$, respectively (King and Minkowski, 1965). These measurements cover a diameter of $\sim 10''$. Bertola (1972) has extended the measurements for NGC 4697 out to $\sim 40''$ (radius) and finds a turnover in the rotation curve near 25''. C. Peterson (1975) also finds rotation in a number of elliptical galaxies including the E1 system NGC 3379. Another example, though possibly special, is the high angular velocity of the nucleus of M32, 78 km s^{-1} per 10 pc (Walker, 1962).

The bulk of the available data are for spiral galaxies. It is only this category of galaxian rotation curves that are discussed further.

As a measure of the distance to which rotation curves extend, the distribution of the ratio: furthest measured point to optical radius for a number of galaxies is shown in Figure 1. Galaxies in which the observations were restricted to the nuclear region are omitted. The bottom histogram in this figure uses diameters from the Reference Catalogue of Galaxies (de Vaucouleurs and de Vaucouleurs, 1964). A number of these galaxies, shown cross hatched, have diameters measured to a fainter isophote (Holmberg, 1958). These galaxies are replotted in the central histogram using the Holmberg radii to form the ratio. For the first, and smaller, photometric radii the median value of the above ratio is $\sim \frac{1}{2}$. For the Holmberg radii, obviously more realistic since the galaxies extend *at least* out to this distance, the ratio has a median value of $\sim \frac{1}{3}$. Only one-half the sample of measured rotation curves extend beyond $\frac{1}{3}$ of their photometric radius. And for these the majority are less than $\sim \frac{1}{2}$ the Holmberg radius.

We thus find that the available rotation curves derived from optical studies refer to only a fraction of the photometric extent of a galaxy. The description of the rotation curve beyond the last measured point can only be assumed. Such assumptions may be based on a particular model, such as a constant mass-to-luminosity ratio or an even more extreme case: that the galaxy essentially ends beyond the peak of the rotation curve and the declining branch will quickly approach a Keplerian curve.

Essentially all rotation data are based on emission lines: [O II], Hα, [N II], [S II], H I. Absorption lines are generally too faint and subject to possible systematic effects although they are used in the measurement of line inclinations for rotation periods of the central section of a galaxy. The 21-cm data are especially valuable for measurable H I radiation is found at a much larger radial distance than optical emission lines. Thus far, only radio measurements are able to supply information on the rotation of the outer part of a galaxy.

In the sample of optically-studied galaxies M31 (Rubin and Ford, 1970) extends the furthest in linear extent, 24 kpc, a value 1.2 the Holmberg radius. Their derived rotation curve shows no indication of a decline in the circular velocity over the outermost 4 kpc (Rubin and Ford, 1970, especially page 389). A similar finding of constant circular velocity, but now out to 30 kpc, appears in 21-cm observations of M31 (Roberts and Rots, 1973; Roberts and Whitehurst, 1975). The ratio of this distance to the Holmberg diameter is 1.5. The greater extent of the neutral hydrogen compared to Holmberg's limiting blue isophote (26^{m}5 per sq arcsec) has long been recognized and is clearly shown by the top histogram in Figure 1 where 11 of 14 galaxies have a ratio greater than one. As noted earlier, it is this more extensive H I distribution that allows rotation curves to be derived to angular distances compa-

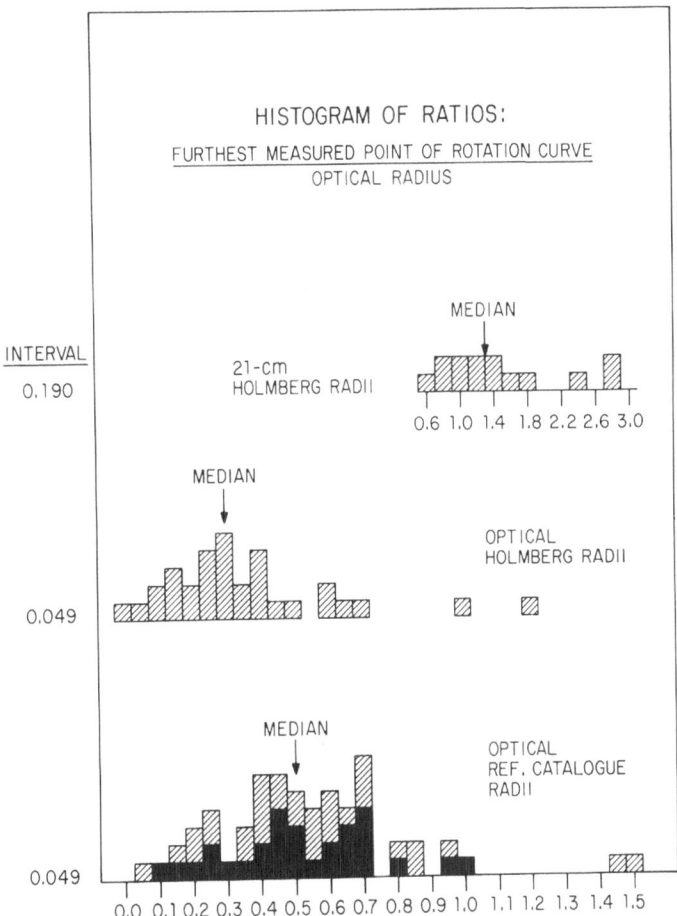

Fig. 1. Distribution of the extent of rotation curves in terms of optical radii. The *bottom histogram* is based on optical determinations of rotation curves and radii taken from the *Reference Catalogue of Bright Galaxies*. Those galaxies which also have radii measured by Holmberg are indicated by hatched lines; they are represented in the *center histogram* using the Holmberg radii to form the ratio. The *top histogram* is based on rotation curves derived from 21-cm measurements and Holmberg radii.

rable to or greater than the Holmberg radius. A limiting factor in such work is the relative resolution: beam size/H I size, available in 21-cm observations. Filled aperture, circular half-power beam widths of 9' (100-m telescope) and 10' (300-ft telescope) are now available. The recently resurfaced Arecibo antenna will have a 4' beam. The

TABLE I

Galaxies studied by 21-cm techniques to large radial distances

NGC, IC	Messier	Last point on rotational curve		Type	Holmberg radius (arcmin)	Description of outer part of rot. curve	Reference
		(arcmin)	(kpc)				
224	31	150	30	Sb	98.5	Flat	Roberts and Whitehurst (1975)
253		11	10.9	Sc	14.2[a]	Flat	Huchtmeier (1972)
598	33	60	12.6	Scd	41.5	Decreasing[c]	Huchtmeier (1973a)
2403		12.8	12.1	Scd	14.5	~Flat	Shostak and Rogstad (1973)
3031	81	33.9	32.1	Sab	17.5	Decreasing slowly	Roberts and Rots (1973)
3109		26.4	16.9	Im	9.2[a]	~Flat	Huchtmeier (1973b)
4236		12.6	11.9	Sdm	13.0	~Flat	Shostak and Rogstad (1973)
4244		10.0	11.1	Scd	9.0	Slight decrease	Huchtmeier (1973c)
5236	83	20.0	51.8	Sc	7.1[a]	Slight decrease	Rogstad et al. (1974)
6946		11.0	32.3	Scd	9.5[b]	~Flat	Rogstad et al. (1973)
5457	101	36.0	72.3	Scd	14.0	Slight decrease	This paper
7640		9.4	12.0	Sc	6.8	Approaching max. V_c	Seielstad and Wright (1973)
IC 342		34.0	44.5	Scd	19.8[b]	Flat	This paper
IC 2574		10.0	9.5	Sm	8.0	Approaching max. V_c	Seielstad and Wright (1973)

[a] Radius converted to Holmberg system from Roberts (1969) Equation (2).

[b] Radius converted to Holmberg system from surface photometry by Ables (1971).

[c] The published data for M33 show an approximately flat rotation curve. Recent unpublished observations by Hutchmeier with the 100-m telescope of positions at larger R show a decrease in V_c.

Nancay telescope has a fan-beam of $4' \times \sim 25'$ (the latter dimension is dependent on the declination of the source). These instruments are particularly sensitive to low surface brightness radiation, i.e., low H I column density, and are therefore particularly suited for such measurements. Their relatively large beams limit the number of galaxies that can be studied for this purpose. Successful observations with synthesized beams as small as 25" have been made at Westerbork. The Cal Tech interferometer has been used to study a number of galaxies with an effective resolution of $2'$ and the Cambridge $\frac{1}{2}$-mile interferometer with comparable resolution. Unless an inordinate amount of observing time is used, these instruments are seriously limited in the surface brightness that is measurable. Instrumental effects may also be greater: in addition to the side lobes of the main beam (common to all telescopes) there are side lobes of the synthesized beam as well as a grating response. However, the relatively high angular resolution that is attainable is ideal for studies of the kinematics and hydrogen distribution in the inner region of a galaxy, i.e., out to ~ 1 Holmberg radius.

   Twenty-one centimeter data based on good relative resolution are available for 14 galaxies listed in Table I. We exclude the Magellanic Clouds where the data are of high quality and extensive but in the present context ambiguous (Hindman, 1967; McGee and Milton, 1966), and IC10 (Shostak, 1974) where only a small velocity gradient is observed. Several of the 14 systems have been studied by more than one group, only the most complete or extensive references are included.

   Examples of some of these rotation curves are shown in Figures 2 (M31), 3 (M101), and 4 (IC 342). The first is taken from Roberts and Whitehurst (1975) and shows for $R > 12$ kpc the optically measured data points (Rubin and Ford, 1970) and the 21-cm data points. For M101 and IC 342 recently derived data obtained with the 300-ft

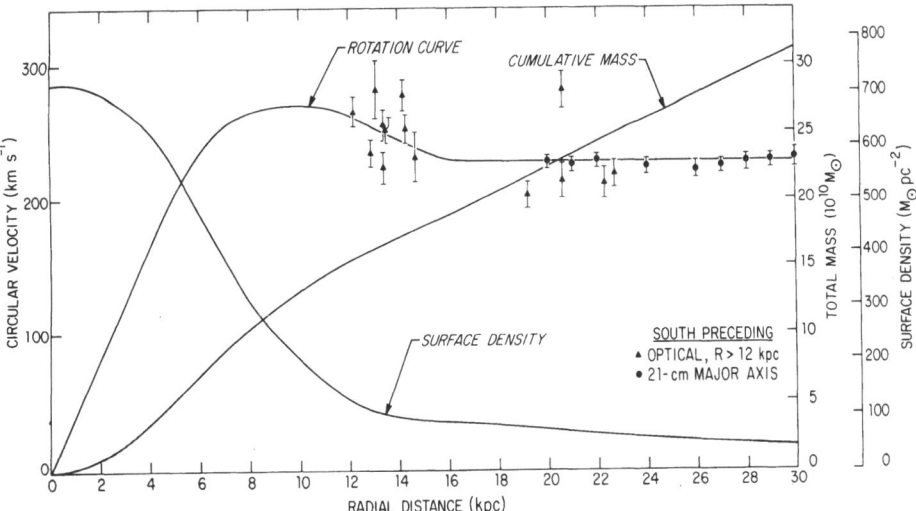

Fig. 2. The rotation curve for M31. For simplicity the inner peak of the rotation curve near 0.8 kpc is omitted. The filled triangles are optically determined rotational velocities from Rubin and Ford (1970) for the outer 12 kpc. The filled circles are 21-cm measurements made with the 300-ft telescope. The surface density and cumulative mass curves are for a thin disk model.

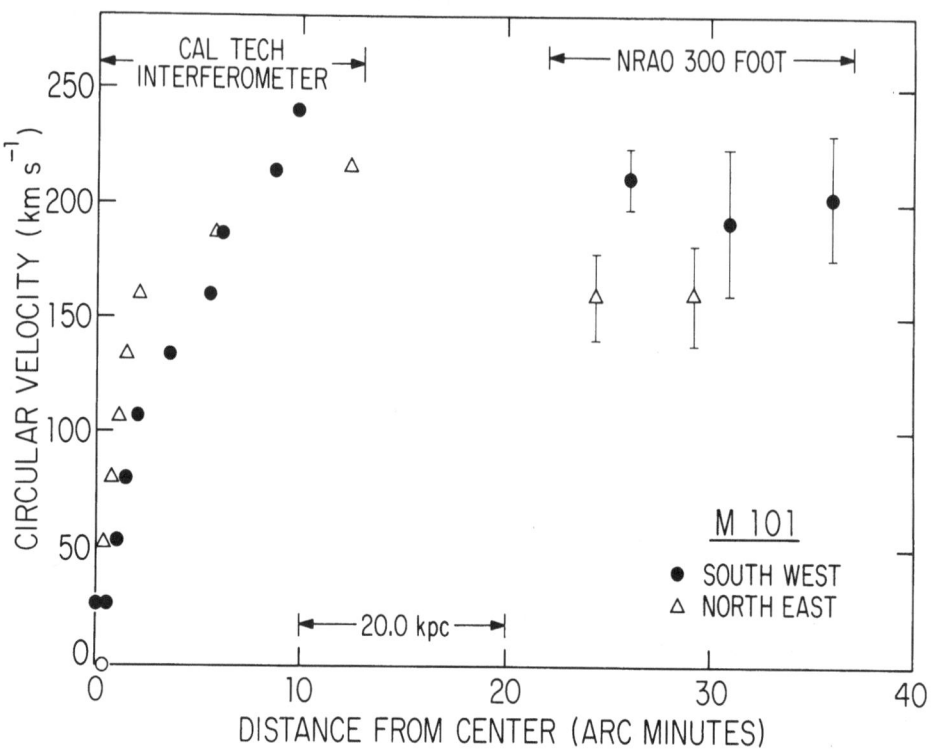

Fig. 3. The rotation curve for M101. All data points are based on 21-cm observations. The inner 12′ are from interferometer measurements by Rogstad and Shostak (1971) and are for major axis values in their Figure 4. The outer points are from recent measurements made with the 300-ft telescope.

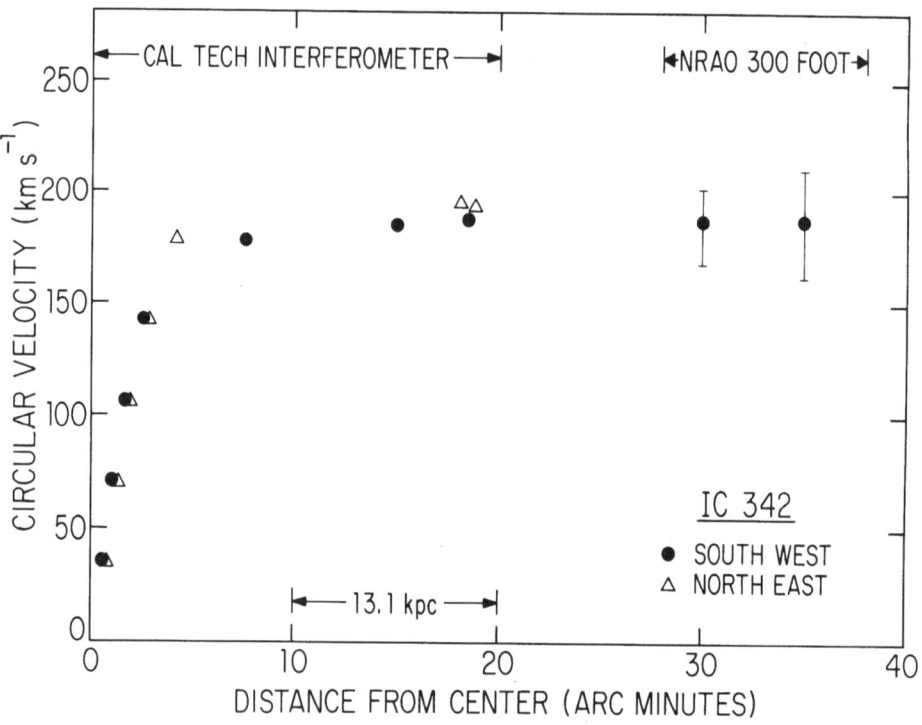

Fig. 4. The rotation curve for IC 342. All data points are based on 21-cm observations. The inner 19′ are from interferometer measurements by Rogstad et al. (1973) and are for major axis values in their Figure 5. The outer points are from recent measurements made with the 300-ft telescope.

telescope are added to inner portions of rotation curves obtained with the Cal Tech interferometer (Rogstad and Shostak, 1971; Rogstad *et al.*, 1973). At least 2 of the 3 curves show a relatively constant circular velocity at large R. Of the 14 galaxies, 7 show an essentially constant rotational velocity at large radius, near or beyond the Holmberg radius. In some instances there is even a suggestion that after the flat portion of the rotation curve there may be an *increase* in V_c with R; M31 is one such example. Of the remaining seven rotation curves, two have not yet reached a turn-over at the maximum velocity, NGC 7640 and IC 2574. The other five rotation curves show, in varying amounts, a gradual decrease. This is most pronounced in M81, an Sab system and the earliest type galaxy in Table I. It is important to note that this sample is heavily weighted to late-type systems. Only 2 of the 14 are earlier than Sc; these are M31 (Sb) and M81 (Sab).

For any specific case, there are a number of possible observational effects that can be invoked to account for the flat rotation curves. The most obvious relates to side lobes sensing radiation from a region away from the direction of the primary beam. However, for the 300-ft observations of M31, the side lobes are carefully measured to -30 dB. At a position 140 arc min along the south-preceding major axis, such side lobe contributions are $\sim 3\%$ of the observed signal. The velocity profile at this position obtained with the 140-ft telescope, the Jodrell Bank 250-ft telescope (Gottes-man and de Jager, 1970), as well as a profile constructed from maps obtained with the Nancay antenna (Guibert, 1970) all yield velocities in substantial agreement with that obtained with the 300-ft telescope. We further note that among the various gal-axies with flat rotation curves, some were derived with the Cal Tech interferometer, some with the Nançay tiltable plane-standing parabola, and some with the NRAO 300-ft antenna. And equally important for possible instrumental effects the sample of galaxies obviously covers a wide range of position angles and angular sizes. Never-theless, the Cambridge group (Emerson and Baldwin, 1973) derive a rotation curve for M31 which is slowly decreasing and thus differs from the rotation curve derived from 300-ft data. It is difficult to reconcile their rotation curve with the one in Fig-ure 2. The outer and most critical points defining their curve fall at the edge and even outside the velocity range in which a signal is measured with the 300-ft telescope. Angular resolution effects due to the different size beams will not account for these differences. The data at the representative point of 140 arcmin along the south-pre-ceding major axis obtained with 4 different filled aperture telescopes are consistent. We conclude that Figure 2 is indeed descriptive of the motion in the outer part of M31.

(Observations of M31 made with the 100-m telescope after this paper was pre-sented at Besançon are in complete agreement with the 300-ft data. The Emerson and Baldwin (1973) rotation curve appears to be based on spurious data even though the 'error' bars on the points defining their rotation curve are remarkably small.)

The 14 rotation curves in Table I are derived from radiation arising from the general interstellar medium. The optical data refer (primarily) to discrete H II regions. Could these two subsystems of a galaxy have different kinematic properties. On a small scale (parsecs) this is possible; on the dimensional scale considered here it is very unlikely. Support for this latter conclusion comes from detailed comparison of H I and H II velocity measurements, e.g., SMC, LMC, M31 and M81. Both neutral and ionized hydrogen may be considered as equivalent test 'particles' of the kinematic field within a galaxy.

The dependence of total mass with radius R, $M(R)$, varies as $\sim R$ for those cases and regions where the rotation curve has a constant circular velocity. But at least $\frac{1}{3}$ of the galaxies studied thus far have V_c decreasing with R, albeit very gradually. In all cases there is a significant contribution to the total mass from large R, a component whose optical surface brightness is very small, indicating a large mass-to-light ratio in the outer parts of the galaxies studied here. Although H I extends well beyond the Holmberg radius for these galaxies, its surface density is only $\sim 1\%$ of the total mass surface density implied by the rotation curves. The most common type of star in the solar neighborhood, intermediate and late dwarf M stars could easily satisfy both the mass surface density requirement and the upper limits of the optical surface brightness. As an example, in M31 at 28 kpc from the center, the surface density (as computed for a thin disk model) is $\sim 50\ M_\odot\ \mathrm{pc}^{-2}$. This location is beyond the limit of de Vaucouleurs's (1958) photometry of M31 but an extrapolation of his data indicate $M/L_{pg} \simeq 400$. The implication of a change in the luminosity function with radial distance from the center of a galaxy need not be alarming. We know this to be the case by the very appearance of spiral arms (see also McCuskey, 1965). In the solar vicinity the shape of the luminosity function changes significantly with height above the galactic plane (Bok and MacRae, 1941). This change is such that at high z (1 kpc) there are relatively more late type stars per volume of space. This is the sense of the change required to explain the shape of the rotation curves at large R.

The picture that emerges from these data is one in which spiral galaxies are significantly more extensive than indicated by blue limiting exposures. The projected surface density in these outer regions decreases approximately linearly with distance. The necessary mass and the implied mass-to-luminosity ratio can be attributed to those stars found near the peak of the local luminosity function, late-type dwarfs. This is not a unique explanation; it is the simplest and is susceptible to observational confirmation through near infrared observations.

References

Ables, H. D.: 1971, *Pub. U.S. Naval Obs.* 2nd Series, **2**, No. 4, Washington, D.C.
Bertola, F.: 1972, *Osservatorio Astronomico di Padova Comunicazioni E Rassegne*, No. 98.
Bok, B. J. and MacRae, D. A.: 1941, in E. M. Schlaikjer (ed.), *The Fundamental Properties of the Galactic System*, *Annals of the New York Academy of Sciences* **42**, 219.
Emerson, D. T. and Baldwin, J. E.: *Monthly Notices Roy. Astron. Soc.* **165**, 9P.
Gottesman, S. T. and de Jager, G.: 1970, *Mem. Roy. Astron. Soc.* **74**, 67.
Guibert, J.: 1973, *Astron. Astrophys. Suppl. Series* **12**, 263.
Hindman, J. V.: 1967, *Australian J. Phys.* **20**, 147.
Holmberg, E.: 1958, *Medd. Lund*, Ser. II, No. 136.
Hubble, E.: 1924, Paper read at the 33rd Meeting of the American Astronomical Society. Abstracted in *Popular Astron.* **33**, 252, 1925.

Huchtmeier, W.: 1972, *Astron. Astrophys.* **17**, 207.

Huchtmeier, W.: 1973a, *Astron. Astrophys.* **22**, 91.

Huchtmeier, W.: 1973b, *Astron. Astrophys.* **22**, 27.

Huchtmeier, W.: 1973c, *Astron. Astrophys.* **23**, 93.

King, I. R. and Minkowski, R.: 1966, *Astrophys. J.* **143**, 1002.

McCuskey, S. W.: 1965, *Stars and Stellar Systems* **5**, 1.

McGee, R. X. and Milton, J. A.: 1966, *Australian J. Phys.* **19**, 343.

Mayall, N. U.: 1960, *Ann. Astrophys.* **23**, 344.

Mayall, N. U. and Lindblad, P. O.: 1970, *Astron. Astrophys.* **8**, 364.

Peterson, C.: 1975, 'Stellar Motions in Elliptical Galaxies', Univ. of Calif., Berkeley (Ph.D. Thesis).

Roberts, M. S.: 1969, *Astron. J.* **74**, 859.

Roberts, M. S. and Rots, A. H.: 1973, *Astron. Astrophys.* **26**, 483.

Roberts, M. S. and Whitehurst, R. N.: 1975, *Astrophys. J.* (submitted).

Rogstad, D. H. and Shostak, G. S.: 1971, *Astron. Astrophys.* **13**, 99.

Rogstad, D. H., Lockhart, I. O., and Wright, M. C. H.: 1974, *Astrophys. J.* (in press).

Rogstad, D. H., Shostak, G. S., and Rots, A. H.: 1973, *Astron. Astrophys.* **22**, 111.

Rubin, V. C. and Ford, K. W., Jr.: 1970, *Astrophys. J.* **159**, 379.

Seielstad, G. A. and Wright, M. C. H.: 1973, *Astrophys. J.* **184**, 343.

Shostak, G. S.: 1974, *Astron. Astrophys.* **31**, 97.

Shostak, G. S. and Rogstad, D. H.: 1973, *Astron. Astrophys.* **24**, 405.

Slipher, V. M.: 1914, *Lowell Obs. Bull.* No. 62.

Vaucouleurs, G. de: 1958, *Astrophys. J.* **128**, 379.

Vaucouleurs, G. de and Vaucouleurs, A. de: 1964, *Reference Catalogue of Bright Galaxies*, Univ. of Texas Press, Austin.

Walker, M. F.: 1962, *Astrophys. J.* **136**, 695.

Wolf, M.: 1914, *Vierteljahrsschieft der Astr. Gesellsch.* **49**, 162.

DISCUSSION

Freeman: Did you say a surface density of 50 $M_\odot$ pc^{-2}?

Roberts: Yes.

Freeman: This means an M/L way over 100 to go below de Vaucouleurs photometry.

Roberts: The mass-to-light ratio is indeed high, several hundred. Note that I use photographic luminosity in these considerations.

Bertola: Did you find any relation between the shape of the rotation curve and the behaviour of M/L within the galaxy?

Roberts: There is a correlation of the general form of a rotation curve with type, earlier galaxies are more centrally concentrated. But there are too few curves as yet, to look for a shape vs M/L relation.

Baldwin: What are the Holmberg radii in NGC 2403 and 4236 relative to the limits of the H I observations at about 12′ radius.

Roberts: For both galaxies the observed rotation curves extend to just less than a Holmberg radius.

Gott: As far as extrapolations of the rotation curves are concerned. The total mass of M31 may be estimated by an examination of the dynamics of our galaxy and M31, namely that they have just turned around and are approaching each other due to their mutual gravitational attraction. This fact together with the distance to M31 gives a total mass for the Milky Way plus M31 system. This indicates a $(M/L)_B \sim 100$ consistent with your values and would allow a continuation of a flat rotation curve to approximately twice the distance shown by the radio data so far.

King: Larger masses for galaxies would help to solve the other great 'missing mass' problem, that in clusters of galaxies. Evidence is accumulating that in the Coma cluster, where about a factor of 8 is needed, the missing mass is actually in the galaxies.

Lynden-Bell: Smart, a student at Cambridge looking at 11 So satellites of one of the giants in Coma points out that they form an isolated system of So's and gives a mass of 2×10^{13} $M_\odot$ for the giant. The So's are essentially embedded in the giant so there is no doubt that at least this mass is in the giant galaxy.

Gott: I might mention two additional pieces of information which also support the viewpoint that the mass resides primarily in the galaxies. Some time ago Dr Gunn and I estimated the mass of the two largest galaxies in the Coma cluster. These galaxies dominate the center of the cluster and therefore can be treated as a binary system and their masses estimated from their separation and relative radial velocities. These estimates also give several times 10^{13} $M_\odot$ for the masses of these galaxies in agreement with your figure. A second point is that Gus Omeler has recently done a study of the degree of central condensation of galaxies in the Coma cluster and found that for the brighter galaxies there was greater central concentration, but for the fainter galaxies there was no effect. Comparison with models by Aarseth showed that this was just the effect expected from two-body relaxation if the brightest galaxies had mass to light ratios equal to the total mass to light ratio for the cluster.

Lecar: Theoretical models by Gott, Larson of formation of galaxies have difficulty obtaining space density fall offs as shallow as $r^{-2.5}$ to r^{-3}. At this time, no model has produced $\varrho \sim r^{-2}$ (isothermal sphere) or $M \sim r$ as indicated by the flat rotation curves Roberts has just shown us.

M/L RATIOS IN GALACTIC DISKS

J. E. BALDWIN

Cavendish Laboratory, Cambridge, U.K.

Abstract. Models of galactic disks based on optical photometry and uniform M/L ratios have rotation curves which are indistinguishable from those derived from 21 cm hydrogen line observations.

What is the distribution of mass in spiral galaxies? On current evidence there are many answers to this question. So it is of interest to consider the much more limited question "Are the present observational data consistent with the view that the mass-to-light ratios in the disks of spiral galaxies are constant throughout any particular galaxy?" I believe that the published data indicate that the answer is yes. Morton Roberts has already explained that we are in disagreement about this conclusion. Our differences are mainly ones of interpretation rather than of observation and it is appropriate to examine first the basis of many reports in recent years of mass-to-light ratios which increase with radius in the outer part of spiral galaxies. These reports fall into three categories:

(a) A rotation curve has been used to derive a model of the mass distribution which, combined with the results of photometry, leads to an M/L ratio increasing with radius. Examples are M33 (Brandt, 1965; Boulesteix and Monnet, 1970), M81 (Brandt *et al.*, 1972), and M31 (Gottesman and Davies, 1970).

(b) The observed rotation curve shows no turnover point. There are many examples in the published literature.

(c) The outer parts of the rotation curve are flat, even though there may be a peak in the curve at a smaller radius. Roberts and Rots (1973) stressed this point for both M31 and M81. The flat rotation curve implies a total mass diverging as r whereas the total luminosity of spiral galaxies converges rapidly with r.

In contrast to these claims, Freeman (1970) found that in disk galaxies with essentially no spheroidal component, the rotation curves are consistent with exponential disks having uniform M/L. Nordsieck (1973) confirmed this result for galaxies for which there are good optical rotation curves. Our radio data from Cambridge on M33 (Warner *et al.*, 1973) and M31 (Emerson and Baldwin, 1973) indicated the same conclusion.

The cases which fall in category (a) arise mainly from fitting model rotation curves of the Brandt type to the observations. For the curve fitting procedure used in the analysis, this type of curve may be as good as any other (although this is discussed at some length by Nordsieck) but it is well known both that the exact curve fitted depends critically on the position of the outermost measured points in the rotation curve and that the fitted curve is often used at radii beyond the last observed point. This has frequently led to very dramatic increases of M/L with radius in the outer parts of galaxies. This is apparent, for instance, in Brandt's work (1965) on M33.

Hayli (ed.), Dynamics of Stellar Systems, 341–348. All Rights Reserved.

There is no evidence for it being a real effect and I shall not discuss claims of class (a) any further.

In discussing (b) and (c) we shall consider mainly those galaxies for which there is good photometry and good observations of radial velocities which extend to large radial distances. In all cases these depend on the 21 cm hydrogen line results and for reasons of angular resolution they are all very nearby. The galaxies are M33, NGC 6946, IC 342, M31 and M81.

We now examine claims of type (b), that some rotation curves do not show turn-over points, with the implied consequence that there must be large amounts of matter in the outer parts of galaxies for which there is no visible counterpart. Without some idea of where the turn-over point is expected to be, this claim is not a strong one. Take a specific model in which M/L is uniform throughout the disk. Freeman (1970) has shown that a thin exponential disk (a very good description of the luminosity distribution in all disks) having a surface mass density σ of the form $\sigma = \mu_0 e^{-\alpha r}$ has a rotational velocity V_{max} at the turn-over point in its rotation curve at radius $r = 2.15$ α^{-1}. At $r = 4.0 \, \alpha^{-1}$ the rotational velocity has dropped only to $0.9 \, V_{max}$. In fact the disks of spiral galaxies are not infinitesimally thin and have axial ratios which are typically 0.2. The rotation curve for a disk of this type, retaining the exponential fall off in projected surface density, is of very nearly the same shape but with different scale factors in r and V_{max}. Analysing such a disk into homogeneous spheroidal shells

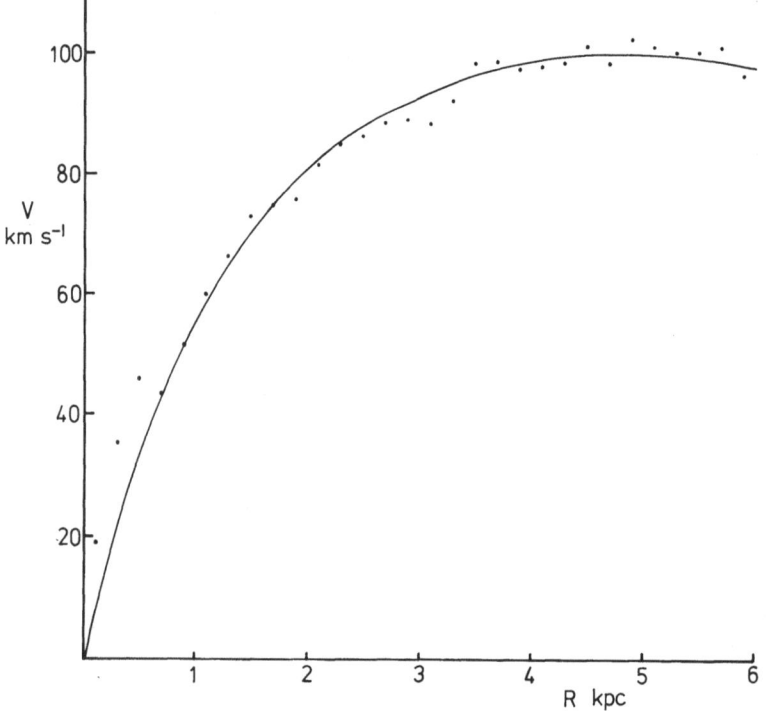

Fig. 1. The rotation curve of M33, uncorrected for inclination. Solid points are the radio observations of Warner et al. (1973). The full line is a model of uniform M/L based on de Vaucouleurs' (1959) photometry.

of axial ratio 0.2 one may calculate its rotation curve numerically. The turn-over radius is 2.42 α^{-1} and the rotational velocity drops to 0.9 V_{max} at 4.5 α^{-1} i.e. the radial scale is about 12% greater than for the thin disk. Freeman (1970) found that the central surface brightness of most disks show a remarkably small scatter about $21\overset{m}{.}6$ (arcsec)$^{-2}$. If M/L is constant in the disk, the surface brightness at the radius where $V = 0.9\ V_{max}$ will be $4\overset{m}{.}9$ ($e^{-4.5}$) fainter at $26\overset{m}{.}5$ (arcsec)$^{-2}$. This value is close to that of the faintest features detectable on all but the most recent long exposure plates. So it is no surprise that, in galaxies where the H I detected so far extends only to the Holmberg radius ($26\overset{m}{.}5$ (arcsec)$^{-2}$) there should be little sign of any turn-over in the rotation curve. Two of the examples given by Roberts (this volume, p. 331) illustrate this point exactly.

This result forces us to make a more quantitative comparison of observations with model predictions and restricts the discussion to the galaxies listed earlier. As Dr King reminded us yesterday, it is best to make comparisons in the observational plane rather than using derived quantities. The data for M33 is presented in Figure 1 which compares the rotation curve of a thick exponential disk of uniform M/L, whose

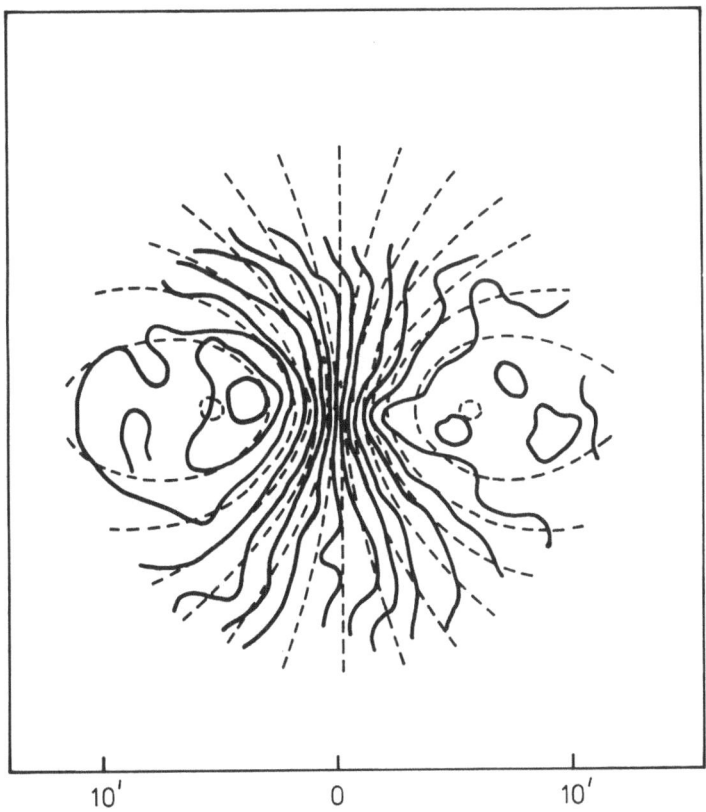

Fig. 2. The radial velocity field in NGC 6946. Full lines are the observations of Rogstad *et al.* (1973). Dotted lines are for a uniform M/L model based on Ables' (1971) photometry.

scale length α fits de Vaucouleurs (1959) photometry, with our radio rotation curve data (Warner *et al.*, 1973). The agreement is very good. H I measurements have been made at radii larger than 6 kpc but the radial velocities show deviations from uniform circular motion. Within 6 kpc the deviations are everywhere less than 10 km s^{-1} and for this reason the mean rotation curve provides the best comparison for model predictions.

For NGC 6946 and IC 342, whose luminosity profiles are also nearly pure exponential disks with only very small central spheroidal components, it is better to make comparisons of models with the measured radial velocity fields since noncircular motions are evidently more important. Figures 2 and 3 show the radio radial

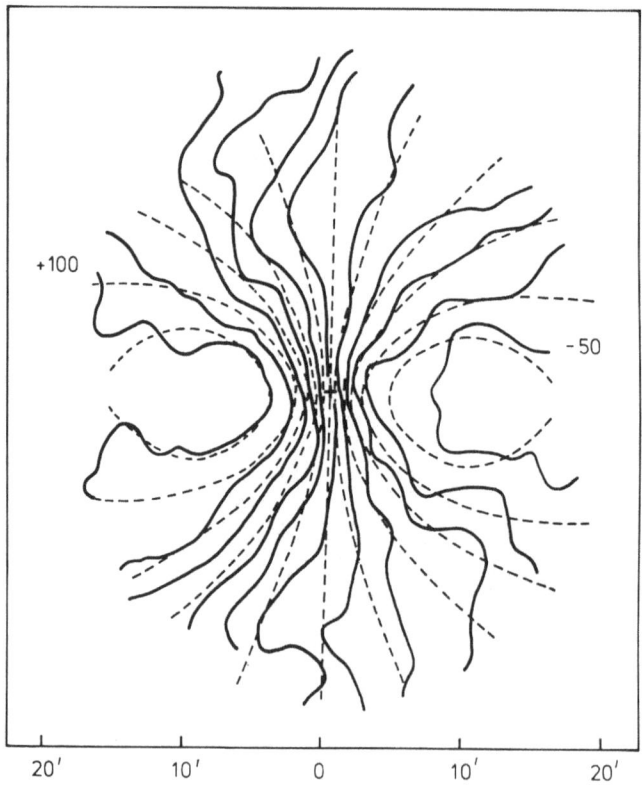

Fig. 3.   The radial velocity field in IC 342. Full lines are the observations of Rogstad *et al.* (1973). Dotted lines are for a uniform M/L model based on Ables' (1971) photometry.

velocity data from Rogstad *et al.* (1973) as full lines compared with exponential disk models (dotted) having scale lengths obtained from Ables' (1972) photometry. Axial ratios of the disks were taken to be 0.2 and the inclinations used for NGC 6946 and IC 342 were 37° and 30° respectively. One criterion for judging the fit to be good or not is whether any large scale differences between the observations and the model are masked by the departures from circular motion shown in the observations. On this basis it seems that the fit is good in both cases. IC 342 suffers from observational limitations in the lack of some data at short interferometer baselines, giving rise to the apparently greater extension of the H I along the minor axis than along the major axis. It is difficult to assess how this may affect the velocity field in the outermost parts on the major axis.

M31 and M81 differ from the other galaxies mentioned in having significant spheroidal components. For galaxies of this type the luminosity distribution can be modelled in most cases by a spheroidal distribution whose surface brightness follows Hubble's Law for elliptical galaxies superposed on an exponential disk of the kind discussed. The division into these two parts is an arbitrary one and may not correspond to a sharp physical distinction. But for purposes of deriving model rotation curves it provides a way of exploring the next most simple mass distribution, in which the disk has one uniform value of *M/L* and the spheroidal population a different value. The rotation curve of M31 has been discussed elsewhere (Emerson and Baldwin, 1973) and it suffices to say that a good fit to the observations was obtained by a model

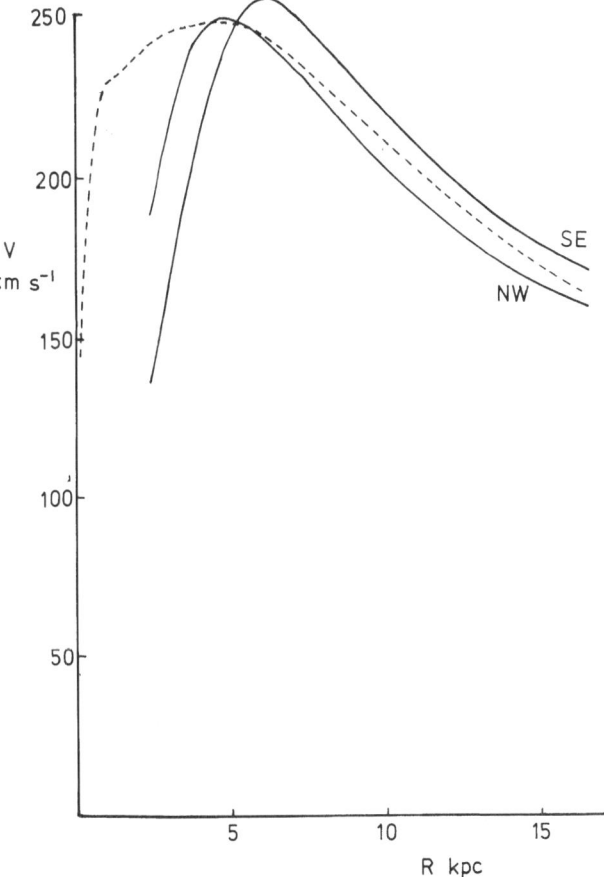

Fig. 4. The rotation curve of M81. Full lines are from the observations of Gottesman and Weliachew (1975) for the two halves of the galaxy. The dotted line is for a model based on the photometry of Brandt *et al.* (1972). See text.

of this kind in which the values of M/L_B, uncorrected for absorption, were 25 for the spheroidal population and 12.5 for the disk. The differences between the model and the observations were smaller than the difference between the rotation curves derived from the two halves of the major axis of M31. In the outer 20 arcmin of the curve there are observational differences from the values obtained by Roberts which reach 20 km s^{-1}. This is the only point on which our disagreement is an observational one. It needs to be cleared up but, whichever way it is resolved, it will not affect the argument presented here that the magnitude of the non-circular motions is larger than any systematic departure from a curve corresponding to a uniform M/L ratio.

The final case is that of M81 and it is of particular interest since the H I extends to large distances beyond the visible limits of the galaxy. Roberts and Rots (1973) give a rotation curve extending to 30 kpc radius and Gottesman and Weliachew (1975) one extending to 16 kpc radius. Good photometry also has been obtained by Brandt *et al.* (1972). The luminosity distribution closely resembles that of M31, the ratios of

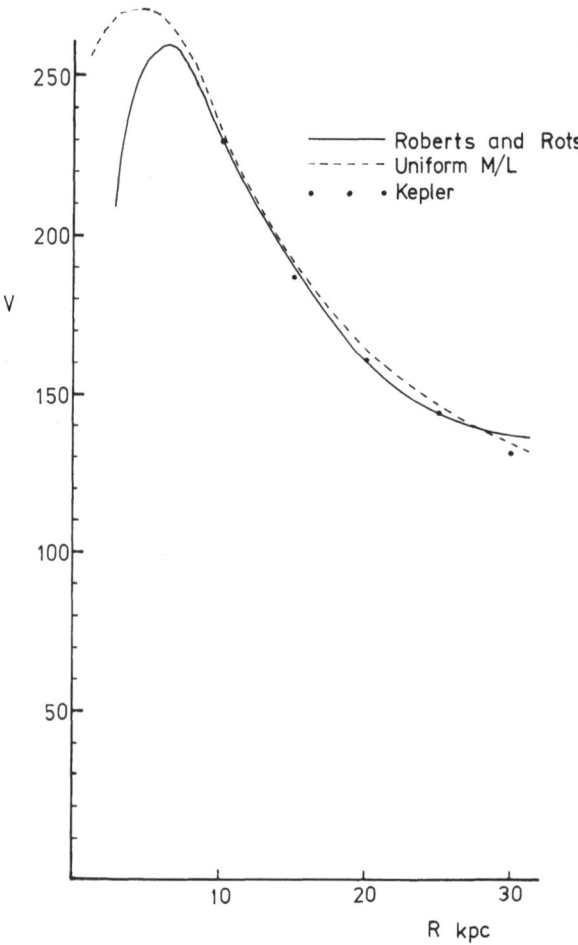

Fig. 5. The rotation curve of M81. The observations of Roberts and Rots (1973) (full line) are compared with model rotation curves described in the text.

the luminosity of the spheroidal distribution to that of the disk being quite similar and the physical scales of the two components in M81 are each about one half of their value in M31. In modelling M81 the exact value of M/L chosen for the spheroidal distribution is not very important since there are no H I measurements closer than 3.5 kpc to the nucleus or optical values closer than 5 kpc. Figure 4 shows Gottesman and Weliachew's (1975) rotation curve compared with a model having $(M/L)_{\rm sph} = = 9$ and $(M/L)_{\rm disk} = 9$. In the range 6 kpc $< r <$ 16 kpc the agreement is better than 10 km s^{-1}. Figure 5 shows Roberts and Rots (1973) curve together with an almost identical model ($M/L = 11$) and also a Keplerian curve. The fit to the model is again satisfactory and the close agreement of that with the Keplerian curve demonstrates that most of its mass (90%) is within 10 kpc radius. The agreement at large r, where Roberts and Rots describe the curve as flat, is excellent but possibly spurious since Weliachew and Gottesman found departures from circular motion as large as 60 km s^{-1} at similar radii in the south following portion of M81. Taken at its face value it suggests that an upper limit to the mass in M81 between 10 kpc and 30 kpc radius is $1.0 \times 10^{10} \, M_{\odot}$.

In conclusion, this discussion of the galaxies for which there are the best data supports the view that no significant variations of mass-to-light ratio with radius in the disks of spiral galaxies have yet been detected. In the case of M81 a useful upper limit can be set to the mass density of any extensive halo at radii near 20 kpc.

References

Ables, H. D.: 1971, *Publ. U.S. Naval Obs.* **20**, Pt. IV.
Brandt, J. C.: 1965, *Monthly Notices Roy. Astron. Soc.* **129**, 309.
Brandt, J. C., Kalinowski, J. K., and Roosen, R. G.: 1972, *Astrophys. J. Suppl.* **24**, 421.
Boulesteix, J. and Monnet, G.: 1970, *Astron. and Astrophys.* **9**, 350.
Emerson, D. T. and Baldwin, J. E.: 1973, *Monthly Notices Roy. Astron. Soc.* **165**, 9P.
Freeman, K. C.: 1970, *Astrophys. J.* **160**, 811.
Gottesman, S. T. and Davies, R. D.: 1970, *Monthly Notices Roy. Astron. Soc.* **149**, 263.
Gottesman, S. T. and Weliachew, L.: 1975, *Astrophys. J.* (in press).
Nordsieck, K. H.: 1973, *Astrophys. J.* **184**, 719.
Roberts, M. S. and Rots, A. H.: 1973, *Astron. and Astrophys.* **26**, 483.
Rogstad, D. H., Shostak, G. S., and Rots, A. H.: 1973, *Astron. and Astrophys.* **22**, 111.
Vaucouleurs, G. de: 1959, *Astrophys. J.* **130**, 728.
Warner, P. J., Wright, M. C. H., and Baldwin, J. E.: 1973, *Monthly Notices Roy. Astron. Soc.* **163**, 163.

DISCUSSION

Roberts: Your examples of the iso-velocity curves derived at Cal Tech and the deviations from pure circular motion that they imply are not germane to the discussion. A constant $V_c(R)$ and one slowly decreasing will show up in the outer parts of your theoretical iso-velocity curves as surprisingly small differences. Thus, for the comparison you wish to make you should also include the constant $V_c(R)$ case.

Further you can not dismiss the real difference in the observed rotation curve for M31 as derived at Cambridge and at Green Bank. These differences cannot be reconciled. You report signals at velocities where no such signal is seen by the 300-ft telescope. The importance in our difference lies in the fact that your decreasing $V_c(R)$ curve is consistent with the model you chose to describe galaxies. However your model does not at all fit the rotation curve derived from Green Bank data for M31. Thus a difference of

~ 30 km s^{-1} may be 'small' but it is a vital discriminant in the description of the mass distribution within a galaxy.

Finally wide band studies of the luminosity distribution within a galaxy do not describe the mass distribution.

Baldwin: It is true that measurements on the major axis are more likely to give good evidence on the rotation curve than ones near the minor axis. But there can be deviations from circular motion even at points on the major axis, both tangential and in the z direction. The latter is particularly likely in the cases of NGC 6946 and IC 342 while both have very low inclinations.

I agree that the differences in the iso-velocity curves of a constant $V_c(R)$ and that of a constant M/L model are quite small. Indeed that was one of the main points of my talk. If we are to say that $V_c(R)$ is constant then we must be very sure that the measured values denote circular motion. If they are disturbed by large non-circular motions, as I illustrated, then the case for $V_c(R)$ being constant is very weak. The differences between our observations of M31 need to be cleared up but regardless of which way this is settled, we still have differences in the two halves of the rotation curve which are large, implying that what is measured is not the true rotation curve.

Pişmiş: The rotation curve, you showed, of M31 exhibits fluctuations resembling waves. Would you consider them physically significant details?

Baldwin: The peaks in the M31 rotation curve correspond with the H I spiral arms in the sense that faster rotation speeds occur on the outside of the arms. But that, I think, does not commit one to any particular theory of such arms!

N-BODY SIMULATIONS OF DISKS

F. HOHL

NASA, Langley Research Center, Hampton, Va., U.S.A.

Abstract. The methods used in the large-scale *n*-body simulations are discussed. However, the present review concentrates on the results already obtained in the *n*-body simulations using systems containing up to 200 000 simulation stars. Results are presented which show that the stability criterion developed for flattened systems applies only to truly axisymmetric instabilities. Purely stellar disks acquire rather large velocity dispersions, generally two or more times the velocity dispersion required by Toomre for axisymmetric stability. In the computer simulations, the bar-forming instability can be prevented only by comparatively large velocity dispersions. However, simulations inclusing the effects of the galactic halo and core as a fixed background field show that bar formation can be prevented for fixed halo components as large or larger than the self-consistent disk component. Experiments performed to determine the collisional relaxation time for the large-scale gravitational *n*-body calculations show that these models are indeed 'collisionless'.

1. Introduction

Large-scale gravitational *n*-body simulations of disks of stars have been in progress for the past seven years. With the introduction of these efficient large-scale computer models for self-gravitating stellar systems, the field of experimental stellar dynamics is providing fresh insights into the structure and the dynamics of galaxies and other stellar systems. These simulations were started by two groups using somewhat different numerical models. The group of Richard Miller and Kevin Prendergast developed a completely discretized model where the forces, star positions, and velocities are allowed only discrete (integer) values less than some given maximum value. This model is described in Miller and Prendergast (1968). Our group developed a model where the forces, star positions, and velocities are not restricted to integer values. This model is described in Hohl and Hockney (1969). To obtain the gravitational potential or force in these models, an $n \times n$ array of cells is superposed over the galactic disk. The mass density in each of the $n \times n$ cells is used to obtain the gravitational force by means of convolution methods making use of fast Fourier transforms (Cooley and Tukey, 1965). Because of the periodic nature of finite Fourier transforms, Miller and Prendergast (1968) and Miller *et al.* (1970) used a doubly periodic configuration space for their earlier force calculations. Hohl and Hockney (1969) developed a modified Fourier transform method to obtain the potential for an isolated disk galaxy. In the simulations the dimension of the array for the force calculations is 128×128 or 256×256 and the number of simulation stars is generally near 100 000. The results obtained with the two models are in general agreement; however, Miller *et al.* (1970) have concentrated their simulations on two-component disks containing both stars and a dissipative 'gas' component. Our own work has been concentrated on simulating purely stellar disks (Hohl, 1971a, 1972a). Some of the interesting problems investigated so far with these models are the 'Jeans instability' and the gravitational two-stream instability (Hohl, 1971b), development of spiral structure

(Miller *et al.*, 1970; Quirk, 1971; Hohl, 1971a), evolution of purely stellar disks (Hohl, 1971a), and tests on stationary self-consistent disks (Hohl, 1972b).

Galaxies are essentially collisionless systems. Thus, the computer models should not cause heating of the 'stars' due to numerical or collisional effects. The collisional relaxation time of the model has been determined to be of the order of 1000 rotations by investigating the rate of energy equipartition of a system with a mass spectrum of stars (Hohl, 1973).

With the exception of computer time and storage, there are no difficulties in extending the simulations to three dimensions. Some initial results of such three-dimensional simulations have been obtained by Hockney and Brownrigg (1974).

2. Evolution of Purely Stellar Disks

Some of the initial simulations were performed for 'cold' (zero velocity dispersion) balanced disks of stars. As expected, such disks were found to be violently unstable. Figure 1 shows the evolution of an initially balanced uniformly-rotating disk con-

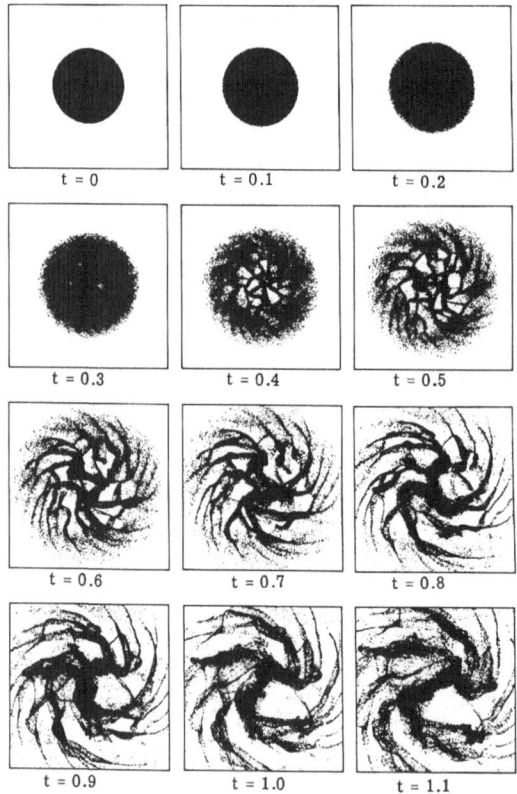

Fig. 1.   Evolutions of an initially 'cold' (zero velocity dispersion) balanced disk containing 200 000 stars. Time is shown in units of the rotational period of the cold balanced disk.

sisting of 200 000 'stars'. The initial surface mass density of the disk is given by

$$\mu(r) = \mu(0)\sqrt{1 - r^2/R^2},$$

where R is the initial radius of the disk. The uniform angular velocity required to balance the cold (zero velocity dispersion) disk is

$$\omega_0 = \tau\sqrt{G\mu(0)/2R},$$

where G is the gravitational constant. As can be seen in Figure 1, the disk quickly develops small-scale instabilities which later develop into a large-scale instability producing a bar with trailing spiral structure. The time shown in Figure 1 and subsequent figures is in units of the rotational period of the cold disk. Figure 2 displays the long-time evolution of another initially cold uniformly-rotating disk containing only 50 000 stars. As can be seen, any spiral-like structure developed during the initial evolution quickly disappears as the system evolves further. The final state of such a

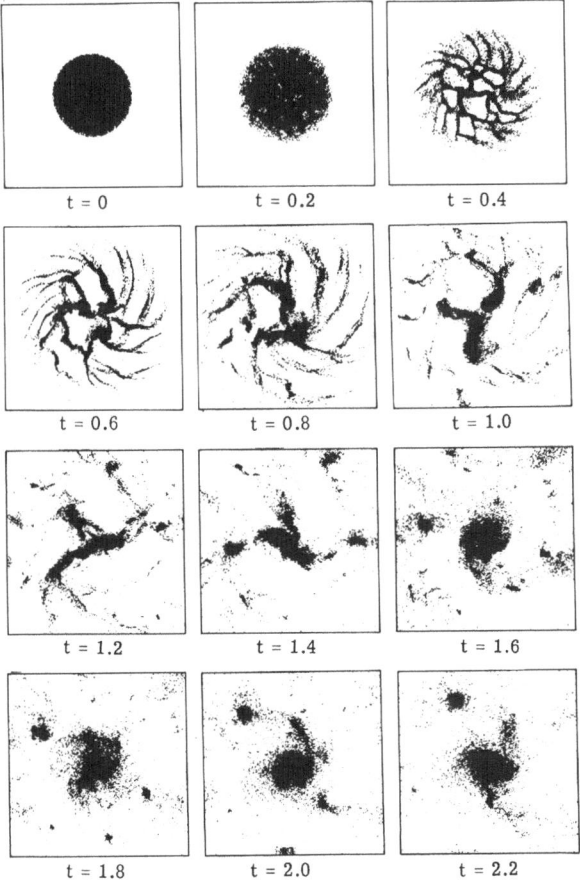

Fig. 2.   Long time behavior of an initially cold balanced disk containing 50 000 stars.

disk has a surface density distribution closely approximated by an exponential variation with the hot core primarily pressure supported. Essentially identical results were obtained for disks with different initial surface density variations (Hohl, 1970).

Toomre (1964) developed a local criterion for the suppression of all axisymmetric instabilities which requires a minimum radial velocity dispersion given by

$$\sigma_{r,\,\min} = 3.36\, G\mu/\kappa,$$

where κ is the local value of the epicyclic frequency. To test Toomre's criterion, an initially balanced uniformly-rotating disk of stars with the velocity dispersion required by Toomre's criterion was investigated to determine whether any axisymmetric instabilities or other misbehavior was present (Hohl, 1971a). This was done by neglecting the azimuthal component of the gravitational field. The evolution of the disk containing 100 000 stars is shown in Figure 3. It remains, of course, axisym-

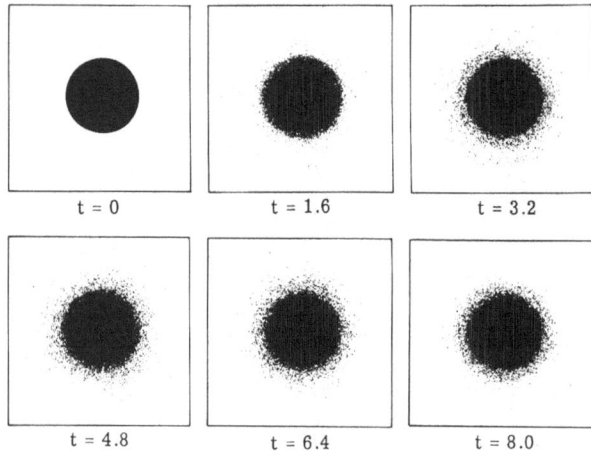

Fig. 3. Axisymmetric evolution of an initially balanced, uniformly rotating disk of 100 000 stars. The stars have an initial velocity dispersion given by Toomre's criterion, and they move under a purely radial gravitational field.

metric. After a few small pulsations it settles down to an essentially steady state with the radial velocity dispersion nearly equal to $\sigma_{r,\,\min}$. Thus, any instabilities the disk may have are not axisymmetric in nature. Similar results by a different method were obtained by Miller (1974). However, when the system at $t=8$ rotations in Figure 3 is allowed to evolve without the axisymmetry constrained, a large-scale bar instability quickly develops.

The totally unrestrained evolution of an initially uniformly rotating disk with Toomre's velocity dispersion is displayed in Figure 4. It shows that the disk is indeed stabilized against the small-scale disturbances which would completely disrupt the disk in less than one rotation, as shown in Figures 1 and 2. However, we again find that the disk is not stabilized against relatively slowly-growing large-scale disturbances which cause the system to assume a very pronounced bar-shaped structure after two rotations. After about four rotations the disk population of stars constitutes a nearly axisymmetric distribution surrounding a dense central oval or bar-shaped core. The oval contains nearly two-thirds of the total mass of the disk and

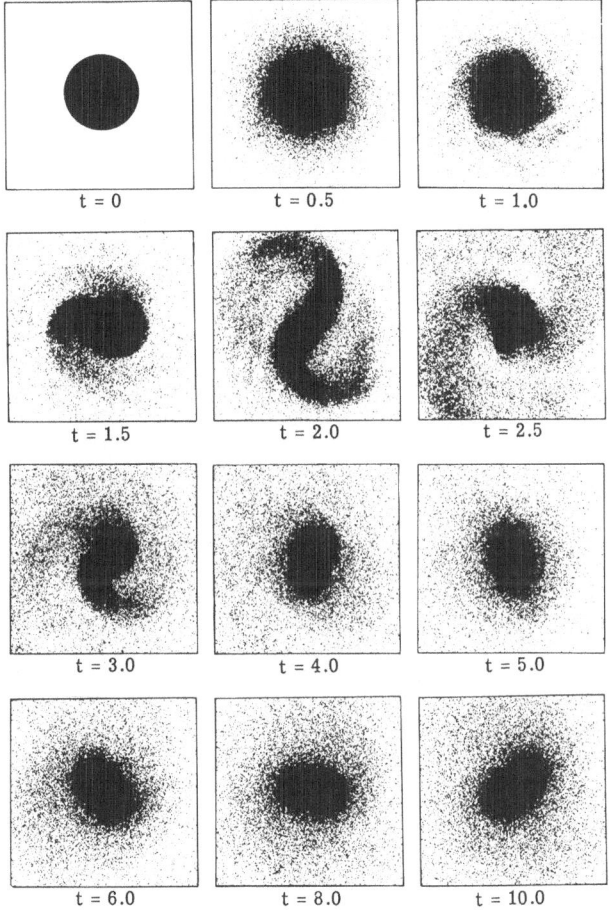

Fig. 4.   Unconstrained evolution of the initially balanced uniformly rotating disk of 100000 stars. The stars have an initial velocity dispersion given by Toomre's criterion.

has an axis ratio of about 3:2. After about four rotations there is little change in the structure of the disk, and the bar rotates with a constant period approximately two times that of the cold balanced disk. To picture the radial variation of parameters describing the disk, the disk was divided into a number of concentric rings each of 0.5 kpc width. The radial dependence of various parameters averaged azimuthally over each ring was then obtained. Figure 5 shows the evolution of the surface mass density obtained in that manner. It can be seen from this figure that the central mass density increases by about a factor of 4 during the 11.6 rotations shown. After eight

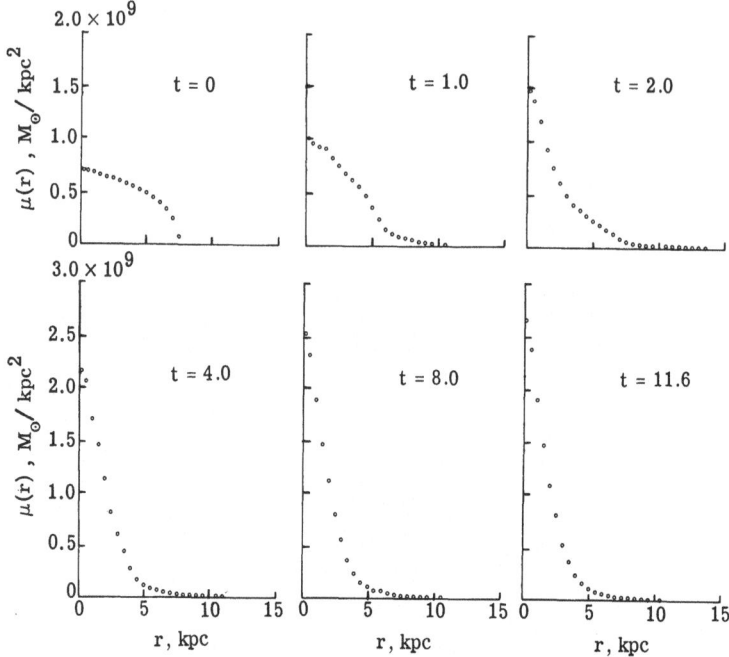

Fig. 5. Evolution of the mass density plotted as a function of radius. The density is given in units of solar masses per kpc². Note that each one of the 100 000 simulation stars has a mass $0.84 \times 10^6\ M_\odot$.

rotations the radial variation in density changes only very little. To determine how closely the final density variation corresponds to an exponential variation the final density is plotted in Figure 6 on a semilog scale. The distribution of the disk population of stars (outside $r = 8$ kpc) is closely approximated by an exponential variation with a scale length of 8.6 kpc. In the high-density central oval the density also closely

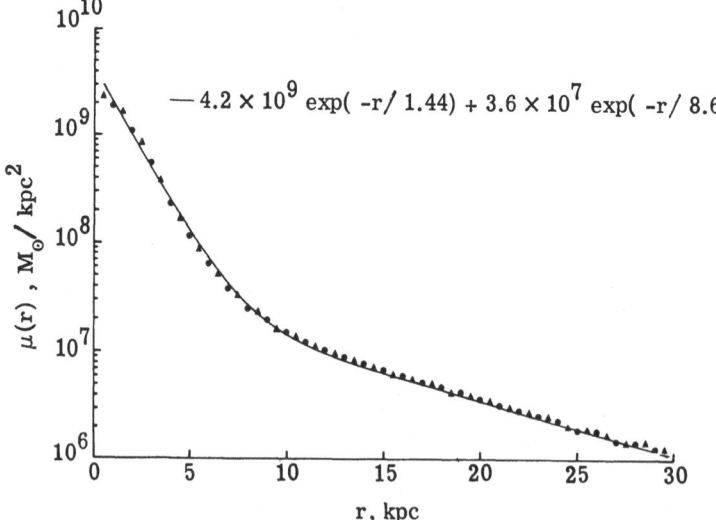

Fig. 6. Dependence of the azimuthally averaged density on radius for the disk after $t = 11.6$.

follows an exponential law, but with the much shorter scale length of 1.44 kpc. The structure of the disk changes very little during the last 5 rotations. Note that the sum of the two exponentials represented by the solid line in Figure 6 closely approximates the density variation.

An interesting end result for all the disks of stars investigated so far which go through the initial unstable evolution is that the final distribution in the radial direction for the 'disk population of stars' is closely approximated by an exponential variation of density. This result may be significant since it agrees with observational evidence (de Vaucouleurs, 1959; Freeman, 1970) which indicates that the luminosity in the outer regions of many spiral and SO galaxies seems also to decrease exponentially with radius.

Figure 7 illustrates the evolution of the angular momentum distribution of the disk.

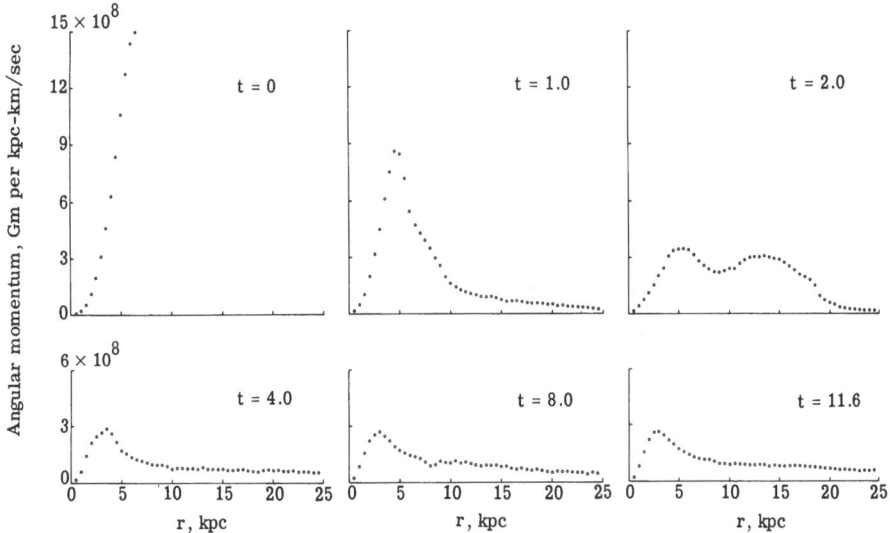

Fig. 7.   Evolution of the azimuthally averaged angular momentum distribution.

As the bar-forming instability is developing at $t=2$ rotations, much of the angular momentum is transferred to large radii as stars are flung outward from the ends of the bar.

The initial radial velocity dispersion is about 120 km s^{-1} at the center of the disk and goes to zero at the edge. The final velocity dispersion at the center is increased to about 180 km s^{-1} and drops quickly to about 60 km s^{-1} at the edge of the bar. The disk population has a nearly constant radial velocity dispersion of about 60 km s^{-1} for r larger than 8 kpc. The ratio of $\sigma_r/\sigma_{r,\,min}$ for the final state of the evolved disk varies from about 1 in the center to about 7 at $r=15$ kpc. Thus the system represents a rather hot disk. This is typical of computer generated galaxies. The radial dependence of the mean circular velocity of the stars $\langle V_\theta \rangle$, of $\omega = \sqrt{K_r/r}$ (K_r=radial gravitational field), and of $r\omega$ are shown in Figure 8. Note that in the central region ($r<8$ kpc) the disk is primarily pressure supported while for large r, rotation is the

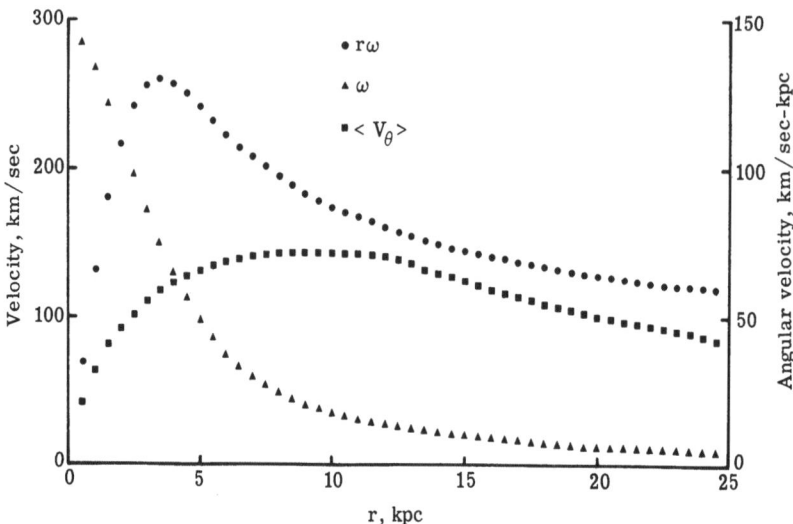

Fig. 8.   Comparison of the azimuthally averaged angular velocity $\omega = \sqrt{K_r/r}$, the circular velocity $r\omega$, and the mean circular velocity of the stars $\langle V_\theta \rangle$.

main supporting factor. Very similar results are obtained for initially Gaussian or exponential surface mass density variations (Hohl, 1972). However, the evolution for the later two density distributions is not quite as violent as indicated by the final values of $\sigma_r/\sigma_{r,\,min}$ of about 4 for the Gaussian variation and of about 2 for the exponential variation (Hohl, 1970). Miller (1971) obtains final states that appear even hotter than those presented here.

3. Cooling of Axisymmetric Disks

The stable axisymmetric disk that we were able to generate was rather hot, especially in the outer regions of the disk. We therefore determined the effects of cooling the disk. Various methods of cooling were tried. For example, during each rotation a certain percentage (we tried from 5 to 30%) of the stars, chosen at random, had a portion of their noncircular velocities removed in proportion to their present radii. Thus stars near the center kept nearly all their random motion whereas stars far away from the center were placed in more nearly circular orbits. Another method, somewhat more analogous to the 'gas collisions' of Miller *et al.* (1970), was simply to place a certain percentage of the stars during each rotation into purely circular orbits. All methods of cooling which we tried gave essentially the same results.

   Figure 9 shows the effects of cooling when for each rotation 10% of the stars are selected at random and are placed in circular orbits. The initial disk has the same density variation and mean circular velocity as shown in Figures 6 and 8. During the first three rotations the appearance of the disk changes little. However, at $t = 3$ there is already a pronounced bar structure at $r = 2.5$ kpc. After $t = 3$ a two-arm structure appears, which during the following 1.5 rotations displays a theta-like structure.

After $t=4.5$ the spiral structure opens up and rotates with a fairly constant speed of about 10 km s^{-1} kpc^{-1}. This compares with a mean circular velocity of the stars of about 15 km s^{-1} kpc^{-1} at $r=10$ kpc. At $t=6$ the value of $\sigma_r/\sigma_{r,\,\text{min}}$ for the disk was between 2 and 3. Any further cooling only caused the collective instabilities to heat up the disk as fast as it was being cooled.

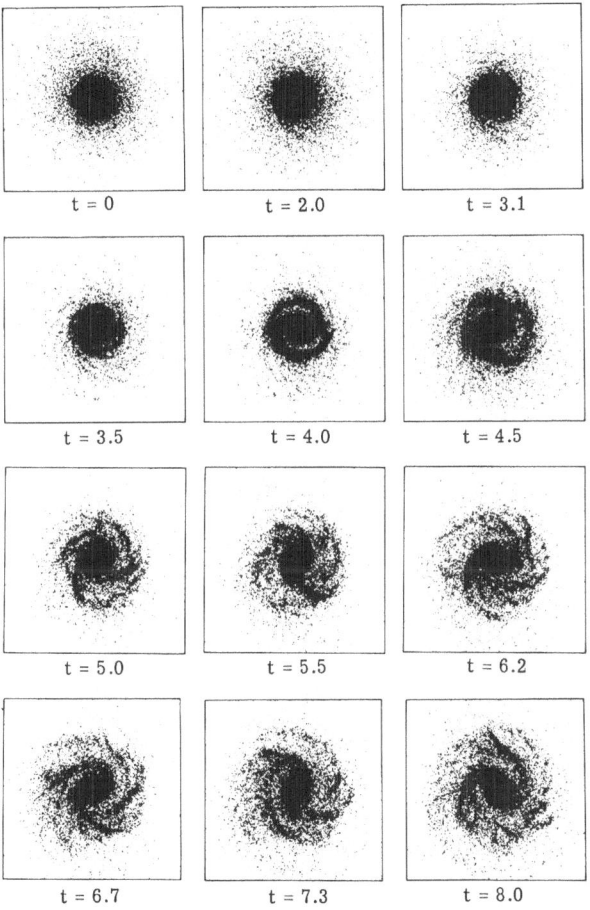

Fig. 9.   Development of spiral structure in a disk of stars which is being cooled.

A much more detailed and somewhat more realistic simulation of spiral structure was performed by Miller *et al.* (1970) and by Quirk (1971) who used a two-component system of stars and a dissipative gas component. Again, this stellar component becomes rather hot while the dissipative 'gas' component is forced to remain cool. Spiral structure which persists for about 3 galactic rotations is found to develop, primarily in the cool gas component. They also find that the spiral pattern has some of the properties of a density wave in that stars move through the pattern at a higher velocity than the pattern speed.

4. Effect of Satellite Galaxy

Toomre and Toomre (1972) have performed a number of simulations of close encounters of galaxies that produced spiral structure and tails and bridges often seen
in multiple galaxies. In these calculations a number of massless points move under
the influence of the two interacting galaxies which are represented as point masses.

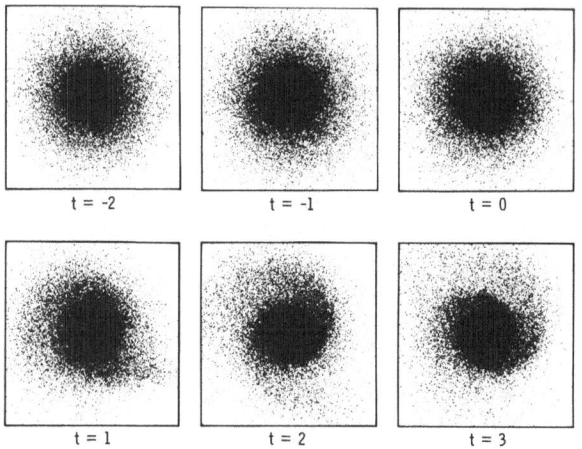

Fig. 10.   Evolution of a relatively 'hot' galaxy under the influence of a companion galaxy.

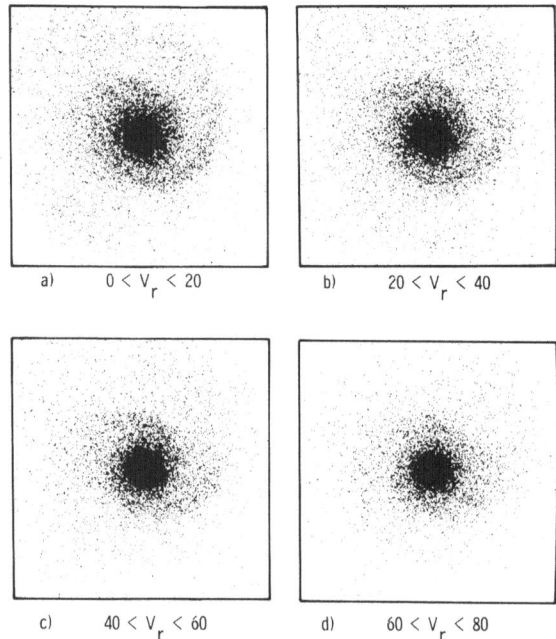

Fig. 11.   Distribution of stars in four velcocity intervals at $t = 3$ for the disk shown in Figure 10.

A simulation of a close passage where the self-gravity of the disk stars is taken into account is of interest. For this purpose a stellar disk was perturbed by the passage of a companion galaxy having one-fourth the mass of the primary galaxy. The parameters for M51 as supplied by Alar Toomre were chosen for the calculations. The orbit was direct, the inclination of the companion's orbit plane to the plane of the primary galaxy was 75°, the angle between the node and periapsis was 20°, and periapsis was 25 kpc. The effect of the passage on the structure of the primary galaxy is shown in Figure 10. The time is given in rotational periods of the primary galaxy; $t=0$ corresponds to the time at periapsis. These results indicate that only a weak two-arm spiral structure developed. This was to be expected since the velocity dispersion of the primary galaxy is rather high. The value of $\sigma_r/\sigma_{r,\,min}$ ranges from about 2 in the central region to 6 in the outer disk. The effect of the large random velocities on the formation of spiral structure can be determined by plotting the distribution of stars in various velocity intervals. This is done in Figure 11. As can be seen, the spiral structure is quite pronounced for the lower velocities, but can hardly be detected for the higher velocities. This was to be expected since the large random velocities have a dispersive effect on the formation of spiral structure. It thus appears that a much cooler primary galaxy is needed for a realistic study of the effects of a close passage.

5. Evolution of Stationary Disks

One of the few stationary solutions of the collisionless Boltzmann equation with velocity dispersion is that for the uniformly rotating disk given by (Hohl, 1972b),

$$f(r, v_r, v_\theta) = \frac{\mu(0)}{2\pi R\sqrt{\omega_0^2 - \omega^2}}\left[(\omega_0^2 - \omega^2)(R^2 - r^2) - v_r^2 - (v_\theta - r\omega)^2\right]^{-1/2},$$

where v_r and v_θ are the radial and azimuthal velocity components, ω_0 is the rotational velocity of the cold balanced disk, and ω is the actual rotational velocity of the disk. Kalnajs (1972) has performed a normal-mode stability analysis and finds that the simple mode corresponding to the bar disturbance becomes unstable for $\omega > 0.508$ ω_0. A series of numerical experiments was performed to test this result.

The evolution of four disks of stars corresponding to equation (1) with (a) $\omega = 0.8\,\omega_0$, (b) $\omega = 0.6\,\omega_0$, (c) $\omega = 0.4\,\omega_0$, and $\omega = 0$ is presented in Figure 12. Each of the 100000 stars in the simulation represents $0.84 \times 10^6\,M_\odot$ so that the total mass of the disk galaxy is $0.84 \times 10^{11}\,M_\odot$. The rectangular border enclosing the disks represents the active 128×128 array of cells used in the calculations. The initial radius of the disks is 16 kpc. Since the disks become progressively more stable as the initial velocity dispersion is increased (or ω is decreased), the evolution of the more stable systems is investigated for longer times. Again the times shown are in units of the rotational period of the cold (zero velocity dispersion) disk $\tau_r = 2\pi/\omega_0$. Figure 12(a) shows that, for $Q = 1$ (or $\omega = 0.8\,\omega_0$), the system is unstable and within two rotations it has formed a bar-shaped structure. After three rotations this structure remains essentially un-

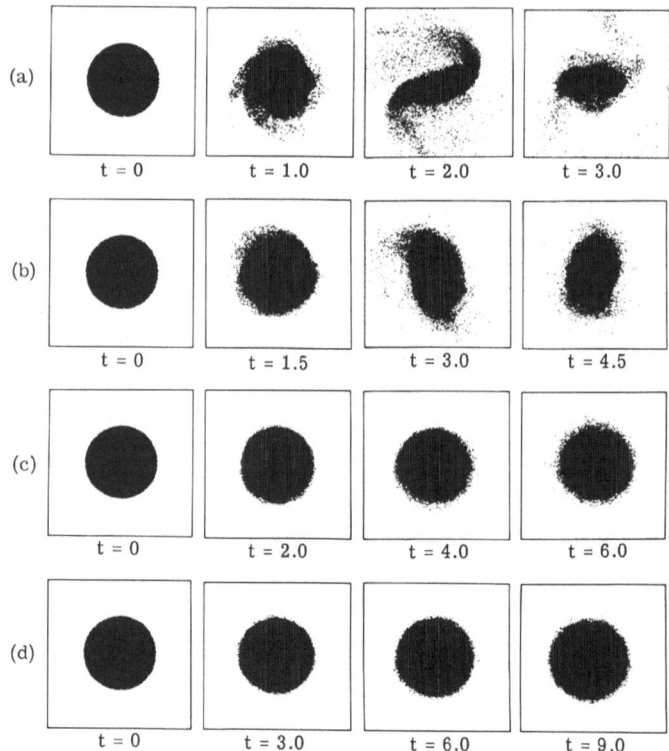

Fig. 12. Evolution of an initially rotating and stationary disk galaxy for four values of the angular velocity given by (a) $\omega = 0.8\,\omega_0$, (b) $\omega = 0.6\,\omega_0$, (c) $\omega = 0.4\,\omega_0$, and (d) $\omega = 0$, where ω_0 is the angular velocity of the cold (zero velocity dispersion) disk.

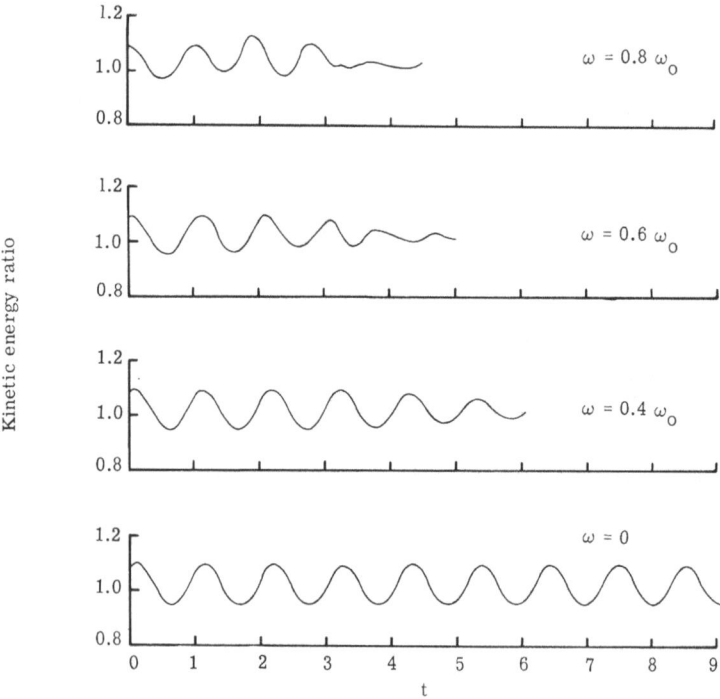

Fig. 13. Oscillation in the total kinetic energy for the four disks shown in Figure 12.

changed. It should be noted that all small-scale instabilities which occurred in the cold disk have been stabilized. Only the large-scale 'bar-making' instability is present. A similar result is shown in Figure 12(b) for $Q = 1.35$. However, the bar structure is now much less pronounced. For $Q = 1.55$, Figure 12(c), the system is essentially stable. Some of the stars near the edge of the disk tend to diffuse to larger radii. This is to be expected since the distribution function $f(r, v_r, v_\theta)$ is singular at the edge and star orbits tend to be unstable there. Similar results are obtained for the nonrotating disk shown in Figure 12(d). Figure 12 indicates that the disk becomes stable for values of Q somewhere between 1.35 and 1.55 or for values of ω between 0.4 ω_0 and 0.6 ω_0. These results are in agreement with the normal mode analysis of Kalnajs. Since a pseudo-random number generator is used to obtain the initial conditions for the disk, small deviations from the stationary solution which result in initial oscillations of some parameters are to be expected. Figure 13 shows the variation of the kinetic energy for the four disks shown in Figure 10. For the two unstable disks these oscillations are quickly damped. For the case $\omega = 0.4$ ω_0 the oscillations show a slow rate of damping whereas for the nonrotating case no damping of the oscillations is noticed.

6. Effects of Fixed Halo Component

A number of simulations have been performed with a model modified to include a fixed central force in addition to the self-consistent disk population of stars (Hohl, 1970). The radial field is taken to represent the hot halo population and central core of the galaxy. Figure 14 gives the evolution of a 50 000-star system with a fixed central potential corresponding to the Schmidt model of the galaxy. The disk stars contain 20% of the total mass of the system and the initial radius of the disk is 15 kpc. The system quickly develops spiral structure which remains quite pronounced up to about four rotations. After this time the spiral structure becomes quite diffuse due to the buildup of random velocities. The buildup of random velocities can be seen in Figure 15 where the evolution of the radial velocity dispersion is shown. After two rotations the velocity dispersion shows little further increase and the value of $\sigma_r/\sigma_{r, \, min}$ is about 2 throughout the system. For systems containing a smaller percentage of disk stars the heating proceeds at a slower rate and the spiral structure remains for a longer time.

The heating rate due to collective effects (collisionless or violent relaxation) can be greatly reduced by modifying the interaction potential of the stars. Thus by using the interaction potential corresponding to a uniform density sphere with a radius of 4 cell dimensions, a system similar to that shown in Figure 14 but with an initial $\sigma_r = \sigma_{r, \, min}$ shows very little further heating (Hohl, 1974). The use of such a modified interaction potential with the resulting low heating rate will be desirable for the study of large-scale galactic stability or density wave propagation since collective heating effects will no longer mask the phenomena under study.

Computer simulations of purely stellar disks have shown that such systems require rather large velocity dispersions to prevent the large-scale bar-making instability. In fact, these experiments indicate that only when the energy in rotation is less than 28% of the total kinetic energy of the stars is the system prevented from developing a bar structure. This result is in agreement with a new stability criterion proposed by Ostriker and Peebles (1973) which states that for stability the total kinetic energy in rotation must not exceed $14 \pm 2\%$ of the absolute value of the potential energy of the system. Indications are that the K or M dwarfs in the solar neighborhood constitute most of the known mass in this region, but their random radial motions are

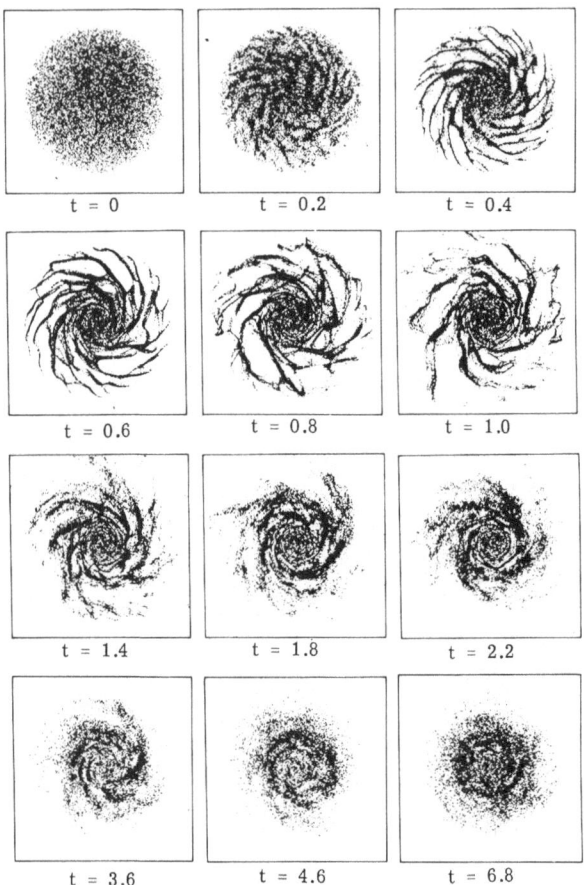

Fig. 14. Evolution of a disk of stars under the influence of a fixed central field corresponding to the Schmidt model of the Galaxy. The 50 000 disk stars contain 20% of the total mass of the system. Time is in rotational periods of the initial balanced disk at $r = 10$ kpc.

much too low to satisfy the stability criterion of Ostriker and Peebles. How then can we account for apparently stable spiral galaxies such as own own? At present the actual mass of the core and/or halo component of our and other spiral galaxies is entirely unknown. Also, there appear to be no difficulties with the assumption that a large portion of the galactic mass is contained in the halo, likely consisting of high-

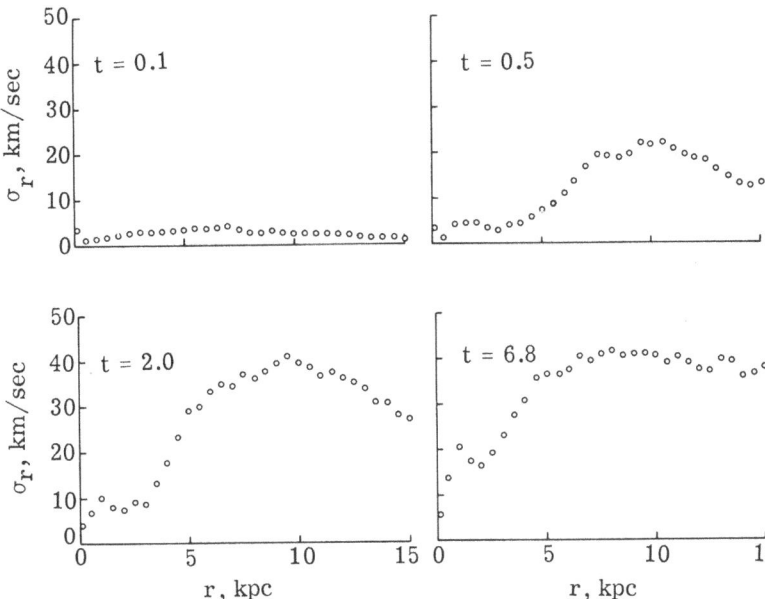

Fig. 15. Evolution of the radial velocity dispersion for the disk shown in Figure 14.

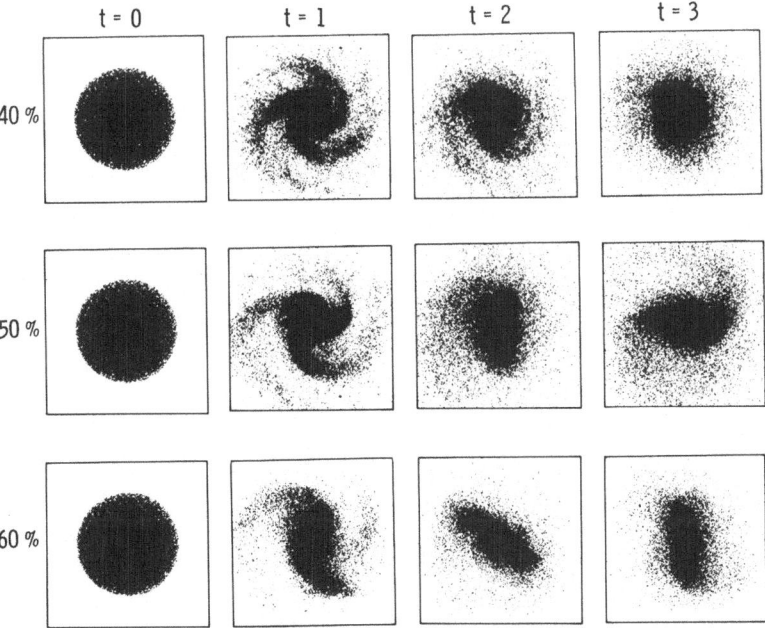

Fig. 16. Evolution of the stellar disk for three different fractions of the total mass contained in the disk stars. Time is in units of the rotational period at $r = 10$ kpc.

velocity dispersion objects. The result of Figure 14 shows that a sufficient large fixed halo component indeed does stabilize a disk galaxy against the bar-making instability. Similar stabilizing effects were found by Ostriker and Peebles (1973). Therefore simulations were performed to determine the amount of halo or fixed mass required to prevent bar formation. Again an axisymmetric fixed radial field corresponding to the Schmidt model was added to simulate the effects of a hot and time-independent core or halo component.

Figure 16 shows the evolution of the disks for three different fractions of total mass contained in the disk stars. The initial radius of the disks is 20 kpc and the time shown is in rotational periods at $r = 10$ kpc. The evolution of the system for 40% is that for a system with the fixed radial field representing 60% of the mass and the 100 000 disk stars containing 40% of the mass. The mass of the disk stars is increased while the fixed mass remains constant to obtain the evolution for systems with larger disk components. The results show that the bar-forming instability is prevented for systems with up to 40% of the mass in the disk stars. For the system with 40% of the mass in the disk stars there is an initial tendency to form a bar-like spiral structure which, however, disappears and the systems become essentially axisymmetric. However, when the fraction of the mass contained in the disk stars becomes 50% or greater, a bar-structure quickly forms and remains for the duration of the calculations.

Additional computer experiments were performed for a more centrally located fixed mass component; namely, a uniform mass density sphere of 6 kpc diameter located at the center of the disk. Again it was found that the system quickly formed a bar when 50% of the mass was contained in the disk stars and the system was stabilized when the disk stars contained 40% of the total mass.

COMPUTER SIMULATION   M 101, Sc-TYPE REGULAR SPIRAL GALAXY

Fig. 17. Comparison of computer generated disk with a normal spiral galaxy.

As was shown in previous simulations (Hohl, 1971a) a purely stellar disk could be stabilized only with very high velocity dispersions such that more than 70% of the total kinetic energy is in random motion with the remaining kinetic energy in rotation. The present computer experiments show that in order to stabilize the large-scale bar-making instability for an initially relatively cool disk with velocity dispersions near Toomre's $\sigma_{r,\,min}$, a central core and/or halo component containing at least 60% of the total galactic mass is required.

In conclusion I would like to show a couple of comparisons of computer simulated galaxies with actual galaxies. This is done in Figures 17 and 18 for a regular and a

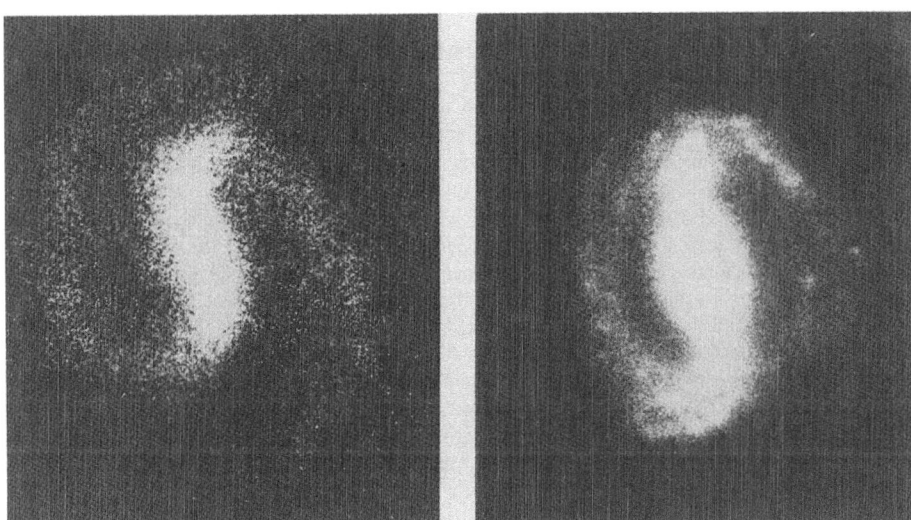

COMPUTER SIMULATION   NGC-175, SBab-TYPE BARRED SPIRAL GALAXY

Fig. 18.   Comparison of computer generated disk with a barred spiral galaxy.

barred spiral galaxy. Such a comparison should not be taken seriously, since for any computer generated galaxy one can find a similar looking object in the various collections of galaxy photographs. Note that the computer generated galaxies considerably change their structure during the following rotation whereas we expect the actual galaxy to preserve its overall structure.

References

Cooley, J. W. and Tukey, J. W.: 1965, *Math. Comp.* **19**, 297.
Freeman, K. C.: 1970, *Astrophys. J.* **160**, 811.
Hockney, R. W. and Brownrigg, D. R. K.: *Monthly Notices Roy. Astron. Soc.* **167**, 351.
Hohl, F.: 1970, NASA Report TR R-343.
Hohl, F.: 1971a, *Astrophys. J.* **168**, 343.
Hohl, F.: 1971b, *Astron. J.* **76**, 202, 207.
Hohl, F.: 1972a, NASA Report TN D-6630.
Hohl, F.: 1972b, *J. Comput. Phys.* **9**, 10.
Hohl, F.: 1972c, *Astrophys. Space Sci.* **14**, 91.
Hohl, F.: 1973, *Astrophys. J.* **184**, 353.
Hohl, F.: 1974, NASA Report TN D-7561.

Hohl, F. and Hockney, R. W.: 1969, *J. Comput. Phys.* **4**, 306.
Kalnajs, A. J.: 1972, *Astrophys. J.* **175**, 63.
Miller, R. H.: 1971, *Astrophys. Space Sci.* **14**, 73.
Miller, R. H.: 1974, *Astrophys. J.* **190**, 539.
Miller, R. H. and Prendergast, K. W.: 1968, *Astrophys. J.* **151**, 699.
Miller, R. H., Prendergast, K. H., and Quirk, W. J.: 1970, *Astrophys. J.* **161**, 903.
Ostriker, J. P. and Peebles, P. J. E.: 1973, *Astrophys. J.* **186**, 467.
Quirk, W. J.: 1971, *Astrophys. J.* **167**, 7.
Toomre, A.: 1964, *Astrophys. J.* **139**, 1217.
Toomre, A. and Toomre, J.: 1972, *Astrophys. J.* **178**, 623.
Vancouleurs, G. de.: 1959, *Handbuch der Physik* **53**, 311.

DISCUSSION

Biermann: If stars evolve, shed gas and this gas dissipates and turns into new stars, doesn't this correspond to a cooling in your calculations already?

Hohl: This process corresponds to the formation of cool young stars and has been included in the simulation of Miller and Prendergast. Of course the problem is that these cool stars quickly heat up.

Lynden-Bell: I notice that all your models have low Q at the middle. I wonder if models with large Q at the centre could satisfie Ostriker's criterium and yet keep Q small far out in the disk as it is observed.

Hohl: I have not yet investigated any models with a large central Q as an initial condition.

Bardeen: In my gas disk calculations a large Q in the center can stabilize the bar mode, but without a substantial halo mass a strong spiral instability remains unless the minimum Q is substantially greater than one. In their gross features the gas disk results are similar to Hohl's for stellar disks; for instance, in that a less centrally condensed halo seems most effective in reducing the value of Q required for stability. I would hope a more detailed comparison will be possible in the future.

Lecar: Before you retire your computer, while you have shown that a massive halo suppresses bar modes, do you think you can kill the bar mode and still retain the small-wavelength (Lin) modes?

Hohl: As you saw from the simulation, even though the bar mode was suppressed by a massive halo, the system still acquired a large velocity dispersion. Thus we still have problems in amplifying the short-wavelength (Lin) modes.

Contopoulos: You said your systems are collisionless. How does this result compare with Rybicki's calculation of the relaxation time, which turns out to be very short for two-dimensional systems?

Hohl: Rybicki's result does not apply to the disk simulation since our forces are cut off at a cell width. Both analytical predictions and computer simulations give relaxation times of several hundred rotations.

Freeman: When you get a persistent bar, is it a large amplitude wave or a material bar?

Hohl: It appears to be more a large amplitude wave since stars in the bar have highly elliptical orbits.

Hunter: This question concerns your numerical simulations of the simple disk models whose stability properties were discussed theoretically by Kalnajs. The slide you showed seemed to indicate that these disks are stable once t is sufficiently small, and the bar instability has been suppressed. According to the theory however, all Kalnajs' elementary disks are unstable to some kind of mode. Did your computations show signs of these other instabilities?

Hohl: No such instabilities were detected in the model. For example, there were small fluctuations in the total kinetic energy for the disks because of the random-number generator used in generating the initial conditions. For a disk with rotation these fluctuations were damped, for the case $\omega = 0$ they showed no damping. Thus any remaining instabilities must have very long growth time and did not show up in the simulations.

Miller: An appeal to the theoreticians present – it could be most useful to have more exact self-consistent models (preferably with stability analysis) along the lines of Kalnajs' models. An exponential disk could be especially helpful. Approximate models won't do, and the entire disk must be self-consistent (including the center).

Hohl: I agree, exact self-consistent model can easily be tested on the computer for stability.

Kalnajs: I can answer Hunter's question: the growth rates of the other unstable modes are small and their spatial structure is less obvious on plots than that of the bar mode.

COMPARISON WITH OBSERVATIONS OF DISK GALAXIES

K. C. FREEMAN

Mount Stromlo and Siding Spring Observatory, Research School of Physical Sciences,
The Australian National University

Abstract. We discuss some observational aspects of the structure and dynamics of disk galaxies. Topics include the disk, the bulge, the lens, and the z-structure of the disk.

1. Introduction

First we recall the two-component (disk + bulge) structure of disk galaxies. Direct photographs and detailed surface photometry show how disk galaxies can have a very wide range in the relative importance of the disk and the bulge. For example, M33 and NGC 5907 have very weak bulges, NGC 4565 has a prominent but not dominant bulge, while the bulge of NGC 4594 is its dominant component. (Photographs of all these systems appear in the *Hubble Atlas*.)

2. The Disk

The disk usually contributes a large part of the total light and angular momentum. For example, in M31 which has a fairly prominent bulge, the disk provides more than 75% of the total light, and probably more than 95% of the total angular momentum. Two comments about the angular momentum:

(1) We know that the disk's surface brightness distribution has the form $I(R) = I_0 \exp(-\alpha R)$, and that for most disks I_0 is roughly constant at 21.65 ± 0.3 B mag. arcsec^{-2}, despite a 5 mag. range in total luminosity (Freeman, 1970). Now assume that the ratio of surface density to surface brightness $\mu(R)/I(R)$ is approximately constant within the disk: some support for this comes from the apparent uniformity of color for the disk and from Warner *et al.* (1973) observations of M33. Then $\mu(R) = \mu_0 \exp(-\alpha R)$. Now the total luminosity $L = 2\pi I_0/\alpha^2$ and the total mass $M = 2\pi\mu_0/\alpha^2$: because M/L appears to be approximately constant for most spirals, it follows that μ_0 is also approximately constant. The total angular momentum for an exponential disk in centrifugal equilibrium is $\mathscr{H} = 1.1 (GM^3/\alpha)^{1/2}$ so $\mathscr{H} \sim M^{7/4}$. This result follows dimensionally from the constancy of μ_0, but is almost unverifiable observationally: see Freeman (1970) and Nordsieck (1973).

(2) Now consider the mass-angular momentum distribution $M(h)$, defined as the mass with angular momentum per unit mass less than h. A real disk will not be cold but will have at least enough velocity dispersion to stabilise it against local axisymmetric instabilities: i.e. $Q = \sigma/\sigma_{min} > 1$, where $\sigma_{min} = 3.36 G\mu/\kappa$ is Toomre's (1964) minimum velocity dispersion for local axisymmetric stability and κ is the local epicyclic frequency. Now a given $\mu(R)$ and $Q(R)$ distribution specifies $M(h)$ (although the con-

Hayli (ed.), Dynamics of Stellar Systems, 367–372. All Rights Reserved.

verse is not true, at least in practice): i.e. if the disks have the exponential surface density distribution and have similar $Q(R)$ distributions, then they have similar $M(h)$ distributions. Small differences in the $Q(R)$ distribution do not change this conclusion significantly. For example, Kalnajs (personal communication) has constructed some selfconsistent disks with the 'isochrone' potential $\Phi = 1/(1 + \sqrt{1 + R^2})$: a 10% change in $Q(R)$ produces an almost imperceptible change in the $M(h)$ distribution. The origin of the $M(h)$ distribution is interesting for real galaxies. If the collapse to the disk was axisymmetric or only weakly barlike, then $M(h)$ was approximately invariant through the collapse and reflects the $M(h)$ for the protocloud, so we need to explain why the protoclouds had such similar $M(h)$ distributions. On the other hand, if the collapse was not axisymmetric, then we need to understand why $M(h)$ has in almost every system relaxed to this particular form.

3. The Bulge

This component has some features in common with elliptical galaxies. (1) Some bulges follow the $R^{1/4}$ law for their surface brightness distributions. (2) Spectrophotometry of the inner parts of some bulges shows that their spectra are very similar to those for the inner parts of some ellipticals, which in turn suggests that their stellar contents are similar. (3) This is reinforced by the apparent similarity of M/L for the bulge of M31 and for normal ellipticals (see Emerson and Baldwin, 1973). (4) Some ellipticals show a change of integrated color with radius which probably reflects the same kind of radial population change as is well known for the bulge of the Milky Way. It seems likely that the bulge component is dynamically like an elliptical, but modified by rotation and the presence of the disk. No good theoretical account of the bulge dynamics is yet published: Carrick's work on this problem at Mt Stromlo will be available soon.

There is much interest now in invoking massive halos to stabilise disk galaxies against barlike deformation (see Ostriker and Peebles, 1973) and there is some indirect evidence that some spirals may have such halos. For example, the flat rotation curves observed for some spirals may imply large amounts of matter outside the visible disk: see M. S. Roberts' talk at this meeting. The total mass of each such spiral could then be of order $10^{12} M_\odot$, which would ease the present problem in binding gravitationally the groups of galaxies (Ostriker et al., 1974). It seems worth putting some observational constraints on such massive coronas. For example, say the halo has a mass of $10^{12} M_\odot$ and a radius of 300 kpc: it is easy to show that this halo would have a significant effect on the dynamics of a typical disk. To make an estimate of the surface brightness of this halo, we can represent it by King's (1966) dynamical models, which give a realistic surface brightness representation for elliptical galaxies and globular clusters. We take the 'tidal' radius $r_t = 300$ kpc and use two models: one with $\log(r_t/r_c) = 2.23$ (concentrated, like an elliptical) and one with $\log(r_t/r_c) = 1.39$ (less concentrated, like a globular cluster of intermediate concentration). The table below gives the projected surface density μ at $R = 20$ kpc for the two models, and then the surface brightness at $R = 20$ kpc, assuming a M/L_B of 200 for the halo.

$\log(r_t/r_c)$	$\mu(20 \text{ kpc})$	$I(20 \text{ kpc})$
2.23	135 $M_\odot$ pc^{-2}	27.5 B mag. arcsec^{-2}
1.39	156	27.3

Although this surface brightness is low, it is possible to test this picture observationally. For example, we have made detailed surface photometry of the nearby almost edge-on spiral NGC 253, by averaging digitally several IIIa-J plates taken with a short focal length wide field Schmidt camera at Siding Spring. We can confidently exclude a surface brightness of 27.5 B mag. arcsec^{-2} at a height of 20 kpc above the galactic plane in this system. Full details of this work will be published later. In summary, while it remains possible that spiral galaxies generally have massive halos with very large M/L ratios, careful surface photometry can certainly put some harsh constraints on these halos.

In our Galaxy, the stellar population near the sun gives some information about the stellar content of the halo. Say the Galaxy, a typical large spiral, has a massive halo like those discussed above. In the solar neighborhood, a sphere of 10 pc radius would then contain about 50 $M_\odot$ of halo matter, which corresponds to at least several hundred faint stars, depending on the appropriate M/L. These stars would have typical velocities of 150 to 200 km s^{-1} (in each coordinate, relative to a non-rotating frame) for the halo to be in equilibrium. Now a velocity of 100 km s^{-1} at a distance of 10 pc corresponds to a proper motion of 2 arcsec yr^{-1}, but there do not appear to be any known stars with such large proper motions, parallaxes greater than 0″.1 and intrinsically faint (say $M_v > 15$). This may put a useful constraint on the stellar content of a massive halo, at least for the Galaxy.

A final point: we know that the halo of our Galaxy certainly extends to at least 100 kpc radius. There are stars in the solar neighborhood with orbital apogalactica of 100 kpc and more, and the globular clusters Pal 1, 3, 4, 13 and NGC 2419 are now about 100 kpc from the galactic center. So it is not the large radial extent of the postulated halos that is unexpected, but rather the large associated mass.

4. The Lens

The disk itself often appears to have two components. This is particularly clear in SO$_1$ lenticulars, where the disk is unconfused by spiral structure. Here is a quote from the *Hubble Atlas*: "To the eye, images of SO$_1$ galaxies present three distinct luminosity zones on the original plate. There is an intense nucleus, an intermediate zone of lower surface brightness, called the lens, and the characteristic faint outer envelope." (This outer envelope is the exponential component in these systems.) Figure 1 shows the luminosity profile along the major axis of the SO system NGC

1553. The lens, which can be seen on the short exposure photograph, also shows clearly as a plateau in the luminosity profile between about 1 and 2 kpc from the nucleus. All three components described in the quote can be seen in the profile; in particular, Figure 1 demonstrates that the lens is not a photographic effect. The *Hubble Atlas* has many examples of systems with lenses: not only SOs but also spirals

Fig. 1a.   The SO galaxy NGC 1553. The nucleus and lens are visible.

like NGC 210 and NGC 1398 show a clear lens. So far, there seems to be no dynamical discussion of this component.

A few more facts: (1) In SBO systems the bar is very often immersed in the lens. See the *Hubble Atlas*. (2) Edge-on SOs, like NGC 4762 and NGC 7332, show that the lens is part of the flat disk component. (3) Not all disk galaxies have lenses. For example, there is no sign of a lens in the luminosity profile of M33 (de Vaucouleurs, 1959).

What is the lens dynamically? Maybe it is the hot inner disk that Ostriker and Peebles (1973) suggest as one way to stabilise the disk against barlike instabilities. However fact (1) above means that the lens is not always hot enough to do this effectively. Hohl's (1971) model disk evolution is interesting here: it shows the abrupt change in the gradient of the radial surface density distribution associated with a

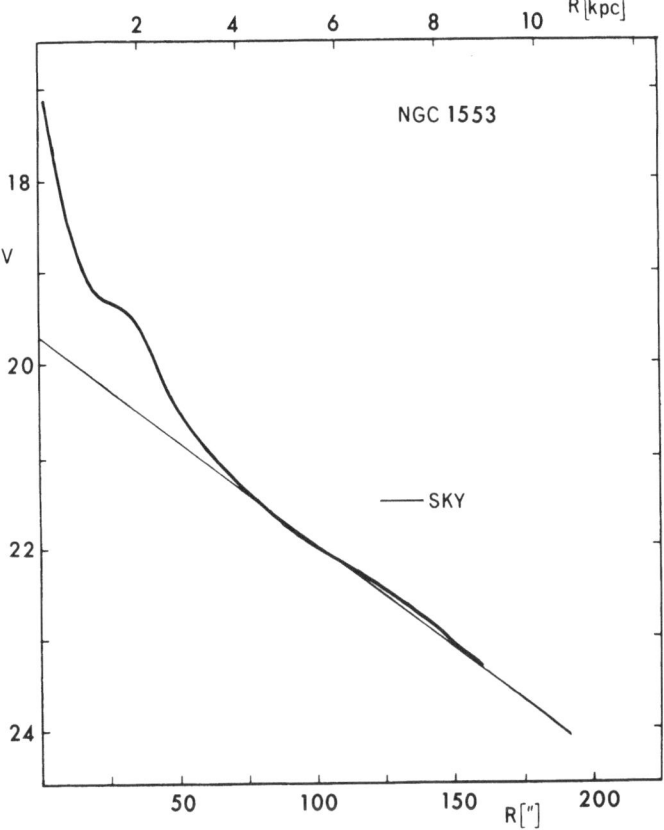

Fig. 1b.   The luminosity profile for NGC 1553 (major axis). The surface brightness is in V mag. arcsec^{-2}.
The lens appears as the plateau between 1 and 2 kpc.

change in the velocity dispersion gradient. It may be then that the lens is the hot inner disk, while the outer envelope may correspond to the cooler disk observed for our Galaxy in the solar neighborhood. This possibility can be tested observationally, at least in principle.

5. The z-Structure of the Disk

We know already how much can be learned about the formation and present structure of elliptical galaxies from their radial luminosity profile: this profile is associated with a particular distribution function (see I. R. King's talk), which itself probably originated through violent relaxation during the Galaxy's collapse phase. In the same way, there may be some information about the z-collapse of the disk, and its present z-structure, to be derived from studying the luminosity distribution in the z-direction for edge-on system, particularly as it seems likely that the z-dynamics of disk stars is almost decoupled from the dynamics in the plane. Edge-on SO galaxies are good candidates for this work, because the interstellar absorption in these systems is low. Mrs Grace is working on this problem at the University of Texas.

Harvey Butcher and I have been studying the z-dependence of the integrated UBV colors in the disk of the edge-on SO NGC 4762. We find a clear change of $U - B$ with z: $U - B$ is smaller for larger z, and this could be analogous to the *radial* color variations in elliptical galaxies, if we believe that metal enrichment, occurring for the disk during its z-collapse, lead to a z-gradient of metal abundance in the disk.

References

Emerson, D. T. and Baldwin, J. E.: 1973, *Monthly Notices Roy. Astron. Soc.* **165**, 9P.
Freeman, K. C.: 1970, *Astrophys. J.* **160**, 811.
Hohl, F.: 1971, *Astrophys. J.* **168**, 343.
King, I. R.: 1966, *Astron. J.* **71**, 64.
Nordsieck, K.: 1973, *Astrophys. J.* **184**, 719 and 735.
Ostriker, J. P. and Peebles, P. J. E.: 1973, *Astrophys. J.* **186**, 467.
Ostriker, J. P., Peebles, P. J. E., and Yahil, A.: 1974, *Astrophys. J. Letters* **193**, 1.
Toomre, A.: 1964, *Astrophys. J.* **139**, 1217.
Warner, P. J., Wright, M. C. H., and Baldwin, J. E.: 1973, *Monthly Notices Roy. Astron. Soc.* **163**, 163.

DISCUSSION

Brosche: With regard to the connection of mass and angular momentum it seems fallacious to condense the information of rotation curves into masses only and neglect the essentially two parameters which they provide (e.g. galaxies with a maximum rotational velocity of 1000 km s^{-1} could exist but are not observed and this has to be explained.)

King: Your remarks about false correlations between mass and angular momentum emphasize the importance of using observations only in the observational domain. To check a presumed relation between M and $\mathcal{H}$, one should figure out what the consequence are for R_{char} and V_{char}, and then look at those quantities. As regards your search for a halo in NGC 253, your model appears to drop off too rapidly to fit the rotation curve. If you had assumed a halo profile that corresponds to the rotation curve, you would expect even more brightness, and thus your negative observational result becomes even stronger.

Freeman: Agree with both comments.

Lecar: Do your observations exclude an Ostriker-Peebles halo which has a very flat surface density $(\sigma \sim 1/r)$?

Freeman: One of the King models I used to estimate the expected surface brightness had $\log(R_t/R_c) = 1.39$ which is a fairly unconcentrated model, and I think we can exclude this sort of distribution for $M/L = 200$, $M \simeq 10^{12} M_\odot$, $R_t = 300$ kpc.

Innanen: A smooth empirical fit of the halo RR Lyrae star distribution to the local Population II data also does not appear to permit a massive halo component for the Galaxy. (See Innanen, K. A.: 1973, *Astrophys. Space Sci.* **22**, 393).

Larson: If galactic halos are made of faint M dwarfs, these stars radiate most of their energy at infrared wavelengths, and one should look for them at red or infrared wavelengths. In this case, I wonder how strong a limit you can set by looking only at blue wavelengths?

Freeman: The number I gave were for $M/L_B = 200$. Going to the red has the advantages you mention but the sky brightness offsets some of this.

Miller: What do the radio astronomers have to say about these edge-on objects? Do any of them show structures like a halo?

Baldwin: In your disk and halo model of our Galaxy, the rotation curve will fit the observed one only if the core radius of the halo is comparable to α^{-1} in the disk. This occurs roughly half way between your models so they do indeed cover the right range of parameters.

DYNAMICS OF EARLY TYPE GALAXIES

F. BERTOLA and M. CAPACCIOLI

Asiago Astrophysical Observatory, University of Padova, Italy

Abstract. An attempt has been made to compute the angular momenta of the early type galaxies NGC 4697 and NGC 4762. Comparison with spiral galaxies indicates that there is a correlation between angular momentum per unit mass and morphological class.

Angular momentum is a crucial parameter in understanding the processes of formation and evolution of galaxies. It is easily derived for spirals when the rotation curve and consequently the density distribution are known. The observed rotation curve of a spiral galaxy, both in optical and radio domain, results from measurements on emission lines produced by the flattest gaseous component, so that the circular velocity is directly derived.

The case of early type galaxies is much more complex from both the observational and theoretical point of view. Their rotation curves are obtained from absorption lines, which are difficult to record and to measure, as discussed elsewhere (Bertola and Capaccioli, 1975). Moreover many problems arise when computing the spatial velocity curve θ_m from the measured rotation curve $\bar{\theta}(x)$. In fact, while it is clearly evident that $\bar{\theta}(x)$ must be some weighted mean of the projected values of θ_m along the line of sight crossing the galaxy at x, we have not complete knowledge about the true nature of the weighting function. This latter must depend on:

(i) tridimensional velocity field geometry of the galaxy;

(ii) space distribution of the stellar populations contributing to the observed absorption lines;

(iii) detailed knowledge of the absorption within the galaxy;

(iv) meaning of what is measured for $\bar{\theta}(x)$, taking into account the effects on the shape of the absorption lines due to the velocity dispersion of the stars and to the observing equipment.

It seems convenient for a first approach to the problem to assume that the weighting function is the space density of light in the galaxy, as deduced from the observed luminosity profile. Even this approximation leaves a difficulty, requiring to extrapolate the rotation curve $\bar{\theta}(x)$ besides the last observed point (Bertola and Capaccioli, 1975).

In addition, in order to compute the angular momentum, the matter density distribution is required. Since the dynamics of an early type galaxy is dominated by the random motions of the stars, the velocity dispersions along three axes should be given. Presently even the most advanced observational techniques allow to obtain only the integrated value of the velocity dispersion in the center. Therefore, rather than using a wide sample of assumptions in deriving the density distribution, it seems here more convenient to have only one, that is the constancy of the mass-to-light ratio, with which the matter density results proportional to the light density.

Hayli (ed.), Dynamics of Stellar Systems, 373–375. All Rights Reserved.

TABLE I

	Morph.T.	D (Mpc)	$\langle b/a \rangle$ adopted	Mass $(10^{10}\,M_\odot)$	Ang. Mom. $(10^{72}\,\mathrm{g\,cm^2\,s^{-1}})$
NGC 4697	E5	14.8	0.6	$2.0 \times \kappa$	$4.2 \times \kappa$
NGC 4762	S0	14.8	0.2	$1.9 \times \kappa$	$10.0 \times \kappa$

With this in mind, we discuss here the determination of the angular momentum for NGC 4697 and NGC 4762, typical examples of E and S0 classes. These galaxies belong to a sample of four objects observed by one of us (F.B.) using the extremely fast image tube nebular spectrograph attached at the Cassegrain focus of the 200″ telescope. A first account of this material has been given by Bertola (1972). The main informations about NGC 4697 and NGC 4762 are summarized in Table I.

In order to compare these early type galaxies with spirals, we computed their masses and angular momenta up to a distance from center $\bar{x} = 250''$ (~ 18 kpc). The

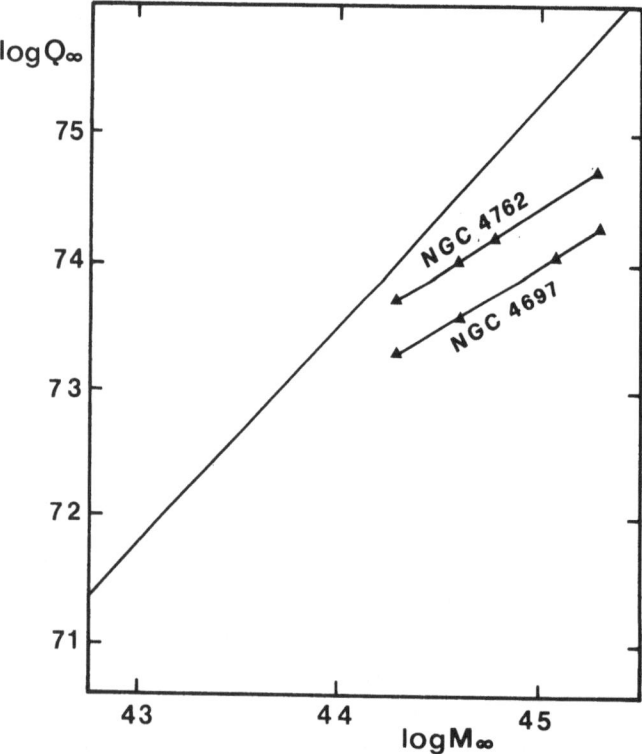

Fig. 1. The straight line represents the mass-angular momentum relation holding for spiral galaxies, from Takase and Kinoshita (1967). Segments refer to the early type galaxies NGC 4697 and NGC 4762, in the range of values of the mass-to-light ratio from 5 to 50. Marks on segments correspond to values of the mass-to-light ratio κ: 5, 10, 30, 50 for NGC 4697 and 5, 10, 15, 50 for NGC 4762. The elliptical NGC 4697 has a much lower angular momentum per unit mass with respect to a spiral of equal mass, while the S0 lies in between. It seems likely that different mass-angular momentum relations should hold for the three morphological classes.

conclusions we reach with this comparison are not modified by increasing $\bar{x}$. Using the observed velocity curves, together with the luminosity profiles $I(x)$ in the B system (Bertola and Capaccioli, 1975; van Houten, 1961) we derived the rotational velocities for the two galaxies. It was assumed, as a working hypothesis, that $\bar{\theta}(x)$, outside the farthest point reached by our observations, is at a constant level equal to that of the last determined velocity. Masses and angular momenta have been computed assuming a zero-thickness model characterized by a surface density $\kappa I(x)$, leaving the mass-to-light ratio κ as a free parameter.

In Figure 1, where the mass is plotted against the angular momentum, we compare NGC 4697 and NGC 4762 with the spirals represented by the relation (straight line) found by Takase and Kinoshita (1967). Representative points for NGC 4697 and NGC 4762 lie on small segments, since κ is assumed to vary in the range from 5 to 50. Typical values of the mass-to-light ratios for these two galaxies could be 30 and 15 respectively. Figure 1 shows that the angular momentum per unit mass of the elliptical galaxy NGC 4697 is lower than that of a spiral of equal mass by a factor from 5 to 30 in the range of κ considered, the most probable value being 20. The SO galaxy NGC 4762 lies in between (factor from 2 to 10, most probable factor 5). Similar computations, carried out on the SO galaxy NGC 4111 (Capaccioli, 1974), give a result close to that on NGC 4762.

The above analysis, although based on very few cases, leads to consider the exhistence of different mass-angular momentum relations for the three classes of spiral, SO and elliptical galaxies. If this result is confirmed by further observations, it will greatly help the understanding of the processes of formation and evolution of galaxies.

References

Bertola, F.: 1972, *Atti XV Riunione SAIt*, 199.
Bertola, F. and Capaccioli, M.: 1975, *Astrophys. J.*, in press.
Capaccioli, M.: 1974, *Atti Cel. V Cent. Coper.*, Univ. Padova, 109.
Houten, C. J. van: 1961, *Bull. Astron. Inst. Neth.* **16**, 1.
Takase, B. and Kinoshita, H.: 1967, *Publ. Astron. Soc. Japan* **19**, 131.

DISCUSSION

Gott: The peak in Bertola's rotation curve is about 65 km s^{-1} while the central velocity dispersion is 310 km s^{-1}. The isophotes at this point are still noticeably elliptical. Reasonable dynamical models would predict the rotation to be more nearly comparable with the velocity dispersion. Previous velocity dispersions by Minkowski have been shown to be in error by a factors of two. Certainly it would be advantageous to have more data both on the velocity dispersion and the rotation curve.

Bertola: I exclude that the velocity of rotation in NGC 4697 can reach a value comparable with that measured for the velocity dispersion, unless the rotation curve increases again outside the observed region.

King: Minkowski's larger velocity dispersions seem to require very little correction, in so far as we have been able to check them. It appears that only his smaller values need serious corrections.

PART III

NUCLEI

GALACTIC NUCLEI AND
COMPACT SUPERMASSIVE OBJECTS

W. C. SASLAW

University of Virginia, and National Radio Astronomy Observatory* Charlottesville, Virginia, U.S.A.
and
Institute of Astronomy, Cambridge, England

Abstract. This is a review of the observational evidence concerning compact supermassive objects, their formation and evolution, and their dynamical interaction with dense stellar systems in galactic nuclei.

> *All these have never yet been seen –*
> *But Scientists, who ought to know,*
> *Assure us that they must be so...*
> *Oh! let us never, never doubt*
> *What nobody is sure about!*
>
> Hilaire Belloc, 1910.

One of the reasons galactic nuclei are so interesting is that they provide an arena for a remarkable range of physical problems. Radiation, gas, dust, magnetic fields and compact objects all interact in a rich variety of ways in different regions of the nucleus. The energy densities of radiation, gas, dust and magnetic fields cover many orders of magnitude; the forms of compact objects may include black holes, neutron stars, white dwarfs, supermassive objects and even ordinary stars.

Among all these interactions, it is probably fair to say that gravitation is the most basic in that it determines the overall structure and evolution of the galactic nucleus. Indeed, it is the gravitational contraction of the whole system, or parts of it, that ultimately drives most other forms of energy in the nucleus. Rather than repeat recent general reviews of gravitational interactions and other problems in galactic nuclei (Saslaw, 1973, 1974), I'd like to try to review one major question which the observations have raised more and more insistently during the last year or two: what is the dynamical relation between galactic nuclei and compact supermassive objects?

As a working definition of compact supermassive objects, we may suppose that their mass is $\gtrsim 10^3 \, M_\odot$, to distinguish them from stars, that they are gravitationally bound, and that their size is much smaller than the size of the galactic nucleus with which they are associated. This leaves open the nature of their structure and lifetime. They may be black holes or black holes surrounded by gaseous disks or satellite systems. They may be supermassive stars supported by gas, radiation pressure, rotation, or magnetic fields. They may be very small (e.g. relativistic) clusters of small mass stars. In principle there is no known reason why matter should not pass through any of these forms. Therefore, let us first ask:

---

* Operated by Associated Universities. Inc. under contract with the National Science Foundation.

Hayli (ed.), Dynamics of Stellar Systems, 379–393. All Rights Reserved.

1. Is There Any Observational Evidence for Compact Supermassive Objects?

There are five main phenomena in which people have suggested looking for compact supermassive objects. These are quasars, galactic nuclei, Lacertids, jets near the centers of galaxies, and the components of extended extragalactic radio sources. Of course, there is also the possibility that all these phenomena, including quasars (Kristian, 1972), are manifestations of active galactic nuclei. If so, the discovery of these objects in one phenomenon would have important implications for all five of them. Let us consider each in turn.

1.1. QUASARS

After Hoyle and Fowler (1963) first proposed the existence of thermally supported supermassive stars in radio galaxies, they were quickly siezed upon as a possible energy source for quasars. After pulsars were discovered, more massive versions of stars supported by rotation or magnetic fields – spinars – were also suggested (Cavaliere *et al.*, 1971). These models could, in principle, be distinguished observationally from models in which quasars were powered by many small explosions (e.g. supernovae or colliding stars) or by multi-pulsar systems, if the massive objects produce periodic variations in intensity. Either a pulsation of the object, or a rotation which produces a pulsar-like beacon could be observed. This gave rise to a still inconclusive debate (e.g. Chertoprud *et al.*, 1973) about the existence of real short term periodicities in quasars, and particularly in 3C 273. Although there is no compelling evidence for periodicity, the absence of evidence does not rule out compact massive objects in the form of black holes surrounded by accretion disks or satellite systems, since these types of supermassive object need not have periodic intensity variations.

There is a second line of evidence in quasars which is also inconclusive at present. This is the possible physical connection between quasars and galaxies which are close together on the celestial sphere. Perhaps the best evidence for a physical connection is the apparent bridge (Arp, 1971) between Markarian 205 ($Z=0.07$) and the spiral galaxy NGC 4319 ($Z=0.006$). However, it is not clear whether this bridge is just a photographic effect of two images close together on the plate (Lynds and Millikan, 1972). A bridge may also join Markarian 205 to a stellar object 3" away (Weedman, 1973). One can also argue for a physical association on the grounds that there are more geometric associations (about a half-dozen) than would be found statistically in a random distribution (Burbidge *et al.*, 1971). However selection effects make the meaning of this result very uncertain (Bahcall *et al.*, 1972; Hazard and Sanitt, 1972; Burbidge *et al.*, 1972). If a physical connection between galaxies and some quasars does turn out to exist, then there could be two types of quasar (Chio *et al.*, 1973). One class would be the more usual type of quasar at cosmological distance; the second, rarer, class would consist of massive objects ejected from their associated galaxies.

1.2. GALACTIC NUCLEI

The most intriguing galactic nucleus from the present point of view is the Seyfert

NGC 1275, which is identified with the strong (~ 50 fu) radio source Perseus A (alias 3C 84). This object is sufficiently peculiar to be interesting and sufficiently close (54 Mpc if $H = 100$ km s^{-1} Mpc^{-1}) to study in detail (Burbidge and Burbidge, 1965; De Young et al., 1973). Very long baseline interferometry of the radio source in the nucleus (Legg et al., 1973) shows it to consist of at least four components. Of course, VLB observations with one baseline do not give a unique picture of the source. In one model fit to the data, there are four sources, each a couple tenths of a parsec in size, stretching along a line about two parsecs long. Each source emits between 5 and 19 fu at ~ 3 cm. The other proposed model contains one strong resolved source, and four weaker sources (0.3–1 fu) with diameters less than 0.05 parsec. While this directly shows that powerful radio emitters ($\sim 10^{42}$ erg s^{-1}) can occupy small volumes, we need still higher resolution before we can understand the nature of these regions.

1.3. LACERTIDS

The Lacertids, named after their prototype BL Lacertae, are a peculiar class of objects characterized by a nearly stellar appearance, a continuous optical spectrum free of lines, optical and radio variability over periods of minutes to months, and a flat or inverted radio spectrum. At various times the Lacertids were suggested to be accreting neutron stars, accreting massive black holes in our Galaxy, and blue shifted quasars. However, the recent measurement (Oke and Gunn, 1974) of a redshift of 0.07 in the fuzz surrounding BL Lacertae suggests that it is associated with the nucleus of a distant elliptical galaxy and radiates $\sim 10^{45}$ erg s^{-1}. These measurements are very difficult and still await confirmation. The Lacertid P1205-008 is separated by $10''$ arc from a galaxy with $Z = 0.1$. (Condon and Jauncey, 1974). If the Lacertid and galaxy have same redshift, their separation is 30 kpc. The size of the Lacertid at 10 GHz, estimated from synchrotron self-absorption ($B \lesssim 1$ G) is $\lesssim 3$ pc, so the ratio of separation to size $\gtrsim 10^4$.

If this interpretation is correct, then the statistical association between Lacertids and galaxies discovered by Condon and Jauncey (1974) becomes very significant. Of the ten Lacertids for which precise radio positions were known, nine are associated with either well defined galaxies or, as in BL Lac, with surrounding nebulosity. Remarkably, several of these Lacertids are not in the center of their galaxies, but rather near the edge of the galaxy or within one or two galactic radii of the center. Thus Condon and Jauncey suggest that they were ejected from the nucleus of their parent galaxy. At present many more potential Lacertids are known, and when their precise radio positions are measured, we will see whether this association improves.

1.4. JETS

In a small number of galaxies a jet containing several optical condensations sticks out from the nucleus. The best studied case is the giant EO galaxy, M87. Even here, there is no answer to the fundamental question whether the jet is colinear, or just several objects scattered in a half-plane seen approximately edge-on. Unfortunately, statistical arguments are not much use because only a few jets are known and selection effects reduce the chances of seeing a disk of objects face-on (although one may be present in the other part of NGC 1275, Sandage, 1971). The optical knots in M87 are $\lesssim 20$ pc in radius (for $H = 55$ m s^{-1} Mpc^{-1}), and their ages are between about 10^4 yr. and 2×10^5 yr (for velocities between the escape velocity and c). The most recent analysis of their structure by Okoye (1973) shows that it is barely possible to stabilize them by inertial confinement if they are gas blobs with a temperature between $\sim 10^4$–10^5 K. If they are very hot ($T \gtrsim 10^7$ K), or not sufficiently dense, they may be confined by ram pressure if they are ejected with $v \approx c$ or if the magnetic field in the blob has its equipartition value. However such a magnetic field would lead to a lifetime of only $\sim 10^3$ yr for electrons generating optical synchrotron radiation, and some mechanism, not presently known, would be necessary to frequently replenish these electrons in the knot. If, on the other hand, the knots contained compact objects of mass $\gtrsim 10^7$ $M_\odot$, the problem of their stability and electron supply would be greatly alleviated, and perhaps even solved.

Perhaps somewhat related to jets, are cases where one galaxy may have ejected a nearby companion. Most of the evidence for this is due to Arp (e.g. Arp, 1972), but few astronomers are very convinced by it. However, there is a recently discovered peculiar object (Arp and O'Connell, 1975) which is rather intriguing from this point of view. The blue compact galaxy CG 1124 + 54 has a main body ~ 1 kpc long ($H = 50$ km s^{-1} Mpc^{-1}) consisting of a central core ~ 200 pc in radius with two fainter disturbed lobes on either side. About 1.6 kpc from the center, approximately along the major axis, is a companion object whose size in about 500 pc $\times$ 250 pc. The edge of this companion farther from the main compact galaxy is rounded and sharp; the closer edge is fainter and less well-defined. Both objects have a redshift of 2895 ± 32 km s^{-1}.

Two explanations for this configuration seem plausible. First, the companion may be gravitationally bound to the main body and we happen to be looking along the polar axis of their orbit. The peculiar structure of the object could then be caused by tidal stresses or by internal dust lanes. Second, the companion may have been ejected from the main body, distorting it in the process. In this case the radial velocity of both objects would be the same even if we are not observing the system perpendicular to the line of ejection. This is because a compact massive object would radiatively ionize the background gas around the main body, but would not drag much of this gas with it. High resolution spectrometry could determine whether the companion shows the turbulence and ionization structure expected for a massive object moving supersonically through surrounding gas (Saslaw and De Young, 1972).

1.5. EXTENDED RADIO SOURCES

In all the previous examples, except for jets, the evidence for ejection is at best circumstantial and at worst merely circumgalactic. In the case of extended extragalactic radio sources, however, there is general agreement that significant amounts of energetic material have been ejected from the nucleus of a galaxy. The main question is the form of the ejecta and the method of ejection. Until the last year or two, models of extended radio sources were dominated by the conventional picture in which an explosion in the nucleus ejects two masses of relativistic plasma in opposite directions, and the escaping plasma is confined by its ram pressure on the intergalactic gas. While this process may sometimes occur, the argument for its general applicability has run into four severe observational problems:

(i) Many radio galaxies contain very compact components. In 3C 390.3, for example, the ratio of the width of the smaller component at 5 GHz to its separation from the central galaxy is greater than 30 (Harris, 1972). In Cygnus A, perhaps the best studied extended source because of its proximity to us, a significant flux ($\gtrsim 20\%$) at 5 GHz from the main double components comes from compact regions $\lesssim 1$ kpc in scale, but the separation of these regions from the central galaxy is ~ 95 kpc. Neither inertial confinement of the plasma by cold gas, nor ram pressure confinement with the intergalactic gas seem adequate to produce such large ratios of separation to size (Hargrave and Ryle, 1974).

(ii) In most double radio sources, the radio components are aligned with the central galaxy to within 5°–10°. In Cygnus A, the compact components are aligned with the central source to within ~ 30 arcmin (Hargrave and Ryle, 1974).

(iii) When the central optical object is an elliptical galaxy, the main radio components have a strong tendency to lie approximately in the plane of the galaxy, rather than near the poles (Mackay, 1971; Bridle and Brandie, 1973). However, in models which eject gas from the galaxy, one would expect the ejection to occur near the poles, along the path of least resistance.

(iv) Perhaps the most severe problem for theories which eject a burst of plasma, is that the lifetime of the relativistic electrons to synchrotron radiation is usually considerably less than the age of the radio source. Examples are Cygnus A (Hargrave and Ryle, 1974) and the giant sources DA 240 and 3C 236 (Willis et al., 1974) which have linear extents of 2.0 and 5.7 Mpc respectively. However this is a difficulty even for many smaller sources.

To meet these objections, two different major theories of radio source structure have arisen. In one, beams of low frequency radiation (Rees, 1971) or relativistic particles (Scheuer, 1974; Blandford and Rees, in press) stream continuously out of the nucleus of the galaxy. This can account for problems (i) and (iv), but has difficulties with (ii) and (iii). The second approach is to suppose that compact massive ($M \gtrsim 10^5$ $M_\odot$) objects are ejected from the nucleus of the galaxy (Burbidge, 1967; Saslaw et al., 1974). The detailed nature of these objects is of secondary importance and there are a number of possibilities. They may be either spinars, supported by thermal, rotational and magnetic energy. They may be black holes surrounded by a disk of gas which is slowly accreted, or by a satellite system of smaller spinars or pulsars. They may even evolve from spinars into black hole systems. Any of these objects provides a natural solution to these four important observational problems. They store energy in compact regions (without adiabatic losses), they can be ejected in close alignment with the central galaxy and near the plane of this galaxy, and they can supply energy fairly continuously over long periods of time.

At present, none of the observational evidence for the existence of compact super-massive objects is really compelling, and a skeptic would probably be justified in doing nothing, or at least working on something else. Nevertheless, the promise of massive objects – and the problems of other explanations – are producing greater interest in their properties and relation to galactic nuclei. We therefore turn first to a brief review of the work which has been done on the formation and evolution of these objects, and then to their dynamical interaction with galactic nuclei.

2. Formation and Evolution of Compact Supermassive Objects

2.1. FORMATION

Not very much is known about the manner in which compact massive objects may form. However, if they form anywhere, it is likely to be in galactic nuclei where the density of gas and stars is high. One possible process is for a dense relativistic rotating disk of gas, released from stars either through normal evolution or collisions, to collect in the center of the nucleus and fragment into several objects (Salpeter, 1971). So far, no detailed calculations have been made to determine the conditions for non-linear fragmentation, the masses of the fragments, or whether fragments fragment. The problem is one of the most intricate in astrophysics; star formation which is a special case seems almost simple by comparison.

A more calculable process for formation of these massive objects is stellar coales-cence. If the nucleus of a galaxy contains a sufficiently dense stellar system, stars may often collide bodily. For a typical star the mean time between collisions is (Spitzer and Saslaw, 1966)

$$\tau_c = \frac{1}{n\sigma v} = \frac{9.7 \times 10^{21} \, R_{(pc)}^{7/2}}{N^{3/2} (m^{1/2} r^2 / m_\odot^{1/2} p_\odot^2)(1 + 8.8 \times 10^7 \, R_{(pc)} r_\odot / Nr)} \, \text{yr}, \tag{1}$$

where r is the radius of the star. In deriving this relation, the virial theorem has been used and the geometrical cross section has been increased by a factor $(1 + 2Gm/rV^2)$ to account for the gravitational attraction, ignoring tidal deformation.

If the collisions are sufficiently energetic, the stars will dissolve into gas. The re-quirement for this to be possible can be estimated approximately in the case of a head-on collision. Most of the kinetic energy is converted into thermal energy, and substantial liberation of gas can occur if the total energy is positive. So if each star is a polytrope of index n,

$$\tfrac{1}{2}m\left(\frac{V}{2}\right)^2 - \frac{3}{2(5-n)}\frac{Gm^2}{r} > 0, \tag{2}$$

where V is their relative velocity at large separation. For two sunlike stars $(n=3)$, this implies $V \gtrsim 1500 \, \text{km s}^{-1}$. This result is approximate since it does not include the effects of shocks and the time-dependent gravitational field. Next we describe the more exact results, first considering what happens when the relative velocities are so small that the bulk of the two stars coalesces (Ulam and Walden, 1964; Colgate, 1967;

Sanders, 1970; Seidl and Cameron, 1972). This problem is somewhat similar to the collision of two gaseous galaxies described by Alladin (this volume, p. 167).

If two similar stars collide at relative velocities exceeding several hundred kilometers per second, most of the gas will interact supersonically relative to the local sound speed, and shocks will convert much of the kinetic energy of stellar motion into thermal energy which is then radiated. Thus the collision is highly inelastic. If, moreover, the positive energy criterion of Equation (2) is not satisfied by a large margin, most of the stars' material will coalesce. The distended, newly-formed object pulsates for some time, and then settles down to a well-defined star. During the collision, the temperature and density are not great enough to generate an important amount of energy by thermonuclear reactions (Spitzer and Saslaw, 1966; Mathis, 1967).

The detailed hydrodynamics of coalescence is very complex; the only extensive treatment is for a star which smashes into its mirror image (Seidl and Cameron, 1972). This is a two-dimensional numerical study of the head-on collision of two polytropes of index 3 with solar mass and radius. As the encounter proceeds, the stars become squashed and a sheet of gas is heated then ejected in the plane perpendicular to the initial relative velocity. Simultaneously a recoil shock forms in the outer layers and ejects gas from the backs of the stars, an effect which is relatively more important at low collision velocities. Experiments for distant relative velocities of zero, 1000, and 2000 km s^{-1} showed that about 5%, 18%, and 60%, respectively, of the gas was liberated. Thus our rough estimate provides a surprisingly good criterion for disruption, especially considering the simple energy argument on which it is based.

Unfortunately, only a small fraction of collisions are head-on, and the rest must be handled gingerly by approximate methods. Sanders (1970) has applied simplified models of undeformed stars to collisions with relative velocities at infinity between 62 and 2356 km s^{-1}, impact parameters from head-on to grazing, and mass ratios from 1:1 $(M_\odot)$ to 1:50 $(M_\odot)$. In these models the two stars are divided into long rectangular tubes of gas parallel to their relative velocity. Each tube collides only with its geometric counterpart, and their changes are not coupled to the rest of the star. The collision converts the kinetic energy of the star's motion into heat, conserving linear momentum, and this thermal energy is divided between the two mass tubes in proportion to their kinetic energy before impact (relative to their own center-of-mass frame). If the thermal energy of a mass element is greater than its binding energy to the star to which it is most strongly bound, the mass element is assumed to escape.

Clearly the accuracy of these assumptions is very uncertain, especially for low-velocity collisions which may transfer substantial momentum perpendicular to the relative velocity of the gas tubes. For a head-on collision, at these low velocities, no mass would be lost on this approximation; therefore we would expect the approximation to become worse as the impact distance, p, becomes smaller. The general shape of the curve of mass loss as a function of p would then be zero for $p=0$ and for $p>2r$, with a maximum in between. As an example, Sanders' computations for two suns colliding with $V_\infty = 62$ km s^{-1} give a total maximum mass loss of $\sim 0.04\ M_\odot$ at $p \approx 0.4\ r_\odot$.

What is the condition that the two stars coalesce? The collision converts kinetic energy from their orbital motion irreversibly into heat. If enough orbital energy is lost, the stars will become bound to each other and successive collisions will reduce the orbit's semimajor axis until most of the mass merges permanently. This occurs if the kinetic energy of motion which is converted into heat exceeds the orbital kinetic energy of the two stars at infinity. Without detailed computations it is not certain how much of the thermal energy is irreversibly lost, i.e. how inelastic the collision is. Sanders has assumed complete inelasticity, which maximizes the chance of coalescence. Then one can compute how much thermal energy is produced in each colliding mass tube, add up the total for all tubes, and see if this is enough to coalesce the stars.

In applying this procedure to stars of different mass, it is important to know the density distribution of the colliding stars. More massive stars will generally have lower average density than the less massive ones. Their density distribution will dominate the question of whether two stars of greatly different mass interact sufficiently strongly to convert enough orbital energy into heat so that they coalesce. Colgate (1967) first suggested that stars of $M \gtrsim 50 \ M_\odot$ would not coalesce with the more numerous field stars of $\sim 1 \ M_\odot$, but would simply have holes punched through them. Thus there would be an upper limit to the mass which could form by such coalescence. However, this estimate assumed that the coalesced star forms with the same binding energy per gram as its progenitors, and retains the same polytrope structure after relaxation so that in the new coalesced star $R \sim M$. On the other hand, Sanders (1970), assumed that $R \sim M^{0.7}$ and the density distribution in the relaxed coalesced star is homologous to the sun. Moreover, he also considered the effect of the gravitational field of the massive star in increasing the relative velocity of collision, resulting in greater heating of the gas tubes. The combined effects of these assumptions is that the ability to coalesce does not decrease so strongly with large mass ratios as in Colgate's calculation. This result is very important for the general evolution of the cluster.

Sanders' main result regarding the physics of coalescence is that, with the assumptions outlined above, two stars can coalesce at sufficiently small impact parameters provided that their relative velocity at infinity is less than a critical value. This critical velocity decreases as the ratio of the more massive to the less massive star increases. For example, with a mass ratio of 1:4, coalescence can occur if $V_{rel} \lesssim 1800$ km s^{-1}, while for a ratio 1:50, $V_{rel} \lesssim 1400$ km s^{-1}.

Stars are very nutritious. A large star can increase its longevity by swallowing a smaller one for two main reasons. First, of course, there is the added hydrogen fuel. Second, more hydrogen of the massive star is mixed throughout the core, increasing the main-sequence lifetime. Thus whether or not a massive star evolves into a supernova depends on the ratio of its coalescence-mixing time scale to its main-sequence lifetime in a given state. If the core of a coalesced star mixes faster than it burns, it may be possible to build up extremely massive stars in the center of the stellar system. This question begs an answer.

Having been swallowed, a star must be digested, then absorbed. The additional heat created in the collision distends the coalesced star beyond the normal size for its total mass. The bloated object pulsates awhile and eventually settles down to mechanical equilibrium after several relaxation times of order $(G\varrho)^{-1/2}$ (about 15 min for the Sun). However, the thermal energy has not yet been absorbed throughout the star, and for this to occur requires several photon diffusion periods, which entails a Kelvin-Helmholtz time scale of order GM/RL (about 10^7 yr for the Sun). The tone of these last two sections indicates that our present knowledge of the structure and evolution of the coalesced star is only qualitative. This represents one of the most important, and one of the most difficult, problems in our understanding of dense stellar systems.

As the stellar system evolves, there is a period during which the time scale for coalescing collisions to involve many stars becomes less than the time scale for these massive stars to evolve off their main sequence. If enough time is spent in this regime (before disrupting collisions take over) stars of extreme mass may form. At first the mass of a typical star is built up by coalescence with smaller stars. Every addition of hydrogen with mixing is assumed to be so effective that it sets the star's evolutionary clock back to zero (an important question for further calculation). In this way stars of $\sim 500\ M_\odot$ may form (Sanders, 1970). They cannot come into equipartition with lighter stars (Spitzer, 1969; Saslaw and De Young, 1971), and so the massive stars sink to the center. There they coalesce one with another and accelerate the building of even more massive stars. Both the limiting mass that can be reached by this process, and the number of supermassive objects which ultimately result are important problems requiring further calculations.

2.2. EVOLUTION

If a hot, thermally supported, supermassive star forms, there are four possible ways it may evolve (Appenzeller and Fricke, 1972; Fricke, 1973, 1974). A non-rotating star with $M \lesssim 4 \times 10^5\ M_\odot$ settles down into thermonuclear equilibrium for $\sim 10^5$–10^6 yr. But if $M \gtrsim 4 \times 10^5\ M_\odot$, it can explode or collapse into a black hole. Explosion occurs, for a given mass, if the initial heavy element abundance is great enough to produce rapid thermonuclear burning. For example, if $M = 10^6\ M_\odot$, explosion occurs if $Z > 0.04$. When Z is too small for a given M, or M is too great for a given Z, thermonuclear energy cannot halt the gravitational collapse. The fourth possibility is that relaxation oscillations occur in which the radius, luminosity, and rate of energy production change periodically in the pulsating star. But this does not seem to occur unless the rate of burning is arbitrarily damped during the explosive phase. The explosive energy of rotating supermassive stars (10^{56}–10^{60} erg) may be several orders of magnitude greater than that of non-rotating ones since rotation stabilizes the star against post-Newtonian instability and increases the upper mass limit for explosions (as against collapse). These explosions may provide an explanation for the optical filaments of NGC 1275, and also for some of the core-halo radio sources.

If the supermassive object which forms is not kept from collapse by gas and radiation pressure, it may be supported mainly by rotation or magnetic fields. No support, however, can last indefinitely, since the disk must eventually cool, and become so thin that it collapses or fragments. The time taken for it to reach this stage depends on details of models, but is usually between 10^5–10^7 yr. When the unstable stage is reached, further collapse and fragmentation occur on a dynamical time scale, which may be only minutes or hours.

The most stable compact supermassive object is a black hole, and often this may be the end result of evolution. However, it is unlikely that all the material goes into the black hole when it first forms; probably much remains behind to form a gaseous disk – or a system of satellites if there is multiple fragmentation – around the black hole. After some initial rearrangements in which gas and fragments close to the hole are swallowed up, and material far away is lost from the system, the total mass surrounding the hole becomes less than the mass of the hole itself. Such a system may be stable for long periods. If it is mostly gas, it evolves on a viscous time scale, during which it is a powerful source of radiation (e.g. Lynden-Bell, 1969). If the fragments have become stars, white dwarfs, neutron stars, small spinars, or small black holes, the system may be stable indefinitely – like our solar system – until secular instabilities destroy the orbits. In this case there may also be strong radiation, especially if the satellites have high magnetic fields. Considerable work has been done on black holes surrounded by accretion disks (e.g. Pringle *et al.*, 1973) but very little is understood about black hole satellite systems. Some of their radiation properties will probably resemble those of multi-pulsar systems (Arons *et al.*, 1974, to be published).

3. Dynamical Interactions between Compact Supermassive Objects and Galactic Nuclei

A compact supermassive object can interact with a galactic nucleus gravitationally and through the effects of its radiation on the surrounding gas. The second interaction is important for models of quasars, Seyfert galaxies, and hydrodynamic explosions in some galaxies. Since the radiation effects have been discussed many times before, and since this symposium is primarily concerned with gravitational dynamics, I'll mostly review the gravitational interactions here.

A dense stellar system loaded with a supermassive object sitting in its middle naturally has a different distribution of stars from an unloaded system. Wolfe and Burbidge (1970) investigated this from the point of view of putting upper limits on the mass of the compact object, by requiring any modification of the stellar density distribution to be consistent with the presently observed projected light and velocity distributions in galactic nuclei. This gave $M_{object} \lesssim 10^{10} \, M_\odot$ from the velocity dispersion, independent of whether the stellar distribution is relaxed. To determine the structure of the system near the massive object, Wolfe and Burbidge assume that the stellar distribution is in isothermal equilibrium, and they superimpose the gravitational potential of the central mass on the potential of a standard *non-singular* isothermal sphere. Although this procedure indicates the main results, it is not quite consistent since one really wants the solution for an isothermal sphere with a singularity at its center. Subsequently, Peebles (1972) assumed that the star distribution function depends only on a single power of the total energy and considered only the region where the massive object dominated the gravitational field. For a steady state distribution this implies that near the compact object $\varrho \sim r^{-9/4}$, in contrast with the density run $\varrho \sim r^{-2}$ in the outer region of an isothermal sphere. This difference is too small to be detected with present observations unless the M/L ratio of the stars also varies strongly with distance from the center.

Recently J. M. Huntley and I (1975) have looked at the effect of a central massive object on distributions of stars satisfying general polytropic or isothermal equations of state, including the gravity of both the stars and the object consistently. The main results are that these loaded polytropes have a steep central cusp in their density distributions, followed at larger radii by a density plateau where the self gravity of the stars becomes comparable to the gravity of the massive object, and then by a further drop at radii where the gravity of the stars dominate and the density approaches that of a normal polytrope. This structure of a central cusp, plateau, and smooth decrease is in contrast to normal polytropes whose density has zero gradient at the center and decreases smoothly to zero further out. One important effect of the central cusp is to decrease the time scale for stars in the core of the nucleus to collide bodily, compared with the average time scale for this in the rest of the system, or in a system with the same number (and mass) of stars, but without a central object. The dynamical relaxation times of stars in the cusp is also strongly modified by the object. Moreover stars venturing too near the center will be consumed by the massive object, whether it is a black hole or a spinar of some sort.

The general conclusion is that the presence of a single massive object will produce more violent activity in a more concentrated region of the center of the nucleus, than if the object were absent. These modifications have yet to be developed in detail. One especially important problem is to work out the rate at which stars flow into the cusp to replace those which collide or are consumed. All the studies so far have assumed an isotropic distribution function, which is probably adequate for understanding the overall structure of the system, but insufficient for explaining the evolution in the center.

Processes of fragmentation or stellar coalescence which create one massive object, may well produce many. We are then faced with the question of how these massive objects interact dynamically with each other, as well as with the rest of the stars. The simplest problem, of course, is to consider the massive objects as point particles exerting Newtonian forces on each other. The orbits of two massive objects interacting in this way are stable, but three or more are usually unstable.

If three massive objects come close together at the center of the nucleus, two of them can give so much kinetic energy to the third that it escapes. The two then form a more compact binary, to conserve total energy. Momentum conservation requires the binary to recoil in the opposite direction. The time scale for this to happen can be very short, ranging from one dynamical crossing time of the initial binary for the case of a flyby on a direct orbit, to (typically) several hundred or (rarely) thousand crossing times for a three body system with all objects initially given random positions.

To understand this problem, Saslaw *et al.* (1974) have numerically computed the orbits of 25000 triple systems and 250 two-binary systems chosen to illustrate a very wide range of initial orbital conditions. In effect, these scattering experiments turn the computer into a high energy accelerator with particles of $\sim 10^{60}$ GeV, in the usual units. The results yield distribution functions for properties of final orbits as a function of the distributions of initial parameters.

All these numerical experiments give a great deal of information about the general three-body problem, which is interesting quite apart from its applications (Valtonen, 1974). Here however, I'll just mention briefly some of the aspects of the gravitational slingshot relevant to the observational problems raised in the first section.

First, the disruption of three bodies with negative total energy always results in a two sided configuration with respect to the center of mass. This effect of momentum conservation results in two components exactly aligned with the central galaxy if both escape and they are not influenced by the galaxy. Thus the alignment of hot spots in Cygnus A is easily accounted for. In order for them to be approximately equal distances from the central galaxy, they must have nearly equal masses, and their initial orbits must satisfy special conditions which Valtonen will describe later in this Symposium. The small amount of misalignment observed in many radio doubles is likely to come from exchange of angular momentum as the asymmetric galaxy perturbs the orbits of massive objects moving through it at different speeds (Saslaw, 1975).

Next, if we consider the finite extent of the massive object, then it must be compact in order to be accelerated to high enough velocity to leave the galaxy ($\gtrsim 0.01\ c$). As a rough rule of thumb, a finite thumb cannot be gravitationally accelerated to a velocity greater than the escape velocity from its surface without being tidally disrupted. Thus $R \lesssim 10^4\ R_{\text{Schwarzschild}}$ which for $10^8\ M_\odot$ is 0.1 pc, and the radio sources will contain compact components (unless they all become unstable and explode). In fact the typical velocity spectrum from the experiments is fairly broad with a low energy cut off and a long high energy tail. However the velocity of ejection must be $\lesssim 10^4$ km s^{-1}, so that gravitational radiation does not destroy the system. There is indeed some evidence (Mackay, 1973) that the velocities of typical radio sources are of this order.

The numerical experiments show that the particles tend to be ejected fairly close to the plane of the total angular momentum, typically within 30°. If we make the fairly natural assumption that this is also approximately the plane of the galaxy's total angular momentum, then we expect most of the ejections to be nearer the plane of the galaxy than the pole, as found observationally. Of course there will always be the occasional exception and, since this is a statistical effect, one can't learn much from one particular observation.

As the massive object moves out through the galaxy, its orbit is altered by interacting both with the mean field of the galaxy and with the fluctuations in this field which the object itself induces as it passes. The deflection produced by the mean field of a static asymmetric galaxy is straightforward to calculate numerically, and Valtonen (1974) has done this for a Schmidt model potential. It is significant if $V_{escape} < V_0 \lesssim 1.2 \, V_{escape}$ where V_0 is the initial ejection velocity of the object and V_{escape} is defined as the velocity necessary to reach 100 kpc starting at the center.

The deflection produced by self-induced time dependent fluctuations can be examined analytically for idealized conditions (Saslaw, 1975). The root mean square angle of deflection varies as $(V_{random}/V_{object})^3$ in this case, showing a strong velocity dependence. Again if $M_{object} \gtrsim 0.1 \, M_{nucleus}$ and $V_{object} \lesssim 1.2 \, V_{random}$ this can also produce a few degrees of misalignment. One prediction of these deflection mechanisms is that the radio sources whose components have the highest ejection velocities should be best aligned. Statistically the largest sources should have some combination of highest velocities and longest lifetimes, but it is not clear how to separate these properties. In this respect it may be significant that the double components of the largest known radio source 3C 236 with a separation 5.7 Mpc are aligned to within $0.5°$, and that the components of other large double sources such as DA 240 (2 Mpc), and Cen A (1.57 Mpc) are also among the best aligned.

If a massive object does not quite have enough velocity to escape from a galaxy it will oscillate back and forth through the center on a dynamical time scale until it is damped. The damping produces disturbances in stellar orbits and gas and the differential rotation of the galaxy shears these disturbances. The result might be a peculiar spiral pattern. As far as I know, the detailed structure produced in this way has not been explored. Since many observers give the impression that galaxies contain more peculiar spiral arms than normal smooth ones, it seems an interesting problem.

In addition to its dynamical interactions, a moving massive object will affect the gas in and around the galaxy as it passes (Saslaw and De Young, 1972). Inside the galaxy, if the object is hot and radiates strongly in the ultraviolet, it will form a large H II region. The effects of the ionization front and increased gas pressure may catalyze star formation in nearby clouds close to instability, leaving a luminous trail of bright young stars or H II regions. The trail may also be heated by high energy particles emitted from the massive object. If the massive object is a black hole surrounded by an accretion disk, the type of radiation it emits will depend strongly on the viscosity and magnetic field in the disk. If the viscosity is low, not much gas will fall into the hole and, for no magnetic field, the radiation will be mainly optical. A high viscosity would enable the object to radiate approximately at its Eddington limit – where radiation pressure prevents further gas from falling into the hole – which is $L \simeq 10^{38}$ $M/M_{\odot}$ erg s^{-1}. Low frequency radio emission could be produced either by rotating magnetic satellites of the black hole, or by magnetic flares in the rotating gaseous disk. This radiation would evacuate a cavity in the intergalactic gas surrounding the galaxy. A number of mechanisms exist in this situation for accelerating particles relativistically and producing radio synchrotron radiation, and the results could well resemble the observed radio structure of extended sources. However, details of this complex situation have not yet been worked out. It will probably be especially important to take into account the inhomogeneity of the surrounding gas (cf. Rees and Saslaw, 1975).

All these observations, calculations, and speculations that I've tried to describe here, suggest that we may be starting to uncover a new set of ideas which relate the formation and evolution of massive objects in galactic nuclei to a wide range of astronomical problems. However, although there are some exciting trends of evidence favoring the existence of massive objects, I think we should still be cautious about believing in them too strongly. In this respect (and perhaps in some others) it is probably good to recall Hilaire Belloc's whimsical verse.

Acknowledgements

Part of this review was written at the Aspen Center for Physics during the summer of 1974, and I am happy to thank the visitors at the Center for many stimulating high altitude discussions.

References

Appenzeller, I. and Fricke, K.: 1972, *Astron. Astrophys.* **21**, 285.
Arp, H. C.: 1971, *Astrophys. Letters* **9**, 1.
Arp, H. C.: 1972, in D. E. Evans (ed.), 'External Galaxies and Quasi-Stellar Sources', *IAU Symp.* **44**, p. 380.
Arp, H. C. and O'Connell, R. W.: 1975, *Astrophys. J.* **197**, 291.
Bahcall, J. N., McKee, C. C., and Bahcall, N. A.: 1972, *Astrophys. Letters*, **10**, 147.
Belloc, H.: 1910, *More Beasts for Worse Children*, Duckworth, London.
Blandford, R. and Rees, M. J.: 1974, *Monthly Notices Roy. Astron. Soc.* (in press).
Bridle, A. H. and Brandie, G. W.: 1973, *Astrophys. Letters* **15**, 21.
Burbidge, E. M. and Burbidge, G. R.: 1965, *Astrophys. J.* **142**, 1351.
Burbidge, E. M., Burbidge, G. R., Solomon, P. M., and Strittmatter, P. A.: 1971, *Astrophys. J.* **170**, 233.
Burbidge, G. R.: 1967, *Nature Phys. Sci.* **216**, 1287.
Burbidge, G. R., O'Dell, S. L., and Strittmatter, P. A.: 1972, *Astrophys. J.* **175**, 601.
Cavaliere, A., Morrison, P., and Wood, K.: 1971, *Astrophys. J.* **170**, 223.
Chertoprud, V. E., Gudzenko, L. I., and Ozernoy, L. M.: 1973, *Astrophys. J. Letters* **182**, L53.
Chio, B. C., Morrison, P., and Sartori, L.: 1973, *Astrophys. J.* **181**, 295.
Colgate, S.: 1967, *Astrophys. J.* **150**, 163.
Condon, J. and Jauncey, D.: 1974, unpublished.
De Young, D. S., Roberts, M. S., and Saslaw, W. C.: 1973, *Astrophys. J.* **185**, 809.
Fricke, K.: 1973, *Astrophys. J.* **183**, 941.
Fricke, K.: 1974, *Astrophys. J.* **189**, 535.
Hargrave, P. J. and Ryle, M.: 1974, *Monthly Notices Roy. Astron. Soc.* **166**, 305.
Harris, A.: 1972, *Monthly Notices Roy. Astron. Soc.* **158**, 1.
Hazard, C. and Sanitt, N.: 1972, *Astrophys. Letters* **11**, 77.
Hoyle, F. and Fowler, W. A.: 1963, *Monthly Notices Roy. Astron. Soc.* **125**, 169.
Huntley, J. M. and Saslaw, W. C.: 1975, *Astrophys. J.*, in press (July 15).
Kristian, J.: 1972, *Astrophys. J. Letters* **179**, L61.
Legg, T. H., Broten, N. W., Fort, D. N., Yen, J. L., Bale, F. V., Barber, P. C., and Quigley, M. J. S.: 1973, *Nature Phys. Sci.* **244**, 18.
Lynden-Bell, D.: 1969, *Nature Phys. Sci.* **223**, 690.
Lynds, C. R. and Millikan, A. G.: 1972, *Astrophys. J. Letters* **176**, L5.
Mackay, C. D.: 1971, *Monthly Notices Roy. Astron. Soc.* **151**, 421.
Mackay, C. D.: 1973, *Monthly Notices Roy. Astron. Soc.* **162**, 1.
Mathis, J. S.: 1967, *Astrophys. J.* **147**, 1050.
Oke, J. B. and Gunn, J. E.: 1974, *Astrophys. J. Letters* **189**, L5.

Okoye, S.: 1973, *Monthly Notices Roy. Astron. Soc.* **165**, 393.

Peebles, P. J. E.: 1972, *Astrophys. J.* **178**, 371.

Pringle, J. E., Rees, M. J., and Pacholczyk, A. G.: 1973, *Astron. Astrophys.* **29**, 179.

Rees, M. J.: 1971, *Nature Phys. Sci.* **229**, 312.

Rees, M. J. and Saslaw, W. C.: 1975, *Monthly Notices Roy. Astron. Soc.* (in press).

Salpeter, E. E.: 1971, *Nature Phys. Sci.* **223**, 5.

Sandage, A. R.: 1971, in D. J. K. O'Connell (ed.), *Nuclei of Galaxies*, North Holland, Amsterdam.

Sanders, R. H.: 1970, *Astrophys. J.* **162**, 791.

Saslaw, W. C.: 1973, *Publ. Astron. Soc. Pacific* **85**, 5.

Saslaw, W. C.: 1974, in J. R. Shakeshaft (ed.), 'The Formation and Dynamics of Galaxies', *IAU Symp.* **58**, 305.

Saslaw, W. C.: 1975, *Astrophys. J.* **195**, 773.

Saslaw, W. C. and De Young, P. S.: 1971, *Astrophys. J.* **170**, 423.

Saslaw, W. C. and De Young, D. S.: 1972, *Astrophys. Letters* **11**, 87.

Saslaw, W. C., Valtonen, M. J., and Aarseth, S. J.: 1974, *Astrophys. J.* **190**, 253.

Scheuer, P. A. G.: 1974, *Monthly Notices Roy. Astron. Soc.* **166**, 513.

Seidl, F. G. P. and Cameron, A. G. W.: 1972, *Astrophys. Space Sci.* **15**, 44.

Spitzer, L.: 1969, *Astrophys. J. Letters* **158**, L139.

Spitzer, L. and Saslaw, W. C.: 1966, *Astrophys. J.* **143**, 400.

Ulam, S. W. and Walden, W. E.: 1964, *Nature Phys. Sci.* **210**, 1202.

Valtonen, M. J.: 1974, Ph. D. Thesis, Cambridge Univ.

Weedman, D.: 1973, *Astrophys. J.* **183**, 29.

Willis, A. G., Strom, R. G., and Wilson, A. S.: 1974, *Nature* **250**, 625.

Wolfe, A. M. and Burbidge, G. R.: 1970, *Astrophys. J.* **161**, 419.

DISCUSSION

Lecar: Why, if the ejected components for, say, massive ellipticals, are black holes: is the primary radiation radio, rather than X-ray?

Saslaw: In the case of a black hole surrounded by a disk of gas, the spectrum of emitted radiation will depend strongly on the form and amount of turbulence in the gas and the magnetic field. The result will be a combination of thermal radiation, bremsstrahlung, synchrotron, and Compton scattering with possible important interactions between photons and plasmons. All this is so complicated, however, that no one has worked out realistic detailed spectra yet. It would be especially interesting to know if the massive object starts optically bright when young – and close to its parent galaxy as Lacertids may be – but becomes optically fainter as it ages and radio emission predominates. If the black hole is surrounded by a satellite system of pulsars, then there is no necessity for any optical radiation since we observe old pulsars only in the radio. When young, these systems might also produce optical and X-ray emission.

Bardeen: The thermal radiation from an accretion disk around a supermassive black hole would be in the ultraviolet rather than the X-ray region of the spectrum, since the area of the emitting region increases faster than the maximum luminosity with mass of the black hole.

King: You have indicated that a massive object at the center of a galaxy would produce a central spike of density. Could you indicate quantitatively what should be observed, so that perhaps it can be looked for observationally?

Saslaw: For galactic nuclei with total masses less than about $10^{10} M_\odot$ in which the mass of the object is $\lesssim 0.1$ of the mass of the nucleus, the central cusp would have an angular diameter much less than one arcsecond. This would put it well within the atmospheric seeing disk, so it would be necessary for high resolution optical observations to be made from above the atmosphere. In the near future, it seems more likely that radio interferometry will be able to measure activity in the cusp.

SYMMETRIC EJECTION OF MASSIVE OBJECTS
FROM GALACTIC NUCLEI

M. J. VALTONEN

Institute of Astronomy, Cambridge, England
and
Research Institute for Theoretical Physics, Helsinki, Finland

Abstract. It is shown that massive objects can be ejected symmetrically in two opposite directions from the nucleus of a giant elliptical galaxy only under special initial conditions. A way of obtaining these conditions is outlined, and the resulting picture of evolution of galactic nuclei, including a quasar or a Seyfert galaxy phase, is shown to be compatible with observations.

1. Introduction

It has recently been demonstrated (Saslaw *et al.*, 1974; subsequently called SVA) that a system of three or four massive objects can in principle be ejected from the nucleus of a giant elliptical galaxy. In particular, a system of three mass points, strongly bound to each other and of similar masses, usually breaks up after a few tens of crossing times to form a close binary, with the third particle escaping from it. The third particle escape velocity V_s, relative to the centre of mass of the triple system, is distributed as shown in Figure 1 (dotted line). The units are such that the initial

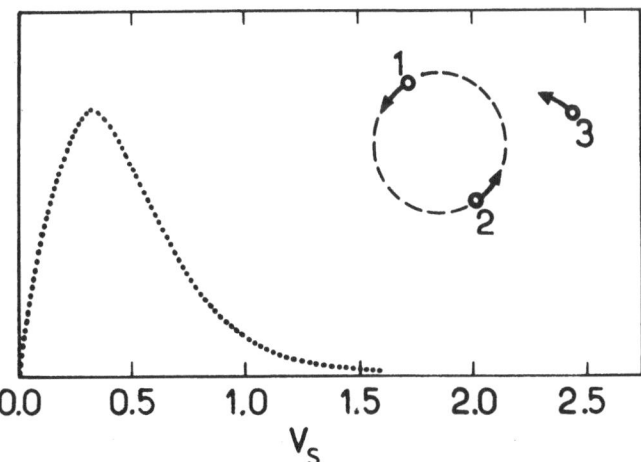

Fig. 1. The escape velocity distribution of the third particle (dotted line). The initial configuration of the three mass points is shown in the corner. For details, see the text.

binary components m_1 and m_2, which are sketched in the figure, have the velocity of 0.5 relative to their centre of mass in their circular initial orbit. This distribution is rather insensitive to the initial conditions of the system and does not depend on the mass of the escaping particle.

The particle which escapes is usually the least massive one and a particle containing approximately half the total mass has less than 10% chance to be the escaper. Therefore the break-up is asymmetric with most of the mass in the binary, whereas the single object acquires most of the velocity. The simplest way to achieve symmetry is to keep the initial binary and the third object, with mass equal to the binary, separate throughout the interaction. We then have a binary interacting with a third particle which is initially on a nearly parabolic orbit relative to the centre of mass of the binary. In Figure 2 the average change in velocity ΔV is shown (full line) as

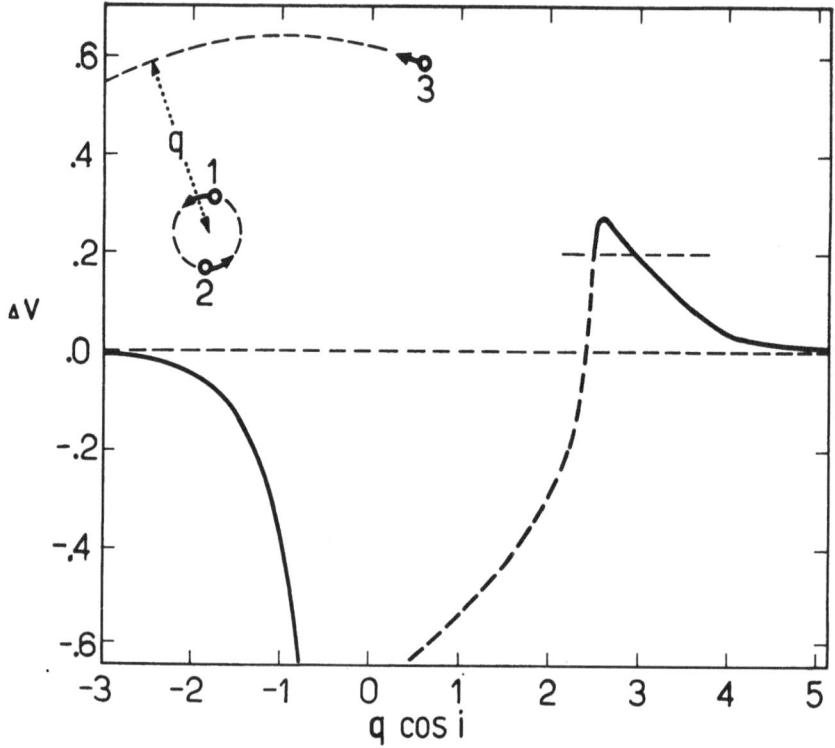

Fig. 2.   The average change in the velocity of the passing particle as a function of pericentre distance q for direct ($\cos i = 1$) and retrograde ($\cos i = -1$) orbits. The initial configuration of the mass points is shown in the corner. For details, see the text.

a function of the pericentre distance q of the outer orbit, which has either zero inclination relative to the binary plane (direct orbit; $\cos i = 1$) or the inclination is 180° (retrograde orbit; $\cos i = -1$). The average change ΔV is not well defined for $-0.8 \leqslant q \times \cos i < 2.5$ since the passing particle is frequently captured and becomes a component of the final binary. The semi-major axis of the outer orbit a_3 was taken to be 300, measured in terms of the semi-major axis a of the initially circular binary ($e = 0$). Also q is measured in units of a, and the velocity unit is such that the initial binary components have a velocity of 0.5 relative to their centre of mass (see the figure). It should be noted that there exists a resonance region when q lies between 2.5 and 3.0 in the case of direct orbits, giving rise to a relatively large acceleration.

The astronomical application in SVA was concerned with the ejection of radio source components, assumed to be massive compact objects ($\sim 10^8\ M_\odot$), from a giant elliptical galaxy. As an example of a radio galaxy we may consider the nearby system Fornax A, for which relatively good information is available (Matthews *et al.*, 1964; Searle, 1965). On both sides of the central object NGC 1316 there are extensive radio emitting regions, with the peak intensity areas lying on a straight line through, and approximately but not exactly equidistant from the nucleus of the galaxy. The galaxy has prominent dust arms and optical extensions lead from the galaxy near to the peak radio intensity regions. The measured rotation velocities and the geometrical configuration suggest that all the optical extensions lie in the plane perpendicular to the rotation axis of the galaxy (Searle, 1965), implying that the ejection of the radio source components has occurred in this plane. The slingshot theory (SVA) predicts preferentially this ejection direction. In the following we denote a system with this type of geometry of radio components a two-sided double.

Apparently there also exists cases when one component of the radio source lies outside the galaxy and the other coincides with the nucleus (e.g. 3C 293; Branson *et al.*, 1972). This situation may arise through an asymmetric ejection, and we call these systems one-sided doubles. The ratio of two-sided to one-sided double radio sources is probably greater then ten. This number is rather uncertain since the present sample of well studied radio sources is small and may be biased by selection effects. There are also a great number of sources completely confined within the parent galaxy. We call these single sources irrespective of whether they show double or some other structure, since they are much too numerous to be components on their way out of the galaxy. The single and double sources are about equally common among strong radio sources. More complex sources usually show an underlying double structure which enables them to be included in the above classification scheme. A similar classification of double sources has been previously used by Miley (1971) and in that notation D2 corresponds to one-sided doubles and D1 to two-sided doubles. However, we include D1 and D2 sources in the class of single sources whenever their extent is smaller than that of the central galaxy.

2. Dynamical Models for Radio Sources

We have considered simple dynamical models to compare with observed radio source structures. A set of identical parent galaxies with a similar triple system of massive objects in their nuclei was assumed and the statistics of the resulting ejections was studied. A square well approximation for the galactic potential was used: any single or binary object possessing an initial velocity greater than V_0 ($\sim 2000\ \mathrm{km\ s^{-1}}$) was assumed to move outside the galaxy, whereas lower velocity objects would remain inside. When an object was ejected outside the galaxy, its distance from the nucleus 10^9 yr later was estimated using a more realistic galactic potential (see SVA) and assuming no interaction with the intergalactic medium.

When a triple system of three massive objects disrupts, both the single object and the binary receive a certain initial velocity V relative to their centre of mass, taken to be the nucleus. If $V < V_0$ for both components, a single source is counted. When the velocity of only one component exceeds V_0, a one-sided double results and when $V > V_0$ for both components, a two-sided double is counted. In the latter case the ratio of the distances of the two components from the nucleus, d_{max}/d_{min}, is calculated. This classification scheme is not so artificial as it may appear. In realistic models of the galactic potential an object having a velocity V only 10% below V_0 does not reach very far from the nucleus and it will return there in a relatively short time, if V_0 is properly chosen (SVA).

Two types of triple configurations were considered: a strongly bound system of three massive objects, and a third particle passing a circular binary on a direct co-planar orbit. In the former case models with various ranges in mass, total angular momentum and energy were tried, and even though this produces many different distributions, all models have a common feature: one-sided structures are by far more common than the two-sided ones. This is understandable considering the basic asymmetry of the three-body instability, and the property that the highest velocities available in such an event are not much greater than V_0 (SVA). As an example we show the distributions for one case in the second line of Table I. The mass range

TABLE I

Structure of radio sources

	Two-sided doubles d_{max}/d_{min}			One-sided doubles	Single sources
	1.0–1.4	1.4–3.0	>3.0		
Observed	22	12	2	~3	~36
Strongly bound	19	15	17	10^3	5×10^3
Fly-by	34	0	27	60	120

for the three objects was taken to be within a factor of ten at most, and the physical scaling was such that no more than 50% of all triple systems would be destroyed by gravitational radiation losses before an ejection occurred (SVA). The present observations are shown in the first line of the table. The number distribution for d_{max}/d_{min} is taken from SVA and the numbers of one-sided doubles and single sources are adjusted according to the estimates made in the previous section. The numbers in the following lines are also normalized to 34 in the first two columns. The difference between the observed and the predicted number of single sources may not be significant because single sources may evolve at a rate different from the double sources.

In the second model an equal-mass circular binary interacts with a single object of similar mass. The relative orbit of the binary centre of mass and the third object has a semi-major axis $a_3 \simeq 200$ and pericentre distance q is randomly distributed between given limits. The resulting structures again show an excess of one-sided doubles; however, these can be decreased by preventing encounters with small values of q. The third line of Table I shows an example with q uniformly distributed between 0.7 and 4.0. To get distributions which correspond to the observations, within the large uncertainties of the latter, a lower limit of $q_{min} \simeq 2.3$ must be introduced.

In conclusion, the following requirements should be satisfied by a two-sided ejection model (SVA): (1) The ejecting system must be composed of a nearly circular binary ($e \sim 0$–0.05) and a third object weakly bound to it ($a_3/a \sim 50$–300). Alternatively, the system could be made out of two approximately equal circular binaries on a weakly bound orbit with $a_3/a \sim 15$–300. (2) The orbits must be direct with inclination $i \sim 0°$–$30°$. The pericentre distance q of the outer orbit must be greater than $\simeq 2.3$, and in most cases it should lie between 2.5 and 2.8. (3) The mass of the binary and the third object or, alternatively, the masses of the binaries, should be approximately equal: $(m_1 + m_2)/m_3 \sim 0.9$–1.1. (4) The diameter of the binary or binaries must be in range $a \sim \frac{1}{4}$–2×10^{-3} $(M/10^8 \, M_\odot)$ pc, where $M = m_1 + m_2$ is the binary mass. Consequently, the lifetime of the binaries due to gravitational radiation losses is $\sim 10^2$ yr in the case of a triple system and $\sim 10^5$ yr in the two binary systems.

3. Formation of Massive Objects

In principle a number of different physical processes may lead to the formation and subsequent ejection of massive objects (SVA). However, the four conditions listed in the previous section may be used to discriminate between various possibilities. The short life-time of the binary seems to exclude any scheme where the third particle is completely unrelated to the binary. The close relation between the binary and third particle masses, and the narrow range of impact distances also seem difficult to achieve in such schemes. Finally, adding the requirement of low inclinations suggests that the objects would most likely originate simultaneously in the fragmentation of a disk-like rotating object.

When a gaseous rotationally supported object cools, it flattens and, after reaching a certain critical degree of flattening, becomes dynamically unstable (Salpeter, 1971). The instability causes the disk to develop a bar-like structure which perhaps subsequently fragments into a few objects (Ostriker and Peebles, 1973). In computer simulations of a rotating disk of stars (Hohl, 1971) a bar develops with an axial ratio of about 3:2 where the length is equal to the disk diameter. It is not clear what the effect of subsequent cooling will be, but here we assume that the bar will finally fragment into two parts forming a circular binary. This may be an essential requirement from the ejection point of view since it is difficult to produce circular binaries from eccentric ones through dissipative processes within a relatively short time. We may take the semi-major axis of the resulting binary to be of the order of the radius of the initial disk.

The narrow range of impact distances and the equality of masses combined with the condition for a_3 suggests that there initially exists a very eccentric binary of two massive disks, which almost touch each other during each pericentre passage. When one or both disks fragment into a binary, an ejection may follow the next pericentre approach. Extremely eccentric binaries may result from an ejection of a third object from a low angular momentum planar triple system (SVA). Such a triple system may arise, for example, through an ejection of one object from a high angular momentum planar system of four bodies. The four-body system could again have been formed in a fragmentation process.

We are thus lead to consider the evolution of a rotating disk under successive stages of fragmentation. In each stage two fragments are produced, and consequently a four-body system results after two stages. Since the life-time for each generation of objects is greater than the dynamical time-scale for their mutual interactions, frequent collisions will take place, transferring angular momentum from the internal rotation of the objects into the relative motion between them (Salpeter, 1971). As a result of loosing their angular momenta the disks will shrink and may become sufficiently small to avoid suffering direct collisions any more. This would happen after contraction by a factor of ~ 100 (SVA). However, two of the objects may escape before this occurs. The remaining binary components might still suffer a few collisions during their first pericentre passage and shrink further until their diameters become smaller than the closest approach distance.

In Figure 3 this process is shown schematically. As an illustration we assume an

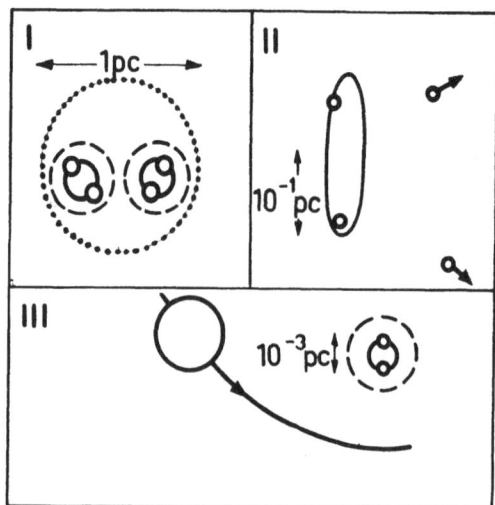

Fig. 3. A schematic representation of the three phases of the fragmentation process leading to an ejection. For details, see the text.

initial disk of diameter 1 pc and total mass $10^9 \, M_\odot$. In the first phase the four objects are formed and interact with each other. In the second phase only the eccentric binary remains near the nucleus while the two other objects are moving independently towards the outer parts of the galaxy. In the third phase one of the binary components fragments, and during the next pericentre passage an ejection of all three objects may occur with a speed somewhat greater than the speed of the first two ejections.

4. Observational Consequences

We now briefly outline the history of a massive rotating disk in a galactic nucleus from an observational point of view. The disk could form in a time much shorter than the life-time of a galaxy from the material shed by evolving stars. It is estimated that approximately ten solar masses per year would be received by the interstellar medium of a giant elliptical galaxy from its stars (Knapp and Kerr, 1974). Due to the small specific angular momentum of a spheroidal galaxy this gas will fall to the centre of the galaxy where it will pile up and form a rotating object perhaps of 10 pc diameter (Shklovskii, 1972). Let us suppose that 10^{10} $M_\odot$ has collected in this object before it becomes unstable, and therefore both the length and mass scale would be of order of magnitude greater than indicated in Figure 3.

In the first phase after the instability sets in frequent collisions between the four fragments take place, each lasting ~ 100 yr initially but becoming shorter and more energetic later on. The total time involved in these collisions before escape takes place may not be more than $\sim 10^3$ yr and thus it is unlikely that we would observe the collisions occurring in any celestial object. The collisions would, however, tear off gas clouds, which would be accelerated in the gravitational field of the massive objects. The number of clouds would build up rapidly until the nuclear region is filled by them. The clouds would then collide among each other and turn their large-scale kinetic energy into electromagnetic radiation through shock waves. The resulting radiation may have many of the properties observed in Seyfert galaxies (Osterbrock, 1971) and in quasi-stellar objects (Daltabuit and Cox, 1972). Also the massive objects would collide frequently with gas clouds. Such events may be observed as variable radio sources (Bignell, 1973). The fact that the best models for many variable radio sources include three or four small components (Kellermann *et al.*, 1974) could mean that they are in their pre-ejection stage of evolution. For example in NGC 1275 the model (Legg *et al.*, 1973) shows four components at a typical separation of a parsec from each other.

The first phase may last $\sim 10^6$ yr after which an eccentric binary remains, and therefore a double radio structure could be observed during the second phase. The cloud collisions would be expected to cease during this phase, which may last 10^5–10^6 yr. This is the cooling time-scale for the disks (Salpeter, 1971). After the symmetric ejection the clouds would find their gravitational binding reduced and they would start to move outwards from the nucleus. Another reminder of the past events would be a halo of relativistic particles, mainly produced during the collisions between the disks, which therefore would be elongated perpendicular to the plane of symmetry of the galaxy. These features may be observable in M87 which shows evidence of a recent symmetrical ejection (Walker and Hayes, 1967; Graham, 1970).

The ejected objects are not presumably stable themselves and may fragment again. Because of strong gravitational radiation losses, a bar instability and resulting binary formation is excluded, and only fragmentation into a large number of smaller objects seems possible. Some of the fragments could escape through three-body interactions, and this may be happening in one of the components of M87. One of the three main knots (perhaps the middle one) would represent the original object and the two others would be ejected objects. This interpretation may also offer an explanation of the peculiar polarization directions in the knots (Hiltner, 1959; Turland, 1975).

Some other peculiar features of M87 are worth noting. The luminous filament which connects the nucleus and the main knots originates from a point slightly north of the nucleus (Felten *et al.*, 1970). If the filament outlines the path of the massive object from the nuclear region, it is not surprising that the path does not originate right from the middle of the nucleus, since in the present theory the eccentric binary should have been displaced by earlier ejections. Also the counterjet (Arp, 1967) and the orthogonal fan-jet (Walker, 1968) may result from interactions between gas clouds and the two independently ejected objects. The peculiar optical and radio extensions originating from the outermost bright knot could also be understood as resulting from further fragmentations and ejections, which have occurred in the ejected knots themselves.

In conclusion, the requirement for a symmetric ejection of massive objects to occur leads to a rather definite picture of the evolution of galactic nuclei which appears to have many connections with the observed forms of their activity.

Acknowledgement

We would like to thank Drs D. Lynden-Bell, J. Bardeen and F. Hohl for discussions on the fragmentation process, and Drs S. J. Aarseth, W. C. Saslaw and N. Sanitt for reading the manuscript and for valuable comments. The author has been supported by an Osk. Huttusen Säätiö-Foundation Fellowship during this work.

References

Arp, H. C.: 1967, *Astrophys. Letters* **1**, 1.
Bignell, R. C.: 1973, *Astron. J.* **78**, 557.
Branson, N. J. B. A., Elsmore, B., Pooley, G. G., and Ryle, M.: 1972, *Monthly Notices Roy. Astron. Soc.* **156**, 377.
Daltabuit, E. and Cox, D.: 1972, *Astrophys. J. Letters* **173**, L13.
Felten, J. E., Arp, H. C., and Lynds, C. R.: 1970, *Astrophys. J.* **159**, 415.
Graham, I.: 1970, *Monthly Notices Roy. Astron. Soc.* **149**, 319.
Hiltner, W. A., 1959, *Astrophys. J.* **130**, 340.
Hohl, F.: 1971, *Astrophys. J.* **168**, 343.
Kellermann, K. I., Clark, B. G., Shaffer, D. B., Cohen, M. H., Jauncey, D. L., Broderick, J. J., and Niell, A. E.: 1974, *Astrophys. J. Letters*, **189**, L19.
Knapp, G. R. and Kerr, F. J.: 1974, *Astron. J.* **79**, 667.
Legg, T. H., Broten, N. W., Fort, D. N., Yen, J. L., Bale, F. V., Barber, P. C., and Quigley, M. J. S.: 1973, *Nature Phys. Sci.* **244**, 18.
Matthews, T. A., Morgan, W. W., and Schmidt, M.: 1964, *Astrophys. J.* **140**, 35.
Miley, G. K.: 1971, *Monthly Notices Roy. Astron. Soc.* **152**, 477.
Osterbrock, D. E.: 1971, *Pontificiae Academiae Scientiarum*, Scripta Varia No. 35, p. 151.
Ostriker, J. P. and Peebles, P. J. E.: 1973, *Astrophys. J.* **186**, 467.
Salpeter, E. E.: 1971, *Nature Phys. Sci.* **233**, 5.
Saslaw, W. C., Valtonen, M. J., and Aarseth, S. J.: 1974, *Astrophys. J.* **190**, 253.

Searle, L.: 1965, *Nature Phys. Sci.* **207**, 1282.
Shklovskii, I. S.: 1972, *Soviet Astron.* **16**, 193.
Turland, B. D.: 1975, in press.
Walker, M. F.: 1968, *Astrophys. Letters* **2**, 65.
Walker, M. F. and Hayes, S.: 1967, *Astrophys. J.* **149**, 481.

DISCUSSION

Bardeen: The size of the objects is not too much smaller than their separation. Have you considered the effects of tidal interactions on the ejection process?

Valtonen: In the binary responsible for the final double ejection, any tidal elongation of the components would be dissipated by gravitational radiation and most likely the components would contract until the tidal effects become negligible.

Baldwin: After an ejection of the kind you describe, how long might it be before the situation could arise once more?

Valtonen: In this over-simplified model two ejections occur with a lower velocity before the double ejection and the massive objects from the former events may return to the nucleus and might perhaps prevent further symmetrical ejections.

EXPÉRIENCES NUMERIQUES SUR L'ÉJECTION DE MASSES DEPUIS LE CENTRE D'UN DISQUE D'ÉTOILES

S. CLAIREMIDI et A. HAYLI

Observatoire de Besançon et Faculté des Sciences, Besançon, France

Abstract. Numerical experiments on simple or symmetric ejections of massive bodies from the nucleus of a disk of stars are presented. The ejected masses range from 0.04 to 0.1 of the total mass. Either spiral features or a barred structure appear but do not last for a long time. Material arms start leading and become trailing.

1. Introduction

Des preuves s'accumulent depuis quelques années qui accréditent la thèse de l'éjection par le noyau de certaines galaxies de nuages gazeux massifs, de particules relativistes et d'objets massifs cohérents. Nous renvoyons pour la bibliographie à O'Connell (1970).

L'une des hypothèses formulées sur la nature du noyau est qu'il serait composé essentiellement d'un objet supermassif capable de fragmentations récurrentes et dont l'activité régirait le comportement de la galaxie.

Les observations montrent un objet brillant à l'extrémité d'un bras de certaines galaxies (Arp, 1969) ou encore des paires de radio-sources alignées avec des galaxies parentes (Van der Kruit, 1971) qui pourraient avoir été éjectées de la galaxie mère.

Les expériences numériques décrites ci-dessous ont été suggérées par cette hypothèse. Leur objet est l'étude du comportement dynamique qu'aurait un disque d'étoiles si son centre éjectait une fraction de sa masse sous la forme d'un ou deux objets cohérents. On a voulu voir si des éjections d'objets massifs pouvaient être responsables de l'amorce d'une structure spirale dans le disque d'étoiles.

2. Expériences numériques

Dans les trois cas présentés ci-dessous on a adopté une représentation du disque d'étoiles analogue à celle de Toomre et Toomre (1972). On a supposé que toute la masse du disque, soit $10^{11}\ M_\odot$ était concentrée en un point. Les étoiles du disque sont des objets tests disposés sur des cercles de rayons 5, 7.5, 10, 12.5 et 15 kpc. Ils ont une vitesse circulaire.

Pour éviter les difficultés numériques dans l'intégration des équations du mouvement, on a supposé que la masse éjectée se trouve à 100 pc du centre à l'instant $T = 0$. L'éjection se fait toujours dans le plan du disque. L'unité de temps vaut 10^8 ans.

On est réduit à intégrer, pour le mouvement d'une étoile test quelconque un problème de trois corps (dans le cas d'une seule éjection) ou de quatre corps (dans le cas de deux éjections) particulièrement simple.

Hayli (ed.), Dynamics of Stellar Systems, 405–419. *All Rights Reserved.*

2.1. Éjection d'une masse unique de 10^{10} $M_\odot$, soit 1/10 de la masse totale

On prend comme vitesse initiale du corps éjecté $v = 2400$ km s^{-1}. Figure 1. Les étoiles ont été suivies pendant 7 unités de temps correspondant à un peu plus d'une révolution complète pour celles qui sont le plus à l'extérieur du disque, et à 7 révolutions pour les plus intérieures. Au temps $T=7$ la masse éjectée est à environ 100 kpc du centre du disque et sa vitesse est de 150 km s^{-1}.

Dès $T=1$ on voit se dessiner l'amorce d'un bras qui devance le disque dans le sens de la rotation. Puis le bras se déforme et finit par traîner.

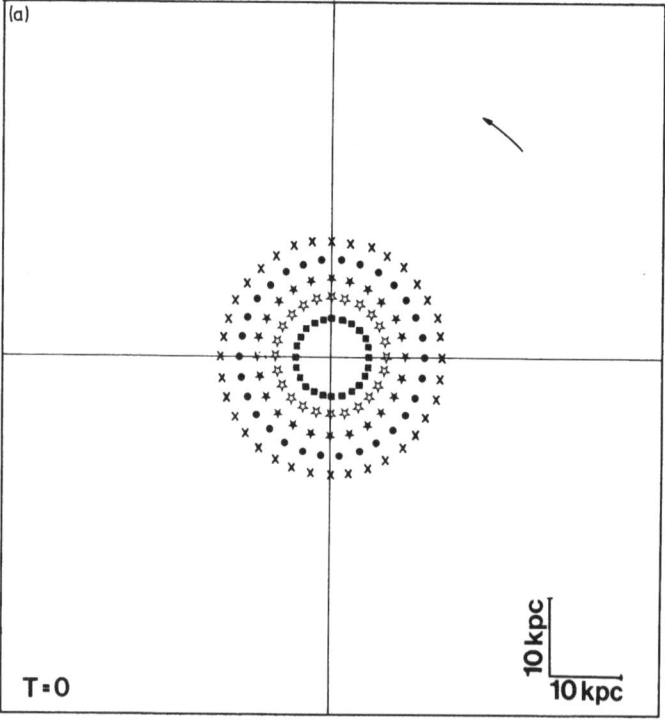

Fig. 1a.

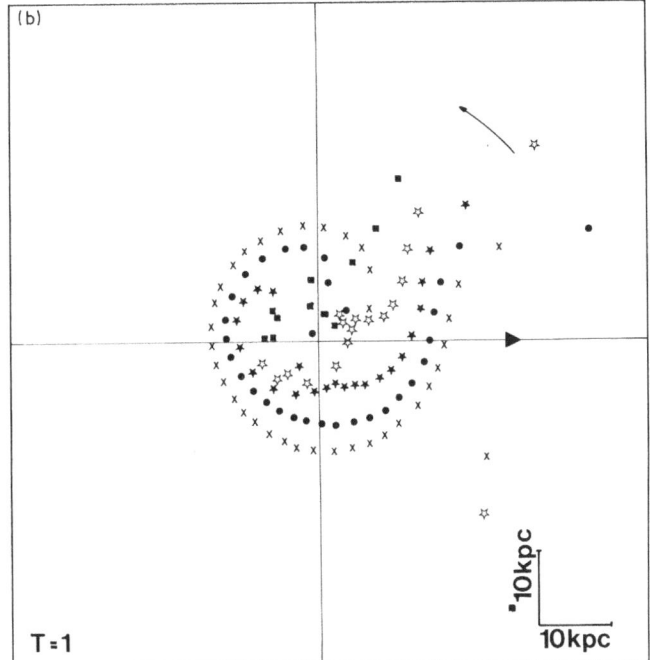

Fig. 1b.

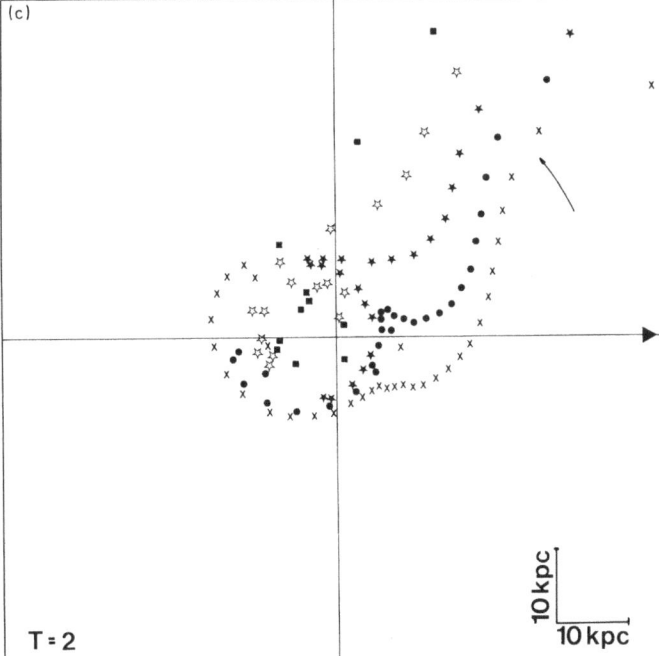

Fig. 1c.

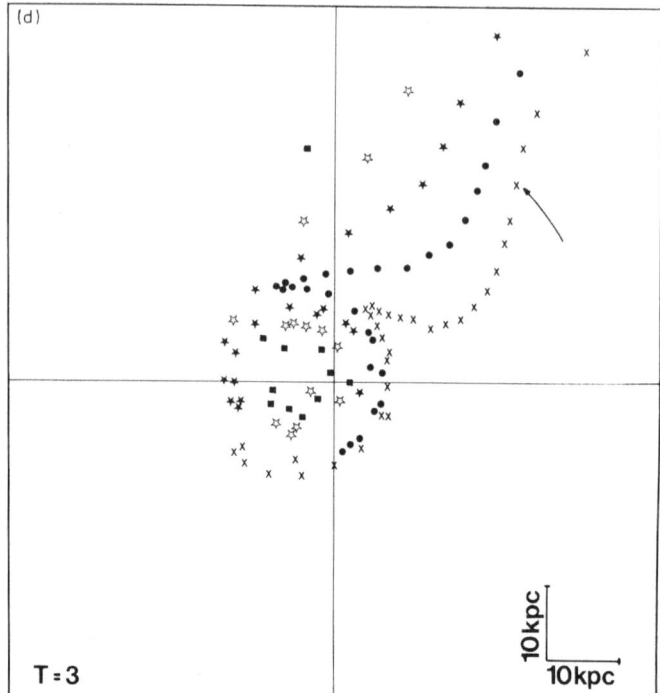

Fig. 1d.

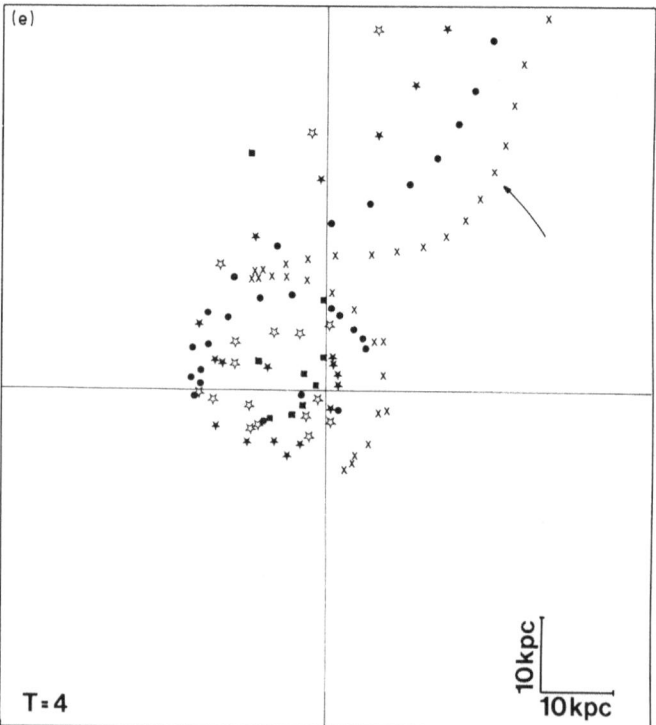

Fig. 1e.

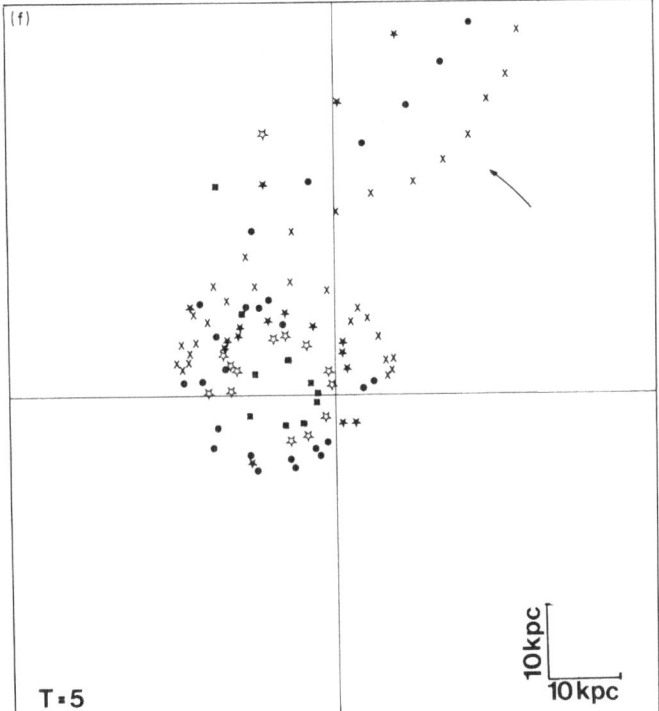

Fig. 1f.

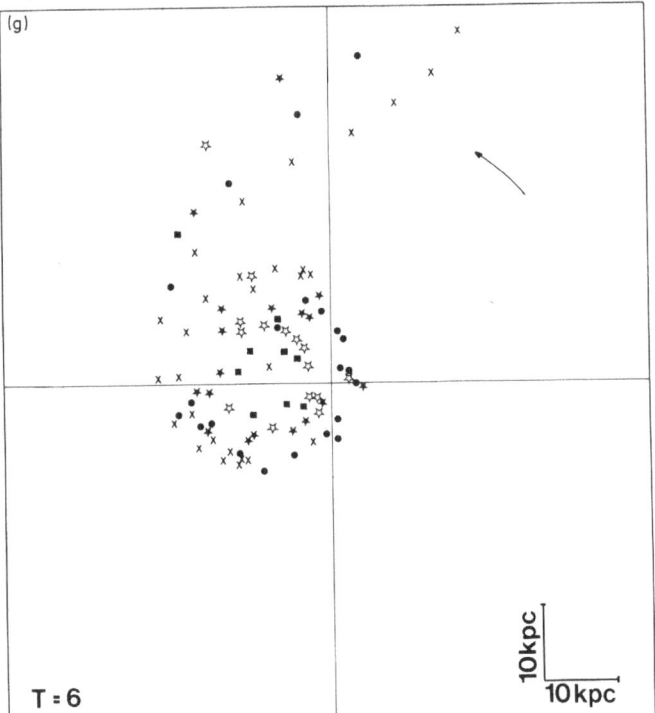

Fig. 1g.

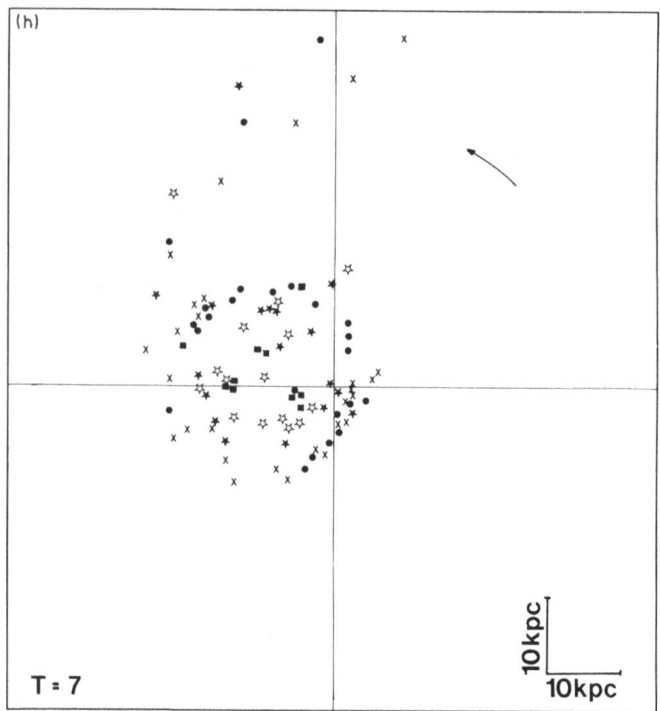

Fig. 1 (a) à (h).   Toute la masse du disque, soit 10^{11} $M_\odot$ est au centre. Les étoiles qui se trouvent à la même distance du centre à l'instant $T=0$ sont représentées par le même symbole. La flèche indique le sens de rotation d'ensemble du disque. Un fragment de 10^{10} $M_\odot$ est éjecté à $T=0$. Il est représenté par le signe ▶ sur les Figures (b) et (c). On a choisi un système de référence lié à la masse centrale. On voit se former un bras qui devient traînant à partir de $T=6$.

2.2. Éjection symétrique de deux masses égales, chacune, à 0.2×10^{10} $M_\odot$

Les vitesses initiales sont de 2700 km s^{-1}. Figure 2. Les étoiles ont été suivies pendant 10 unités de temps. La figure 2 ne montre l'évolution du système que jusqu'à $T=7$. A la fin du calcul les masses éjectées sont à 120 kpc de part et d'autre de la masse centrale et leur vitesse est de 120 km s^{-1} environ.

Pendant les 4 premières unités de temps on voit apparaître deux bras qui devancent le disque dans sa rotation, puis les bras disparaissent et la configuration ressemble davantage à celle d'une galaxie spirale barrée. Au delà de $T=8$ la structure du système devient très confuse.

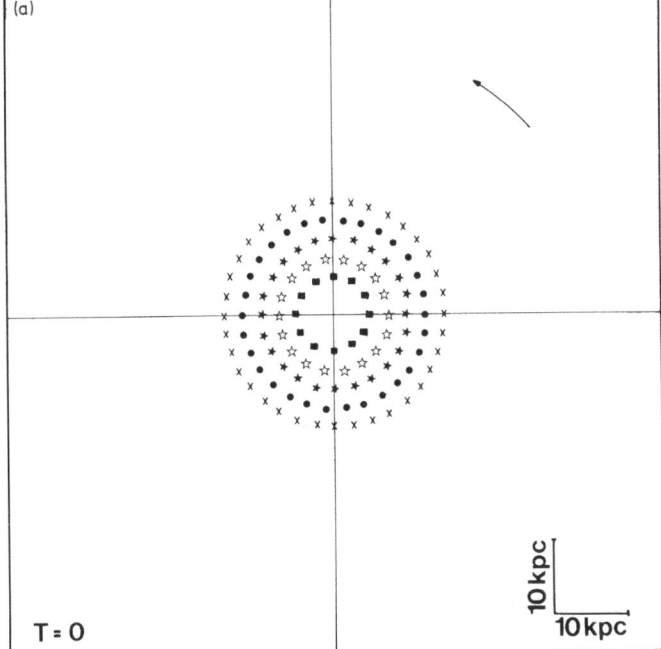

Fig. 2a.

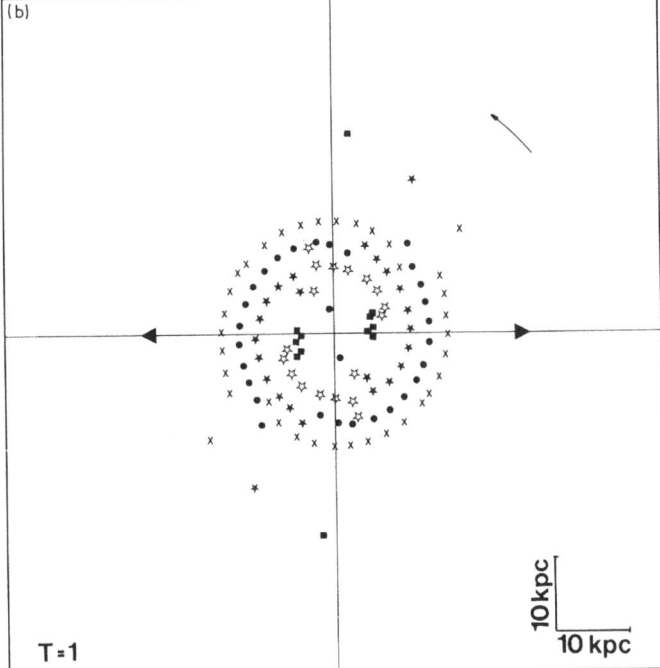

Fig. 2b.

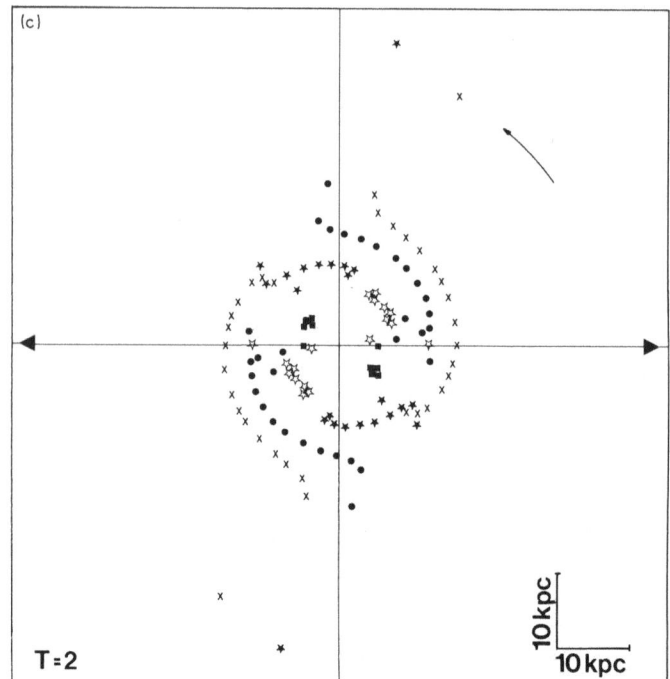

Fig. 2c.

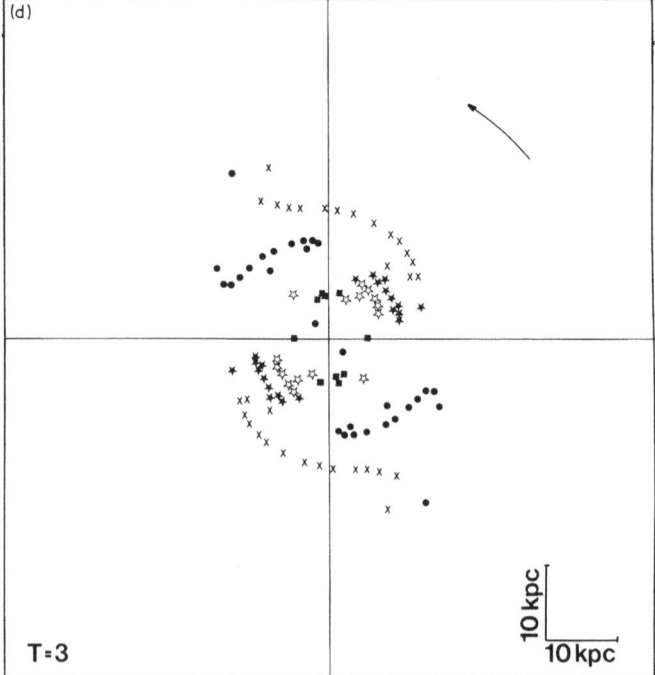

Fig. 2d.

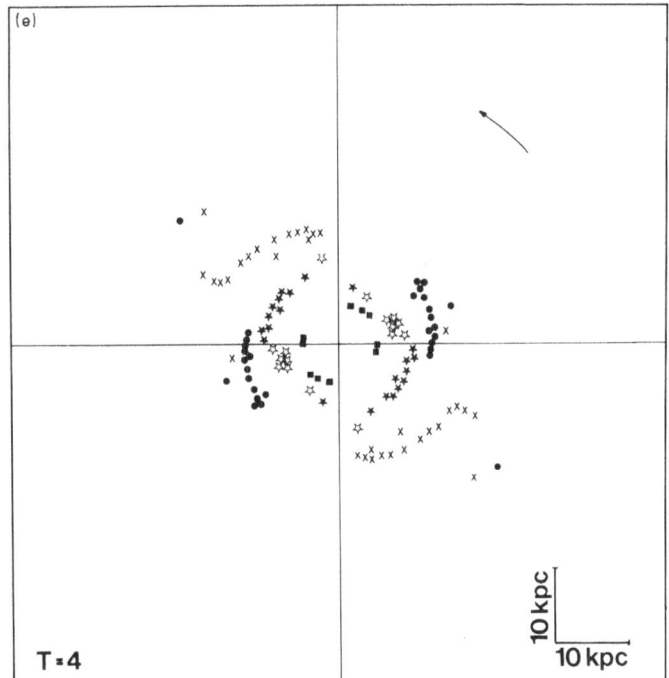

Fig. 2e.

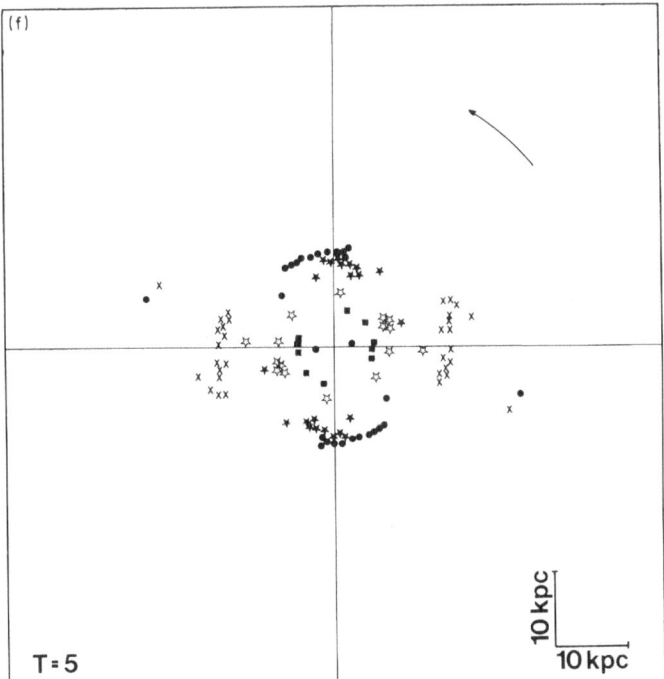

Fig. 2f.

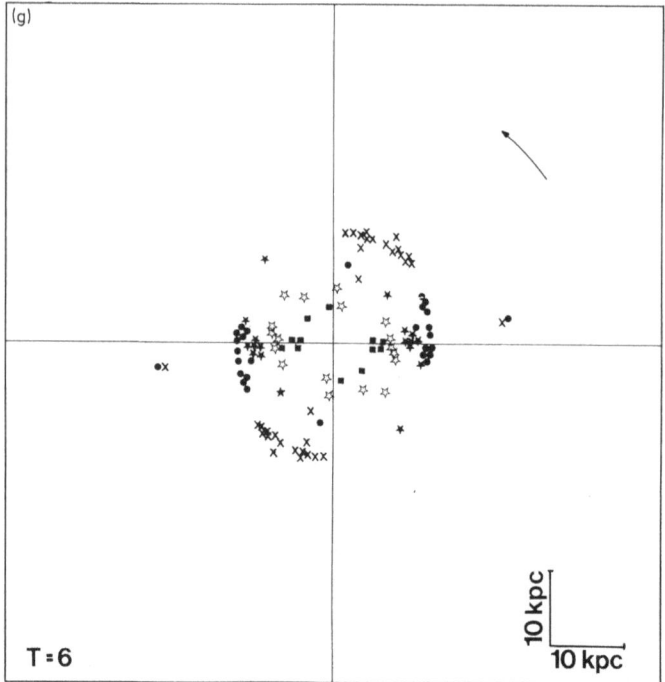

Fig. 2g.

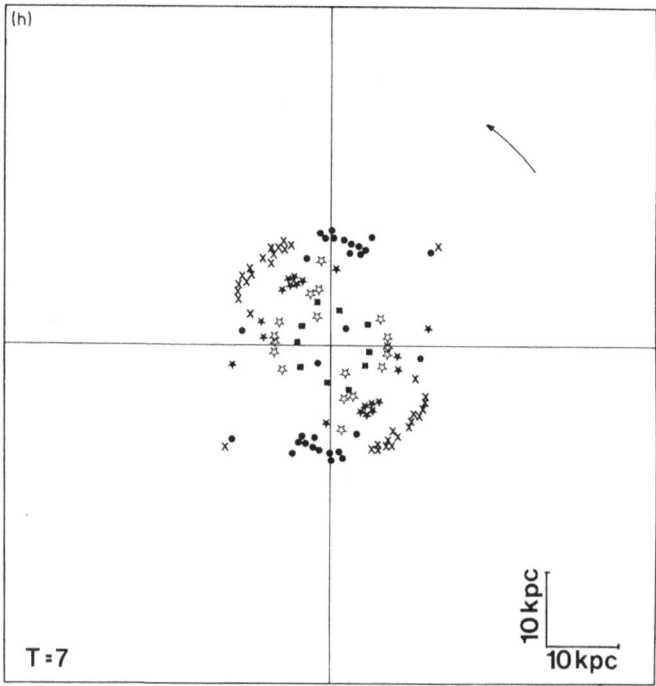

Fig. 2 (a) à (h). Deux fragments de même masse, $0.2 \times 10^{10}\ M_{\odot}$, sont éjectés dans deux directions opposées. On voit apparaître des bras, Figures (b) à (e) qui cèdent la place à une structure qui rappelle une spirale barrée, (g) et (h).

2.3. ÉJECTION SYMÉTRIQUE DE DEUX MASSES ÉGALES, CHACUNE, À $0.5 \times 10^{10}\ M_{\odot}$

Les vitesses initiales sont de 2600 km s^{-1}. Figure 3. Les étoiles ont été suivies pendant 10 unités de temps. Dès $T = 1$ on voit apparaître un bras qui devance d'abord la rotation du disque puis se déforme et suit le disque pendant près de 4 unités de temps. Au delà de ce temps la structure du système devient confuse.

D'autres expériences numériques sont en cours, pour examiner notamment si l'on peut obtenir une structure spirale qui persiste plus longtemps.

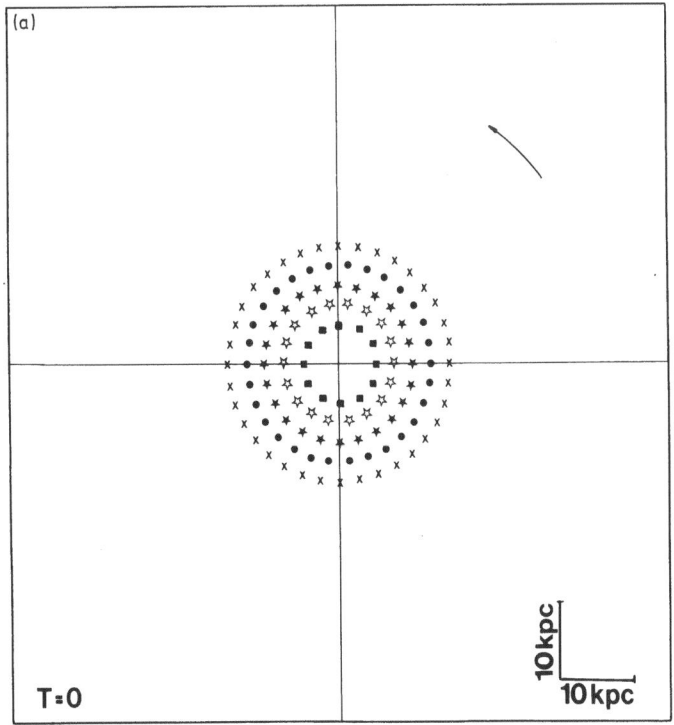

Fig. 3a.

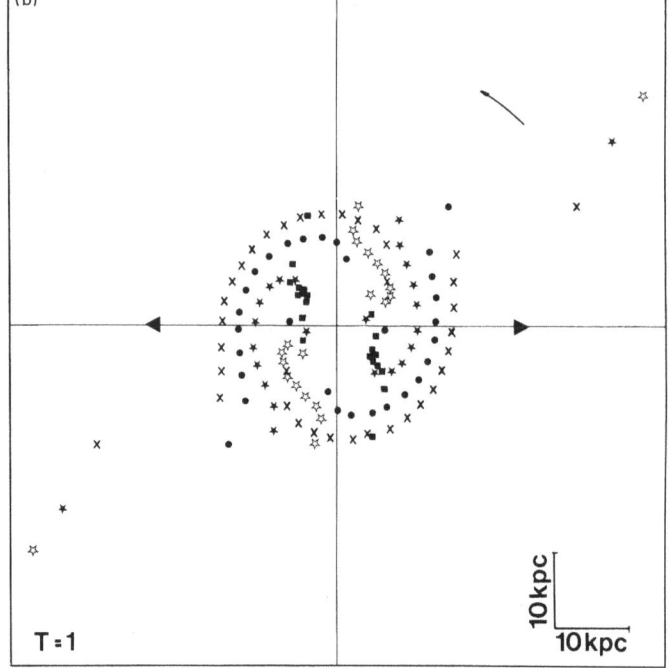

Fig. 3b.

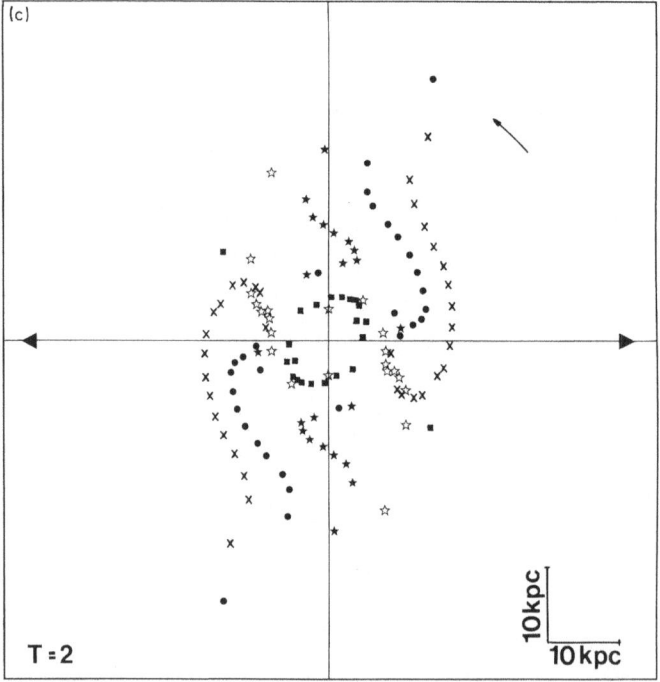

Fig. 3c.

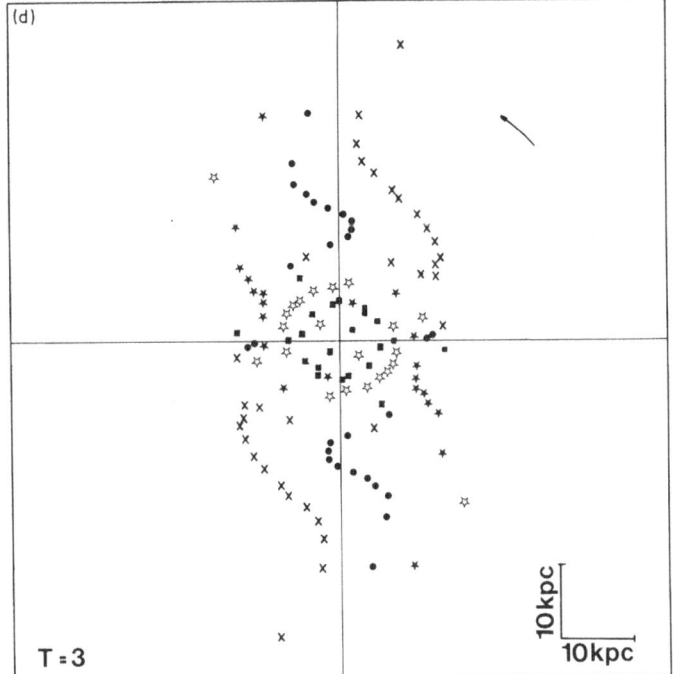

(d)

T = 3

10 kpc

10 kpc

Fig. 3d.

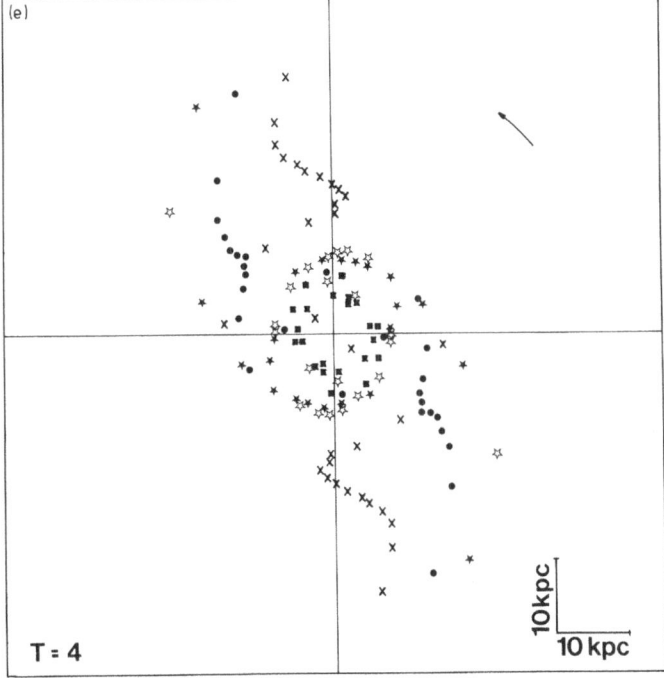

(e)

T = 4

10 kpc

10 kpc

Fig. 3e.

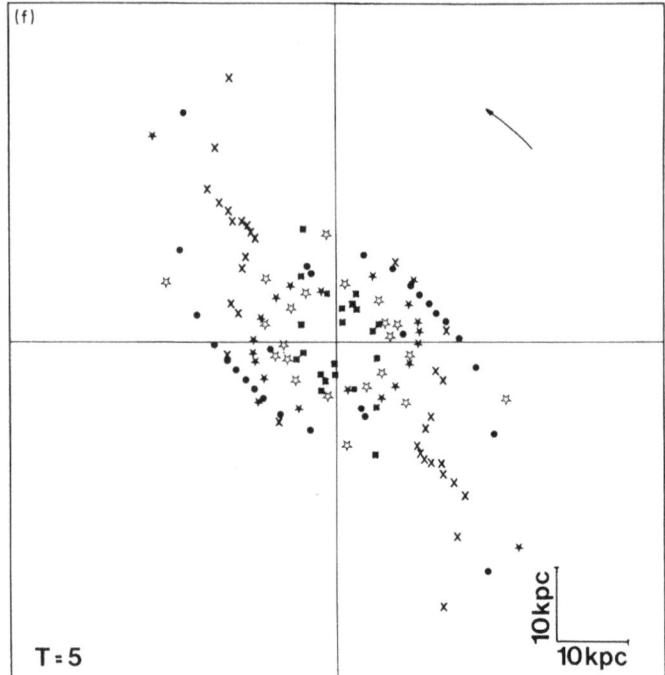

Fig. 3f.

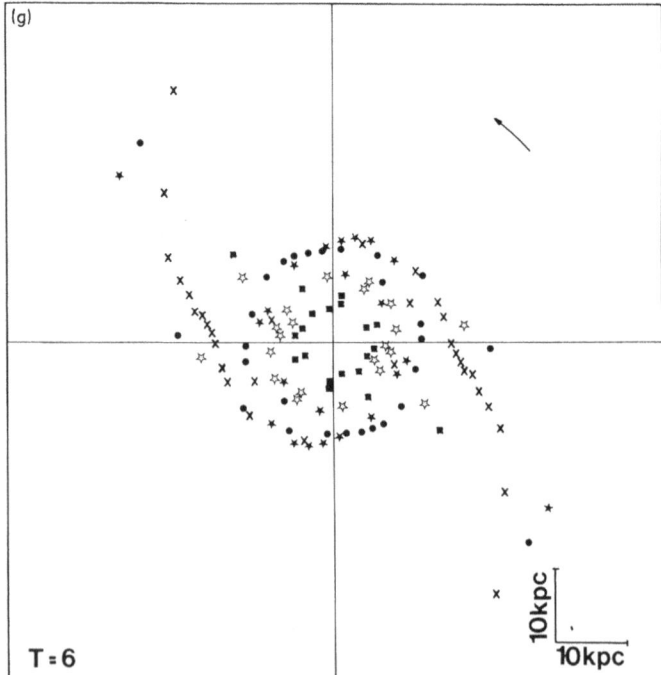

Fig. 3g.

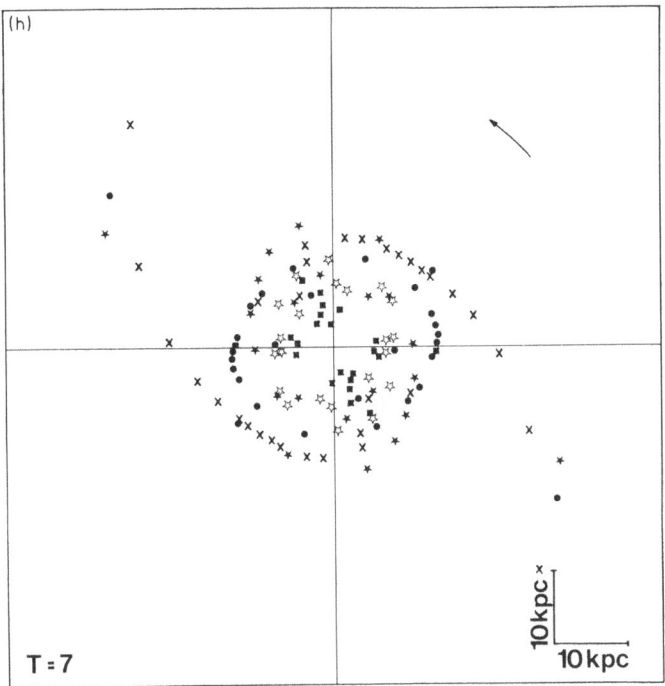

Fig. 3 (a) à (h).   Deux fragments de même masse 0.5×10^{10} $M_{\odot}$, sont éjectés dans deux directions opposées. Une structure de type spiral apparaît sur (b); elle est remplacée ensuite par des bras traînant, (g) et (h).

Bibliographie

Arp, H.: 1969, *Astron. Astrophys.* **3**, 418.
O'Connell, D. J. K. (ed.): 1971, *Nuclei of Galaxies*, North-Holland Publ. Co., Amsterdam, London.
Toomre, A. and Toomre, Y.: 1972, *Astrophys. J.* **178**, 623.
Van der Kruit, P. C.: 1971, *Astron. Astrophys.* **15**, 110.

DISCUSSION

Contopoulos: What is the life-time of the spiral feature?
   Hayli: About 5×10^{8} yr.
   King: Why did you choose this particular ejection velocity?
   Hayli: For the chosen velocities ejected bodies arrive at infinity with zero velocity.

PART IV

RELATIVISTIC STELLAR DYNAMICS

DYNAMICS OF RELATIVISTIC STELLAR SYSTEMS*

J. R. IPSER

University of Chicago, Chicago, Ill., U.S.A.

Abstract. Relativistic stellar-dynamical systems and their possible occurrence in nature are discussed. Features of the equilibrium models that have been constructed for spherical star clusters in general relativity are delineated. The results of studies of the stability of relativistic spherical clusters are reviewed. It is noted that the results, while not conclusive, indicate that realistic spherical clusters are stable against gravitational collapse if their central redshifts $z_c \lesssim 0.55$ and unstable if $z_c \gtrsim 0.55$. More work is needed on this and other problems.

1. Introduction

About a decade ago, exciting developments in astronomy, involving, for example, the cosmic blackbody radiation, quasars, radio sources, and galaxies exhibiting violent activity, helped to nurture a renewed interest in possible astrophysical applications of general relativity. This interest flowered into a sustained outburst of work on relativistic models for the universe and for individual systems such as stars, clusters of stars, and black holes.

The fact that no individual relativistic systems had yet been discovered – or course, it had long been generally accepted that the overall structure of the universe requires a relativistic description – did not deter speculation about them. And one might say rightly so, if only because of the present experimental situation, which associates relativistic neutron stars with pulsars and black holes with the unseen components in certain binary-star systems.

There is as yet still no direct evidence for the existence of relativistic stellar systems (relativistic clusters of stars). And, depending on one's prejudices, he may or may not hope that the relativist's luck has run out as far as the arrival of evidence for the existence of yet another class of relativistic objects is concerned. In any case, at least a few relativists have expended considerable effort trying to discern the properties of relativistic stellar systems, and we shall here review the current theoretical situation.

2. General Features of Relativistic Stellar Systems

Relativistic effects should be important for a star cluster if the gravitational redshift of a photon emitted from the center of the cluster and received at infinity is greater than, say, a hundredth or so. Generally, except for a rather special class of clusters discussed below, such a significant redshift requires the ratio (we use units in which $G = c = 1$)

$$\frac{2M}{R} \approx 0.01 \, \frac{M/(10^{11} \, m_\odot)}{R/(1 \text{ pc})} \tag{1}$$

* Supported in part by the National Science Foundation (MPS 74-17456).

Hayli (ed.), Dynamics of Stellar Systems, 423–431. All Rights Reserved.

to exceed about a hundredth, where M and R are the total mass-energy and the mean size of the system. This clearly requires either a very large mass density or, if the volume is moderate or large, a very large mass.

If one supposes that clusters are not born in a relativistic state, the question immediately arises whether they have been around long enough to permit evolution to such a state. This is a nagging question. It seems clear, on the one hand, that evolution pictured as driven by 2-body encounters will not do the trick: the associated evolution time-scales are simply too long, at least for the types of stellar systems that are known to exist (see, for example, Fackerell *et al.*, 1969). On the other hand, computer simulations of the evolution of stellar systems (see, for example, the contributions to this volume by Hénon and by Spitzer) often point to a scenario involving rapid evolution – compared with evolution expected on the basis of 2-body encounters – of a stellar system's central regions toward ever increasing ratio (1). It has been suggested that what one is seeing here is a phenomenon driven by many-particle interactions and analogous to a thermal runaway, or gravothermal catastrophe, of the type proposed by Antonov (1962) and Lynden-Bell and Wood (1968).

We shall return to the question of thermal runaways later, but for the moment let us suppose that nature does in fact have some way of making relativistic clusters. Once a cluster becomes relativistic, it should evolve rapidly. If the stars are of normal size, direct collisions between stars can be expected to occur rather frequently and to dominate the evolution. (For example, a relativistic cluster with 10^{12} solar-type stars in 1 pc will produce $\sim 10^8$ collisions per year!) One is then faced with the possibility that the center of the cluster is turned into one supermassive object. If the cluster 'stars' are neutron stars or black holes, then energy-loss via emission of gravitational radiation during close binary encounters dominates the evolution (Greenstein, 1969), and it is interesting to speculate about the possibility of observing such radiation.

If a relativistic cluster is not to evolve catastrophically rapidly on a dynamical time-scale, the evolution time-scale must be large compared with the time a typical star takes to traverse the system. For a given value of the ratio (1), this requirement places a lower limit on the size, and hence on the mass, of the system (Fackerell *et al.*, 1969). For example, if the ratio $2M/R \sim 0.1$, the lower limit on $R \sim 0.01$ pc for a cluster of solar-type stars. For $2M/R \sim 0.1$ and $R \sim 1$ pc, $M \sim 10^{12} \, m_\odot$ and the collision time-scale $\sim 10^4$ yr. If the evolution time-scale is sufficiently greater than the star crossing-time, then the cluster should evolve quasistatically; and at any moment it should be able to be approximated by a collisionless near-equilibrium state.

3. Fundamental Equations of Relativistic Stellar Dynamics

To date, almost all formal studies of relativistic stellar systems have adopted descriptions in terms of a one-particle distribution function, $\mathcal{N}$. An observer at a spacetime event $(\mathbf{x}, t)$ determines $\mathcal{N}$ there by measuring, in his local Lorentz frame, the number dN of stars occupying a volume d^3x in physical space, occupying a volume d^3p in

3-momentum space and having rest masses in the range dm:

$$\mathcal{N} = \mathrm{d}N/\mathrm{d}m \; \mathrm{d}^3 p \; \mathrm{d}^3 x. \tag{2}$$

If the geometry of spacetime is described by the line element (latin indices run over 0, 1, 2, 3)

$$\mathrm{d}s^2 = g_{ab} \; \mathrm{d}x^a \; \mathrm{d}x^b \tag{3}$$

in a particular curvilinear coordinate system (x^0, x^1, x^2, x^3), it can be shown that Equation (2) is equivalent to

$$\mathcal{N} = \mathrm{d}N/\mathrm{d}\mathscr{V}_p \; \mathrm{d}\mathscr{V}_x, \tag{4}$$

with

$$\mathrm{d}\mathscr{V}_p = -\,\mathrm{d}p_0 \; \mathrm{d}p_1 \; \mathrm{d}p_2 \; \mathrm{d}p_3 / \sqrt{-g}, \qquad \mathrm{d}\mathscr{V}_x = (p^0/m)\sqrt{-g} \; \mathrm{d}x^1 \; \mathrm{d}x^2 \; \mathrm{d}x^3. \tag{5}$$

Here p_a and p^a are the covariant and contravariant components of a star's 4-momentum, g is the determinant of the metric tensor, and m is a star's rest mass. The distribution function $\mathcal{N}$ is an invariant in that all observers at a given event agree on its value for a given group of stars. The distribution function determines a smoothed-out stress-energy tensor

$$T^{ab} = \int (\mathcal{N}/m) \; p^a p^b \; \mathrm{d}\mathscr{V}_p. \tag{6}$$

The stress-energy tensor in turn determines the geometry via Einstein's field equations,

$$G^{ab} = 8\pi T^{ab}, \tag{7}$$

where G^{ab} is the Einstein field tensor. If, for one reason or another, a cluster can be approximated as collisionless, the circle begun by Equations (6) and (7) is completed by demanding that the geometry determines the distribution function via the Boltzmann-Liouville, or collisionless Boltzmann, equation

$$\begin{aligned} \mathscr{D}\mathcal{N} &\equiv \frac{\mathrm{d}x^a}{\mathrm{d}s} \frac{\partial \mathcal{N}}{\partial x^a} + \frac{\mathrm{d}p_a}{\mathrm{d}s} \frac{\partial \mathcal{N}}{\partial p_a} \\ &= \frac{p^a}{m} \frac{\partial \mathcal{N}}{\partial x^a} - \frac{1}{2m} \frac{\partial g^{bc}}{\partial x^a} p_b p_c \frac{\partial \mathcal{N}}{\partial p_a} = 0. \end{aligned} \tag{8}$$

The operator $\mathscr{D}$ is the derivative with respect to proper time along the path of a star through phase space.

4. Relativistic Equilibrium Configurations

Generalizations of familiar Newtonian methods enable one to break the circle formed by Equation (6)–(8) and to construct collisionless equilibrium configurations, for which there is no explicit dependence on time $t = x^0$ (see, for example, Fackerell, 1968; and Zel'dovich and Novikov, 1971). To date, only spherically symmetric configurations have been constructed – almost always numerically.

Under the assumption of spherical symmetry, the line element (3) takes the form

$$ds^2 = -e^{\nu(r,\,t)}\,dt^2 + e^{\lambda(r,\,t)}\,dr^2 + r^2(d\theta^2 + \sin^2\theta\,d\varphi^2) \qquad (9)$$

in Schwarzschild coordinates (t, r, θ, φ). At equilibrium, when the metric functions ν and λ are time independent, the Boltzmann-Liouville Equation (8) implies that $\mathcal{N}$ is generally (there are contrived exceptions – see Zel'dovich and Novikov, 1971) a function of the isolating integrals

$$m, \quad E = p_0, \quad \text{and} \quad J = [p_\theta^2 + (p_\varphi/\sin\theta)^2]^{1/2} \qquad (10)$$

along stellar orbits. The integral E is the energy of a star as measured at infinity, and J is the magnitude of its angular momentum. If $\mathcal{N}$ does not depend on J, the cluster has an isotropic velocity distribution and an effective isotropic pressure

$$P = T_r^r = T_\theta^\theta = T_\varphi^\varphi. \qquad (11)$$

Zel'dovich and Podurets (1965) constructed numerical equilibrium models by taking $\mathcal{N}$ to be that appropriate to a truncated isothermal cluster of identical stars,

$$\mathcal{N} = Ae^{-E/kT}\,\delta(m - m_0)\,H(E_{\mathrm{CUT}} - E). \qquad (12)$$

Here A is a normalization constant, k is Boltzmann's constant, T is the constant temperature as measured from infinity (the temperature as measured locally in the cluster is not constant but increases toward the center) and m_0 is the rest mass of a star; $H(x)$ is the step function

$$\begin{aligned} H(x) &= 1 \qquad x \geqslant 0 \\ &= 0 \qquad x < 0, \end{aligned} \qquad (13)$$

so that the constant E_{CUT} is the maximum energy of a star in the cluster. Fackerell (1968, 1970, 1971) and Ipser (1969) have also constructed isothermal models as well as a variety of other isotropic models that obey polytropic relations of the form

$$-T_0^0 = \alpha P^{n/(n+1)} + P/(\Gamma - 1), \qquad (14)$$

where α, n, and Γ are constants and P is the isotropic pressure (4.3).

Certain features appear to be common to all these types of models. For example, as the central redshift z_c – a convenient measure of the importance of general relativity that can take any value in the range $(0, \infty)$ – increases from zero, the ratio (1) at first increases to a maximum value typically ~ 0.3 and thereafter oscillates about its maximum. More importantly (and we shall shortly see why), the fractional binding energy,

$$\mathscr{E}_B \equiv (M_0 - M)/M_0, \qquad (15)$$

$M_0 =$ rest mass-energy of cluster, $M =$ total mass-energy of cluster, also at first increases to a maximum value and thereafter oscillates about its maximum. Remarkably, the maximum value of $\mathcal{E}_B$ and the corresponding z_c show little variation from one type of model to another: for all isothermal and polytropic models that have been studied, $\mathcal{E}_B$ has a maximum value $\sim 0.0355 \pm 0.005$ at a central redshift $z_c \sim 0.53 \pm 0.03$.

5. The Stability of Relativistic Stellar Systems

Hoyle and Fowler (1967) proposed a quasar model in which the quasar lies at the center of a massive relativistic star cluster and derives its redshift from the gravitational field of the cluster. The viability of such a model clearly depends crucially on whether equilibrium clusters become unstable against gravitational collapse before their redshifts reach values as large as $z_c \sim 3$. The type of instability referred to here involves a gross overall instability of the gravitational field of a cluster. Having evolved to a point at which such an instability sets in, a cluster would subsequently collapse on a time scale of the order of the star transit-time across the cluster. One is consequently dealing here with a dynamical instability.

Methods for studying the dynamical stability of relativistic spherical clusters to small spherical perturbations were developed by Ipser and Thorne (1968). They used the fact that collisions between stars can be neglected in a study of dynamical stability if the evolution time-scale is assumed to be large compared with the dynamical time-scale. Briefly, their analysis begins with the expression

$$ds^2 = -\exp\left[v_A(r) + v_B(r, t)\right] dt^2 + \exp\left[\lambda_A(r) + \lambda_B(r, t)\right] dr^2 + $$
$$+ r^2 (d\theta^2 + \sin^2 \theta) \, d\varphi^2 \qquad (16)$$

for the line element of a slightly perturbed spherical cluster, where the subscripts A and B refer to equilibrium quantities and small perturbations of those quantities, respectively. The distribution function is split in a particular way into an equilibrium part, $\mathcal{N}_A = F(m, E_A = e^{v_A} p_0, J)$, and a perturbed part, $\mathcal{N}_B = f(x^a, p^a)$, so that

$$\mathcal{N} = \mathcal{N}_A + \mathcal{N}_B = F(m, E_A, J) + f(x^a, p^a). \qquad (17)$$

Equations (6)–(8) are then linearized in v_B, λ_B, and f. The perturbation f is split into its even and odd parts,

$$f_{\pm} \equiv \tfrac{1}{2}\left[f(x^a, p^0, p^\alpha) \pm f(x^a, p^0, -p^\alpha) \right], \qquad (18)$$

as a function of spatial momenta p^α ($\alpha = 1, 2, 3$), and a dynamical equation of the form

$$(1/-F_E) \, \partial^2 f_-/\partial t^2 = \mathcal{T} f_- \qquad (19)$$

is derived. In this equation

$$F_E \equiv [\partial F(m, E_A, J)/\partial E_A]_{m, J}, \qquad (20)$$

and $\mathcal{T}$ is an integro-differential operator in phase space. It is self-conjugate for well-behaved odd functions h, g which vanish outside the phase space of the equilibrium cluster; that is

$$\int h \mathscr{T} g \, d\mathscr{V}_p \, d\mathscr{V}_x = \int g \mathscr{T} h \, d\mathscr{V}_p \, d\mathscr{V}_x. \tag{21}$$

Attention is restricted to clusters satisfying the physically reasonable condition

$$F_E \leqslant 0, \tag{22}$$

which implies that there are fewer stars at high energies than at low energies. It then follows from Equations (19) and (21) that a cluster is dynamically stable to spherical perturbations if and only if $\mathscr{T}$ is positive-definite,

$$\int g \mathscr{T} g \, d\mathscr{V}_p \, d\mathscr{V}_x > 0, \tag{23}$$

for all well-behaved odd perturbation functions g.

One can attempt determine a cluster's stability by inserting various trial functions in the integral in Equation (23). Ipser (1969) followed this path for trial functions of the simple form

$$g = F_E p^r C(r). \tag{24}$$

His study of a wide variety of isothermal and polytropic clusters reveals a pattern that changes strikingly little from one cluster type to another: at small redshift $\mathscr{T}$ is positive-definite over the subspace of functions (24); but $\mathscr{T}$ ceases to be positive-definite at a redshift very near that at which the fractional binding energy peaks. Consequently, the clusters studied are unstable if $z_c \gtrsim 0.55$.

This behavior, which tends to be supported by Fackerell's (1970) stability calculations, tempts one to adopt the general picture that a spherical cluster born at low z_c will evolve in the direction of increasing z_c and binding energy along a sequence of initially stable states, and will become unstable near $z_c = 0.55$. There is an impediment to adopting this picture, however. It arises from work by Bisnovatyi-Kogan and Zel'dovich (1969), who studied the so-called γ-law cluster models. Such a model consists of a nearly constant-density core surrounded by an extended mantle in which the density of total mass-energy, $-T_0^0(r)$, and the mass energy inside radius r, $M(r)$, have the behavior

$$-T_0^0(r) \sim \frac{\gamma}{1+6\gamma+\gamma^2} \frac{1}{2\pi r^2}, \qquad M(r) \sim \frac{2\gamma r}{1+6\gamma+\gamma^2}. \tag{25}$$

The constant γ is the ratio of the isotropic pressure to the density in the core and mantle. The join between the core and mantle occurs at a radius

$$r_{core} \sim (\gamma/2\pi\varrho_c)^{1/2}, \qquad \varrho_c \equiv -T_0^0(r=0). \tag{26}$$

Surrounding the mantle is an envelope in which the sensity drops to zero. The redshift a photon experiences in traveling from the center to the outer edge, R_m, of the mantle is

$$z_c \sim (R_m/r_{\mathrm{core}})^{2\gamma} - 1. \tag{27}$$

Bisnovatyi-Kogan and Zel'dovich (1969) pointed out that, in principle, Equations (25)–(27) for sufficiently small γ allow the construction of models with arbitrarily large redshifts and arbitrarily *small* ratios $2M(r)/r$ everywhere. Hence, they argued, by choosing γ very small, one could construct large-redshift models that must be stable, since the models have very small ratios $P/(-T_0^0)$ and $2M(r)/r$ and hence are locally nearby Newtonian everywhere.

Bisnovatyi-Kogan and Thorne (1970) showed that the operator $\mathcal{T}$ is in fact positive-definite for all trial functions of the form (24) if and only if $\gamma \lesssim 0.117$. Add to this the fact that the fractional binding energy of the pure-mantle models (infinite ϱ_c, z_c, and R_m) peaks as a function of γ at $\gamma \sim 0.025$. Then, remembering the way isothermal and polytropic clusters behave, one is strongly drawn to the conjecture that γ-law models are stable if $\gamma \lesssim 0.025$.

What will be the properties of stable γ-law models? Note that for each value of γ and ϱ_c, a minimum size R, and hence total mass M, of the system is needed to produce a desired redshift. For example, for $\gamma = 0.025$,

$$R \gtrsim (1 + z_c)^{20}/(15\varrho_c^{1/2}), \qquad M \gtrsim 0.05\,R. \tag{28}$$

Hence if $z_c = 1$, then $R \gtrsim 3 \times 10^5$ pc for a massive object with $\varrho_c \sim 10^{12}\, m_\odot$ pc^{-3}; and $R \gtrsim 3 \times 10^4$ pc for a less massive object with the large density $\varrho_c \sim 10^{14}\, m_\odot$ pc^{-3} (corresponding to, say, 10^8 stars in ~ 0.01 pc). Though these radii may not be unmanageable, the corresponding total masses, $M \gtrsim 3 \times 10^{17}\, m_\odot$ and $3 \times 10^{16}\, m_\odot$, might be. And matters only get worse as z_c is increased. At very small γ one runs into problems with the very large radii ($\gg$ Hubble radius) needed to produce large redshifts. Conversely, if models can be stable for γ near 0.1, then the possibility of large redshifts becomes more attractive.

Obviously, more work is needed to conclusively demonstrate whether reasonable spherical clusters can be stable at central redshifts significantly larger than $z_c = 0.55$.

Up until now we have not raised the question of dynamical stability to nonspherical perturbations. Nonspherical perturbations, in contrast with spherical perturbations, take on an added significance in general relativity in that the associated motion of the cluster's overall gravitational field generates gravitational radiation. If a cluster reached a point of onset of instability to nonspherical perturbations, its subsequent unstable motion might produce copious amounts of gravitational waves. Unfortunately, the gravity-wave experimentalist is out of luck here, at least regarding clusters with $F_E \leqslant 0$. One can show (Ipser, 1975) that, just as in Newtonian theory, so also in general relativity, a spherical isotropic cluster with $F_E \leqslant 0$ is dynamically stable to nonspherical perturbations.

On occasion, there appear analyses expressing interest in some sort of secular, as opposed to dynamical, stability of stellar systems. One type of such analyses seeks to determine whether there exist equilibrium states for which the system's entropy, defined in one way or another, is a maximum subject to certain constraints, usually those of fixed mass-energy and stars. For example, Antonov (1962) and Lynden-Bell and Wood (1968) studied the Boltzmann entropy of Newtonian systems for which there are no limits on the velocities of the constituent particles, but which are held in by boxes of finite size in physical space. They found that a Newtonian spherical equilibrium configuration is a local entropy-maximum, when compared with all neighboring configurations having the same energy, mass, and confining radius, if and only if it is isothermal and the confining radius is less than the critical value $-0.335 \, M_0^2/\mathscr{E}$, where M_0 and $\mathscr{E}$ are the mass and energy of the configuration. They argued that confined systems with radii greater than the critical value have no states of locally maximum entropy to which they can evolve, and should experience a thermal runaway, or gravothermal catastrophe. Presumably, a thermal runaway is driven by encounters – perhaps collective in nature – and involves evolution to states of ever increasing entropy and density contrast, on a time scale that is fairly rapid and is determined by the driving mechanism but is nevertheless long compared with the dynamical time-scale (hence the term secular).

Even if some sort of thermal runaway is actually important for confined systems, it is not immediately clear what the situation is for realistic star clusters, which are not held in by boxes. In fact, recent calculations (Ipser, 1974, 1975) in Newtonian theory and in general relativity suggest that perhaps secular-stability criteria involving entropy maxima are not very useful for realistic self-bound stellar systems. The calculations derive a criterion for the Boltzmann entropy,

$$S \equiv -k \int \mathscr{N} \ln \mathscr{N} \, d\mathscr{V}_p \, d\mathscr{V}_x, \tag{29}$$

of a self-bound spherical equilibrium cluster to be a maximum, when compared with the entropy of all neighboring configurations having the same energy and stars. According to the criterion, S is a maximum if and only if the cluster has a truncated isothermal distribution function and the corresponding isentropic gas sphere – which has the same radial distribution of density and pressure as the cluster – is dynamically stable to spherical perturbations. One finds, however, that this criterion cannot be satisfied for a cluster unless its distribution function is so heavily truncated that the cluster is unrealistic. Hence one is led to conclude that no realistic clusters are Boltzmann-entropy maxima; and, further, that at present it is not clear how the occurrence of any sort of thermal runaway can be predicted or pinpointed by appealing to entropy arguments.

6. Conclusion

In this paper we have reviewed aspects of the current theoretical situation pertaining to relativistic stellar systems. Several important problems remain at an unsatisfactory stage of solution, and others have not even been tackled yet. It seems clear that relativistic stellar dynamics offers fertile ground for future study.

References

Antonov, V. A.: 1962, *Vestnik Leningrad. gos. Univ.* **7**, 135.

Bisnovatyi-Kogan, G. S. and Thorne, K. S.: 1970, *Astrophys. J.* **160**, 875.

Bisnovatyi-Kogan, G. S. and Zel'dovich, Ya. B.: 1969, *Astrofizika* **5**, 223.

Fackerell, E. D.: 1968, *Astrophys. J.* **153**, 643.

Fackerell, E. D.: 1970, *Astrophys. J.* **160**, 859.

Fackerell, E. D.: 1971, *Astrophys. J.* **165**, 489.

Fackerell, E. D., Ipser, J. R., and Thorne, K. S.: 1969, *Comments Astrophys. Space Phys.* **1**, 134.

Greenstein, G.: 1969, *Astrophys. J.* **158**, L145.

Hoyle, F. and Fowler, W. A.: 1967, *Nature* **213**, 373.

Ipser, J. R.: 1969, *Astrophys. J.* **158**, 17.

Ipser, J. R.: 1974, *Astrophys. J.* **193**, 463.

Ipser, J. R.: 1975, 'On Using Entropy and Energy Arguments to Study the Evolution and Stability of Relativistic Stellar-Dynamical Systems', *Astrophys. J.*, to be published.

Ipser, J. R. and Thorne, K. S.: 1968, *Astrophys. J.* **154**, 251.

Lynden-Bell, D. and Wood, R.: 1968, *Monthly Notices Roy. Astron. Soc.* **138**, 495.

Zel'dovich, Ya. B. and Novikov, I. D.: 1971, *Relativistic Astrophysics, Vol. 1, Stars and Relativity*, University of Chicago Press, Chicago.

Zel'dovich, Ya. B. and Podurets, M. A.: 1965, *Astron. Zh.* **42**, 963 (English transl. in 1966, *Soviet Astron.* **9**, 742).

DISCUSSION

Contopoulos: What is the role of gravitational radiation in your systems?

(1) How large is the gravitational radiation due to the graininess of the system?

(2) What is the role of non-spherical perturbations of spherical clusters?

Ipser: For spherical perturbations of spherical systems there is no gravitational radiation. Nonspherical perturbations do involve gravitational radiation. I have recently been able to show that, just as in Newtonian theory, so also in general relativity, spherical systems are stable to nonradial perturbations if the distribution function is a decreasing function of the energy of a star. Also, the incoherent gravitational radiation emitted due to the graininess of the cluster is not important compared with other processes.

Bardeen: There is one example of a rotating relativistic 'stellar-dynamical' system – the relativistic version of the cold MacLaurin disk. The structure of such disks was calculated numerically for models with all central redshifts from zero to infinity by Bardeen and Wagoner a few years ago. The binding energy increases monotonically with redshift, so one suspects the disks are stable against overall collapse even when infinitely relativistic. However, they are certainly violently unstable to fragmentation. They do exhibit such interesting phenomena as the ergotoroids mentioned by Ipser.

Ipser: I agree that your rapidly rotating disks are probably stable to overall collapse.

THE CONCENTRIC SHELL METHOD FOR RELATIVISTIC
STAR CLUSTERS

E. D. FACKERELL

Dept. of Applied Mathematics, University of Sydney, Australia

Abstract. The analytic aspects of the Campbell-Hénon method of concentric spherical shells are generalized for application to relativistic spherically symmetric star clusters.

1. Introduction

Hoyle and Fowler (1967) gave considerable impetus to the study of the properties of relativistic star clusters when they introduced their relativistic star cluster model for quasistellar sources. In this model the redshift is gravitational in origin, since the emitting region is postulated to be at the centre of a highly relativistic star cluster. The main difficulty with this model is to find reasonable models of relativistic star clusters which are stable and yet which have large central redshifts (z_c). Ipser (1969b) found that wide classes of reasonable models of relativistic spherically symmetric collisionless star clusters become unstable against radial perturbations at values of the central redshift greater than about 0.5. Using a refinement of Ipser's method, Fackerell (1970) showed that a relativistic polytropic star cluster model with a core-halo structure was unstable for $z_c \gtrsim 0.73$. Because of these results interest in the Hoyle-Fowler model waned after 1970.

However, in a recent publication Horwitz *et al.* (1974) have claimed, on the basis of a thermodynamic stability criterion, that there exist stable truncated isothermal relativistic star clusters with values of z_c as high as 4. These clusters have a pronounced core-halo structure, and so the analysis of the dynamical stability of these models requires the introduction of more powerful methods than those available at present, since Ipser's methods are usually inconclusive for models with pronounced core-halo structures (Ipser, 1969a, b). A further reason why it is desirable to develop new methods for the analysis of the dynamical stability of relativistic clusters is that all existing methods require that the equilibrium distribution function should be a monotonic decreasing function of energy (Ipser and Thorne, 1968). Another problem which is as yet intractable and which nevertheless is of considerable interest is the investigation of the effect on a cluster of a central black hole.

In Newtonian theory the most significant method developed to date for the analysis of the stability against radial disturbances of collisionless spherically symmetric stellar systems is the concentric shell method due to Campbell (1962) and developed for full numerical application by Hénon (1964. 1967). Hénon has demonstrated the power of this method in his investigation of the stability of a wide class of generalized polytropic models (Hénon, 1973). A compact summary of the method has been given by Janin (1972).

Hayli (ed.), Dynamics of Stellar Systems, 433–439. All Rights Reserved.

The purpose of this paper is to begin the development of a general relativistic version of the Campbell-Hénon method. This requires the solution of three problems, namely, (1) the analysis of the motion of a spherical shell with a given 'equation of state' in a spherically symmetric gravitational field (possibly due to a black hole), (2) the determination of the 'equation of state' of a shell of collisionless particles, and (3) the determination of the configuration that arises when two collisionless shells pass through one another.

2. Motion of a Spherical Shell in General Relativity

The motion of a spherical surface layer in general relativity has been investigated by a number of authors. Israel (1966, 1967) investigated the motion of a pressure-free shell collapsing under its own self gravitation. Chase (1970) extended Israel's formalism to deal with a charged shell with internal pressure moving in the combined field due to itself and to an internal central charged mass. The equation of motion that we require may be deduced from Chase's results by setting all charges equal to zero.

The spherical shell, which we denote by Σ, has the parametric equations

$$r = R(\tau),$$ (1)

$$t_+ = T_+(\tau),$$ (2)

and

$$t_- = T_-(\tau),$$ (3)

where τ is the proper time on the shell, t_+ and t_- are the Schwarzschild times on the exterior and the interior of Σ and r is the radius of the shell in Schwarzschild coordinates. Note that, while r is continuous across the shell, the Schwarzschild time is discontinuous across the shell. If the mass interior or the shell is m_1 and the mass of the whole system is m_2 (both in geometrized units), the equation of motion of the shell is given by

$$(\dot{R}^2/c^2 + 1 - 2m_1/R)^{1/2} - (\dot{R}^2/c^2 + 1 - 2m_2/R)^{1/2} = \frac{4\pi G}{c^2} R\sigma,$$ (4)

where

$$\dot{R} = dR/d\tau$$ (5)

and σ is the surface energy density of the shell. If the surface pressure of the shell is denoted by p, the surface energy tensor S_{ab} of the shell is given by

$$S_{ab} = \left(\sigma + \frac{p}{c^2}\right) U_a U_b + \frac{p}{c^2} g_{ab} \qquad (a, b = 0, 1, 2),$$ (6)

where g_{ab} is the intrinsic metric tensor of the shell and U_a is the intrinsic velocity of an observer moving with the shell. If we take our intrinsic coordinates ξ^a on the shell as

$$(\xi^0, \xi^1, \xi^2) = (\tau, \theta, \phi)$$ (7)

we find that

$$(U^0, U^1, U^2) = (c, 0, 0) \tag{8}$$

and

$$(g_{ab}) = \mathrm{diag}(-c^2, R^2(\tau), R^2(\tau) \sin^2 \theta). \tag{9}$$

S_{ab} satisfies the conservation law

$$S^{ab}{}_{;b} = 0$$

which in our case becomes

$$d(R^2\sigma)/d\tau + 2R\dot{R}\frac{p}{c^2} = 0. \tag{10}$$

The Schwarzschild times on the two sides of Σ may be determined from the differential equations

$$\frac{dT_+}{d\tau} = (1 - 2m_2/R)^{-1}(\dot{R}^2/c^2 + 1 - 2m_2/R)^{1/2} \tag{11}$$

and

$$\frac{dT_-}{d\tau} = (1 - 2m_1/R)^{-1}(\dot{R}^2/c^2 + 1 - 2m_1/R)^{1/2}. \tag{12}$$

Equations (4), (5), (11) and (12) are sufficient to determine the motion of the shell provided σ and hence p are given as functions of R.

3. Equation of State of a Shell of Collisionless Particles

The kinetic theory of surface layers has been developed in general relativity by Voorhees (1972) whose results may be stated as follows. The surface energy tensor is given by

$$S_{ab} = \int \frac{1}{\mu} f p_a p_b \, dv_{(p)}, \tag{13}$$

where f is the distribution function in the surface layer, μ is the conserved rest mass of a particle with intrinsic momentum p_a and $dv_{(p)}$ is the natural element of volume in the momentum space associated with the surface layer. The distribution function f is to be a suitable function of the constants of the motion of a particle in the surface layer. These constants can be obtained from the intrinsic Hamilton-Jacobi equation

$$g^{ab} \frac{\partial S}{\partial \zeta^a} \frac{\partial S}{\partial \zeta^b} = -\mu^2 c^2, \tag{14}$$

which in our case becomes

$$\frac{1}{c^2} \left(\frac{\partial S}{\partial \tau}\right)^2 - \frac{1}{R^2(\tau)} \left[\left(\frac{\partial S}{\partial \theta}\right)^2 + \frac{1}{\sin^2 \theta} \left(\frac{\partial S}{\partial \phi}\right)^2 \right] = \mu^2 c^2, \tag{15}$$

so that

$$p_0 = -c[\mu^2 c^2 + \alpha^2/R^2(\tau)]^{1/2} \tag{16}$$

$$p_1 = \pm[\alpha^2 - J^2/\sin^2\theta]^{1/2} \tag{17}$$

$$p_2 = J, \tag{18}$$

where J is the axial component of angular momentum and α^2 is the squared angular momentum.

If all the particles have the same rest mass μ_0 and the same squared angular momentum α_0^2, a suitable distribution function is

$$f = \frac{1}{2\pi^2} A\delta(\mu - \mu_0)\,\delta(\alpha^2 - \alpha_0^2), \tag{19}$$

where the constant A is in fact the total number of particles in the shell, since the intrinsic particle flux vector

$$N_a = \int \frac{1}{\mu} f p_a \, dv_{(p)} \tag{20}$$

has the sole surviving proper component

$$N_{(0)} = A/(4\pi R^2). \tag{21}$$

From Equations (13) and (19) we find that

$$\sigma = \frac{A}{4\pi R^2 c}[\mu_0^2 c^2 + \alpha_0^2/R^2]^{1/2} \tag{22}$$

and

$$p = \frac{A\alpha_0^2 c}{8\pi R^4}[\mu_0^2 c^2 + \alpha_0^2/R^2]^{-1/2}. \tag{23}$$

These expressions differ from those given by Papapetrou and Hamoui (1968). However, they are entirely compatible with Gerlach's results on the motion of a thin shell (Gerlach, 1970) so we are compelled to reject the results of Papapetrou and Hamoui together with those of Urbantke (1972) who followed their derivation.

If we substitute Equation (22) into Equation (4) and solve for $\dot{R}^2$ we obtain the equation of motion in the form

$$\frac{1}{c^2}\left(\frac{dR}{d\tau}\right)^2 = -1 + \frac{m_1 + m_2}{R} + \frac{(m_2 - m_1)^2}{K^2} + \frac{1}{4}\frac{K^2}{R^2} \tag{24}$$

where the variable K is defined by

$$K = \frac{GA}{c^3}(\mu_0^2 c^2 + \alpha_0^2/R^2)^{1/2}. \tag{25}$$

4. The Crossing of Two Shells

We now consider the problem of determining the configuration that results when two collisionless spherical shells pass through each other at radius R say. Let the velocity of the inner shell just before crossing be specified by $\dot{R}_1^- = dR_1/d\tau_1$, where τ_1 is the proper time on this shell and the derivative is evaluated just before crossing. Let the total number of particles in this shell be A_1, let their common rest mass be μ_1, let their common angular momentum have magnitude α_1 and let the variable K_1 be defined by

$$K_1 = \frac{GA_1}{c^3} (\mu_1^2 c^2 + \alpha_1^2/R^2)^{1/2}. \tag{26}$$

Let $\dot{R}_2^-$, τ_2, A_2, μ_2, α_2 and K_2 have similar meanings for the outer shell. Let the Schwarzschild mass of the region interior to the inner shell be m_1, let the Schwarzschild mass of the region between the two shells be m_2 before crossing takes place and let the Schwarzschild mass of the total system be m_3. We need to determine the Schwarzschild mass m_4 of the region between the shells after crossing occurs together with the new velocity $\dot{R}_1^+$ of the new outer shell and the new velocity $\dot{R}_2^+$ of the new inner shell.

Because A_1, μ_1, α_1, A_2, μ_2, α_2 and therefore K_1 and K_2 are all conserved we obtain the two relations

$$(\dot{R}_1^+/c)^2 = -1 + \frac{m_3 + m_4}{R} + \frac{(m_3 - m_4)^2}{K_1^2} + \tfrac{1}{4}\frac{K_1^2}{R^2} \tag{17}$$

and

$$(\dot{R}_2^+/c)^2 = -1 + \frac{m_4 + m_1}{R} + \frac{(m_4 - m_1)^2}{K_2^2} + \tfrac{1}{4}\frac{K_2^2}{R^2} \tag{28}$$

between the three unknowns $\dot{R}_1^+$, $\dot{R}_2^+$ and m_4. A third relation comes from the fact that the relative velocity of the shells before crossing is equal to their relative velocity after crossing. We find that

$$[(\dot{R}_1^+/c)^2 + 1 - 2m_4/R]^{1/2} [(\dot{R}_2^+/c)^2 + 1 - 2m_4/R]^{1/2} - \dot{R}_1^+ \dot{R}_2^+/c^2 =$$
$$= (1 - 2m_4/R)\, Y, \tag{29}$$

where Y is defined by

$$Y = \{[(\dot{R}_1^-/c)^2 + 1 - 2m_2/R]^{1/2} [(\dot{R}_2^-/c)^2 + 1 - 2m_2/R]^{1/2} - \dot{R}_1^- \dot{R}_2^-/c^2\}/(1 - 2m_2/R). \tag{30}$$

A tedious algebraic manipulation yields the elegant result

$$m_4 = m_1 + m_3 - m_2 + Y K_1 K_2/R, \tag{31}$$

which shows that the mass energy of a given shell is not conserved during the crossing of that shell with some other shell. The values of $\dot{R}_1^+$ and $\dot{R}_2^+$ may then be determined from Equations (27) and (28), the ambiguities of sign being resolved by the fact that we require $\dot{R}_1^+ > \dot{R}_2^+$ and we also require that (29) should be satisfied.

5. Problems in Integration of the Equations

It is clear that the problem of numerical integration of the basic equations for relativistic shells will be considerably more difficult than the corresponding Newtonian problem. The difficulties arise principally because of the fact that each shell needs its own time coordinate and because the $\dot{R}_i$ are discontinuous at the crossing of shells, together with the fact that the mass-energy of a given shell is also discontinuous at shell crossings. The last two facts require that the precise spatial location of each shell crossing has to be determined, which will add considerably to computing time. The problem of different times for each shell may be overcome by the expedient of slicing space-time into sections determined by taking constant Schwarzschild time between shells, but it needs to be noted that the difference in Schwarzschild time between successive slices will not be constant in between different shells. Finally we note that the collapse of a shell inside its event horizon does not cause any serious computing problem since we have used proper time on the various shells as our basic time coordinates. Once the innermost shell has begun an irreversible journey towards its event horizon it may be ignored in further computation, apart from the fact that its mass is then to be added to the mass of the central black hole.

6. Discussion

Two main areas of research will be opened up by the development of a successful numerical integration procedure for relativistic shells. First, it will be possible to investigate the full non-linear stability of the truncated isothermal clusters introduced by Zel'dovich and Podurets (1965) and Fackerell (1966). This will enable a decisive answer to be given to the claim by Horwitz et al. (1974) that there exist stable clusters with high central redshifts. Second, the interesting problem of the effect of a central black hole on the evolution of a spherical cluster will be numerically analyzable. Both of these problems are being pursued by Mr. K. G. Suffern at the University of Sydney.

7. Acknowledgement

I wish to acknowledge the helpful and constructive criticism of K. G. Suffern during the course of this work.

References

Campbell, P. M.: 1962, *Proc. Nat. Acad. Sci.* **48**, 1993.
Chase, J. E.: 1970, *Nuovo Cimento* **67B**, 136.
Fackerell, E. D.: 1966, 'Relativistic Stellar Dynamics', University of Sydney (Ph.D. Thesis).
Fackerell, E. D.: 1970, *Astrophys. J.* **160**, 859.
Gerlach, U. H.: 1970, *Phys. Rev. Letters* **25**, 1771.
Hénon, M.: 1964, *Ann. Astrophys.* **27**, 83.
Hénon, M.: 1967, *14e Colloque de Liège*, p. 243.
Hénon, M.: 1973, *Astron. Astrophys.* **24**, 229.
Horwitz, G., Katz, J., and Klapisch, M.: 1974, 'Relativistic Star Clusters with High Gravitational Redshifts?', Racah Institute of Physics, Hebrew University of Jerusalem (preprint).

Hoyle, F. and Fowler, W. A.: 1967, *Nature* **213**, 373.

Ipser, J. R.: 1969a, *Astrophys. J.* **156**, 509.

Ipser, J. R.: 1969b, *Astrophys. J.* **158**, 17.

Ipser, J. R. and Thorne, K. S.: 1968, *Astrophys. J.* **154**, 251.

Israel, W.: 1966, *Nuovo Cimento* **44B**, 1.

Israel, W.: 1967, *Phys. Rev.* **153**, 1388.

Janin, G.: 1972, in M. Lecar (ed.), *Gravitational N-Body Problem (IAU Colloq.* **10***)*, D. Reidel Publ. Co., Dordrecht-Holland, p. 311.

Papapetrou, A. and Hamoui, A.: 1968, *Ann. Inst. H. Poincaré* **A9**, 179.

Urbantke, H.: 1972, *Acta Phys. Aust.* **35**, 1.

Voorhees, B. H.: 1972, *Phys. Rev.* **D5**, 2413.

Zel'dovich, Y. B. and Podurets, M. A.: 1965, *Astron. Zh.* **43**, 963.

PART V

GENERAL CONCLUSION AND FINAL DISCUSSION

DO STELLAR SYSTEMS AGREE WITH THE SINGULAR
EVOLUTION OF *N*-BODY SYSTEMS?

D. LYNDEN-BELL

Institute of Astronomy, The Observatories, Cambridge, U.K.

One of the important jobs of scientists is to formulate new laws and principles, and I have attempted to summarise those we have learnt in this most enjoyable and fruitful symposium in the laws stated below. However, recognising that laws derived from stellar dynamics may lead to important discoveries in other fields, I have translated these laws into the language of sociology. The key concept of hard and soft binaries has been introduced into stellar dynamics by Heggie. If $\langle \frac{1}{2}mv^2 \rangle$ at some point in a stellar system is defined to be $\frac{3}{2}\beta^{-1}$ then Heggie defines a soft binary to be one whose internal energy has $\beta|\varepsilon| < 1$ and a hard binary to be one with $\beta|\varepsilon| > 1$. Heggie's laws cover the interaction of these binaries with the field stars and his first basic discovery is that on interacting with this society hard binaries become harder while soft binaries become softer and may even cease to be binaries at all. The same may be true of criminals in their interaction with society.

Heggie's first law. The rate of hardening of a criminal is independent of his hardness once he has become a hard criminal.

Heggie's second law. Very hard criminals are locked up (by adiabatic invariants) most of the time so they only harden by committing relatively few but very violent crimes, while less hard criminals commit minor crimes more often.

Aarseth's law. Small societies evolve until they are dominated by a central very hard criminal who may expel other members of society, but he may eventually be displaced from his central position or even ejected altogether after a violent interaction with another member of society.

Hénon's law. Every society eventually evolves a central core or clique that dominates it, but when the clique gets too small it may become dominated by a hard criminal just as Aarseth's law describes.

Spitzer's law. Heavyweight individuals jostle their way through society until they join the ruling clique which is the ultimate source of power. Unfortunately power corrupts, and they are also the most likely to become hard criminals.

Hénon's conjecture is that a certain amount of crime is needed by society and if insufficient crime is committed then the ruling clique will shrink until a fresh hard criminal is created by the more prickly interactions among the smaller agitated clique.

The agreement between the rates of evolution calculated by *N* body computations and by the Monte Carlo approaches of Hénon and Spitzer is impressive as Wielen showed, and the refinement of this agreement by Hénon should convince us that the Monte Carlo method is accurate, at least until the number of stars in the central core

Hayli (ed.), Dynamics of Stellar Systems, 443–446. All Rights Reserved.

becomes so few that two body encounters leading to escape become frequent. When the number of stars in the core is only 100 or so the N body integrations become more powerful in giving us a detailed description of what happens. An area that needs far more study is the correct boundary condition to apply at the junction between this small core and the rest of the cluster. Is Hénon's conjecture – that the rest of the cluster needs a certain energy supply and that a central binary in the core will be formed and supply energy at just this rate – correct? If so, how should this energy demand by the rest of the cluster be represented when the N body calculation of the core is performed? Perhaps the boundary conditions adopted may be open to check by the powerful new N body method of Ahmad and Cohen.

If we take King's advice and turn to the great book of Nature for final confirmation that these cores have developed and on the expected timescale, we find our difficulties are not over. In globular clusters where we expect that the cores must have formed quite long ago, he sees no sign of a central nucleus, while in galaxies where there is no time for the stellar dynamical processes to act we see the sort of nuclei that we might have expected for the globulars. Of course, we may be able to wriggle out of this – the predicted cores should not contain a large proportion of the mass or light, only a significant fraction of the binding energy, so perhaps the core radius as defined by King has little to do with the cores as predicted by Monte Carlo calculations which have shrunk to be of negligible mass. My present belief is that some sort of funnel going up to a core of high density with one or more binary stars at its centre ought to be seen. However, we must bear in mind that globulars must have had massive stars in them at one time and their whole history may have been quite complicated. Let us turn back to the theories and simulations to see what we expect to happen once the core has developed to very great densities etc. There is little doubt that the contraction of the core leads to the creation of Aarseth's central hard binary. However, for a globular cluster as it is to-day a single pair probably cannot hold an energy of $\frac{1}{10}$ of the cluster energy, even with the two stars almost touching. Furthermore, such a binary will wander, due to the recoil whenever it gives up energy to a third star. The critical question is whether this recoil is sufficient to send it out of the cluster, as often happens in the numerical simulations of 100 stars when the stars have equal mass. As it hardens due to such encounters one might expect its escape but for the fact that the surrounding core will be getting deeper and deeper, and so the internal energy of the binary may only keep pace with the increasing temperature of the few stars left in the core. Thus the central binary may merely wander with translational energy $\frac{3}{2}kT$ like the remaining stars. In any case the central binary will occasionally eject a star through a violent encounter so there will be mass loss from the centre. In clusters of unequal mass the central hard binary will be heavy, and this will help to keep it fixed at the centre, and it will eject those lighter stars that have close encounters. However, at this point stellar evolution probably intervenes for the heavy stars will blow up; their debris will leave the cluster and the way will be open for another heavy binary to form. It is an in-teresting question to know whether the core evolution has to start all over again once its dominating binary leaves, or whether a sufficient funnel leading to the core has remained so that a new binary can form almost at once. The only heavy bodies that we expect to be stable for long times will be the black holes. If massive ones are made we must expect dynamical friction to bring them back to the central regions, provided that they have remained bound to the cluster.

Turning now to galaxies and clusters of galaxies, we have our first difference in that the galaxies are relatively large and fragile, so that their close encounters are dissipative. Work by Alladin, and more recently by Toomre, shows clearly the possibility of coalescence and the nuclei will wind their way down to the centre by dynamical friction. A number of people have speculated that the stars thrown off in these processes may be the origin of the envelopes of the cD galaxies. One wonders whether the two nuclei will eventually form a strong central binary and if so what will happen as more nuclei collect at the centre. If we were to apply our N body experience to a giant elliptical ignoring the wrong time-scale, it is interesting to note that a core should develop with a binding energy $\frac{1}{10}$ of the whole. This core energy will be concentrated into few degrees of freedom and must be 10^{60} erg for a large galaxy, sufficient for a large radio outburst. It was a bit sad to see that the 3 massive body model of Valtonen and Saslaw was in difficulty with the rarity of the correct initial conditions.

I turn now to haloes. I was impressed by Freeman's negative result on NGC 253 and his emphasis on the heavy lens components of some SO galaxies. Could it be that we have central disc components with large $Q = \langle V_R^2 \rangle^{1/2} K/(3.36 \, G\mu)$. Here $\mu(R)$ is the surface density of the galaxy and $K(R)$ is the epicyclic frequency. I note that Hohl's thorough discussion of bar instabilities in his numerical experiments did not include experiments with large Q in the middle, and that large Q there would be most useful in satisfying Ostriker and Peeble's criterion. Furthermore, that is just the place where every astronomer expects high random motion. The M dwarf problems discussed by Schmidt and Biermann seems to be more and more intriguing. We would probably patch things up if high central Q would allow us a Q of just over 1 out here, but even then we have the basic problem of the very flat layer of stars that are probably old. King's report on velocities of five stars and their lack of H and K emission certainly adds weight to the case already made by Sanduleak, Murray and Weistrop. It seems to me that we need haloes very badly in the system with large central bulges and the ellipticals and that large haloes may not be necessary to stabilise the flat Sc's. In the clusters of galaxies large haloes accompanying all galaxies with strong central bulges are probably all that is necessary.

We will hear more of the importance of viscosity in the formation of the galaxy as introduced by Larson, and all of us will remember Brahic's fascinating talk on the role of collisions and dissipation in all of astrophysics and in Saturn's ring in particular. While Larson was criticised for the many free parameters in his models, I think they are all introduced for good physical reasons, and his work is important in defining what sorts of star formation rates and friction coefficients are necessary. These are the food for other theorists to taste and think about, because we shall need an explanation of why the rates take values in Larson's range.

I agree with King that the problems of star formation are so little understood that the rest of the subject will be held back without more work theoretically and observationally on this problem.

It was very nice to see the advance of Technology making possible now what was almost impossible 20 years back. Thus Illingworth's velocity dispersions in globular clusters and Bertola's fine rotation curves for ellipticals and SOs are hints of what is to come. Lastly, it is always fun to mention the wilder speculations that can be the spice of life in science. Contopoulos in his review referred to Galgani's enthusiastic espousal of the theory of third and up to the Nth integral in a system of N degrees of freedom at low energy. The N adiabatic invariants there do give a basis for a sort of classical zero point energy. Can we still tell our students that classical mechanics has no explanations for the lowering of specific heats at low energy? If we still do so, we may well be telling less than the truth. It was also fun to see the close analogies between the classical theory and relativistic stellar dynamics represented by Ipser and Fackerell. It is sad that such a beautiful theory may not prove useful for real systems, but there is hope yet. The evolution of the core of an N body system as described on the constant energy hypothesis does have the central potential behaving like $\psi_0 \propto (t_0 - t)^{-2/7}$. If we can find a large enough system of small enough stars there may yet be room in the world for relativistic clusters.

FINAL DISCUSSION

King: I should like to emphasize the problem of the cores of globular clusters. A cluster core – within one core radius of the center – has about 10^4 stars, and the N-body calculations have not yet reached this domain. Yet the Monte Carlo results disturb me very much because they are adjusted to represent the right N and they predict a central behavior that we do not observe.

Contopoulos: Is there any evidence whether we have collisions, actual collisions, in the centre of clusters?

King: I don't know how you would observe a collision. There has been a nova observed at the center of a globular cluster; but as far as I know, it was a perfectly authentic nova.

Spitzer: Dr King remarked that there was a discrepancy between the Monte Carlo computations, which predicted a collapsing core, and the observations of globular clusters, which do not show such a collapse. It is not obvious to me that such a discrepancy is necessarily present. We do not yet understand the collapse in detail, but it is conceivable that such a collapse may be intermittent. For example, a hard binary may eject stars from the core and finally get ejected itself by recoil. The remaining stars would then resume their contraction, with another collapse occuring in due course. We cannot predict that this occurs, but we cannot exclude this possibility.

Lynden-Bell: I agree that a cycle of increasing relaxation time can occur with hard binaries forming and being ejected in turn but I would speculate that we might still see the core of high density even when the central binary was removed and that this core might quite rapidly make another binary. Of course this core might be much smaller than the observed core but I am still worried that it is not seen.

Contopoulos: Do we have any evidence from numerical experiments whether a very hard binary can explode or be ejected?

Aarseth: Close binaries can certainly be ejected from small clusters. The recoil effect is particularly strong if the particle masses are similar.

Hénon to Lynden-Bell: You pointed out that in my models, the halo expands faster than the core, so that the contrast increases and there is an apparent contradiction with observations. These models, however, are isolated; while real clusters are subjected to the tidal field of the Galaxy, which will stop the expansion of the halo at some point.

Lecar: Isn't the formation of energetic binaries considerably enhanced in a system with a mass spectrum?

Aarseth: Equal-mass systems do form energetic binaries and at least in small systems the time-scale is not much longer than for unequal masses.

Freeman: The most massive stars in globular clusters are probably about $0.8\ M_\odot$:

Hayli (ed.), Dynamics of Stellar Systems, 447–451. All Rights Reserved.
Copyright © 1975 by the IAU.

Illingworth's M/L ratios mean that the lightest stars are around 0.1 $M_\odot$ and maybe less, so the mass range is not so small.

Severne to Lynden-Bell: In view of the remarkable agreement now obtained between the N-body calculations and the Monte Carlo method based on the standard Fokker-Planck equation, do you think that it is still worth investing much effort into improving the description of encounters, in particular, retaining the curvature of the trajectories?

King: To expand somewhat on Severne's remark, there is a serious gap that I would like to bring up. In the domain of the N-body problem, we have simulations and we have observation, but we have no theory. For stellar encounters, we have the Fokker-Planck equation, but its validity breaks down just in the range where the N-body calculations are done. In this domain the simulations clearly show the importance of large energy changes, and these are just what invalidates the Fokker-Planck equation. Hénon has shown how to calculate the statistics of these large changes, but we lack a way of following their effect in the form of a differential equation. I know of no way of following a diffusion process in which the individual steps are large, except for some recent papers in the Soviet literature that used the Kolmogorov-Feller equation. Can anyone present say how relevant that work is to this problem?

Heggie: In these papers the Kolmogorov-Feller equation is solved numerically and a variety of problems are treated: 'equipartition' between different masses, escape, and so on.

Lynden-Bell: The numerical method is not equivalent to the Monte Carlo approach?

Heggie: No, in this work the distribution function is treated directly, not individual particles.

Hénon: One possible improvement would be to use the full Boltzmann equation instead of the Fokker-Planck equation, which is a limiting case valid only for small velocity changes.

King: I have tried to use the Boltzmann equation directly, by making a Taylor expansion and integrating term by term. It behaves as you would expect: beyond the second order, none of the terms contains the logarithmic factor. The trouble is that the series seems to converge quite poorly; so I gave up.

Hénon: Is it necessary to expand the Boltzmann equation? One should rather use it in its closed form, as an integral.

King: I don't know how. I hope that someone who is here today will go home and do it.

Hénon: Numerically at least, one can imagine how it could be done.

Spitzer to Lynden-Bell: One step would be possible in this direction. In the Princeton computations it would be possible to perturb the stellar velocities in accordance with the exact probability distribution function for two-body encounters, taking into account both the close and distant encounters. Not all features of the problem would be taken into account, but some would be.

Lynden-Bell: I have two general worries about the Fokker-Planck equation and its coefficients.

(1) They are calculated in the absence of the curvature in the cluster orbits although the effect extends to longer distances.

(2) The far field effects neighbouring regions of phase space similarly and should not lead to any change in phase density whereas the Fokker-Planck equation seems to me to assume that they contribute to the changes in phase density so important in cluster evolution.

Severne: A related problem which remains open and which is possibly of some importance is the description of the dynamics of the violent relaxation phase. Taken globally, violent relaxation gives a very satisfying picture of the overall evolution of clusters and galaxies. While the end state is well specified, we have no operational characterization of the evolution during violent relaxation.

Lynden-Bell: Several people have done simulations but I an very doubtful that one can do much analytically; even the theory of the equilibrium does not agree at all perfectly.

Lecar: I understand that contrary to early expectations, the structure of a stellar system with a cut-off Maxwellian distribution of velocities is quite sensitive to the cut-off. Equilibrium statistical mechanics provides no prescription for the cut-off.

Lynden-Bell: The cut off in the Maxwellian does not come out of equilibrium theory without special assumption. However the assumption is the same as one uses in cluster cores that the relaxation is mainly confined to the well bound stars in a central core.

Lecar to Wielen: A comparison of systems with a mass spectrum would provide a sensitive check on the validity of extrapolating N-body simulations to large N. Do you have such a comparison?

Wielen: I made a comparison of the mass segregation found in Monte Carlo models and in N-body simulations. Unfortunately, a detailed comparison is hampered at present, because our computer outputs do not provide the same quantities up to now. A rather global comparison, using the mean stellar mass as a function of radius, indicates no severe discrepancies between Monte Carlo and N-body results.

Spitzer: We have similar results from Monte Carlo computations by Shull and myself for a system with three components. These will be available shortly for comparison with the results by Dr Wielen.

Lynden-Bell: I would like to make the speculative remark that the binding energies of giant elliptical galaxies are around 10^{61} erg or more and in the gravothermal catastrophe the evolution causes the concentration of the energy into few degrees of freedom in the cluster core. Typically this may lead to 10% or so of the energy in the central core or in small systems the central binary. It is remarkable that this would give us 10^{60} erg in giant E nuclei, very much the order of magnitude needed for their radio source explosions. Have you any remarks on this, Bill?

Saslaw: The coincidence between the energy required for radio source and the energy of a massive binary in a galactic nucleus could occur if there is approximate equipartition between the binding energy of the binary and the rotational and magnetic energies of its individual components.

Contopoulos to Lynden-Bell: I was puzzled by your remark that the energy goes into a few degrees of freedom. If one would apply ergodic arguments, one would expect the energy to go to all degrees of freedom, not be concentrated in a few of them.

Lynden-Bell: The gravothermal catastrophe concentrates the kinetic energy into very few degrees of freedom because the entropy increases as one goes further from equipartition in this problem due to the open phase volume of the system.

Miller: The matter of concentration of energy in a few degrees of freedom is easily understood as a state of maximum phase volume in the microcanonical ensemble. The maximum phase volume is attained in a state in which all the (negative) energy is concentrated in a single binary with all the remaining stars at rest at very large distance. Those remaining stars should be at rest because non-zero velocities for those stars requires more negative energy for the binary, which reduces the phase volume more than enough to compensate the increase of phase volume because of the velocities of the single stars.

I prefer this formulation to those based on f_1's because it avoids the logical inconsistencies inherent in the f_1 description. These mainly center about the requirement that correlations be generated at higher order, and the experimental result that the correlation energy is a substantial fraction of the total cluster energy.

Lynden-Bell: I think to make your phase volume statement you need a confining sphere around the system.

Miller: A confining volume is not required.

Bardeen: I would like to bring up the question of halos vs warm disks, and whether a large velocity dispersion in the center can stabilize an otherwise cool disk. The large N numerical simulations carried out so far do not help, but my gas disk calculations seem to indicate that a large Q near the center is not sufficient to stabilize the disk by itself, particularly if a substantial fraction of the mass has a low Q. Much more investigation of this point is required. However, I have been impressed during this meeting by how little observational evidence there is for low Q, since Q is only known for the solar neighborhood in our Galaxy. In the outer parts of a galactic disk large Q is consistent with a velocity dispersion very small compared with the circular velocity, so direct measurements of the velocity dispersion may be only possible near the center, if there.

Innanen: The work of Van Flandern at Washington seems to indicate that G may be at least very slowly time-variable. Are there any comments on this?

Gott: I believe it was Dr I. I. Shapiro's opinion that that result was incorrect.

Freeman: The words 'hot' and 'disk' mean a highly anisotropic velocity dispersion: σ_z must be small for the disk to be fairly thin, yet σ_R, σ_φ must be large. My question is to large N-body computors: would the process that heat the disk keep the heating to the plane, while keeping the σ_z small?

King: One way of converting motions parallel to the plane into z-motions might be the Spitzer-Schwarzschild mechanism, provided we still think that that mechanism is relevant to the circumstances in the Galaxy. The original analysis considered only motions parallel to the plane, and I'm not sure that the full 3-dimensional case has ever been treated in this sense.

Freeman: There is the difficulty, when measuring velocity dispersions in the disks of edge-on galaxies, that differential rotation along the line of sight induces some extra dispersion which would be as large maybe as the dispersion one is trying to measure.

King: There is one galaxy where observations of a hot disk might be available. Many years ago, when Oort discussed NGC 3115, he remarked that either Humason or Minkowski had observed a considerable velocity dispersion. I believe that Schmidt (who is no longer here this afternoon) has some newer and better spectra, and perhaps from them he can help to answer this question.

Schmidt: Image-tube spectra of NGC 3115 were obtained in 1969. No analysis of the velocity dispersion has been undertaken. T. Williams has determined the rotation velocity from these spectra. He finds no evidence for the broad secondary minimum in the rotation curve previously found by Minkowski*.

Gott: The importance of star formation in the formation of galaxies has been mentioned. I think this is a point that can not be overemphasized. It is my own feeling that this is the key factor in determining whether a spiral or an elliptical galaxy is formed, of what disk and halo components are produced. Another important question is the dynamics of the early gaseous component. It is important to know what the early gas clouds are like, (i.e. what are their mean free paths between collision) so that one may know how dissipative and how viscous the early gas may be.

Freeman: I want to mention again the young globular clusters in the Magellanic Clouds – these are only a few collapse times old, have stars in the mass range at least 10 to 0.5 $M_{\odot}$, and appear from their brightness distribution to be dynamically like the old Milky Way clusters. There is cluster formation going on here before our eyes and we could maybe ask why. We can easily now do obvious things like comparison of distributions for different mass classes in these clusters. If there is anything N-body computors would like us to look for, please let us know.

Larson: It is certainly clear that further progress will require a better understanding of the gas dynamics and star formation processes in forming galaxies. As a theoretician, it is my impression that further improvements in our understanding of star formation in galaxiws will have to come from observations, so I would like to encourage observers to try to identify and study carefully any galaxies which may still be in the process of formation. A possible example is NGC 5253, a small galaxy of elliptical outline whose interior structure is very irregular and shows evidence of concentrations of gas and young stars. This galaxy has recently achieved notoriety as a prolific producer of supernovae, which indicates very active recent or ongoing star formation. Perhaps this is the place to look to understand more about star formation in forming galaxies.

---

* This answer was sent later on by M. Schmidt to the editor.

INDEX OF NAMES

INDEX OF SUBJECTS